Lecture Notes in Computer Science   5747

Commenced Publication in 1973
Founding and Former Series Editors:
Gerhard Goos, Juris Hartmanis, and Jan van Leeuwen

Christophe Clavier   Kris Gaj (Eds.)

Cryptographic Hardware and Embedded Systems – CHES 2009

11th International Workshop
Lausanne, Switzerland, September 6-9, 2009
Proceedings

 Springer

Volume Editors

Christophe Clavier
Université de Limoges
Département de Mathématiques et d'Informatique
83 rue d'Isle, 87000 Limoges, France
E-mail: christophe.clavier@unilim.fr
and
Institut d'Ingénierie Informatique de Limoges (3iL)
42 rue Sainte Anne, 87000 Limoges, France
E-mail: christophe.clavier@3il.fr

Kris Gaj
George Mason University
Department of Electrical and Computer Engineering
Fairfax, VA 22030, USA
E-mail: kgaj@gmu.edu

Library of Congress Control Number: 2009933191

CR Subject Classification (1998): E.3, D.4.6, K.6.5, E.4, C.2, H.2.7

LNCS Sublibrary: SL 4 – Security and Cryptology

ISSN	0302-9743
ISBN-10	3-642-04137-X Springer Berlin Heidelberg New York
ISBN-13	978-3-642-04137-2 Springer Berlin Heidelberg New York

springer.com

© International Association for Cryptologic Research 2009
Printed in Germany

Typesetting: Camera-ready by author, data conversion by Scientific Publishing Services, Chennai, India
Printed on acid-free paper   SPIN: 12753017   06/3180   5 4 3 2 1 0

Preface

CHES 2009, the 11th workshop on Cryptographic Hardware and Embedded Systems, was held in Lausanne, Switzerland, September 6–9, 2009. The workshop was sponsored by the International Association for Cryptologic Research (IACR).

The workshop attracted a record number of 148 submissions from 29 countries, of which the Program Committee selected 29 for publication in the workshop proceedings, resulting in an acceptance rate of 19.6%, the lowest in the history of CHES. The review process followed strict standards: each paper received at least four reviews, and some as many as eight reviews. Members of the Program Committee were restricted to co-authoring at most two submissions, and their papers were evaluated by an extended number of reviewers.

The Program Committee included 53 members representing 20 countries and five continents. These members were carefully selected to represent academia, industry, and government, as well as to include world-class experts in various research fields of interest to CHES. The Program Committee was supported by 148 external reviewers. The total number of people contributing to the review process, including Program Committee members, external reviewers, and Program Co-chairs, exceeded 200.

The papers collected in this volume represent cutting-edge worldwide research in the rapidly growing and evolving area of cryptographic engineering. The submissions were sought in several general areas, including, but not limited to, cryptographic hardware, cryptographic software, attacks against implementations and countermeasures against these attacks, tools and methodologies of cryptographic engineering, and applications and implementation environments of cryptographic systems. Ten years after its first workshop, CHES is now very firmly established as the premier international forum for presenting scientific and technological advances in cryptographic engineering research, the event that bridges the gap between theoretical advances and their practical application in commercial products.

In order to further extend the scope of CHES, this year's CHES included for the first time a special Hot Topic Session. The goal of this session was to attract new authors and attendees to CHES by highlighting a new area, not represented at CHES before, but of potential interest to CHES participants. The topic of this year's Hot Topic Session was: Hardware Trojans and Trusted ICs. The session was chaired by Anand Raghunathan from Purdue University, USA, who prepared a separate call for papers, and oversaw an evaluation of papers submitted to this session. This evaluation was supported by a special Hot Topic Session Committee, composed of six experts in the field of trusted integrated circuit manufacturing. The session included two regular presentations, and an invited talk, entitled "The State-of-the-Art in IC Reverse Engineering," delivered by Randy Torrance from Chipworks, Inc., Canada.

Additionally, the workshop included two other excellent invited talks. Christof Paar from Ruhr-Universität Bochum, one of the two founders of CHES, discussed his vision of cryptographic engineering, and its evolution over years, in a talk entitled "Crypto Engineering: Some History and Some Case Studies." Srini Devadas, MIT, an inventor of PUF (Physical Unclonable Function), and a founder of a company that develops practical products based on this new technology, described his experiences in a talk entitled "Physical Unclonable Functions and Secure Processors."

The workshop also included two special sessions. Elisabeth Oswald chaired a session on the DPA contest, which included an introduction and discussion of the contest by one of the primary contest organizers, Sylvain Guilley from Telecom ParisTech. Following the introduction was a short presentation by the winners of the contest and a panel discussion devoted to the current and future rules of the contest and the ethical issues associated with inadvertently facilitating through the contest practical attacks against implementations of cryptography. The second special session, chaired by Patrick Schaumont from Virginia Tech, was on benchmarking of cryptographic hardware. The session included several interesting short talks on problems and solutions related to fair and comprehensive evaluation of the performance of cryptographic hardware. This session was of particular significance in light of the ongoing evaluation of the SHA-3 candidates competing to become a new American, and a de-facto worldwide, hash function standard. Additionally, the workshop included two traditional events: a rump session, chaired by Guido Bertoni from STMicroelectronics, Italy, and a poster session chaired by Stefan Mangard, Infineon Technologies, Germany. Our great thanks go to all Special Session Chairs for their initiative, enthusiasm, commitment, innovative spirit, and attention to every detail of their respective sessions.

Through a nomination process and a vote, the Program Committee awarded three CHES 2009 Best Paper Awards. The selected papers represent three distinct areas of cryptographic engineering research: efficient hardware implementations of public key cryptography, efficient and secure software implementations of secret key cryptography, and side-channel attacks and countermeasures. The winners of the three equivalent awards were: Jean-Luc Beuchat, Jérémie Detrey, Nicolas Estibals, Eiji Okamoto, and Francisco Rodríguez-Henríquez for their paper "Hardware Accelerator for the Tate Pairing in Characteristic Three Based on Karatsuba-Ofman Multipliers," Emilia Käsper and Peter Schwabe for their paper "Faster and Timing-Attack Resistant AES-GCM," and Thomas Finke, Max Gebhardt, and Werner Schindler for their paper "A New Side-Channel Attack on RSA Prime Generation."

The selection of 29 best papers out of 148 predominantly very strong submissions was a very challenging and difficult task. The Program Committee members dedicated a very significant amount of time and effort in order to comprehensively and fairly evaluate all submitted papers and provide useful feedback to the authors. Our deepest thanks go to the members of the Program Committee for their hard work, expertise, dedication, professionalism, fairness, and team spirit.

We deeply thank Marcelo Kaihara, the General Chair of CHES 2009, for his excellent and always timely work on managing the local organization and orchestrating conference logistics. Only because of his tireless effort, flexibility, and team spirit were we able to fit so many additional events and special sessions in the program of this year's CHES. We would like to also thank EPFL for providing an excellent venue for holding the workshop, and for assisting with many local arrangements. Our gratitude also goes to the generous sponsors of CHES 2009, namely, Cryptography Research, Inc., Nagravision Kudelski Group, Oberthur Technologies, RCIS AIST Japan, Riscure, and Telecom ParisTech.

We are also very grateful to Çetin Kaya Koç for managing conference announcements and advertising as the Publicity Chair, and to Jens-Peter Kaps for diligently maintaining the CHES website. The review and discussion process was run using an excellent Web Submission and Review System developed and maintained by Shai Halevi, who was always very quick and precise in addressing our questions and concerns regarding the operation of the system.

We would like to deeply thank the Steering Committee of CHES, for their trust, constant support, guidance, and kind advice on many occasions. Special thanks go to Jean-Jacques Quisquater and Colin Walter, who were always first to respond to our questions and concerns, and often volunteered the advice and support needed to resolve a wide array of challenging issues associated with the fair, firm, and transparent management of the evaluation process.

Finally, we would like to profoundly thank and salute all the authors from all over the world who submitted their papers to this workshop, and entrusted us with a fair and objective evaluation of their work. We appreciate your creativity, hard work, and commitment to push forward the frontiers of science. All your submissions, no matter whether accepted or rejected at this year's CHES, represent the vibrant field of research that CHES is proud to exemplify.

September 2009   Christophe Clavier
                                                              Kris Gaj

CHES 2009

Workshop on Cryptographic Hardware and Embedded Systems
Lausanne, Switzerland, September 6–9, 2009

Sponsored by *International Association for Cryptologic Research*

General Chair

Marcelo Kaihara   EPFL, Switzerland

Program Co-chairs

Christophe Clavier   Université de Limoges, France
Kris Gaj   George Mason University, USA

Publicity Chair

Çetin Kaya Koç   University of California Santa Barbara, USA

Program Committee

Lejla Batina   Katholieke Universiteit Leuven, Belgium
Daniel J. Bernstein   University of Illinois at Chicago, USA
Guido Bertoni   STMicroelectronics, Italy
Jean-Luc Beuchat   University of Tsukuba, Japan
Luca Breveglieri   Politecnico di Milano, Italy
Ernie Brickell   Intel, USA
Dipanwita Roy
  Chowdhury   Indian Institute of Technology, Kharagpur,
                      India
Jean-Sébastien Coron   University of Luxembourg, Luxembourg
Joan Daemen   STMicroelectronics, Belgium
Ricardo Dahab   University of Campinas, Brazil
Markus Dichtl   Siemens AG, Germany
Benoît Feix   Inside Contactless, France
Viktor Fischer   Université de Saint-Étienne, France
Pierre-Alain Fouque   ENS, France
Catherine H. Gebotys   University of Waterloo, Canada
Christophe Giraud   Oberthur Technologies, France

Program Committee Advisory Member

Hot Topic Session Committee

Anand Raghunathan
(Chair)   Purdue University, USA

Farinaz Koushanfar	Rice University, USA
Jim Plusquellic	University of New Mexico, USA
Pankaj Rohatgi	IBM T.J. Watson Research Center, USA
Patrick Schaumont	Virginia Tech, USA
Berk Sunar	Worcester Polytechnic Institute, USA

External Reviewers

Onur Aciicmez	Guerric Meurice	Markus Kasper
Guido Costa Souza	de Dormale	Chang Hoon Kim
de Araújo	Emmanuelle Dottax	Chong Hee Kim
Kubilay Atasu	Saar Drimer	Minkyu Kim
Alain Aubert	Milos Drutarovsky	Mario Kirschbaum
Maxime Augier	Sylvain Duquesne	Ilya Kizhvatov
Jean-Claude Bajard	Thomas Eisenbarth	Thorsten Kleinjung
Brian Baldwin	Nicolas Estibals	Heiko Knospe
Alessandro Barenghi	Martin Feldhofer	Sandeep S. Kumar
Florent Bernard	Wieland Fischer	Yun-Ki Kwon
Alex Biryukov	Georges Gagnerot	Tanja Lange
Andrey Bogdanov	Berndt Gammel	Cédric Lauradoux
Simone Borri	Pierrick Gaudry	Hee Jung Lee
Joppe Bos	Willi Geiselmann	Arjen K. Lenstra
Arnaud Boscher	Benedikt Gierlichs	Gaëtan Leurent
Lilian Bossuet	Guy Gogniat	Yann L'Hyver
Nicolas Brisebarre	Michael Gora	Wei-Chih Lien
Marco Bucci	Aline Gouget	Feng-Hao Liu
Philippe Bulens	Johann Großschädl	Pierre Loidreau
Anne Canteaut	Sylvain Guillet	Raimondo Luzzi
Nathalie Casati	Eric Xu Guo	François Mace
Mathieu Chartier	Ghaith Hammouri	Abhranil Maiti
Sanjit Chatterjee	Dong-Guk Han	Theo Markettos
Chien-Ning Chen	Neil Hanley	Marcel Medwed
Zhimin Chen	Guillaume Hanrot	Filippo Melzani
Ray Cheung	Christoph Herbst	Nele Mentens
Fred Chong	Clemens Heuberger	Giacomo de Meulenaer
Baudoin Collard	Michael Hutter	Bernd Meyer
René Cumplido	Laurent Imbert	Atsushi Miyamoto
Jean-Luc Danger	Seyyd Hasan Mir Jalili	Amir Moradi
Pascal Delaunay	Pascal Junod	Csaba Andras Moritz
Elke De Mulder	Marcelo Kaihara	Sergey Morozov
Jérémie Detrey	Deniz Karakoyunlu	Andrew Moss

Debdeep Mukhopadhyay
David Oswald
Siddika Berna Ors Yalcin
Young-Ho Park
Hervé Pelletier
Gerardo Pelosi
Christophe Petit
Gilles Piret
Thomas Plos
Thomas Popp
Axel Poschmann
Bart Preneel
Emmanuel Prouff
Carine Raynaud
Francesco Regazzoni
Mathieu Renauld
Matthieu Rivain
Bruno Robisson

Mylène Roussellet
Christian Rust
Akashi Satoh
Werner Schindler
Jörn-Marc Schmidt
Michael Scott
Christian Scwarz
Hermann Seuschek
Chang Shu
Hervé Sibert
Sergei Skorobogatov
Takeshi Sugawara
Ruggero Susella
Daisuke Suzuki
Yannick Teglia
Nicolas Thériault
Hugues Thiebeauld
Stefan Tillich

Kris Tiri
Arnaud Tisserand
Lionel Torres
Michael Tunstall
Jonny Valamehr
Gilles Van Assche
Jérôme Vasseur
Ihor Vasyltsov
Vincent Verneuil
Nicolas Veyrat-
Charvillon
Karine Villegas
Hassan Wassel
Christopher Wolf
Tugrul Yanik
Yu Yu

Table of Contents

Side Channel and Fault Analysis Countermeasures

Invited Talk 2

Pairing-Based Cryptography

New Ciphers and Efficient Implementations

TRNGs and Device Identification

Invited Talk 3

Hot Topic Session: Hardware Trojans and Trusted ICs

Theoretical Aspects

Fault Analysis

Faster and Timing-Attack Resistant AES-GCM

Emilia Käsper[1] and Peter Schwabe[2,*]

[1] Katholieke Universiteit Leuven, ESAT/COSIC
Kasteelpark Arenberg 10, B-3001 Leuven-Heverlee, Belgium
emilia.kasper@esat.kuleuven.be
[2] Department of Mathematics and Computer Science
Technische Universiteit Eindhoven, P.O. Box 513, 5600 MB Eindhoven, Netherlands
peter@cryptojedi.org

Abstract. We present a bitsliced implementation of AES encryption in counter mode for 64-bit Intel processors. Running at 7.59 cycles/byte on a Core 2, it is up to 25% faster than previous implementations, while simultaneously offering protection against timing attacks. In particular, it is the only cache-timing-attack resistant implementation offering competitive speeds for stream as well as for packet encryption: for 576-byte packets, we improve performance over previous bitsliced implementations by more than a factor of 2. We also report more than 30% improved speeds for lookup-table based Galois/Counter mode authentication, achieving 10.68 cycles/byte for authenticated encryption. Furthermore, we present the first constant-time implementation of AES-GCM that has a reasonable speed of 21.99 cycles/byte, thus offering a full suite of timing-analysis resistant software for authenticated encryption.

Keywords: AES, Galois/Counter mode, cache-timing attacks, fast implementations.

1 Introduction

While the AES cipher has withstood years of scrutiny by cryptanalysts, its *implementations* are not guaranteed to be secure. Side-channel attacks have become the most promising attacks, and cache-timing attacks pose a security threat to common AES implementations, as they make heavy use of lookup tables. Countermeasures against cache-timing attacks on software implementations include hardware-based defenses to limit cache leakage; or obscuring timing data, e.g.,

---

* The first author was supported in part by the European Commission through the ICT Programme under Contract ICT-2007-216646 ECRYPT II, the IAP–Belgian State–Belgian Science Policy BCRYPT and the IBBT (Interdisciplinary institute for BroadBand Technology) of the Flemish Government, and by the FWO-Flanders project nr. G.0317.06 Linear Codes and Cryptography. The second author was supported by the European Commission through the ICT Programme under Contract ICT-2007-216499 CACE, and through the ICT Programme under Contract ICT-2007-216646 ECRYPT II. Permanent ID of this document: cc3a43763e7c5016ddc9cfd5d06f8218. Date: June 15, 2009.

C. Clavier and K. Gaj (Eds.): CHES 2009, LNCS 5747, pp. 1–17, 2009.

via adding dummy instructions. However, both approaches are generally deemed impractical due to a severe performance penalty.

This leaves us with the third option: writing dedicated constant-time software. While several cryptographic algorithms such as the Serpent block cipher [8] have been designed with a lookup-table-free implementation in mind, it is generally extremely difficult to safeguard a cipher against side-channel attacks *a posteriori*.

Matsui and Nakajima were the first to show a constant-time implementation of AES on an Intel Core 2 processor faster than any other implementation described before [24]. However, the reported speed of 9.2 cycles/byte[1] is only achieved for chunks of 2 KB of input data that are transposed into a dedicated bitsliced format. Including format conversion, this implementation thus runs at around 10 cycles/byte for stream encryption. On the other hand, encrypting, say, 576-byte packets would presumably cause a slowdown by more than a factor of 3, making the approach unsuitable for many network applications.

Könighofer presents an alternative implementation for 64-bit platforms that processes only 4 input blocks in parallel [22], but at 19.8 cycles/byte, his code is even slower than the reference implementation used in OpenSSL.

Finally, Intel has announced a new AES-NI instruction set [17] that will provide dedicated hardware support for AES encryption and thus circumvent cache leaks on future CPUs. However, processors rolled out to the market today do not yet support these instructions, so cache-timing attacks will continue to be a threat to AES for several years until all current processors have been replaced.

This paper presents a constant-time implementation of AES which only needs 7.59 cycles/byte on an Intel Core 2 Q9550, including costs for transformation of input data into bitsliced format and transformation of output back to standard format. On the newer Intel Core i7, we show even faster speeds of 6.92 cycles/byte, while lookup-table-based implementations on the same platform are still behind the 10 cycles/byte barrier. Not only is our software up to 30% faster than any previously presented AES software for 64-bit Intel processors, it also no longer needs input chunks of 2 KB but only of 128 bytes to achieve optimal speed and is thus efficient for packet as well as stream encryption.

Secondly, we propose a fast implementation of Galois/Counter mode (GCM) authentication. Combined with our fast AES encryption, we demonstrate speeds of 10.68 cycles per encrypted and authenticated byte on the Core 2 Q9550. Our fast GCM implementation, however, uses the standard method of lookup tables for multiplication in a finite field. While no cache-timing attacks against GCM have been published, we acknowledge that this implementation might be vulnerable to cache leaks. Thus, we also describe a new method for implementing GCM without lookup tables that still yields a reasonable speed of 21.99 cycles/byte. The machine-level strategies for implementing AES-GCM in constant time might be of independent interest to implementors of cryptographic software.

Note. All software presented in this paper is in the public domain and is available online on the authors' websites [19, 31] to maximize reusability of results.

---

[1] From here on, we consider only AES-128. All results extend straightforwardly to other key sizes, with an appropriate downscaling in performance.

Organization of the paper. In Section 2, we analyze the applicability of cache-timing attacks to each component of AES-GCM authenticated encryption. Section 3 gives an overview of the target platforms. In Sections 4 and 5, we describe our implementations of AES and GCM, respectively. Finally, Section 6 gives performance benchmarks on three different platforms.

2   Cache Timing Attacks against AES and GCM

Cache-timing attacks are software side-channel attacks exploiting the timing variability of data loads from memory. This variability is due to the fact that all modern microprocessors use a hierarchy of caches to reduce load latency. If a load operation can retrieve data from one of the caches (*cache hit*), the load takes less time than if the data has to be retrieved from RAM (*cache miss*).

Kocher [21] was the first to suggest cache-timing attacks against cryptographic algorithms that load data from positions that are dependent on secret information. Initially, timing attacks were mostly mentioned in the context of public-key algorithms until Kelsey et al. [20] and Page [30] considered timing attacks, including cache-timing attacks, against secret-key algorithms. Tsunoo et al. demonstrated the practical feasibility of cache-timing attacks against symmetric-key ciphers MISTY1 [33] and DES [32], and were the first to mention an attack against AES (without giving further details).

In the rest of this section, we analyze separately the cache-timing vulnerability of three components of AES-GCM: encryption, key schedule, and authentication.

2.1   Attacks against AES Encryption

A typical implementation of AES uses precomputed lookup tables to implement the S-Box, opening up an opportunity for a cache-timing attack. Consider, for example, the first round of AES: the indices of the table lookups are then defined simply by the xor of the plaintext and the first round key. As the attacker knows or even controls the plaintext, information about the lookup indices directly leaks information about the key.

Bernstein [3] was the first to implement a cache-timing key-recovery attack against AES. While his attack relies on the attacker's capability of producing reference timing distributions from known-key encryptions on a platform identical to the target platform and has thus been deemed difficult to mount [29,9], several improved attack strategies have subsequently been described by Bertoni et al. [6], Osvik et al. [29], Acıiçmez et al. [18], Bonneau and Mironov [9], and Neve et al. [28,27].

In particular, Osvik et. al. [29] propose an attack model where the attacker obtains information about cache access patterns by manipulating the cache between encryptions via user-level processes. Bonneau and Mironov [9] further demonstrate an attack detecting cache hits in the encryption algorithm itself, as opposed to timing a process controlled by the attacker. Their attack requires no active cache manipulation, only that the tables are (partially) evicted from cache prior to the encryption. Finally, Acıiçmez et. al. [18] note that if the encrypting machine is running multiple processes, workload on the target machine achieves the desired cache-cleaning effect, and provide simulation results suggesting that it is possible to recover an AES encryption key via a passive remote timing attack.

2.2 Attacks against AES Key Expansion

The expansion of the 128-bit AES key into 11 round keys makes use of the SUB-BYTES operation which is also used for AES encryption and usually implemented through lookup tables. During key schedule, the lookup indices are dependent on the secret key, so in principle, ingredients for a cache-timing attack are available also during key schedule.

However, we argue that mounting a cache-timing attack against AES key-expansion will be very hard in practice. Common implementations do the key expansion just once and store either the fully expanded 11 round keys or partially-expanded keys (see e.g. [2]); in both cases, table lookups based on secret data are performed just once, precluding statistical timing attacks, which require multiple timing samples.

We nevertheless provide a constant-time implementation of key expansion for the sake of completeness. The cycle count of the constant-time implementation is however inferior to the table-based implementation; a performance comparison of the two methods is given in Section 6.

2.3 Attacks against Galois/Counter Mode Authentication

The computationally expensive operations for GCM authentication are multiplications in the finite field $\mathbb{F}_{2^{128}}$. More specifically, each block of input requires multiplication with a secret constant factor H derived from the master encryption key. As all common general-purpose CPUs lack support for multiplication of polynomials over $\mathbb{F}_2$, the standard way of implementing GCM is through lookup tables containing precomputed multiples of H.

The specification of GCM describes different multiplication algorithms involving tables of different sizes allowing to trade memory for computation speed [25]. The basic idea of all of these algorithms is the same: split the non-constant factor of the multiplication into bytes or half-bytes and use these as indices for table lookups.

For the first block of input P_1, this non-constant factor is C_1, the first block of ciphertext. Assuming the ciphertext is available to the attacker anyway, the indices of the first block lookups do not leak any secret information. However, for the second ciphertext block C_2, the non-constant input to the multiplication is $(C_1 \cdot H) \oplus C_2$. An attacker gaining information about this value can easily deduce the secret value H necessary for a forgery attack.[2]

---

[2] The authentication key H is derived from the master key via encrypting a known constant. Thus, learning H is equivalent to obtaining a known plaintext-ciphertext pair and should pose no threat to the master encryption key itself.

The lookup tables used for GCM are usually at least as large as AES lookup tables; common sizes include 4 KB, 8 KB and 64 KB. The values retrieved from these tables are 16 bytes long; knowledge of the (64-byte) cache line thus leaves only 4 possibilities for each lookup index. For example, the 64-KB implementation uses 16 tables, each corresponding to a different byte of the 128-bit input. Provided that cache hits leak the maximum 6 bits in each byte, a 2^{32} exhaustive search over the remaining unknown bits is sufficient to recover the authentication key.

We conclude that common implementations of GCM are potentially vulnerable to authentication key recovery via cache timing attacks. Our software thus includes two different versions of GCM authentication: a fast implementation based on 8-KB lookup tables for settings where timing attacks are not considered a threat; and a slower, constant-time implementation offering full protection against timing attacks. For a performance comparison of these two implementations, see Section 6.

3   The Intel Core 2 and Core i7 Processors

We have benchmarked our implementations on three different Intel microarchitectures: the 65-nm Core 2 (Q6600), the 45-nm Core 2 (Q9550) and the Core i7 (920). These microarchitectures belong to the amd64 family, they have 16 128-bit SIMD registers, called XMM registers.

The 128-bit XMM registers were introduced to Intel processors with the "Streaming SIMD Extensions" (SSE) on the Pentium III processor. The instruction set was extended (SSE2) on the Pentium IV processor, other extensions SSE3, SSSE3 and SSE4 followed. Starting with the Core 2, the processors have full 128-bit wide execution units, offering increased throughput for SSE instructions.

Our implementation mostly uses bit-logical instructions on XMM registers. Intel's amd64 processors are all able to dispatch up to 3 arithmetic instructions (including bit-logical instructions) per cycle; at the same time, the number of simultaneous loads and stores is limited to one.

Virtually all instructions on the amd64 operate on two registers; that is, a two-operand instruction, such as an XOR, overwrites one of the inputs with the output. This introduces an overhead in register-to-register moves whenever both inputs need to be preserved for later reuse.

Aside from these obvious performance bottlenecks, different CPUs have specific limitations:

The pshufb instruction: This instruction is part of the SSSE3 instruction-set extension and allows to shuffle the bytes in an XMM register arbitrarily. On a 65-nm processor, pshufb is implemented through 4 μops; 45-nm Core 2 and Core i7 CPUs need just 1 μop (see [15]). This reduction was achieved by the introduction of a dedicated shuffle-unit [12]. The Core i7 has two of these shuffle units, improving throughput by a factor of two.

Choosing between equivalent instructions: The SSE instruction set includes three different logically equivalent instructions to compute the xor of two 128-bit registers: xorps, xorpd and pxor; similar equivalences hold for other bit-logical instructions: andps/andpd/pand, orps/orpd/por.

While xorps/xorpd consider their inputs as floating point values, pxor works on integer inputs. On Core 2 processors, all three instructions yield the same performance. On the Core i7, on the other hand, it is crucial to use integer instructions: changing all integer bit-logical instructions to their floating-point equivalents results in a performance penalty of about 50% on our benchmark Core i7 920.

What about AMD processors? Current AMD processors do not support the SSSE3 pshufb instruction, but an even more powerful SSE5 instruction pperm will be available for future AMDs. It is also possible to adapt the software to support current 64-bit AMD processors. The performance of the most expensive part of the computation—the AES S-box—will not be affected by this modification, though the linear layer will require more instructions.

4   Bitsliced Implementation of AES in Counter Mode

Bitslicing as a technique for implementing cryptographic algorithms was proposed by Biham to improve the software performance of DES [7]. Essentially, bitslicing simulates a hardware implementation in software: the entire algorithm is represented as a sequence of atomic Boolean operations. Applied to AES, this means that rather than using precomputed lookup tables, the 8×8-bit S-Box as well as the linear layer are computed on-the-fly using bit-logical instructions. Since the execution time of these instructions is independent of the input values, the bitsliced implementation is inherently immune to timing attacks.

Obviously, representing a single AES byte by 8 Boolean variables and evaluating the S-Box is much slower than a single table lookup. However, collecting equivalent bits from multiple bytes into a single variable (register) allows to compute multiple S-Boxes at the cost of one. More specifically, the 16 XMM registers of the Core 2 processors allow to perform packed Boolean operations on 128 bits. In order to fully utilize the width of these registers, we thus process 8 16-byte AES blocks in parallel. While our implementation considers 8 consecutive blocks of AES in counter mode, the same technique could be applied equally efficiently

Table 1. Instruction count for one AES round

	xor/and/or	pshufd/pshufb	xor (mem-reg)	mov (reg-reg)	TOTAL
SUBBYTES	128	–	–	35	163
SHIFTROWS	–	8	–	–	8
MIXCOLUMNS	27	16	–	–	43
ADDROUNDKEY	–	–	8	–	8
TOTAL	155	24	8	35	222

to other modes, as long as there is sufficient parallelism. For example, while the CBC mode is inherently sequential, one could consider 8 parallel independent CBC encryptions to achieve the same effect.

Table 1 summarizes the instruction count for each component of AES. In total, one full round of AES requires 222 instructions to process 8 blocks, or 1.73 instructions/byte. In comparison, a typical lookup-table-based implementation performs 1 lookup per byte per round. As the Core 2 can issue up to 3 arithmetic instructions per clock cycle, we are able to break the fundamental 1 cycle/byte barrier of lookup-table-based implementations.

Several AES implementations following a similar bitslicing approach have been reported previously [22, 23, 24]. However, compared to previous results, we have managed to further optimize every step of the round function. Our implementation of SUBBYTES uses 15% fewer instructions than previously reported software implementations. Also, replacing rotates with the more general byte shuffling instructions has allowed us to design an extremely efficient linear layer (see Section 4.3 and 4.4). In the rest of this section, we describe implementation aspects of each step of the AES round function, as well as the format conversion algorithm.

4.1   Bitsliced Representation of the AES State

The key to a fast bitsliced implementation is finding an efficient bitsliced representation of the cipher state. Denote the bitsliced AES state by $a[0], \ldots, a[7]$, where each $a[i]$ is a 128-bit vector fitting in one XMM register. We take 8 16-byte AES blocks and "slice" them bitwise, with the least significant bits of each byte in $a[0]$ and the most significant bits in the corresponding positions of $a[7]$. Now, the AES S-Box can be implemented equally efficiently whatever the order of bits within the bitsliced state. The efficiency of the linear layer, on the other hand, depends crucially on this order.

In our implementation, we collect in each byte of the bitsliced state 8 bits from identical positions of 8 different AES blocks, assuring that bits within each byte are independent and all instructions can be kept byte-level. Furthermore, in order to simplify the MIXCOLUMNS step, the 16 bytes of an AES state are collected in the state row by row. Figure 1 illustrates the bit ordering in each 128-bit state vector $a[i]$.

Several solutions are known for converting the data to a bitsliced format and back [22, 24]. Our version of the conversion algorithm requires 84 instructions to bitslice the input, and 8 byte shuffles to reorder the state row by row.

row 0				....	row 3		
column 0	column 1	column2	column 3	....	column 0	......	column 3
block 0 block 1 ...	block 7 block 0 block 1 ...	block 7 block 0 block 1 ...	block 7 block 0 block 1 ... block 7	....	block 0 block 1 ... block 7	......	block 0 block 1 ... block 7

Fig. 1. Bit ordering in one 128-bit vector of the bitsliced state

Table 2. Instruction count for the AES S-Box

	xor	and/or	mov	TOTAL
Hardware	82	35	–	117
Software	93	35	35	163

4.2   The SUBBYTES Step

The SUBBYTES step of AES transforms each byte of the 16-byte AES state according to an 8×8-bit S-Box S based on inversion in the finite field $\mathbb{F}_{2^8}$. We use well-known hardware implementation strategies for decomposing the S-Box into Boolean instructions. The starting point of our implementation is the compact hardware S-Box proposed by Canright [11], requiring 120 logic gates, and its recent improvements by Boyar and Peralta [10], which further reduce the gate count to 117. Our implementation of the SUBBYTES step is obtained by converting each logic gate (xor, and, or) in this implementation to its equivalent CPU instruction. All previous bitsliced implementations use a similar approach, nevertheless, by closely following hardware optimizations, we have improved the software instruction count by 15%, from 199 instructions [24] to 163.

We omit here the lengthy description of obtaining the Boolean decomposition; full details can be found in the original paper [11]. Instead, we highlight differences between the hardware approach and our software "simulation", as the exchange rate between hardware gates and instructions on the Core 2 is not one-to-one.

First, the packed Boolean instructions of the Core 2 processors have one source and one destination; that is, one of the inputs is always overwritten by the result. Thus, we need extra move instructions whenever we need to reuse both inputs. Also, while the compact hardware implementation computes recurring Boolean subexpressions only once, we are not able to fit all intermediate values in the available 16 XMM registers. Instead, we have a choice between recomputing some values, or using extra load/store instructions to keep computed values on the stack. We chose to do away without the stack: our implementation fits entirely in the 16 registers and uses 128 packed Boolean instructions and 35 register-to-register move instructions. Table 2 lists the instruction/gate counts for the S-Box in software and hardware.

4.3   The SHIFTROWS Step

Denote the 4×4-byte AES state matrix by $[a_{ij}]$. SHIFTROWS rotates each row of the matrix left by 0, 1, 2 and 3 bytes, respectively:

$$
\begin{bmatrix}
a_{00} & a_{01} & a_{02} & a_{03} \\
a_{10} & a_{11} & a_{12} & a_{13} \\
a_{20} & a_{21} & a_{22} & a_{23} \\
a_{30} & a_{31} & a_{32} & a_{33}
\end{bmatrix}
\mapsto
\begin{bmatrix}
a_{00} & a_{01} & a_{02} & a_{03} \\
a_{11} & a_{12} & a_{13} & a_{10} \\
a_{22} & a_{23} & a_{20} & a_{21} \\
a_{33} & a_{30} & a_{31} & a_{32}
\end{bmatrix} .
$$

Since each byte of the bitsliced state contains 8 bits from identical positions of 8 AES blocks, SHIFTROWS requires us to permute the 16 bytes in each 128-bit vector according to the following permutation pattern:

$$[a_{00}|a_{01}|a_{02}|a_{03}|a_{10}|a_{11}|a_{12}|a_{13}|a_{20}|a_{21}|a_{22}|a_{23}|a_{30}|a_{31}|a_{32}|a_{33}] \mapsto$$
$$[a_{00}|a_{01}|a_{02}|a_{03}|a_{11}|a_{12}|a_{13}|a_{10}|a_{22}|a_{23}|a_{20}|a_{21}|a_{33}|a_{30}|a_{31}|a_{32}].$$

Using the dedicated SSSE3 byte shuffle instruction `pshufb`, the whole SHIFTROWS step can be done in 8 XMM instructions.

4.4   The MIXCOLUMNS Step

MIXCOLUMNS multiplies the state matrix $[a_{ij}]$ by a fixed 4×4 matrix to obtain a new state $[b_{ij}]$:

$$\begin{bmatrix} b_{00} & b_{01} & b_{02} & b_{03} \\ b_{10} & b_{11} & b_{12} & b_{13} \\ b_{20} & b_{21} & b_{22} & b_{23} \\ b_{30} & b_{31} & b_{32} & b_{33} \end{bmatrix} = \begin{bmatrix} 02_x & 03_x & 01_x & 01_x \\ 01_x & 02_x & 03_x & 01_x \\ 01_x & 01_x & 02_x & 03_x \\ 03_x & 01_x & 01_x & 02_x \end{bmatrix} \cdot \begin{bmatrix} a_{00} & a_{01} & a_{02} & a_{03} \\ a_{11} & a_{12} & a_{13} & a_{10} \\ a_{22} & a_{23} & a_{20} & a_{21} \\ a_{33} & a_{30} & a_{31} & a_{32} \end{bmatrix}.$$

Owing to the circularity of the multiplication matrix, each resulting byte b_{ij} can be calculated using an identical formula:

$$b_{ij} = 02_x \cdot a_{ij} \oplus 03_x \cdot a_{i+1,j} \oplus a_{i+2,j} \oplus a_{i+3,j},$$

where indices are reduced modulo 4.

Recall that each byte a_{ij} is an element of $\mathbb{F}_{2^8} = \mathbb{F}_2[X]/X^8 + X^4 + X^3 + X + 1$, so multiplication by 02_x corresponds to a left shift and a conditional masking with 00011011_b whenever the most significant bit $a_{ij}[7] = 1$. For example, the least significant bit $b_{ij}[0]$ of each byte is obtained as

$$b_{ij}[0] = a_{ij}[7] \oplus a_{i+1,j}[0] \oplus a_{i+1,j}[7] \oplus a_{i+2,j}[0] \oplus a_{i+3,j}[0].$$

As the bitsliced state collects the bits of an AES state row by row, computing $a_{i+1,j}[0]$ from $a_{ij}[0]$ for all 128 least significant bits in parallel is equivalent to rotating $a[0]$ left by 32 bits:

$$[a_{00}|a_{01}|a_{02}|a_{03}|a_{10}|a_{11}|a_{12}|a_{13}|a_{20}|a_{21}|a_{22}|a_{23}|a_{30}|a_{31}|a_{32}|a_{33}] \mapsto$$
$$[a_{10}|a_{11}|a_{12}|a_{13}|a_{20}|a_{21}|a_{22}|a_{23}|a_{30}|a_{31}|a_{32}|a_{33}|a_{00}|a_{01}|a_{02}|a_{03}].$$

Similarly, computing $a_{i+2,j}$ ($a_{i+3,j}$) requires rotation by 64 (resp. 96) bits. To obtain the new bitsliced state vector $b[0]$, we can now rewrite the above equation as

$$b[0] = (a[7] \oplus (rl^{32}a[7])) \oplus (rl^{32}a[0]) \oplus rl^{64}(a[0] \oplus (rl^{32}a[0])).$$

Similar equations can be obtained for all state vectors $b[i]$ (see App. A for a complete listing). By observing that $rl^{64}a[i] \oplus rl^{96}a[i] = rl^{64}(a[i] \oplus rl^{32}a[i])$, we are able to save a rotation and we thus only need to compute two rotations per register, or 16 in total. There is no dedicated rotate instruction for XMM registers; however, as all our rotations are in full bytes, we can use the pshufd 32-bit-doubleword permutation instruction. This instruction allows to write the result in a destination register different from the source register, saving register-to-register moves. In total, our implementation of MixColumns requires 43 instructions: 16 pshufd instructions and 27 xors.

4.5 The AddRoundKey Step

The round keys are converted to bitsliced representation during key schedule. Each key is expanded to 8 128-bit values, and a round of AddRoundKey requires 8 xors from memory to the registers holding the bitsliced state. The performance of the AddRoundKey step can further be slightly optimized by interleaving these instructions with the byte shuffle instructions of the ShiftRows step.

4.6 AES Key Schedule

The AES key expansion algorithm computes 10 additional round keys from the initial key, using a sequence of SubBytes operations and xors. With the input/output transform, and our implementation of SubBytes, we have all the necessary components to implement the key schedule in constant time. The key schedule performs 10 unavoidably sequential SubBytes calls; its cost in constant time is thus roughly equivalent to the cost of one 8-block AES encryption. The performance results in Section 6 include an exact cycle count.

5   Implementations of GCM Authentication

Galois/Counter mode is a NIST-standardized block cipher mode of operation for authenticated encryption [25]. The 128-bit authentication key H is derived from the master encryption key K during key setup as the encryption of an all-zero input block. The computation of the authentication tag then requires, for each 16-byte data block, a 128-bit multiplication by H in the finite field $\mathbb{F}_{2^{128}} = \mathbb{F}_2[X]/(X^{128} + X^7 + X^2 + X + 1)$. Figure 2 illustrates the mode of operation; full details can be found in the specification [25].

The core operation required for GCM authentication is thus Galois field multiplication with a secret constant element H. This section describes two different implementations of the multiplication—first, a standard table-based approach, and second, a constant-time solution. Both implementations consist of a one-time key schedule computing H and tables containing multiples of H; and an online phase which performs the actual authentication. Both implementations accept standard (non-bitsliced) input.

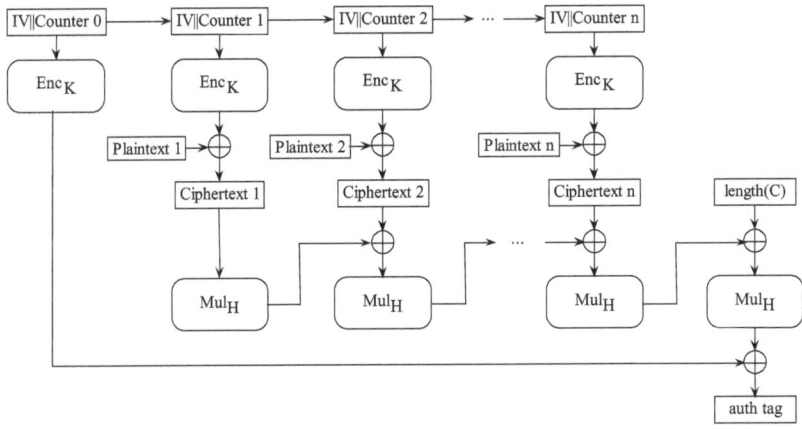

Fig. 2. Galois/Counter Mode Authenticated Encryption

5.1 Table-Based Implementation

Several flavors of Galois field multiplication involving lookup tables of different sizes have been proposed for GCM software implementation [25]. We chose the "simple, 4-bit tables method", which uses 32 tables with 16 precomputed multiples of H each, corresponding to a memory requirement of 8 KB.

Following the ideas from [13], we can do one multiplication using 84 arithmetic instructions and 32 loads.

The computation is free of long chains of dependent instructions and the computation is thus mainly bottlenecked by the number of 32 loads per multiplication yielding a performance of 10.68 cycles/byte for full AES-GCM on a Core 2 Q9550.

5.2 Constant-Time Implementation

Our alternative implementation of GCM authentication does not use any table lookups or data-dependent branches and is thus immune to timing attacks. While slower than the implementation described in Section 5.1, the constant-time implementation achieves a reasonable speed of 21.99 cycles per encrypted and authenticated byte and, in addition, requires only 2 KB of memory for precomputed values, comparing favorably to lookup-table-based implementations.

During the offline phase, we precompute values $H, X \cdot H, X^2 \cdot H, \ldots, X^{127} \cdot H$. Based on this precomputation, multiplication of an element D with H can be computed using a series of xors conditioned on the bits of D, as shown in Algorithm 1.

For a constant-time version of this algorithm we have to replace the conditional statements by a sequence of deterministic instructions. Suppose that we want to xor register %xmm3 into register %xmm4 if and only if bit b_0 of register

---

Algorithm 1. Multiplication in $\mathbb{F}_{2^{128}}$ of D with a constant element H.

Require: Input D, precomputed values $H, X \cdot H, X^2 \cdot H, \ldots, X^{127} \cdot H$
Ensure: Output product $DH = D \cdot H$
  $DH = 0$
  for $i = 0$ to 127 **do**
    if $d_i == 1$ **then**
        $DH = DH \oplus X^i \cdot H$
    end if
  end for

Listing 1. A constant-time implementation of conditional xor

```
1: movdqa  %xmm0, %xmm1        # %xmm1 - tmp
2: pand    BIT0 , %xmm1        # BIT0  - bit mask in memory
3: pcmpeqd BIT0 , %xmm1
4: pshufd  $0xff, %xmm1, %xmm1 #
5: pand    %xmm3, %xmm1        #
6: pxor    %xmm1, %xmm4        #
```

%xmm0 is set. Listing 1 shows a sequence of six assembly instructions that implements this conditional xor in constant time. Lines 1–4 produce an all-zero mask in register %xmm1 if $b_0 = 0$ and an all-one mask otherwise. Lines 5–6 mask %xmm3 with this value and xor the result. We note that the precomputation described above is also implemented in constant time, using the same conditional-xor technique.

In each 128-bit multiplication in the online phase, we need to loop through all 128 bits of the intermediate value D. Each loop requires $6 \cdot 128$ instructions, or 48 instructions per byte. We managed to further optimize the code in Listing 1 by considering four bitmasks in parallel and only repeating lines 1–3 of the code once every four bits, yielding a final complexity of 3.75 instructions per bit, or 30 instructions/byte. As the Core 2 processor can issue at most 3 arithmetic instructions per cycle, a theoretical lower bound for a single Galois field multiplication, using our implementation of the conditional xor, is 10 cycles/byte. The actual performance comes rather close at around 14 cycles/byte for the complete authentication.

6   Performance

We give benchmarking results for our software on three different Intel processors. A description of the computers we used for benchmarking is given in Table 3; all benchmarks used just one core.

To ensure verifiability of our results, we used the open eSTREAM benchmarking suite [14], which reports separate cycle counts for key setup, IV setup, and for encryption.

Table 3. Computers used for benchmarking

	latour	berlekamp	dragon
CPU	Intel Core 2 Quad Q6600	Intel Core 2 Quad Q9550	Intel Core i7 920
CPU frequency	2404.102 MHz	2833 MHz	2668 MHz
RAM	8 GB	8 GB	3 GB
OS	Linux 2.6.27.11 x86_64	Linux 2.6.27.19 x86_64	Linux 2.6.27.9 x86_64
Affiliation	Eindhoven University of Technology	National Taiwan University	National Taiwan University

Table 4. Performance of AES-CTR encryption in cycles/byte

Packet size	4096 bytes	1500 bytes	576 bytes	40 bytes	Simple Imix
latour					
This paper	9.32	9.76	10.77	34.36	12.02
[5]	10.58	10.77	10.77	19.44	11.37
Cycles for key setup (this paper), table-based: 796.77					
Cycles for key setup (this paper), constant-time: 1410.56					
Cycles for key setup [5]: 163.25					
berlekamp					
This paper	7.59	7.98	8.86	28.71	9.89
[5]	10.60	10.77	10.75	19.34	11.35
Cycles for key setup (this paper), table-based: 775.14					
Cycles for key setup (this paper), constant-time: 1179.21					
Cycles for key setup [5]: 163.21					
dragon					
This paper	6.92	7.27	8.08	26.32	9.03
[5]	10.01	10.24	10.15	18.01	10.72
Cycles for key setup (this paper), table-based: 763.38					
Cycles for key setup (this paper), constant-time: 1031.11					
Cycles for key setup [5]: 147.70					

Benchmarking results for different packet sizes are given in Tables 4 and 5. The "simple Imix" is a weighted average simulating sizes of typical IP packages: it takes into account packets of size 40 bytes (7 parts), 576 bytes (4 parts), and 1500 bytes (1 part).

For AES-GCM authenticated encryption, the eSTREAM benchmarking suite reports cycles per encrypted and authenticated byte without considering final computations (one 16-byte AES encryption and one multiplication) necessary to compute the authentication tag. Cycles required for these final computations are reported as part of IV setup. Table 5 therefore gives performance numbers as reported by the eSTREAM benchmarking suite (cycles/byte and cycles required for IV setup) and "accumulated" cycles/byte, illustrating the "actual" time required for authenticated encryption.

For AES in counter mode, we also give benchmarking results of previously fastest software [5], measured with the same benchmarking suite on the same computers. Note however that this implementation uses lookup tables. The previous fastest

Table 5. Cycles/byte for AES-GCM encryption and authentication

Packet size	4096 bytes	1500 bytes	576 bytes	40 bytes	Simple Imix
latour					
Table-based (eSTREAM)	12.22	13.73	16.12	76.82	19.41
Table-based (accumulated)	12.55	14.63	18.49	110.89	23.41
Constant-time (eSTREAM)	27.13	28.79	31.59	99.90	35.25
Constant-time (accumulated)	27.52	29.85	34.36	139.76	39.93
Cycles for precomputation and key setup, table-based: 3083.31					
Cycles for precomputation and key setup, constant-time: 4330.94					
Cycles for IV setup and final computations for authentication, table-based: 1362.98					
Cycles for IV setup and final computations for authentication, constant-time: 1594.39					
berlekamp					
Table-based (eSTREAM)	10.40	11.64	13.72	65.95	16.54
Table-based (accumulated)	10.68	12.39	15.67	94.24	19.85
Constant-time (eSTREAM)	21.67	23.05	25.34	82.79	28.44
Constant-time (accumulated)	21.99	23.92	27.62	115.57	32.30
Cycles for precomputation and key setup, table-based: 2786.79					
Cycles for precomputation and key setup, constant-time: 3614.83					
Cycles for IV setup and final computations for authentication, table-based: 1131.97					
Cycles for IV setup and final computations for authentication, constant-time: 1311.21					
dragon					
Table-based (eSTREAM)	9.86	10.97	12.87	59.05	15.34
Table-based (accumulated)	10.12	11.67	14.69	85.24	18.42
Constant-time (eSTREAM)	20.00	21.25	23.04	73.95	25.87
Constant-time (accumulated)	20.29	22.04	25.10	103.56	29.36
Cycles for precomputation and key setup, table-based: 2424.50					
Cycles for precomputation and key setup, constant-time: 3429.55					
Cycles for IV setup and final computations for authentication, table-based: 1047.49					
Cycles for IV setup and final computations for authentication, constant-time: 1184.41					

bitsliced implementation [24] is not available for public benchmarking; based on the results in the paper, we expect it to perform at best equivalent for stream encryption; and significantly slower for all packet sizes below 2 KB.

For AES-GCM, there exist no benchmarking results from open benchmarking suites such as the eSTREAM suite or the successor eBASC [4]. The designers of GCM provide performance figures for 128-bit AES-GCM measured on a Motorola G4 processor which is certainly not comparable to an Intel Core 2 [26]. Thus, we only give benchmarks for our software in Table 5. As a frame of reference, Brian Gladman's implementation needs 19.8 cycles/byte using 64-KB GCM lookup tables and 22.3 cycles/byte with 8-KB lookup tables on a non-specified AMD64 processor [16]. LibTomCrypt needs 25 cycles/byte for AES-GCM on an Intel Core 2 E6300 [1]. Our implementation of AES-CTR achieves up to 30% improved performance for stream encryption, depending on the platform. Compared to previous bitsliced implementations, packet encryption is *several times* faster. Including also lookup-table-based implementations, we still improve speed for all packet sizes except for the shortest, 40-byte packets.

Similarly, our lookup-table-based implementation of AES-GCM is more than 30% faster than previously reported. Our constant-time implementation is the first of its kind, yet its performance is comparable to previously published software, confirming that it is a viable solution for protecting GCM against timing attacks.

Finally, our benchmark results show a solid improvement from the older 65nm Core 2 to the newer i7, indicating that bitsliced implementations stand to gain more from wider registers and instruction set extensions than lookup-table-based implementations. We conclude that bitslicing offers a practical solution for safeguarding against cache-timing attacks: several of the techniques described in this paper extend to other cryptographic algorithms as well as other platforms.

Acknowledgements

Emilia Käsper thanks the Computer Laboratory of the University of Cambridge for hosting her. The authors are grateful to Dan Bernstein, Joseph Bonneau, Wei Dai, George Danezis, Samuel Neves, Jing Pan, and Vincent Rijmen for useful comments and suggestions.

References

1. LTC benchmarks (accessed 2009-03-07), http://libtomcrypt.com/ltc113.html
2. Bernstein, D.J.: AES speed (accessed 2009-03-07),
   http://cr.yp.to/aes-speed.html
3. Bernstein, D.J.: Cache-timing attacks on AES (2005),
   http://cr.yp.to/antiforgery/cachetiming-20050414.pdf
4. Bernstein, D.J., Lange, T. (eds.): eBACS: ECRYPT benchmarking of cryptographic systems, accessed -03-07 (2009), http://bench.cr.yp.to
5. Bernstein, D.J., Schwabe, P.: New AES software speed records. In: Chowdhury, D.R., Rijmen, V., Das, A. (eds.) INDOCRYPT 2008. LNCS, vol. 5365, pp. 322–336. Springer, Heidelberg (2008)
6. Bertoni, G., Zaccaria, V., Breveglieri, L., Monchiero, M., Palermo, G.: AES power attack based on induced cache miss and countermeasure. In: ITCC 2005: Proceedings of the International Conference on Information Technology: Coding and Computing (ITCC 2005), Washington, DC, USA, vol. I, pp. 586–591. IEEE Computer Society, Los Alamitos (2005)
7. Biham, E.: A fast new des implementation in software. In: Biham, E. (ed.) FSE 1997. LNCS, vol. 1267, pp. 260–272. Springer, Heidelberg (1997)
8. Biham, E., Anderson, R.J., Knudsen, L.R.: Serpent: A new block cipher proposal. In: Vaudenay, S. (ed.) FSE 1998. LNCS, vol. 1372, pp. 222–238. Springer, Heidelberg (1998)
9. Bonneau, J., Mironov, I.: Cache-collision timing attacks against AES. In: Goubin, L., Matsui, M. (eds.) CHES 2006. LNCS, vol. 4249, pp. 201–215. Springer, Heidelberg (2006)
10. Boyar, J., Peralta, R.: New logic minimization techniques with applications to cryptology. Cryptology ePrint Archive, Report 2009/191 (2009),
    http://eprint.iacr.org/

11. Canright, D.: A very compact s-box for AES. In: Rao, J.R., Sunar, B. (eds.) CHES 2005. LNCS, vol. 3659, pp. 441–455. Springer, Heidelberg (2005)
12. Coke, J., Baliga, H., Cooray, N., Gamsaragan, E., Smith, P., Yoon, K., Abel, J., Valles, A.: Improvements in the Intel© Core™ 2 Penryn processor family architecture and microarchitecture. Technical report, Intel Corporation (2008), http://download.intel.com/technology/itj/2008/v12i3/Paper2.pdf
13. Dai, W.: Crypto++ library (accessed 2009-06-14), http://www.cryptopp.com
14. De Cannière, C.: The eSTREAM project: software performance (2008), http://www.ecrypt.eu.org/stream/perf
15. Fog, A.: How to optimize for the Pentium family of microprocessors (2009), http://www.agner.org/assem/
16. Gladman, B.: AES and combined encryption/authentication modes (2008), http://fp.gladman.plus.com/AES/ (accessed, 2009-03-07)
17. Gueron, S.: Advanced encryption standard (AES) instructions set. Technical report, Intel Corporation (2008), http://softwarecommunity.intel.com/isn/downloads/intelavx/AES-Instructions-Set_WP.pdf
18. Acıiçmez, O., Schindler, W., Koç, Ç.K.: Cache based remote timing attack on the AES. In: Abe, M. (ed.) CT-RSA 2007. LNCS, vol. 4377, pp. 271–286. Springer, Heidelberg (2006)
19. Käsper, E.: AES-GCM implementations (2009), http://homes.esat.kuleuven.be/~ekasper
20. Kelsey, J., Schneier, B., Wagner, D., Hall, C.: Side channel cryptanalysis of product ciphers. Journal of Computer Security 8(2-3), 141–158 (2000)
21. Kocher, P.C.: Timing attacks on implementations of Diffie-Hellman, RSA, DSS, and other systems. In: Koblitz, N. (ed.) CRYPTO 1996. LNCS, vol. 1109, pp. 104–113. Springer, Heidelberg (1996)
22. Könighofer, R.: A fast and cache-timing resistant implementation of the AES. In: Malkin, T.G. (ed.) CT-RSA 2008. LNCS, vol. 4964, pp. 187–202. Springer, Heidelberg (2008)
23. Matsui, M.: How far can we go on the x64 processors? In: Robshaw, M.J.B. (ed.) FSE 2006. LNCS, vol. 4047, pp. 341–358. Springer, Heidelberg (2006), http://www.iacr.org/archive/fse2006/40470344/40470344.pdf
24. Matsui, M., Nakajima, J.: On the power of bitslice implementation on Intel Core2 processor. In: Paillier, P., Verbauwhede, I. (eds.) CHES 2007. LNCS, vol. 4727, pp. 121–134. Springer, Heidelberg (2007), http://dx.doi.org/10.1007/978-3-540-74735-2_9
25. McGrew, D.A., Viega, J.: The Galois/Counter Mode of operation (GCM), http://www.cryptobarn.com/papers/gcm-spec.pdf
26. McGrew, D.A., Viega, J.: The security and performance of the Galois/Counter Mode (GCM) of operation. In: Canteaut, A., Viswanathan, K. (eds.) INDOCRYPT 2004. LNCS, vol. 3348, pp. 343–355. Springer, Heidelberg (2004)
27. Neve, M., Seifert, J.-P.: Advances on access-driven cache attacks on AES. In: Biham, E., Youssef, A.M. (eds.) SAC 2006. LNCS, vol. 4356, pp. 147–162. Springer, Heidelberg (2007)
28. Neve, M., Seifert, J.-P., Wang, Z.: A refined look at Bernstein's AES side-channel analysis. In: ASIACCS 2006: Proceedings of the 2006 ACM Symposium on Information, computer and communications security, pp. 369–369. ACM Press, New York (2006)
29. Osvik, D.A., Shamir, A., Tromer, E.: Cache attacks and countermeasures: the case of AES. In: Pointcheval, D. (ed.) CT-RSA 2006. LNCS, vol. 3860, pp. 1–20. Springer, Heidelberg (2006)

30. Page, D.: Theoretical use of cache memory as a cryptanalytic side-channel. Technical report, Department of Computer Science, University of Bristol (June 2002), http://www.cs.bris.ac.uk/Publications/Papers/1000625.pdf
31. Schwabe, P.: AES-GCM implementations (2009), http://cryptojedi.org/crypto/#aesbs
32. Tsunoo, Y., Saito, T., Suzaki, T., Shigeri, M., Miyauchi, H.: Cryptanalysis of DES implemented on computers with cache. In: Walter, C.D., Koç, Ç.K., Paar, C. (eds.) CHES 2003. LNCS, vol. 2779, pp. 62–76. Springer, Heidelberg (2003)
33. Tsunoo, Y., Tsujihara, E., Minematsu, K., Miyauchi, H.: Cryptanalysis of block ciphers implemented on computers with cache. In: Proceedings of the International Symposium on Information Theory and Its Applications, ISITA 2002, pp. 803–806 (2002)

A   Equations for MixColumns

We give the full equations for computing MixColumns as described in Section 4.4. In MixColumns, the bits of the updated state are computed as follows:

$$b_{ij}[0] = a_{ij}[7] \oplus a_{i+1,j}[0] \oplus a_{i+1,j}[7] \oplus a_{i+2,j}[0] \oplus a_{i+3,j}[0]$$
$$b_{ij}[1] = a_{ij}[0] \oplus a_{ij}[7] \oplus a_{i+1,j}[0] \oplus a_{i+1,j}[1] \oplus a_{i+1,j}[7] \oplus a_{i+2,j}[1] \oplus a_{i+3,j}[1]$$
$$b_{ij}[2] = a_{ij}[1] \oplus a_{i+1,j}[1] \oplus a_{i+1,j}[2] \oplus a_{i+2,j}[2] \oplus a_{i+3,j}[2]$$
$$b_{ij}[3] = a_{ij}[2] \oplus a_{ij}[7] \oplus a_{i+1,j}[2] \oplus a_{i+1,j}[3] \oplus a_{i+1,j}[7] \oplus a_{i+2,j}[3] \oplus a_{i+3,j}[3]$$
$$b_{ij}[4] = a_{ij}[3] \oplus a_{ij}[7] \oplus a_{i+1,j}[3] \oplus a_{i+1,j}[4] \oplus a_{i+1,j}[7] \oplus a_{i+2,j}[4] \oplus a_{i+3,j}[4]$$
$$b_{ij}[5] = a_{ij}[4] \oplus a_{i+1,j}[4] \oplus a_{i+1,j}[5] \oplus a_{i+2,j}[5] \oplus a_{i+3,j}[5]$$
$$b_{ij}[6] = a_{ij}[5] \oplus a_{i+1,j}[5] \oplus a_{i+1,j}[6] \oplus a_{i+2,j}[6] \oplus a_{i+3,j}[6]$$
$$b_{ij}[7] = a_{ij}[6] \oplus a_{i+1,j}[6] \oplus a_{i+1,j}[7] \oplus a_{i+2,j}[7] \oplus a_{i+3,j}[7].$$

In our bitsliced implementation, this translates to the following computation on the 8 128-bit state vectors:

$$b[0] = (a[7] \oplus (rl^{32}a[7])) \oplus (rl^{32}a[0]) \oplus rl^{64}(a[0] \oplus (rl^{32}a[0]))$$
$$b[1] = (a[0] \oplus (rl^{32}a[0])) \oplus (a[7] \oplus (rl^{32}a[7])) \oplus (rl^{32}a[1]) \oplus rl^{64}(a[1] \oplus (rl^{32}a[1]))$$
$$b[2] = (a[1] \oplus (rl^{32}a[1])) \oplus (rl^{32}a[2]) \oplus rl^{64}(a[2] \oplus (rl^{32}a[2]))$$
$$b[3] = (a[2] \oplus (rl^{32}a[2])) \oplus (a[7] \oplus (rl^{32}a[7])) \oplus (rl^{32}a[3]) \oplus rl^{64}(a[3] \oplus (rl^{32}a[3]))$$
$$b[4] = (a[3] \oplus (rl^{32}a[3])) \oplus (a[7] \oplus (rl^{32}a[7])) \oplus (rl^{32}a[4]) \oplus rl^{64}(a[4] \oplus (rl^{32}a[4]))$$
$$b[5] = (a[4] \oplus (rl^{32}a[4])) \oplus (rl^{32}a[5]) \oplus rl^{64}(a[5] \oplus (rl^{32}a[5]))$$
$$b[6] = (a[5] \oplus (rl^{32}a[5])) \oplus (rl^{32}a[6]) \oplus rl^{64}(a[6] \oplus (rl^{32}a[6]))$$
$$b[7] = (a[6] \oplus (rl^{32}a[6])) \oplus (rl^{32}a[7]) \oplus rl^{64}(a[7] \oplus (rl^{32}a[7])).$$

Accelerating AES with Vector Permute Instructions

Mike Hamburg

Computer Science Dept., Stanford University
mhamburg@cs.stanford.edu

Abstract. We demonstrate new techniques to speed up the Rijndael (AES) block cipher using vector permute instructions. Because these techniques avoid data- and key-dependent branches and memory references, they are immune to known timing attacks. This is the first constant-time software implementation of AES which is efficient for sequential modes of operation. This work can be adapted to several other primitives using the AES S-box such as the stream cipher LEX, the block cipher Camellia and the hash function Fugue. We focus on Intel's SSSE3 and Motorola's Altivec, but our techniques can be adapted to other systems with vector permute instructions, such as the IBM Xenon and Cell processors, the ARM Cortex series and the forthcoming AMD "Bulldozer" core.

Keywords: AES, AltiVec, SSSE3, vector permute, composite fields, cache-timing attacks, fast implementations.

1   Introduction

Since the 2001 selection of the Rijndael block cipher [6] as the Advanced Encryption Standard (AES), optimization of this cipher in hardware and in software has become a topic of significant interest.

Unfortunately, fast implementations of AES in software usually depend on a large table – 4kiB in the most common implementation – to perform the S-box and the round mixing function. While the table's size is problematic only on the most resource-constrained embedded platforms, the fact that the lookups are key- and data-dependent leads to potential vulnerabilities [2]. In an extreme case, Osvik et al. demonstrate how to extract the key from the Linux dm-crypt encrypted disk implementation with 65 milliseconds of measurements and 3 seconds of analysis [11].

This weakness seems to be intrinsic to the Rijndael algorithm itself. Except in bit-sliced designs, no known technique for computing the S-box is remotely competitive with table lookups on most processors, so that constant-time implementations are many times slower than table-based ones except in parallelizable modes of operation. Despite this issue, the Rijndael S-box' excellent cryptographic properties have led to its inclusion in other ciphers, including the LEX stream cipher [5], the Fugue hash function [7] and the Camellia block cipher [10].

C. Clavier and K. Gaj (Eds.): CHES 2009, LNCS 5747, pp. 18–32, 2009.

Several processors — including the VIA C3 and higher, the AMD Geode LX and the forthcoming Intel "Sandy Bridge" and AMD "Bulldozer" cores — support hardware acceleration of Rijndael. This acceleration both speeds up the cipher and reduces its vulnerability to timing attacks. However, such hardware accelerators are processor-specific, and may not be useful in accelerating and protecting Fugue, Camellia or LEX.

We examine another hardware option for accelerating and protecting Rijndael: vector units with permutation instructions, such as the PowerPC AltiVec unit or Intel processors supporting the SSSE3 instruction set. Such units allow us to implement small, constant-time lookups with considerable parallelism. These vector units have attracted attention from AES implementors before: Bhaskar et al. considered a permutation-based implementation in 2001 [4], and Osvik et al. mention such an implementation as a possible defense against cache-based attacks [11].

To implement the S-box, we take advantage of its algebraic structure using composite-field arithmetic [12], that is, by writing $\mathbb{F}_{2^8}$ as a degree-2 field extension of $\mathbb{F}_{2^4}$. This allows efficient computation of the AES S-box without a large lookup table, and so is commonly used in hardware implementations of AES [14]. Käsper and Schwabe's software implementation in [8] takes this approach in a bit-sliced software implementation; this implementation holds the current PC processor speed record of 7.08 cycles/byte on the Intel Core i7 920 ("Nehalem") processor. Our technique achieves fewer cycles/byte on the PowerPC G4, but not on Intel processors.

As usual, hardware-specific optimizations are necessary to achieve optimal performance. This paper focuses on the PowerPC G4e[1] and Intel Core i7 920 "Nehalem", but the techniques can be used in other processors. To this end, we demonstrate techniques that are not optimal on the G4e or Nehalem, but might be preferable on other processors.

2 Preliminaries

2.1 Notation

Because we are working with fields of characteristic 2, addition of field elements amounts to a bitwise exclusive or. We will still write it as "+".

Over subfields of $\mathbb{F}_{2^8}$, we will write x/y for xy^{254}, which are equal when $y \neq 0$ because $y^{255} = 1$. This extension of $\cdot/\cdot$ adds some corner cases when dividing by 0. We will note such corner cases as they arise, and write $\approx$ instead of $=$ for formulae which are incorrect due to these corner cases.

The most frequent remedy for division by zero will be to set an "infinity flag". When dividing by a number with the infinity flag set, we will return 0 instead of the normal value. The flag is bit 4 on AltiVec and bit 7 in SSSE3, for simplicity

---

[1] "G4e" is an unofficial designation of the PowerPC 744x and 745x G4 processors, commonly used to distinguish them from the earlier and considerably different line of G4 processors, the PowerPC 7400 and 7410.

we simply set all 4 high bits. On AltiVec, use of an infinity flag requires extra masking to prevent the high bits of the input from interfering with the flag; in SSSE3, this masking is required anyway.

We write $a||b$ for the concatenation of a and b.

If v is a vector, then v_i is its ith component. By $\langle f(i) \rangle_{i=0}^{15}$, we mean the 16-element vector whose ith element is $f(i)$. For example, if v has 16 elements, then $v = \langle v_i \rangle_{i=0}^{15}$.

We number bits from the right, so that bit 0 is the 2^0 place, bit 1 is the 2^1 place, and so on.

2.2   The Galois Fields $\mathbb{F}_{2^8}$ and $\mathbb{F}_{2^4}$

AES is expressed in terms of operations on the Galois field $\mathbb{F}_{2^8}$. This field is written as

$$\mathbb{F}_{2^8} \cong \mathbb{F}_2[x]/(x^8 + x^4 + x^3 + x + 1)$$

When we write a number in hexadecimal notation, we mean to use this representation of $\mathbb{F}_{2^8}$. For example, $0x63 = x^6 + x^5 + x + 1$.

Because $\mathbb{F}_{2^8}$ is too large for convenient multiplication and division using AltiVec, we will work with the field $\mathbb{F}_{2^4}$, which we will write cyclotomically:

$$\mathbb{F}_{2^4} \cong \mathbb{F}_2[\zeta]/(\zeta^4 + \zeta^3 + \zeta^2 + \zeta + 1)$$

For this generator ζ, we will express $\mathbb{F}_{2^8}$ as

$$\mathbb{F}_{2^8} \cong \mathbb{F}_{2^4}[t]/(t^2 + t + \zeta)$$

The obvious way to represent elements of $\mathbb{F}_{2^8}$ is as $a + bt$ where $a, b \in \mathbb{F}_{2^4}$. A more symmetric, and for our purposes more convenient, representation is to set $\bar{t} := t + 1$ to be the other root of $t^2 + t + \zeta$, so that $t + \bar{t} = 1$ and $t\bar{t} = \zeta$. Then we may write elements of $\mathbb{F}_{2^8}$ uniquely as $xt + y\bar{t}$ with $x, y \in \mathbb{F}_{2^4}$ (here $y = a$ and $x = a + b$ from above). We will use these representations throughout this paper, and they will be reflected at the bit level: our implementations will compute with either $x||y$ or $y||(x + y)$, depending on timing constraints.

2.3   AltiVec and the PowerPC G4

We implemented AES on the PowerPC G4's AltiVec SIMD architecture, specifically the PowerPC 7447a G4e. We will be treating its 128-bit vectors as vectors of 16 bytes. In addition to bytewise arithmetic instructions, this processor has a vector permute instruction:

$$\mathrm{vperm}(a, b, c) := \langle (a||b)_{c_i \bmod 32} \rangle_{i=0}^{15}$$

That is, $\mathrm{vperm}(a, b, c)$ replaces each element c_i of c with the element of the concatenation of a and b indexed by c_i's 5 low-order bits.

We will find two uses for vperm. The first is to permute the block for the ShiftRows and MixColumns steps of AES. In this case, c is a fixed permutation

Table 1. Intel SSE configurations

Core	SSE units	pshufb units	pshufb throughput	pshufb latency
Conroe	3	1	2 cycles	3 cycles
Harpertown	3	1	1 cycle	1 cycle
Nehalem	3	2	1 cycle	1 cycle

and $a = b$ is the input block. The second use is 16 simultaneous lookups in a 32-element table, or a 16-element lookup table when $a = b$.

The processor can dispatch and execute any two vector instructions of different types[2] per cycle, plus a load or store. The arithmetic operations that we will use have a 1-cycle effective latency, and the permute operations have a 2-cycle effective latency; both types have a 1/cycle throughput. Because we won't be saturating either the dispatcher or the load-store unit, loads and stores are effectively free in moderation.

2.4 Intel SSSE3

Intel's SSSE3 instruction set includes a weaker vector permute operation called pshufb. It differs from vperm in three ways. First, it only implements a 16-way shuffle, implicitly taking $a = b$. Second, if the highest-order bit of c_i is set, then the ith output will be 0 instead of $a_{c_i \mod 16}$. This is useful for implementing an infinity flag. Third, its operands follow a CISC 2-operand convention: its destination register is always the same register as a, but c can be loaded from memory instead of from a register.

We will show benchmarks on three different Intel processors: a Core 2 Duo L7600 "Conroe", a Xeon E5420 "Harpertown" and a Core i7 920 "Nehalem". These processors have much more complicated pipelines than the G4e. All three can execute up to 3 instructions per cycle, all of which can be SSE logical instructions. Their shuffle units have different configurations as shown in Table 1 [1].

2.5 Log Tables

Implementing multiplication and division with log tables is a well-known technique. However, it is not trivial to apply it in SIMD. AltiVec's vperm instruction only uses the low-order 5 bits of the permutation, so we must ensure that these 5 bits suffice to determine the result. Furthermore, when dividing we may wish to distinguish between $0/1, 0/0$ and $1/0$ in order to implement an infinity bit. Within these constraints, we worked out the following tables largely by trial and error.

---

[2] There are 4 types of vector operations: floating-point operations; simple integer operations; complex integer operations such as multiplies; and permutations including whole-vector shifts, repacks and immediate loads.

For log tables over $\mathbb{F}_{2^4}$, we set

$$\log_{\text{num}}(x) = \begin{cases} \log(x) + 97, & x \neq 0 \\ -64 \equiv 192, & x = 0 \end{cases} \quad \text{and} \quad \log_{\text{denom}}(y) = \begin{cases} \log(1/y) - 95, & y \neq 0 \\ 65, & y = 0 \end{cases}$$

For multiplication, we perform an *unsigned addition with saturation*, defined as $a \uplus b := \min(a + b, 255)$ so that

$$\log_{\text{num}}(x) \uplus \log_{\text{num}}(y) = \begin{cases} 194 + \log(xy) \equiv 2 + \log(xy) \in [2, 30], & xy \neq 0 \\ 255 \equiv 31, & xy = 0 \end{cases}$$

For division, we perform a *signed addition with saturation*, defined as $a \barwedge b := \min(\max(a + b, -128), 127)$ so that

$$\log_{\text{num}}(x) \barwedge \log_{\text{denom}}(y) = \begin{cases} -128 \equiv 0, & x = 0 \neq y \\ 1, & x = 0 = y \\ 2 + \log(x/y) \in [2, 30], & x \neq 0 \neq y \\ 127 \equiv 31, & x \neq 0 = y \end{cases}$$

Because these sums' residues mod 32 depend only on xy or x/y (and in the same way for both), a lookup table on the output can extract xy or x/y. Furthermore, these log tables allow us to distinguish between $0/1, 1/0$ and $0/0$.

2.6   Cubic Multiplication

Because pshufb operates on tables of size at most 16, it does not appear to admit a efficient implementation of log tables. It would be desirable to multiply instead using the "quarter-squares" identity $xy = (x+y)^2/4 - (x-y)^2/4$. Unfortunately, this identity does not work over fields of characteristic 2. We can instead set ω to a cube root of unity (so that $\omega^2 = \omega + 1$) and use an "omega-cubes" formula such as

$$xy^2 = \omega(x + \omega y)^3 + \omega^2(\omega x + y)^3 + (\omega^2 x + \omega^2 y)^3$$

which is not as horrible as it looks because the map $(x, y) \to (x + \omega\sqrt{y}, \omega x + \sqrt{y})$ is linear. If x and y are given in this basis, xy can be computed with 3 table lookups and 3 xors, but transforming into and out of this basis will cost 4-6 instructions. Alternatively, the above formula can be used to compute x/y^2 and x^2/y in the high and low nibbles of a single register, but the lack of room for an infinity flag makes this strategy less useful.

On the processors we studied, cubic multiplication does not appear to be optimal technique for implementing AES, but it might be useful in other algorithms.

3   Implementing Inversion

This section deals with algorithms for inverting an element $xt + y\bar{t}$ of $\mathbb{F}_{2^8}$.

3.1 Classical Inversion

The simplest way to compute $1/(xt + y\bar{t})$ is to rationalize the denominator: multiplying top and bottom by $x\bar{t} + yt$ gives

$$\frac{1}{xt + y\bar{t}} = \frac{x\bar{t} + yt}{x^2 t\bar{t} + xyt^2 + xy\bar{t}^2 + y^2 t\bar{t}} = \frac{x\bar{t} + yt}{xy + (x^2 + y^2)\zeta} = \frac{x\bar{t} + yt}{(\sqrt{xy/\zeta} + x + y)^2 \zeta}$$

This last, more complicated expression is how we actually perform the computation: the multiplications are computed with log tables; the squares, square roots and multiplications and divisions by ζ come for free. This technique requires few operations, but it has many lookups on the critical path. Therefore it is optimal on the G4e in parallel modes, but not in sequential modes.

3.2 Symmetric Inversion

To improve parallelism, we can rearrange the above to the near-formula:

$$\frac{1}{xt + y\bar{t}} \approx \frac{t}{x + \zeta(x + y)^2/y} + \frac{\bar{t}}{y + \zeta(x + y)^2/x}$$

This formula is incorrect when $x = 0$ or $y = 0$, but this is easily fixed using an infinity flag. Since this technique can be computed in parallel, it is faster than classical inversion in sequential modes.

3.3 Nested Inversion

The most efficient formula that we found for Intel processors uses the fact that

$$\frac{1}{1/x + 1/\zeta(x + y)} + y \approx \frac{xy + \zeta(x^2 + y^2)}{(1 + \zeta)x + \zeta y}$$

which leads to the monstrous near-formula

$$\frac{1}{xt + y\bar{t}} \approx \frac{t + \zeta}{\frac{1}{1/y + 1/\zeta(x+y)} + x} + \frac{\bar{t} + \zeta}{\frac{1}{1/x + 1/\zeta(x+y)} + y}$$

Division by zero is handled by using an infinity flag, which remarkably makes this formula correct in all cases. This technique performs comparably to symmetric inversion on the G4e[3], but much better on Intel because it does not require multiplication.

Figure 1 compares the parallelism of nested inversion and classical inversion. This diagram omits setup and architectural details, but is nonetheless representative: nested inversion completes faster despite having more instructions.

---

[3] In fact, their code is almost exactly the same. Nested inversion has different (and fewer) lookup tables, and some xors replacing adds, but its dependencies are identical. It is due entirely to momentum that our G4e implementation uses symmetric inversion.

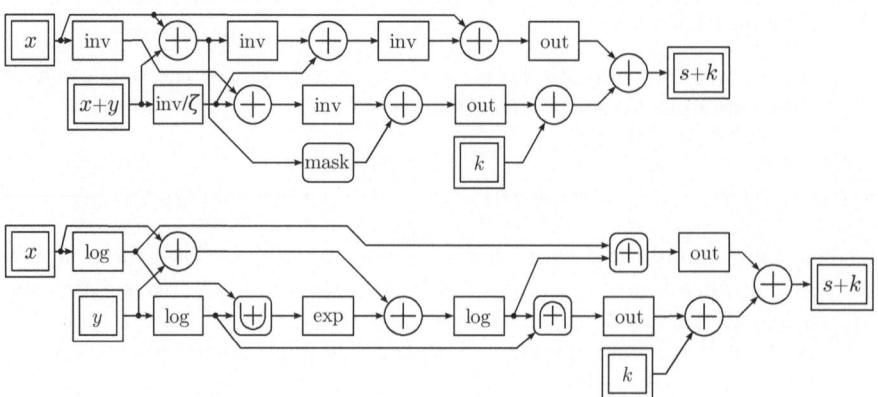

Fig. 1. Nested inversion (top) has more parallelism than classical inversion (bottom)

3.4   Factored Inversion

Another approach is to separate the variables, for example:

$$\frac{1}{xt + y\bar{t}} \approx \frac{1}{x} \cdot \frac{1}{t + (y/x)\bar{t}}$$

Once we compute $\log(y/x)$, we can compute (the log of) the right term in the form $at + b\bar{t}$ with a pair of lookups. The formula is wrong when $x = 0$, but we can look up a correction for this case in parallel with the rest of the computation. This technique combines the low latency of symmetric inversion with the high throughput of classical inversion. However, its many different lookup tables cause register pressure, so we prefer to use the more specialized formulas above.

3.5   Brute Force

The implementation in [4] uses a brute-force technique: a lookup in each of 8 tables of size 32 can emulate a lookup in a table of size 256. This technique is less efficient than symmetric or nested inversion on all the processors we tested. For example, nested inversion requires 7 lookups into 4 tables of size 16 (with an infinity flag) and 6 xors.

4   Implementing AES

4.1   The S-Box and the Multiplication by 0x02

Every inversion algorithm described above (other than brute force) ends by computing $f(a)+g(b)$ for some (f, g, a, b) using two shuffles and an xor. Therefore the S-box's linear skew can be folded into the tables for f and g. However, the use of infinity flags (which may force a lookup to return 0) prevents folding in the addition. Therefore, we use tables for $\texttt{skew}(f(a))$ and $\texttt{skew}(g(b))$; we add 0x63 to the key schedule instead. Similarly, we can multiply by 0x02 by computing $2\,\mathsf{skew}(f(a)) + 2\,\mathsf{skew}(g(b))$.

On the G4e, we make another modification to accomodate classical inversion. It happens that in the basis we use for classical inversion, $\mathsf{skew}(at)$ and $\mathsf{skew}(b\overline{t})$ are functions of the low-order 4 bits of $\mathtt{0x2} \cdot \mathsf{skew}(at)$ and $\mathtt{0x2} \cdot \mathsf{skew}(b\overline{t})$ but not vice-versa. As a result, we use fewer registers if we compute $\mathtt{0x2} \cdot \mathsf{skew}(at)$ first.

4.2   ShiftRows and MixColumns

We have three options to perform the ShiftRows step and the MixColumns rotations.

1. We could keep AES' natural alignment, with each column in 4 contiguous bytes. This would allow us to use an unmodified key schedule. On the G4e, this technique makes MixColumns fast at the expense of ShiftRows. On Intel, both require permutations.
2. We could align each row into 4 contiguous bytes. On the G4, this makes ShiftRows fast at the expense of MixColumns, but relieves register pressure. On Intel, it allows the use of pshufd for MixColumns, but the lack of SIMD rotations means that ShiftRows requires a permutation. Also, an both an input and an output permutation are required.
3. We could use permutations for the MixColumns step. We conjugate by the ShiftRows permutation, so we need not physically perform ShiftRows at all. There will be a different forward and backward permutation each round, with a period of 4 rounds. For 128- and 256-bit keys, the number of rounds is not a multiple of 4, so this technique requires either an input or an output permutation, but not both. (With the MixColumns technique used for classical inversion on the G4e, this method will always require an input or output permutation.) This technique seems to be the fastest both on the G4e and on Intel.

In any case, it is not advantageous to compute ShiftRows directly before computing MixColumns, because there are at least 2 registers live at all times. If ShiftRows is to be physically computed at all, this should be done at the beginning or end of the round, when only 1 register is live.

Let r_k denote a left rotation by k elements. To compute MixColumns, we compute $\boldsymbol{x} := (a, b, c, d)$ and $\mathtt{0x02} \cdot \boldsymbol{x}$ as above. We compute

$$\boldsymbol{y} := r_1(\boldsymbol{x}) + \mathtt{0x02} \cdot \boldsymbol{x} = (\mathtt{0x02} \cdot a + b, \mathtt{0x02} \cdot b + c, \mathtt{0x02} \cdot c + d, \mathtt{0x02} \cdot d + a)$$

We then compute the desired output

$$r_1(\boldsymbol{y}) + \boldsymbol{y} + r_3(\boldsymbol{x}) = (\mathtt{0x02} \cdot a + \mathtt{0x03} \cdot b + c + d,\ \ldots)$$

When using classical inversion, we compute $\mathtt{0x02} \cdot \boldsymbol{x}$ before $\boldsymbol{x}$, so we use a similar addition chain that rotates $\mathtt{0x02} \cdot \boldsymbol{x}$ first.

4.3 `AddRoundKey` and the Modified Key Schedule

Naïvely we would add the round key either at the end of the round, or just before the end of the round, while waiting for a `MixColumns` permutation to finish executing. However, for some implementations, it is more convenient to add the round key during the S-box. Since every implementation of the S-box other than brute force ends with $f(a) + g(b)$ for some (f, g, a, b), we can compute $f(a) + k' + g(b)$, with the addition of k' in parallel with the computation of $g(b)$. This technique is more efficient on Intel and on the original G4, while performing the same on the G4e.

However, this trick makes the key schedule more complicated: the vector k' will be multiplied by

$$M := \begin{pmatrix} 0\ 1\ 1\ 1 \\ 1\ 0\ 1\ 1 \\ 1\ 1\ 0\ 1 \\ 1\ 1\ 1\ 0 \end{pmatrix}$$

during `MixColumns`. Since M is its own inverse, this means that we should set $k' = Mk$. As a result, this trick may not be desirable when very high key agility is required.

Our key schedule also differs from the standard schedule in that keys must be transformed into the $(t, \bar{t})$ basis for $\mathbb{F}_{2^8}$ and rotated by the `ShiftRows` permutation.

5   Parallelism and Related Optimizations

5.1 Interleaving

On the G4e, the classical inversion algorithm gives few instructions, but a high latency. What is more, the instructions are balanced: they put an equal toll on the simple integer and permute units. As a result, they interleave very well: we can run 2 rounds in 24 cycles, compared to 1 round in 20 cycles using classical inversion or 17 cycles using symmetric inversion. For modes which allow parallel encryption, this boosts encryption speed by about 40%. Similiarly, we can interleave 4 rounds in 46 cycles, which should boost encryption speed by another 4% or so.

5.2 Byte-Slicing

A more extreme transformation is to byte-slice encryption. This technique performs 16 encryptions in parallel. Instead of holding state for each of the 16 bytes in one encryption, a vector holds state for one byte in each of the 16 different encryptions. This allows two speedups. First, since each byte is in a separate register, no permutations are needed for `ShiftRows` and `MixColumns`. Second, the same key byte is added to every position of the vector. As a result, we can simply add this key byte to the exponentiation tables each round. Of course, to add this byte every round would save no time at all, but if we schedule each round's exponentiation table ahead of time, we will save an xor instruction.

We implemented this technique on the G4e, and found two difficulties. One is that we must split the data to be encrypted into byte-sliced form, and merge it back into ordinary form at the end of the encryption, which costs about 1/8 cycle per byte in each direction. A second problem is that the byte-sliced round function becomes limited by integer operations, so that saving permutations doesn't help. To alleviate this problem, we can replace a vsrb (element shift right) instruction with a vsr (whole-vector shift right) instruction, which uses the permutation unit on the G4e. In conjunction with double-triple mixing, this optimization reduces the cipher to 9.5 cycles per main round and 6.5 cycles for the final round.

Conveniently, Rogaway et al's OCB mode for authenticated encryption with associated data [13] is simpler in byte-sliced form. The offset Δ may be stored for each slice; then most of the multiplication by 2^{16} – that is, shifting left by 2 bytes – consists of renumbering slice n as slice $n - 2$. The lowest three slices will be a linear function of the upper two slices and the lowest slice; this can be computed with two shifts, 8 permutations and 6 xors. This representation is highly redundant; it is possible that a more concise representation would allow a more efficient computation of the OCB offset Δ.

5.3 Double-Triple Mixing

Let (a, b, c, d) be the output of the S-box. Our implementation of mixing for standard round functions computed $2a$ then a. For highly parallel modes, we can do better by computing $2a$ and then $3a$. If we let

$$(\alpha, \beta, \gamma, \delta) := (3a + 2b, 3b + 2c, 3c + 2d, 3d + 2a)$$

then the output of the mixing function is

$$(\beta + \gamma + \delta, \alpha + \gamma + \delta, \alpha + \beta + \delta, \alpha + \beta + \gamma)$$

This output is easily computed using 10 xors instead of 12. The technique we use computes

$$
\begin{aligned}
&c_0 := \alpha, \quad c_1 := \alpha + \beta, \quad c_2 := \alpha + \beta + \gamma, \quad c_3 := \alpha + \beta + \delta, \\
&c_4 := \beta + \gamma = c_2 + c_0, \\
&c_5 := \alpha + \gamma + \delta = c_3 + c_4 \\
&c_6 := \beta + \gamma + \delta = c_1 + c_5
\end{aligned}
$$

and outputs (c_6, c_5, c_3, c_2). In addition to taking only 10 xors, this method takes its inputs in order $(\alpha, \beta, \gamma, \delta)$ and immediately xors something into them. These features lead to significant savings in register usage, latency and complexity. We suspect that the savings would be even better on Intel hardware.

5.4   Counter-Mode Caching

Bernstein and Schwabe [3] call attention to a useful optimization in Hongjun Wu's eStream implementation of counter mode. Except in every 256th block, only the last byte of the input changes. As a result, only one S-box needs to be computed the first round. Its output affects only 4 bytes, so only 4 S-boxes need to be computed the second round. In a 10-round, 128-bit AES encryption, this saves about 1.7 rounds on average, or about 17% (slightly more, because the last round is shorter). What is more, it allows us to avoid transforming the input into byte-sliced form.

5.5   Scalar-Unit Assistance

Following [4], we considered using the G4e's scalar unit to perform two of the vector xors, reducing the cipher to 9 cycles per round. However, the expense of shuttling data between the vector and scalar units nullifies any advantage from this technique.

6   Decryption

Decryption is more difficult than encryption, because the MixColumns step is more complicated: the coefficients are $(0x0E, 0x09, 0x0D, 0x0B)$, which are linearly independent over $\mathbb{F}_2$. As a result, all four coefficients must be looked up separately. This requires 4 more tables than encryption (minus one for permutations, because we can use only forward permutations with this method), and on Intel means that the lookup tables spill to memory.

7   Benchmarks

We initially tested several experimental implementations on the G4e. They include heavily optimized implementations of several modes several modes, but are also somewhat incomplete; in particular, we did not implement encryption of unaligned data or any sort of decryption.

After realizing that the same techniques are applicable to Intel using SSSE3, we set out to build a practical AES library for Intel machines. However, optimization on x86 processors is much more difficult than on the PowerPC, so our library does not yet approach its theoretical maximum performance. Our Intel library also does not yet implement as many modes or techniques, but it does implement encryption and decryption on aligned and unaligned data.

We tested our implementation on four machines, whose specifications are listed in Table 2.

Our byte-sliced implementations are experimental and so far incomplete. We benchmarked each individual component of the algorithm and added together the times. Similarly, on the G4e we benchmarked only the streaming steps of OCB mode, not the nonce generation and finalization. These consist of one

Table 2. Bechmark machine specifications

Machine	Processor	Core	Speed
altacaca	Motorola PowerPC G4 7447a	Apollo 7	1.67 GHz
peppercorn	Intel Core 2 Duo L7500	Conroe	1.60 GHz
WhisperMoon	Intel Xeon E5420	Harpertown	2.50 GHz
lahmi	Intel Core i7 920	Nehalem	2.67 GHz

Table 3. Encryption timings on altacaca in cycles per byte

Implementation	Par	Mode	128-bit key			192-bit key			256-bit key		
			32	512	long	32	512	long	32	512	long
Symmetric	1	ECB	11.3	10.6	10.7	14.1	12.7	12.8	17.0	14.8	14.9
	1	CBC	11.3	10.8	10.8	14.1	12.9	12.9	17.0	15.0	15.1
Classical	2	ECB	8.5	7.9	7.8	11.3	9.4	9.3	11.3	10.8	10.8
	2	CTR	9.4	7.9	7.9	11.3	9.4	9.3	11.3	10.8	10.8
	2	OCB	19.5	8.6	7.8	25.4	10.2	9.3	25.4	12.0	10.8
Classical (sliced)	16	CTR			5.4			6.7			7.9
	16	OCB			6.6			7.8			9.0
openssl speed	1	CBC			32.6			36.4			40.5

encryption each, so we have added the time required for two encryptions. We expect that a complete implementation would be slightly more efficient due to function call overhead.

We tested encryption and decryption on messages of size 32, 512 and 4096 bytes, with 128-, 192- and 256-bit keys.

Our classical encryption code was optimized for OCB mode; we expect that its ECB and CTR timings could be improved by 1-2% with further tuning. Due to cache effects, encryption of long messages is slightly slower than encryption of short messages in some cases.

7.1   Architecture-Specific Details

Alignment. We tested with 16-byte-aligned input, output and key. Our Intel code supports unaligned input and output; our G4e code does not. Both implementations require 16-byte-aligned round keys, but this is enforced by our library. Our code currently only supports messages which are an integral number of blocks; we are intending to change this before release.

Loop unrolling. We did not unroll the round function at all, except in the byte-sliced case, in which we unrolled it 4 ways. Experiments showed that unrolling was generally unnecessary for optimum performance on the G4e, and our Intel code is still largely unoptimized.

Table 4. Encryption timings on `peppercorn` in cycles per byte

Implementation	Mode	128-bit key			192-bit key			256-bit key		
		32	512	long	32	512	long	32	512	long
Nested	ECB	22.0	21.6	21.5	26.3	25.3	25.5	30.6	29.9	30.1
	ECB^{-1}	27.0	26.5	26.3	32.3	31.7	31.4	37.8	37.1	37.0
	CBC	22.3	21.6	21.4	26.5	25.8	25.6	31.0	30.0	30.0
	CBC^{-1}	27.4	26.3	25.9	32.4	31.7	31.4	37.8	37.3	36.9
	CTR	22.2	21.5	21.8	26.5	25.8	25.8	30.7	30.1	29.9
	OCB	44.2	23.6	22.3	52.8	28.2	26.5	61.4	32.5	30.8
	OCB^{-1}	54.7	28.7	27.4	64.5	34.4	32.8	75.4	40.0	37.8
`openssl speed`	CBC			18.8			21.4			24.1

Table 5. Encryption timings on `WhisperMoon` in cycles per byte

Implementation	Mode	128-bit key			192-bit key			256-bit key		
		32	512	long	32	512	long	32	512	long
Nested	ECB	11.8	11.1	11.0	13.9	13.2	13.3	16.1	15.4	15.4
	ECB^{-1}	14.7	14.3	14.4	17.7	17.0	17.1	20.4	19.9	19.9
	CBC	11.6	11.1	11.2	14.3	13.3	13.5	16.1	15.4	15.8
	CBC^{-1}	14.8	14.2	14.2	17.6	17.0	17.0	20.4	20.2	20.2
	CTR	11.9	11.1	11.1	14.1	13.3	13.4	16.4	15.4	15.8
	OCB	23.3	12.3	11.7	27.6	14.5	13.7	31.8	17.0	16.1
	OCB^{-1}	29.4	15.4	14.6	35.0	18.5	17.5	40.9	21.5	20.4
`openssl speed`	CBC			18.7			21.3			23.9

Table 6. Encryption timings on `lahmi` in cycles per byte

Implementation	Mode	128-bit key			192-bit key			256-bit key		
		32	512	long	32	512	long	32	512	long
Nested	ECB	10.3	10.0	9.9	12.4	11.9	11.9	14.8	13.9	13.9
	ECB^{-1}	12.9	12.4	12.4	15.3	15.0	15.0	18.0	17.6	17.6
	CBC	10.8	10.3	10.3	12.9	12.4	12.3	14.6	14.4	14.2
	CBC^{-1}	13.0	12.6	12.5	16.1	15.1	15.2	18.2	17.8	17.8
	CTR	10.4	10.0	10.0	12.4	12.0	11.9	14.2	13.9	13.9
	OCB	21.4	11.1	10.5	25.5	13.2	12.5	29.5	15.3	14.5
	OCB^{-1}	26.4	13.8	13.1	31.4	16.5	15.6	36.6	19.2	18.2
`openssl speed`	CBC			17.6			20.2			22.6

8   Other Processors

Our techniques are applicable to the PowerPC e600 (modern, embedded G4) with essentially no modification. Other processors have different instruction sets, pipelines and numbers of registers, and so our techniques will require modification for optimimum implementation.

The earlier PowerPC 7410 (original G4) can only issue 2 instructions per cycle instead of three. Because we no longer have "free" loads and branches, more unrolling and caching is necessary. However, the 7410's vperm instruction has an effective latency for only 1 cycle. After accounting for these differences, performance should be slightly faster than on the 7447 when using symmetric inversion, and about the same speed when using interleaved classical inversion. As a result, the interleaved case is less desirable.

The PowerPC 970 (G5) has much higher instruction latencies than the G4, and penalties for moving data between functional units. As a result, more parallelism is required to extract reasonable performance from the G5. It is possible that brute force is the best way to compute the S-box, due to its very high parallelism.

The IBM Cell's SPEs have many more registers than the G4e. Furthermore, their spu_shuffle instruction differs from vperm in that it assigns a meaning to the top 3 bits of their input, so inputs need to be masked before permuting. Furthermore, the SPEs lack a vector byte add with saturation, so a different log-table technique needs to be used. For multiplication, we suggest mapping 0 to 0x50 and nonzero x to $0x30 + \log x$, so that $\log(0 \cdot 0) \rightarrow 0xA0$ and $\log(0 \cdot x) \rightarrow [0x80, 0x8E]$, all of which code for 0x00 in the spu_shuffle instruction. We estimate that with a byte-sliced 128-bit implementation, a Cell SPU would require approximately 8 clock cycles per byte encrypted.

The forthcoming AMD "Bulldozer" core will feature an SSE5 vector permute instruction similar to AltiVec's. Like the Cell's spu_shuffle, this instruction assigns additional meaning to the 3 bits of the input field, which means that more masking and different log tables will be needed. SSE5 has fewer registers than AltiVec, but its ability to take arguments from memory instead of from registers may make up for this if the latency penalty is low enough.

ARM's NEON vector instruction set features a vector permute instruction, but its performance is significantly worse than SSSE3 or AltiVec. Nested inversion is probably the most practical technique due to its smaller tables.

9   Conclusions and Future Work

We have presented a technique for accelerating AES using vector permute units, while simultaneously thwarting known timing- and cache-based attacks. Our technique is the first software design to yield a fast, constant time implementation for sequential modes of operation. Our results include some 150%

improvement over current implementations on the G4e[4]. On recent x86-64 processors, it is some 41% slower than Käsper and Schwabe's bitsliced implementation [8], but doesn't require a parallel mode to attain this speed.

We expect that microarchitectural optimization can improve the speed of our code significantly. This will be a major focus of future work. We also expect that this work can be applied to other primitives; it would be interesting to see if Camellia, Fugue or LEX can be implemented as efficiently.

References

1. Intel 64 and ia-32 architectures optimization reference manual (2009)
2. Bernstein, D.: Cache-timing attacks on AES. Technical report (2005)
3. Bernstein, D.J., Schwabe, P.: New AES software speed records (2008)
4. Bhaskar, R., Dubey, P., Kumar, V., Rudra, A., Sharma, A.: Efficient Galois field arithmetic on SIMD architectures. In: Proceedings of the 15th ACM Symposium on Parallelism in Algorithms and Architectures, pp. 256–257 (2003)
5. Biryukov, A.: A new 128-bit-key stream cipher: LEX. In: eSTREAM, ECRYPT Stream Cipher Project, Report 2005/013 (2005)
6. Daemen, J., Rijmen, V.: Aes proposal: Rijndael (1999)
7. Halevi, S., Hall, W., Jutla, C.: The hash function fugue (2008)
8. Käsper, E., Schwabe, P.: Faster and timing-attack resistant aes-gcm. In: Clavier, C., Gaj, K. (eds.) CHES 2009. LNCS, vol. 5747, pp. 1–17. Springer, Heidelberg (2009)
9. Lipmaa, H.: AES ciphers: speed (2006)
10. Nakajima, J., Aoki, K., Kanda, M., Matsui, M., Moriai, S., Ichikawa, T., Tokita, T.: Camellia: A 128-bit block cipher suitable for multiple platforms — design and analysis (2000)
11. Osvik, D.A., Shamir, A., Tromer, E.: Cache attacks and countermeasures: the case of aes. In: Pointcheval, D. (ed.) CT-RSA 2006. LNCS, vol. 3860, pp. 1–20. Springer, Heidelberg (2006)
12. Rijmen, V.: Efficient implementation of the rijndael s-box (2000)
13. Rogaway, P.: Authenticated-encryption with associated-data. In: Proc. 9th CCS, pp. 98–107. ACM Press, New York (2002)
14. Rudra, A., Dubey, P.K., Jutla, C.S., Kumar, V., Rao, J.R., Rohatgi, P.: Efficient Rijndael encryption implementation with composite field arithmetic. In: Koç, Ç.K., Naccache, D., Paar, C. (eds.) CHES 2001. LNCS, vol. 2162, pp. 171–184. Springer, Heidelberg (2001)

---

[4] Bernstein and Schwabe claim 14.57 cycles/byte for CTR mode on a G4. We have not benchmarked their code on a G4e; Denis Ahrens claims a 4% speedup for AES on the G4e [9], so we estimate that Bernstein and Schwabe's code will also gain 4% on the G4e.

SSE Implementation of Multivariate PKCs on Modern x86 CPUs

Anna Inn-Tung Chen[1], Ming-Shing Chen[2], Tien-Ren Chen[2],
Chen-Mou Cheng[1], Jintai Ding[3], Eric Li-Hsiang Kuo[2],
Frost Yu-Shuang Lee[1], and Bo-Yin Yang[2]

[1] National Taiwan University, Taipei, Taiwan
{anna1110,doug,frost}@crypto.tw
[2] Academia Sinica, Taipei, Taiwan
{mschen,trchen1103,lorderic,by}@crypto.tw
[3] University of Cincinnati, Cincinnati, Ohio, USA
ding@math.uc.edu

Abstract. Multivariate Public Key Cryptosystems (MPKCs) are often
touted as future-proofing against Quantum Computers. It also has been
known for efficiency compared to "traditional" alternatives. However, this
advantage seems to erode with the increase of arithmetic resources in
modern CPUs and improved algorithms, especially with respect to Elliptic Curve Cryptography (ECC). In this paper, we show that *hardware advances do not just favor ECC*. Modern commodity CPUs also
have many small integer arithmetic/logic resources, embodied by SSE2
or other vector instruction sets, that are useful for MPKCs. In particular, Intel's SSSE3 instructions can speed up both public and private
maps over prior software implementations of Rainbow-type systems up
to 4×. Furthermore, *MPKCs over fields of relatively small odd prime
characteristics* can exploit SSE2 instructions, supported by most modern 64-bit Intel and AMD CPUs. For example, Rainbow over $\mathbb{F}_{31}$ can
be up to 2× faster than prior implementations of similarly-sized systems
over $\mathbb{F}_{16}$. Here a key advance is in using Wiedemann (as opposed to
Gauss) solvers to invert the small linear systems in the central maps. We
explain the techniques and design choices in implementing our chosen
MPKC instances over fields such as $\mathbb{F}_{31}$, $\mathbb{F}_{16}$ and $\mathbb{F}_{256}$. We believe that
our results can easily carry over to modern FPGAs, which often contain
a large number of small multipliers, usable by odd-field MPKCs.

Keywords: multivariate public key cryptosystem (MPKC), TTS, rainbow, ℓIC, vector instructions, SSE2, SSSE3, Wiedemann.

1 Introduction

Multivariate public-key cryptosystems (MPKCs) [35, 13] is a genre of PKCs
whose public keys represent multivariate polynomials over a small field $\mathbb{K} = \mathbb{F}_q$:

$$\mathcal{P} : \mathbf{w} = (w_1, w_2, \ldots, w_n) \in \mathbb{K}^n \mapsto \mathbf{z} = (p_1(\mathbf{w}), p_2(\mathbf{w}), \ldots, p_m(\mathbf{w})) \in \mathbb{K}^m.$$

C. Clavier and K. Gaj (Eds.): CHES 2009, LNCS 5747, pp. 33–48, 2009.

Polynomials $p_1, p_2, \ldots$ have (almost always) been quadratic since MPKCs came to public notice [30]. Since this is public-key cryptography, we can let $\mathcal{P}(\mathbf{0}) = \mathbf{0}$.

Of course, a random $\mathcal{P}$ would not be invertible by the legitimate user, so almost always $\mathcal{P} = T \circ \mathcal{Q} \circ S$ with two affine maps $S : \mathbf{w} \mapsto \mathbf{x} = \mathrm{M}_S \mathbf{w} + \mathbf{c}_S$ and $T : \mathbf{y} \mapsto \mathbf{z} = \mathrm{M}_T \mathbf{y} + \mathbf{c}_T$, and an "efficiently invertible" quadratic map $\mathcal{Q} : \mathbf{x} \mapsto \mathbf{y}$. The public key then comprise the polynomials in $\mathcal{P}$, while the private key is $\mathrm{M}_s^{-1}, \mathbf{c}_s, \mathrm{M}_T^{-1}, \mathbf{c}_T$, plus information to determine the *central map* $\mathcal{Q}$.

MPKCs have been touted as (a) potentially surviving future attacks using quantum computers, and (b) faster than "traditional" competitors — in 2003, SFLASH was a finalist for the NESSIE project signatures, recommended for embedded use. *We seek to evaluate whether (b) is affected by the evolution of computer architecture.* Without going into any theory, we will discuss the implemention of MPKCs on today's commodity CPUs. *We will conclude that modern single-instruction-multiple-data (SIMD) units also make great cryptographic hardware for MPKCs, making them stay competitive speed-wise.*

1.1   History and Questions

Conventional wisdom used to be: "MPKCs replace arithmetic operations on large units (e.g., 1024+-bit integers in RSA, or 160+-bit integers in ECC) by faster operations on many small units." But the latter means many more memory accesses. People came to realize that eventually the memory latency and bandwidth would become the bottleneck of the performance of a microprocessor [7, 36].

The playing field is obviously changing. When MPKCs were initially proposed [25, 30], commodity CPUs computed a 32-bit integer product maybe every 15–20 cycles. When NESSIE called for primitives in 2000, x86 CPUs could compute one 64-bit product every 3 (Athlon) to 10 (Pentium 4) cycles. The big pipelined multiplier in an AMD Opteron today can produce one 128-bit integer product every 2 cycles. ECC implementers quickly exploited these advances.

In stark contrast, a MOSTech 6502 CPU or an 8051 microcontroller from Intel multiplies in $\mathbb{F}_{256}$ in a dozen instruction cycles (using three table look-ups) — not too far removed from the latency of multiplying in $\mathbb{F}_{256}$ in modern x86.

This striking disparity came about because the number of gates available has been doubling every 18 to 24 months ("Moore's Law") for the last few decades. Compared to that, memory access speed increased at a snail's pace. Now the width of a typical arithmetic/logic unit is 64 bits, vector units are everywhere, and even FPGAs have hundreds of multipliers built in. On commodity hardware, the deck has never seemed so stacked against MPKCs or more friendly to RSA and ECC. Indeed, ECC over $\mathbb{F}_{2^k}$, the only "traditional" cryptosystem that has been seemingly left behind by advances in chip architectures, will get a new special struction from Intel soon — the new carryless multiplication [27].

Furthermore, we now understand attacks on MPKCs much better. In 2004, traditional signature schemes using RSA or ECC are much slower than TTS/4 and SFLASH [1, 10, 37], but the latter have both been broken [17, 18]. Although TTS/7 and 3IC-p seem ok today [8], the impending doom of SHA-1 [33] will force longer message digests and thus slower MPKCs while leaving RSA untouched.

The obvious question is, then: *Can all the extras on modern commodity CPUs be put to use with MPKCs as well? If so, how do MPKCs compare to traditional PKCs today, and how is that likely going to change for the future?*

1.2   Our Answers and Contributions

We will show that advances in chip building also benefit MPKCs. First, vector instructions available on many modern x86 CPUs can provide significant speed-ups for MPKCs over binary fields. Secondly, we can derive an advantage for MPKCs by using as the base field $\mathbb{F}_q$ for q equal to a small odd prime such as 31 on most of today's x86 CPUs. This may sound somewhat counter-intuitive, since for binary fields addition can be easily accomplished by the logical exclusive-or (XOR) operation, while for odd prime fields, costly reductions modulo q are unavoidable. Our reasoning and counter arguments are detailed as follows.

1. Virtually all x86 CPUs today support SSE2 instructions, which can pack eight 16-bit integer operands in its 128-bit xmm registers and hence dispatch eight simultaneous integer operations per cycle in a SIMD style.
   - Using MPKCs with a small odd prime base field $\mathbb{F}_q$ (say $\mathbb{F}_{31}$, as opposed to the usual $\mathbb{F}_{256}$ or $\mathbb{F}_{16}$) enables us to take advantage of vector hardware. Even with an overhead of conversion between bases, schemes over $\mathbb{F}_{31}$ is usually faster than an equivalent scheme over $\mathbb{F}_{16}$ or $\mathbb{F}_{256}$ without SSSE3.
   - MPKCs over $\mathbb{F}_q$ can still be faster than ECC or RSA. While q can be any prime power, it pays to tune to a small set of carefully chosen instances. In most of our implementations, we specialize to $q = 31$.
2. Certain CPUs have simultaneous look-ups from a small, 16-byte table:
   - all current Intel Core and Atom CPUs with SSSE3 instruction PSHUFB;
   - all future AMD CPUs, with SSE5 PPERM instruction (superset of PSHUFB);
   - IBM POWER derivatives — with AltiVec/VMX instruction PERMUTE.
   Scalar-to-vector multiply in $\mathbb{F}_{16}$ or $\mathbb{F}_{256}$ can get around a $10\times$ speed-up; MPKCs like TTS and Rainbow get a $4\times$ factor or higher speed-up.

In this work, we will demonstrate that the advances in chip architecture *do not* leave MPKCs behind while improving traditional alternatives. Furthermore, we list a set of *counter-intuitive* techniques we have discovered during the course of implementing finite field arithmetic using vector instructions.

1. When solving a small and dense matrix equation in $\mathbb{F}_{31}$, iterative methods like Wiedemann may still beat straight Gaussian elimination.
2. $X \mapsto X^{q-2}$ may be a fast way to component-wise invert a vector over $\mathbb{F}_q^*$.
3. For big-field MPKCs, some fields (e.g., $\mathbb{F}_{31^{15}}$) admit *very fast arithmetic representations* — in such fields, inversion is again by raising to a high power.
4. It is important to manage numerical ranges to avoid overflow, for which certain instructions are unexpectedly useful. For example, the PMADDWD (*packed multiply-add word to double word*) instruction, which computes from two 8-long vectors of 16-bit signed words $(x_0, \ldots, x_7)$ and $(y_0, \ldots, y_7)$ the 4-long vector of 32-bit signed words $(x_0 y_0 + x_1 y_1, x_2 y_2 + x_3 y_3, x_4 y_4 + x_5 y_5, x_6 y_6 + x_7 y_7)$, avoids many carries when evaluating a matrix-vector product (mod q).

Finally, we reiterate that, like most implementation works such as the one by
Bogdanov et al [6], we only discuss implementation issues and do not concern
ourselves with the security of MPKCs in this paper. Those readers interested in
the security and design of MPKCs are instead referred to the MPKC book [13]
and numerous research papers in the literature.

2   Background on MPKCs

In this section, we summarize the MPKC instances that we will investigate. Using
the notation in Sec. 1, we only need to describe the central map $\mathcal{Q}$ (M_S and M_T
are square and invertible matrices, usu. resp. of dim $= n$ and m, respectively.
To execute a private map, we replace the "minus" components if needed, invert
T, invert $\mathcal{Q}$, invert S, and if needed verify a prefix/perturbation.

Most small-field MPKCs — TTS, Rainbow, oil-and-vinegar [11, 12, 17, 29]
seem to behave the same over small odd prime fields and over $\mathbb{F}_{2^k}$. Big-field
MPKCs in odd-characteristic were mentioned in [35], but not much researched
until recently. In some cases e.g., ℓIC-derivatives, an odd-characteristic version
is inconvenient but not impossible. Most attacks on and their respective defenses
of MPKCs are fundamentally independent of the base field. Some attacks are
known or conjectured to be easier over binary fields than over small odd prime
fields [5, 19, 9, 15], but never vice versa.

2.1   Rainbow and TTS Families of Digital Signatures

Rainbow($\mathbb{F}_q, o_1, \ldots, o_\ell$) is characterized as follows as a u-stage UOV [14, 17].

- The segment structure is given by a sequence $0 < v_1 < v_2 < \cdots < v_{u+1} = n$.
  For $l = 1, \ldots, u + 1$, set $S_l := \{1, 2, \ldots, v_l\}$ so that $|S_l| = v_l$ and $S_0 \subset S_1 \subset$
  $\cdots \subset S_{u+1} = S$. Denote by $o_l := v_{l+1} - v_l$ and $O_l := S_{l+1} \setminus S_l$ for $l = 1 \cdots u$.
- The central map $\mathcal{Q}$ has components $y_{v_1+1} = q_{v_1+1}(\mathbf{x})$, $y_{v_1+2} = q_{v_1+2}(\mathbf{x}), \ldots,$
  $y_n = q_n(\mathbf{x})$, where $y_k = q_k(\mathbf{x}) = \sum_{i=1}^{v_l} \sum_{j=i}^{n} \alpha_{ij}^{(k)} x_i x_j + \sum_{i<v_{l+1}} \beta_i^{(k)} x_i$, if $k \in$
  $O_l := \{v_l + 1 \cdots v_{l+1}\}$.
- In every q_k, where $k \in O_l$, there is no cross-term $x_i x_j$ where both i and j are
  in O_l. So given all the y_i with $v_l < i \le v_{l+1}$, and all the x_j with $j \le v_l$, we can
  easily compute $x_{v_l+1}, \ldots, x_{v_{l+1}}$. So given $\mathbf{y}$, we guess $x_1, \ldots x_{v_1}$, recursively
  solve for all x_i's to invert $\mathcal{Q}$, and repeat if needed.

Ding et al. suggest Rainbow/TTS with parameters ($\mathbb{F}_{2^4}, 24, 20, 20$) and ($\mathbb{F}_{2^8}, 18,$
$12, 12$) for 2^{80} design security [8, 17]. According to their criteria, the former
instance should not be more secure than Rainbow/TTS at ($\mathbb{F}_{31}, 24, 20, 20$) and
roughly the same as ($\mathbb{F}_{31}, 16, 16, 8, 16$). Note that in today's terminology, TTS is
simply a Rainbow with sparse coefficients, which is faster but less understood.

2.2   Hidden Field Equation (HFE) Encryption Schemes

HFE is a "big-field" variant of MPKC. We identify $\mathbb{L}$, a degree-n extension of the
base field $\mathbb{K}$ with $(\mathbb{F}_q)^n$ via an implicit bijective map $\phi : \mathbb{L} \to (\mathbb{F}_q)^n$ [34]. With

$\mathbf{y} = \sum_{0 \le i,j < \rho} a_{ij}\mathbf{x}^{q^i + q^j} + \sum_{0 \le i < \rho} b_i\mathbf{x}^{q^i} + c$, we have a quadratic $\mathcal{Q}$, invertible via the Berlekamp algorithm with $\mathbf{x}, \mathbf{y}$ as elements of $(\mathbb{F}_q)^n$.

Solving HFE directly is considered to be sub-exponential [22], and a "standard" HFE implementation for 2^{80} security works over $\mathbb{F}_{2^{103}}$ with degree $d = 129$. We know of no timings below 100 million cycles on a modern processor like a Core 2. Modifiers like vinegar or minus cost extra.

The following multi-variable HFE appeared in [5]. First, randomly choose a $\mathbb{L}^h \to \mathbb{L}^h$ quadratic map $\overline{\mathcal{Q}}(X_1, ..., X_h) = (Q_1(X_1, ..., X_h), \cdots, Q_h(X_1, ..., X_h))$ where each $Q_\ell = Q_\ell(X_1, \ldots, X_h) = \sum_{1 \le i \le j \le h} \alpha_{ij}^{(\ell)} X_i X_j + \sum_{j=1}^h \beta_j^{(\ell)} X_j + \gamma^{(\ell)}$ is also a randomly chosen quadratic for $\ell = 1, \ldots, h$. When h is small, this $\overline{\mathcal{Q}}$ can be easily converted into an equation in one of the X_i using Gröbner basis methods at degree no higher than 2^h, which is good since solving univariate equations is cubic in the degree. The problem is that the authors also showed that these schemes are equivalent to the normal HFE and hence are equally (in-)secure.

It was recently conjectured that for odd characteristic, Gröbner basis attacks on HFE does not work as well [15]. Hence we try to implement multivariate HFEs over $\mathbb{F}_q$ for an odd q. We will be conservative here and enforce one prefixed zero block to block structural attacks at a q-time speed penalty.

2.3 C^*, ℓ-Invertible Cycles (ℓIC) and Minus-p Schemes

C^* is the original Matsumoto-Imai scheme [30], also a big-field variant of MPKC. We identify a larger field $\mathbb{L}$ with $\mathbb{K}^n$ with a $\mathbb{K}$-linear bijection $\phi : \mathbb{L} \to \mathbb{K}^n$. The central map $\mathcal{Q}$ is essentially $\overline{\mathcal{Q}} : \mathbf{x} \longmapsto \mathbf{y} = \mathbf{x}^{1+q^\alpha}$, where $\mathbb{K} = \mathbb{F}_q$. This is invertible if $\gcd(1 + q^\alpha, q^n - 1) = 1$.

The ℓ-Invertible Cycle (ℓIC) can be considered as an improved extension of C^* [16]. Here we use the simple case where $\ell = 3$. In 3IC we also use an intermediate field $\mathbb{L} = \mathbb{K}^k$, where $k = n/3$. The central map is $\mathcal{Q} : (X_1, X_2, X_3) \in (\mathbb{L}^*)^3 \mapsto (Y_1, Y_2, Y_3) := (X_1 X_2, X_2 X_3, X_3 X_1)$. 3IC and C^* maps have a lot in common [16,20,11]. To sign, we do "minus" on r variables and use s prefixes (set one or more of the variables to zero) to defend against all known attacks against C^* schemes [11]. This is written as $C^{*-}\mathrm{p}(q, n, \alpha, r, s)$ or 3IC-p(q, n, r, s). Ding et al. recommend $C^{*-}\mathrm{p}(2^4, 74, 22, 1)$, also known as the "pFLASH" [11].

To invert 3IC-p over a field like $\mathbb{F}_{31^{18}}$, from (Y_1, Y_2, Y_3) we do the following.

1. Compute $A = Y_1 Y_2$ [1 multiplication].
2. Compute $B = A^{-1}$ [1 inverse].
3. Compute $C = Y_3 B = X_2^{-2}$ [1 multiplication].
4. Compute $D = C^{-1} = X_2^2$ and $\pm\sqrt{C} = X_2^{-1}$ [1 sqrt+inverse].
5. Multiply X_2^{-1} to Y_1, Y_2, and D [3 multiplications].

We note that for odd q, square roots are non-unique and slow.

3 Background on x86 Vector Instruction Set Extensions

The use of vector instructions to speed up MPKCs is known since the seminal Matsumoto-Imai works, in which bit slicing is suggested for MPKCs over $\mathbb{F}_2$ as a form of SIMD [30]. Berbain et al. pointed out that bit slicing can be extended appropriately for $\mathbb{F}_{16}$ to evaluate public maps of MPKCs, as well as to run the QUAD stream cipher [2]. Chen et al. extended this further to Gaussian elimination in $\mathbb{F}_{16}$, to be used for TTS [8].

To our best knowledge, the only mention of more advanced vector instructions in the MPKC literature is T. Moh's suggestion to use AltiVec instructions (only available then in the PowerPC G4) in his TTM cryptosystem [31]. This fell into obscurity after TTM was cryptanalyzed [21].

In this section, we describe one of the most widely deployed vector instruction sets, namely, the x86 SIMD extensions. The assembly language mnemonics and code in this section are given according Intel's naming convention, which is supported by both gcc and Intel's own compiler icc. We have verified that the two compilers give similar performance results for the most part.

3.1   Integer Instructions in the SSE2 Instruction Set

SSE2 stands for Streaming SIMD Extensions 2, i.e., doing the same action on many operands. It is supported by all Intel CPUs since the Pentium 4, all AMD CPUs since the K8 (Opteron and Athlon 64), as well as the VIA C7/Nano CPUs. The SSE2 instructions operate on 16 architectural 128-bit registers, called the xmm registers. Most relevant to us are SSE2's integer operations, which treat xmm registers as vectors of 8-, 16-, 32- or 64-bit *packed* operands in Intel's terminology. The SSE2 instruction set is highly non-orthogonal. To summarize, there are the following.

Load/Store: To and from xmm registers from memory (both aligned and un-aligned) and traditional registers (using the lowest unit in an xmm register and zeroing the others on a load).

Reorganize Data: Various permutations of 16- and 32-bit packed operands (Shuffle), and Packing/Unpacking on vector data of different densities.

Logical: AND, OR, NOT, XOR; Shift (packed operands of 16, 32, and 64 bits) Left, Right Logical and Right Arithmetic (copies the sign bit); Shift entire xmm register byte-wise only.

Arithmetic: Add/Subtract on 8-, 16-, 32- and 64-bits; Multiply of 16-bit (high and low word returns, signed and unsigned, and fused multiply-adds) and 32-bit unsigned; Max/Min (signed 16-bit, unsigned 8-bit); Unsigned Averages (8/16-bit); Sum-of-differences on 8-bits.

3.2   SSSE3 (Supplementary SSE3) Instructions

SSSE3 adds a few very useful instructions to assist with our vector programming.

PALIGNR ("packed align right"): "PALIGNR xmm (i), xmm (j), k" shifts xmm (j) right by k bytes, and insert the k rightmost bytes of xmm (i) in the space vacated by the shift, with the result placed in xmm (i). Can be used to rotate an xmm register by bytes.

PHADDx,PHSUBx H means horizontal. E.g., consider PHADDW. If destination register xmm (i) starts out as $(x_0, x_1, \ldots, x_7)$, the source register xmm (j) as $(y_0, y_1, \ldots, y_7)$, then after "PHADDW xmm (i), xmm (j)", xmm (i) will hold: $(x_0 + x_1, x_2 + x_3, x_4 + x_5, x_6 + x_7, y_0 + y_1, y_2 + y_3, y_4 + y_5, y_6 + y_7)$. From 8 vectors $\mathbf{v}_0, \mathbf{v}_1, \ldots, \mathbf{v}_7$, after seven invocations of PHADDW, we can obtain $\left(\sum_j v_j^{(0)}, \sum_j v_j^{(1)}, \ldots, \sum_j v_j^{(7)} \right)$ arranged in the right order.

PSHUFB From a source $(x_0, x_1, \ldots, x_{15})$, and destination register $(y_0, y_1, \ldots, y_{15})$, the result at position i is $x_{y_i \bmod 32}$. Here x_{16} through x_{31} are taken to be 0.

PMULHRSW This gives the *rounded* higher word of the product of two signed words in each of 8 positions. [SSE2 only has PMULHW for higher word of the product.]

The source register xmm (j) can usually be replaced by a 16-byte-aligned memory region. The interested reader is referred to Intel's manual for further information on optimizing for the x86-64 architecture [26]. To our best knowledge SSE4 do not improve the matter greatly for us, so we skip their descriptions here.

3.3 Speeding Up in $\mathbb{F}_{16}$ and $\mathbb{F}_{256}$ via PSHUFB

PSHUFB enables us to do 16 simultaneous look-ups at the same time in a table of 16. The basic way it helps with $\mathbb{F}_{16}$ and $\mathbb{F}_{256}$ arithmetic is by speeding up multiplication of a vector $\mathbf{v}$ by a scalar a.

We will use the following notation: if i, j are two bytes (in $\mathbb{F}_{256}$) or nybbles (in $\mathbb{F}_{16}$), each representing a field element, then $i * j$ will be the byte or nybble representing their product in the finite field.

$\mathbb{F}_{16}$, **v is unpacked, 1 entry per byte:** Make a table TT of 16 entries, each 128 bits, where the i-th entry contains $i * j$ in byte j. Load TT$[a]$ into xmm (i), and do "PSHUFB xmm (i), **v**".

$\mathbb{F}_{16}$, **v 2-packed per byte or $\mathbb{F}_{16}$, a 2-packed:** Similar with shifts and ORs.

$\mathbb{F}_{256}$: Use two 256×128-bit tables, for products of any byte-value by bytes [0x00, 0x10, ..., 0xF0], and [0x00, 0x01, ..., 0x0F]. One AND, one shift, 2 PSHUFBs, and one OR dispatches 16 multiplications.

Solving a Matrix Equation: We can speed up Gaussian elimination a lot on fast row operations. Note: Both SSSE3 and bit-slicing require column-first matrices for matrix-vector multiplication and evaluating MPKCs' public maps.

Evaluating public maps: We can do $z_k = \sum_i w_i \left[P_{ik} + Q_{ik} w_i + \sum_{i<j} R_{ijk} w_j \right]$. But on modern processors it is better to compute $\mathbf{c} := [(w_i)_i, (w_i w_j)_{i \leq j}]^T$, then $\mathbf{z}$ as a product of a $m \times n(n+3)/2$ matrix (public key) and $\mathbf{c}$.

In theory, it is good to bit-slice in $\mathbb{F}_{16}$ when multiplying a scalar to a vector that is a multiple of 64 in length. Our tests show bit-slicing a $\mathbb{F}_{16}$ scalar-to-64-long-vector to take a tiny bit less than 60 cycles on a core of a newer (45nm) Core 2 CPU. The corresponding PSHUFB code takes close to 48 cycles. For 128-long vectors, we can still bit-slice using xmm registers. It comes out to around 70 cycles with bit-slicing, against 60 cycles using PSHUFB. This demonstrate the usefulness of SSSE3 since these should be optimal cases for bit-slicing.

4   Arithmetic in Odd Prime Field $\mathbb{F}_q$

4.1   Data Conversion between $\mathbb{F}_2$ and $\mathbb{F}_q$

The first problem with MPKCs over odd prime fields is the conversion between binary and base-q data. Suppose the public map is $\mathcal{P} : \mathbb{F}_q^n \to \mathbb{F}_q^m$. For digital signatures, we need to have $q^m > 2^\ell$, where ℓ is the length of the hash, so that all hash digests of the appropriate size fit into $\mathbb{F}_q$ blocks. For encryption schemes that pass an ℓ-bit session key, we need $q^n > 2^\ell$.

Quadword (8-byte) unsigned integers in $[0, 2^{64} - 1]$ fit decently into 13 blocks in $\mathbb{F}_{31}$. So to transfer 128-, 192-, and 256-bit AES keys, we need at least 26, 39, and 52 $\mathbb{F}_{31}$ blocks, respectively.

Packing $\mathbb{F}_q$-blocks into binary can be more "wasteful" in the sense that one can use more bits than necessary, as long as the map is injective and convenient to compute. For example, we have opted for a very simple packing strategy in which every three $\mathbb{F}_{31}$ blocks are fit in a 16-bit word.

4.2   Basic Arithmetic Operations and Inversion Mod q

$\mathbb{F}_q$ operations for odd prime q uses many modulo-q. We almost always replace slow division instructions with multiplication as follows.

Proposition 1 ([23]). If $2^{n+\ell} \leq Md \leq 2^{n+\ell} + 2^\ell$ for $2^{\ell-1} < d < 2^\ell$, then $\left\lfloor \frac{X}{d} \right\rfloor = \left\lfloor 2^{-\ell} \left\lfloor \frac{XM}{2^n} \right\rfloor \right\rfloor = \left\lfloor 2^{-\ell} \left(\left\lfloor \frac{X(M-2^n)}{2^n} \right\rfloor + X \right) \right\rfloor$ for $0 \leq X < 2^n$.

An instruction giving "top n bits of product of n-bit integers x, y" achieves $\left\lfloor \frac{xy}{2^n} \right\rfloor$ and thus can be used to implement division by multiplication. E.g.,when we take $n = 64$, $\ell = 5$, and $d = 31$, $Q = \left\lfloor \frac{1}{32} \left(\left\lfloor \frac{595056260442243601\, x}{2^{64}} \right\rfloor + x \right) \right\rfloor = x$ div 31, $R = x - 31\, Q$, for an unsigned integer $x < 2^{64}$. Note often $M > 2^n$ as here.

Inverting one element in $\mathbb{F}_q$ is usually via a look-up table. Often we need to invert simultaneously many $\mathbb{F}_q$ elements. As described later, we vectorize most arithmetic operations using SSE2 and hence need to store the operands in xmm registers. Getting the operands between xmm and general-purpose registers for table look-up is very troublesome. Instead, we can use a $(q-2)$-th power ("patched inverse") to invert a vector. For example, the following raises to the 29-th to find multiplicative inverses in $\mathbb{F}_{31}$ using 16-bit integers (`short int`):

$y = x{*}x{*}x \mod 31$; $y = x{*}y{*}y \mod 31$; $y = y{*}y \mod 31$; $y = x{*}y{*}y \mod 31$.

Finally, if SSSE3 is available, inversion in a $\mathbb{F}_q$ for $q < 16$ is possible using one PSHUFB, and for $16 < q \leq 31$ using two PSHUFB's and some masking.

Overall, the most important optimization is *avoiding unnecessary modulo operations* by delaying them as much as possible. To achieve this goal, we need to carefully track operand sizes. SSE2 uses fixed 16- or 32-bit operands for most of its integer vector operations. In general, the use of 16-bit operands, either signed or unsigned, gives the best trade-off between modulo reduction frequency (wider operands allow for less frequent modulo operations) and parallelism (narrower operands allow more vector elements packed in an xmm register).

4.3 Vectorizing Mod q Using SSE2

Using vectorized integer add, subtract, and multiply instructions provided by SSE2, we can easily execute multiple integer arithmetic operations simultaneously. A problem is how to implement vectorized modulo operations (cf. Sec. 4.2). While SSE2 does provide instructions returning the upper word of a 16-by-16-bit product, there are no facilities for carries, and hence it is difficult to guarantee a range of size q for a general q. It is then important to realize that *we do not always need the tightest range*. Minus signs are okay, as long as the absolute values are relatively small to avoid non-trivial modulo operations.

- If IMULHIb returns "the upper half in a signed product of two b-bit words", $y = x - q \cdot$ IMULHI$b\left(\left\lfloor \frac{2^b}{q} \right\rfloor, \left(x + \lfloor \frac{q-1}{2} \rfloor\right)\right)$ will return a value $y \equiv x \pmod q$ such that $|y| \leq q$ for b-bit word arithmetic, where $-2^{b-1} \leq x \leq 2^{b-1} - (q-1)/2$.
- For $q = 31$ and $b = 16$, we do better finding $y \equiv x \pmod{31}, -16 \leq y \leq 15$, for any $-32768 \leq x \leq 32752$ by $y = x - 31 \cdot$ IMULHI16 $(2114, x + 15)$. Here IMULHI16 is implemented via the Intel intrinsic of __mm_mulhi_epi16.
- For I/O in $\mathbb{F}_{31}$, the principal value between 0 and 30 is $y' = y - 31 \,\&\, (y \ggg 15)$, where $\&$ is the logical AND, and $\ggg$ arithmetically shifts in the sign bit.
- When SSSE3 is available, *rounding* with PMULHRSW is faster.

4.4 Matrix-Vector Multiplication and Polynomial Evaluation

Core 2 and newer Intel CPUs have SSSE3 and can add horizontally within an xmm register, c.f., Sec. 3.2. Specifically, the matrix M can be stored row-major. Each row is multiplied component-wise to the vector $\mathbf{v}$. Then PHADDW can add horizontally and arrange the elements at the same time. Surprisingly, this convenience only makes at most a 10% difference for $q = 31$.

If we are restricted to using just SSE2, then it is advisable to store M in the column-major order and treat the matrix-to-vector product as taking a linear combination of the column vectors. For $q = 31$, each 16-bit component in $\mathbf{v}$ is copied eight times into every 16-bit word in an xmm register using an __mm_set1 intrinsic, which takes three data-moving (shuffle) instructions, but still avoids the penalty for accessing the L1 cache. Finally we multiply this register into one column of M, eight components at a time, and accumulate.

Public maps are evaluated as in Sec. 3.3, except that we may further exploit PMADDWD as mentioned in Sec. 1.2, which computes $(x_0 y_0 + x_1 y_1, x_2 y_2 + x_3 y_3, x_4 y_4 + x_5 y_5, x_6 y_6 + x_7 y_7)$ given $(x_0, \ldots, x_7)$ and $(y_0, \ldots, y_7)$. We interleave one xmm with two monomials (32-bit load plus a single __mm_set1 call), load a 4×2 block in another, PMADDWD, and *continue in 32-bits until the eventual reduction mod q*. This way we are able to save a few mod-q operations.

The Special Case of $\mathbb{F}_{31}$: We also pack keys (c.f., Sec. 4.1) so that the public key is roughly $mn(n + 3)/3$ bytes, which holds $mn(n + 3)/2$ $\mathbb{F}_{31}$ entries. For $\mathbb{F}_{31}$, we avoid writing the data to memory and execute the public map on the fly as we unpack to avoid cache contamination. It turns out that it does not slow things down too much. Further, we can do the messier 32-bit mod-q reduction *without* __mm_mulhi_epi32 via shifts as $2^5 = 1 \mod 32$.

4.5   Solving Systems of Linear Equations

Solving systems of linear equations are involved directly with TTS and Rainbow, as well as indirectly in others through taking inverses. Normally, one runs a Gaussian elimination, where elementary row operations can be sped up by SSE2.

However, during a Gaussian elimination, one needs frequent modular reductions, which rather slows things down from the otherwise expected speed. Say we have an augmented matrix $[A|\mathbf{b}]$ modulo 31 in row-major order. Let us do elimination on the first column. Each entry in the remaining columns will now be of size up to about 1000 (31^2), or 250 if representatives are between ± 16.

To eliminate on the second column, we must reduce that column mod 31 before looking up the correct multipliers. Note that reducing a single column by table look-up is no less expensive than reducing the entire matrix when the latter is not too large due to the overhead associated with moving data in and out of the xmm registers, so we end up reducing the entire matrix many times.

We can switch to an iterative method like Wiedemann or Lanczos. To solve by Wiedemann an $n \times n$ system $A\mathbf{x} = \mathbf{b}$, one computes $\mathbf{z}A^i\mathbf{b}$ for $i = 1 \ldots 2n$ for some given $\mathbf{z}$. Then one computes the minimal polynomial from these elements in $\mathbb{F}_q$ using the Berlekamp-Massey algorithm.

It looks very counter-intuitive, as a Gaussian elimination does around $n^3/3$ field multiplications but Wiedemann takes $2n^3$ for a dense matrix for the matrix-vector products, plus extra memory/time to store the partial results and run Berlekamp-Massey. Yet in each iteration, we only need to reduce a single vector, not a whole matrix. That is the key observation and the tests show that Wiedemann is significantly faster for convenient sizes and odd q. Also, Wiedemann outperforms Lanczos because the latter fails too often.

5   Arithmetic in $\mathbb{F}_{q^k}$

In a "big-field" or "two-field" variant of MPKC, we need to handle $\mathbb{L} = \mathbb{F}_{q^k} \cong \mathbb{F}_q[t]/(p(t))$, where p is an irreducible polynomial of degree k. It is particularly efficient if $p(t) = t^k - a$ for a small positive a, which is possible $k|(q-1)$ and in a few other cases. With a convenient p, the map $X \mapsto X^q$ in $\mathbb{L}$, becomes an easy precomputable linear map over $\mathbb{K} = \mathbb{F}_q$. Multiplication, division, and inversion all become much easier. See some example timing for such a tower field in Tab. 1.

Table 1. Cycle counts for various $\mathbb{F}_{31^{18}}$ arithmetic operations using SSE2

Microarchitecture	MULT	SQUARE	INV	SQRT	INV+SQRT
C2 (65nm)	234	194	2640	4693	6332
C2+ (45nm)	145	129	1980	3954	5244
K8 (Athlon 64)	397	312	5521	8120	11646
K10 (Phenom)	242	222	2984	5153	7170

5.1 Multiplication and the S:M (Square:Multiply) Ratio

When $\mathbb{F}_{q^k} \cong \mathbb{F}_q[t]/(t^k - a)$, a straightforward way to multiply is to copy each x_i eight times, multiply by the correct y_i's using PMULLW, and then shift the result by the appropriate distances using PALIGNR (if SSSE3 is available) or unaligned load/stores/shifts (otherwise), depending on the architecture and compiler. For some cases we need to tune the code. E.g., for $\mathbb{F}_{31^9}$, we multiply the x-vector by y_8 and the y-vector by x_8 with a convenient 8×8 pattern remaining.

For very large fields, we can use Karatsuba [28] or other more advanced multiplication algorithms. E.g. $\mathbb{F}_{31^{30}} := \mathbb{F}_{31^{15}}[u]/(u^2 - t)\mathbb{F}_{31^{15}} = \mathbb{F}_{31}[t]/(t^{15} - 3)$. Then $(a_1 u + a_0)(b_1 u + b_0) = [(a_1 + a_0)(b_1 + b_0) - a_1 b_1 - a_0 b_0]u + [a_1 b_1 t + a_0 b_0]$. Similarly, we treat $\mathbb{F}_{31^{54}}$ as $\mathbb{F}_{31^{18}}[u]/(u^3 - t)$, where $\mathbb{F}_{31^{18}} = \mathbb{F}_{31}[t]/(t^{18} - 3)$. Then

$$(a_2 u^2 + a_1 u + a_0)(b_2 u^2 + b_1 u + b_0) = [(a_2 + a_0)(b_2 + b_0) - a_2 b_2 - a_0 b_0 + a_1 b_1] u^2$$
$$+ [(a_1 + a_0)(b_1 + b_0) - a_1 b_1 - a_0 b_0 + t a_2 b_2] u + [t ((a_2 + a_1)(b_2 + b_1) - a_1 b_1 - a_2 b_2) + a_0 b_0].$$

For ECC, often the rule-of-thumb is "**S=0.8M**". Here the **S:M** ratio *ranges from 0.75 to 0.92* for fields in the teens of $\mathbb{F}_{31}$-blocks depending on architecture.

5.2 Square and Other Roots

Today there are many ways to compute square roots in a finite field [3]. For field sizes $q = 4k + 3$, it is easy to compute the square root in $\mathbb{F}_q$ via $\sqrt{y} = \pm y^{\frac{q+1}{4}}$. Here we implement the Tonelli-Shanks method for $4k + 1$ field sizes, as working with a fixed field we can include pre-computed tables with the program "for free." To recap, assume that we want to compute square roots in the field $\mathbb{L}$, where $|\mathbb{L}| - 1 = 2^k a$, with a being odd.

0. Compute a primitive solution to $g^{2^k} = 1$ in $\mathbb{L}$. We only need to take a random $x \in \mathbb{L}$ and compute $g = x^a$, and it is almost even money (i.e., x is a non-square) that $g^{2^{k-1}} = -1$, which means we have found a correct g. *Start with a pre-computed table of (j, g^j) for $0 \le j < 2^k$.*

1. We wish to compute an x such that $x^2 = y$. First compute $v = y^{\frac{a-1}{2}}$.

2. Look up in our table of 2^k-th roots $yv^2 = y^a = g^j$. If j is odd, then y is a non-square. If j is even, then $x = \pm vyq^{\frac{-j}{2}}$ because $x^2 = y(yv^2 g^{-j}) = y$.

Since we implemented mostly mod 31, for $\mathbb{F}_{31^k}$ taking a square root is easy when k is odd and not very hard when k is even. For example, via fast 31^k-th powers, in $\mathbb{F}_{31^9}$ we take square roots by raising to the $\frac{1}{4}(31^9 + 1)$-th power

i.   `temp1 := (((input)²)²)²`,   ii. `temp2:=(temp1)² * ((temp1)²)²`,
iii. `temp2 := [temp2 * ((temp2)²)²]³¹`,   iv. `temp2:= temp2 * (temp2)³¹`,
v.   `result := temp1 * temp2 * ((temp2)³¹)³¹`;

5.3   Multiplicative Inverse

There are several ways to do multiplicative inverses in $\mathbb{F}_{q^k}$. The classical one is an extended Euclidean Algorithm; another is to solve a system of linear equations; the last one is to invoke Fermat's little theorem and raise to the power of $q^k - 2$.

For our specialized tower fields of characteristic 31, the extended Euclidean Algorithm is slower because after one division the sparsity of the polynomial is lost. Solving every entry in the inverse as a variable and running an elimination is about 30% better. *Even though it is counter-intuitive to compute* $X^{31^{15}-2}$ *to get* $1/X$, *it ends up fastest by a factor of 2 to 3.*

Finally, we note that when we compute $\sqrt{X}$ and $1/X$ as high powers at the same time, we can share some exponentiation and save 10% of the work.

5.4   Equation Solving in an Odd-Characteristic Field $\mathbb{L} = \mathbb{F}_{q^k}$

Cantor-Zassenhaus solves a univariate degree-d equation $u(X) = 0$ as follows. The work is normally cubic in $\mathbb{L}$-multiplications and quintic in $(d, k, \lg q)$ overall.

1. Replace $u(X)$ by $\gcd(u(X), X^{q^k} - X)$ so that u factors completely in $\mathbb{L}$.
   (a) Compute and tabulate $X^d \bmod u(X), \ldots, X^{2d-2} \bmod u(X)$.
   (b) Compute $X^q \bmod u(X)$ via square-and-multiply.
   (c) Compute and tabulate $X^{qi} \bmod u(X)$ for $i = 2, 3, \ldots, d - 1$.
   (d) Compute $X^{q^i} \bmod u(X)$ for $i = 2, 3, \ldots, k$, then $X^{q^k} \bmod u(X)$.
2. Compute $\gcd\left(v(X)^{(q^k-1)/2} - 1, u(X)\right)$ for a random $v(X)$, where $\deg v = \deg u - 1$; half of the time we find a nontrivial factor; repeat till u is factored.

6   Experiment Results

Clearly, we need to avoid too large q (too many reductions mod q) and too small q (too large arrays). The choice of $q = 31$ seems the best compromise, since it also allows us several convenient tower fields and easy packing conversions (close to $2^5 = 32$). This is verified empirically.

Some recent implementations of MPKCs over $\mathbb{F}_{2^k}$ are tested by Chen et al. [8] We choose the following well-known schemes for comparison: HFE (an encryption scheme); pFLASH, 3IC-p, and Rainbow/TTS (all signature schemes). We summarize the characteristics and performances, measured using SUPERCOP-20090408 [4] on an Intel Core 2 Quad Q9550 processor running at 2.833 GHz, of these MPKCs and their traditional competitors in Tab. 2. The current MPKCs are over odd-characteristic fields except for pFLASH, which is over $\mathbb{F}_{16}$. The table is divided into two regions: top for encryption schemes and bottom for signature schemes, with the traditional competitors (1024-bit RSA and 160-bit ECC) listed first. The results clearly indicate that MPKCs can take advantage of the latest x86 vector instructions and hold their speeds against RSA and ECC.

Tab. 3 shows the speeds of the private maps of the MPKCs over binary vs. odd fields on various x86 microarchitectures. As in Tab. 1, the C2 microarchitecture

Table 2. Current MPKCs vs. traditional competitors on an Intel C2Q Q9550

Scheme	Result	PubKey	PriKey	KeyGen	PubMap	PriMap
RSA (1024 bits)	128 B	128 B	1024 B	27.2 ms	26.9 μs	806.1 μs
4HFE-p (31,10)	68 B	23 KB	8 KB	4.1 ms	6.8 μs	659.7 μs
3HFE-p (31,9)	67 B	7 KB	5 KB	0.8 ms	2.3 μs	60.5 μs
RSA (1024 bits)	128 B	128 B	1024 B	26.4 ms	22.4 μs	813.5 μs
ECDSA (160 bits)	40 B	40 B	60 B	0.3 ms	409.2 μs	357.8 μs
C^*-p (pFLASH)	37 B	72 KB	5 KB	28.7 ms	97.9 μs	473.6 μs
3IC-p (31,18,1)	36 B	35 KB	12 KB	4.2 ms	11.7 μs	256.2 μs
Rainbow (31,24,20,20)	43 B	57 KB	150 KB	120.4 ms	17.7 μs	70.6 μs
TTS (31,24,20,20)	43 B	57 KB	16 KB	13.7 ms	18.4 μs	14.2 μs

Table 3. MPKC private map timings in kilocycles on various x86 microarchitectures

Scheme	Atom	C2	C2+	K8	K10
4HFE-p (31,10)	4732	2703	2231	8059	2890
3HFE-p (31,9)	528	272	230	838	259
C^*-p (pFLASH)	7895	2400	2450	5010	3680
3IC-p (31,18,1)	2110	822	728	1550	1410
3IC-p (16,32,1)	1002	456	452	683	600
Rainbow (31,16,16,8,16)	191	62	51	101	120
Rainbow (16,24,24,20)	147	61	48	160	170
Rainbow (256,18,12,12)	65	27	22	296	211
TTS (31,24,20,20)	78	38	38	65	72
TTS (16,24,20,20)	141	61	65	104	82
TTS (256,18,12,12)	104	31	36	69	46

refers to the 65 nm Intel Core 2, C2+ the 45 nm Intel Core 2, K8 the AMD Athlon 64, and K10 the AMD Phenom processors. The results clearly indicate that even now MPKCs in odd-characteristic fields hold their own against prior MPKCs that are based in $\mathbb{F}_{2^k}$, if not generally faster, on various x86 microarchitectures.

7   Concluding Remarks

Given the results in Sec. 6 and the recent interest into the theory of algebraic attacks on odd-characteristic HFE, we believe that odd-field MPKCs merit more investigation. Furthermore, today's FPGAs have many built-in multipliers and intellectual properties (IPs), as good integer multipliers are common for application-specific integrated circuits (ASICs). One excellent example of using the multipliers in FPGAs for PKCs is the work of Güneysu and Paar [24]. We believe our results can easily carry over to FPGAs as well as any other specialized hardware with a reasonable number of small multipliers. There are also a variety of massively parallel processor architectures, such as NVIDIA, AMD/ATI, and

Intel [32] graphics processors coming. The comparisons herein must of course be re-evaluated with each new instruction set and new silicon implementation, but we believe that the general trend stands on our side.

Acknowledgements. CC thanks the National Science Council of Taiwan for support under grants NSC 97-2628-E-001-010- and Taiwan Information Security Center (grant 98-2219-E-011-001), BY for grant NSC 96-2221-E-001-031-MY3.

References

1. Akkar, M.-L., Courtois, N.T., Duteuil, R., Goubin, L.: A fast and secure implementation of SFLASH. In: Desmedt, Y.G. (ed.) PKC 2003. LNCS, vol. 2567, pp. 267–278. Springer, Heidelberg (2002)
2. Berbain, C., Billet, O., Gilbert, H.: Efficient implementations of multivariate quadratic systems. In: Biham, E., Youssef, A.M. (eds.) SAC 2006. LNCS, vol. 4356, pp. 174–187. Springer, Heidelberg (2007)
3. Bernstein, D.J.: Faster square roots in annoying finite fields. In: High-Speed Cryptography (2001) (to appear), http://cr.yp.to/papers.html#sqroot
4. Bernstein, D.J.: SUPERCOP: System for unified performance evaluation related to cryptographic operations and primitives (April 2009), http://bench.cr.yp.to/supercop.html
5. Billet, O., Patarin, J., Seurin, Y.: Analysis of intermediate field systems. Presented at SCC 2008, Beijing (2008)
6. Bogdanov, A., Eisenbarth, T., Rupp, A., Wolf, C.: Time-area optimized public-key engines: MQ-cryptosystems as replacement for elliptic curves? In: Oswald, E., Rohatgi, P. (eds.) CHES 2008. LNCS, vol. 5154, pp. 45–61. Springer, Heidelberg (2008)
7. Burger, D., Goodman, J.R., Kägi, A.: Memory bandwidth limitations of future microprocessors. In: Proceedings of the 23rd annual international symposium on Computer architecture, pp. 78–89 (1996)
8. Chen, A.I.-T., Chen, C.-H.O., Chen, M.-S., Cheng, C.-M., Yang, B.-Y.: Practical-sized instances of multivariate pkcs: Rainbow, TTS, and ℓIC-derivatives. In: Buchmann, J., Ding, J. (eds.) PQCrypto 2008. LNCS, vol. 5299, pp. 95–108. Springer, Heidelberg (2008)
9. Courtois, N.: Algebraic attacks over $GF(2^k)$, application to HFE challenge 2 and SFLASH-v2. In: Bao, F., Deng, R., Zhou, J. (eds.) PKC 2004. LNCS, vol. 2947, pp. 201–217. Springer, Heidelberg (2004)
10. Courtois, N., Goubin, L., Patarin, J.: SFLASH: Primitive specification (second revised version), Submissions, Sflash, 11 pages (2002), https://www.cosic.esat.kuleuven.be/nessie
11. Ding, J., Dubois, V., Yang, B.-Y., Chen, C.-H.O., Cheng, C.-M.: Could SFLASH be repaired? In: Aceto, L., Damgard, I., Goldberg, L.A., Halldórsson, M.M., Ingólfsdóttir, A., Walukiewicz, I. (eds.) ICALP 2008, Part II. LNCS, vol. 5126, pp. 691–701. Springer, Heidelberg (2008)
12. Ding, J., Gower, J.: Inoculating multivariate schemes against differential attacks. In: Yung, M., Dodis, Y., Kiayias, A., Malkin, T.G. (eds.) PKC 2006. LNCS, vol. 3958, pp. 290–301. Springer, Heidelberg (2006), http://eprint.iacr.org/2005/255

13. Ding, J., Gower, J., Schmidt, D.: Multivariate Public-Key Cryptosystems. In: Advances in Information Security. Springer, Heidelberg (2006) ISBN 0-387-32229-9
14. Ding, J., Schmidt, D.: Rainbow, a new multivariable polynomial signature scheme. In: Ioannidis, J., Keromytis, A.D., Yung, M. (eds.) ACNS 2005. LNCS, vol. 3531, pp. 164–175. Springer, Heidelberg (2005)
15. Ding, J., Schmidt, D., Werner, F.: Algebraic attack on hfe revisited. In: Wu, T.-C., Lei, C.-L., Rijmen, V., Lee, D.-T. (eds.) ISC 2008. LNCS, vol. 5222, pp. 215–227. Springer, Heidelberg (2008)
16. Ding, J., Wolf, C., Yang, B.-Y.: ℓ-invertible cycles for multivariate quadratic public key cryptography. In: Okamoto, T., Wang, X. (eds.) PKC 2007. LNCS, vol. 4450, pp. 266–281. Springer, Heidelberg (2007)
17. Ding, J., Yang, B.-Y., Chen, C.-H.O., Chen, M.-S., Cheng, C.-M.: New differential-algebraic attacks and reparametrization of rainbow. In: Bellovin, S.M., Gennaro, R., Keromytis, A.D., Yung, M. (eds.) ACNS 2008. LNCS, vol. 5037, pp. 242–257. Springer, Heidelberg (2008), http://eprint.iacr.org/2008/108
18. Dubois, V., Fouque, P.-A., Shamir, A., Stern, J.: Practical cryptanalysis of SFLASH. In: Menezes, A. (ed.) CRYPTO 2007. LNCS, vol. 4622, pp. 1–12. Springer, Heidelberg (2007)
19. Faugère, J.-C., Joux, A.: Algebraic cryptanalysis of Hidden Field Equations (HFE) using Gröbner bases. In: Boneh, D. (ed.) CRYPTO 2003. LNCS, vol. 2729, pp. 44–60. Springer, Heidelberg (2003)
20. Fouque, P.-A., Macario-Rat, G., Perret, L., Stern, J.: Total break of the ℓIC- signature scheme. In: Cramer, R. (ed.) PKC 2008. LNCS, vol. 4939, pp. 1–17. Springer, Heidelberg (2008)
21. Goubin, L., Courtois, N.T.: Cryptanalysis of the TTM cryptosystem. In: Okamoto, T. (ed.) ASIACRYPT 2000. LNCS, vol. 1976, pp. 44–57. Springer, Heidelberg (2000)
22. Granboulan, L., Joux, A., Stern, J.: Inverting HFE is quasipolynomial. In: Dwork, C. (ed.) CRYPTO 2006. LNCS, vol. 4117, pp. 345–356. Springer, Heidelberg (2006)
23. Granlund, T., Montgomery, P.: Division by invariant integers using multiplication. In: Proceedings of the SIGPLAN 1994 Conference on Programming Language Design and Implementation, pp. 61–72 (1994), http://www.swox.com/~tege/divcnst-pldi94.pdf
24. Güneysu, T., Paar, C.: Ultra high performance ecc over nist primes on commercial fpgas. In: Oswald, E., Rohatgi, P. (eds.) CHES 2008. LNCS, vol. 5154, pp. 62–78. Springer, Heidelberg (2008)
25. Imai, H., Matsumoto, T.: Algebraic methods for constructing asymmetric cryptosystems. In: Calmet, J. (ed.) AAECC 1985. LNCS, vol. 229, pp. 108–119. Springer, Heidelberg (1986)
26. Intel Corp. Intel 64 and IA-32 architectures optimization reference manual (November 2007), http://www.intel.com/design/processor/manuals/248966.pdf
27. Intel Corp. Carryless multiplication and its usage for computing the GCM mode. (2008), http://software.intel.com/en-us/articles/carry/less-multiplication-and-%its-usage-for-computing-the-gcm-mode
28. Karatsuba, A., Ofman, Y.: Multiplication of many-digital numbers by automatic computers. Doklady Akad. Nauk SSSR 145, 293–294 (1962); Translation in Physics-Doklady. 7, 595–596 (1963)
29. Kipnis, A., Patarin, J., Goubin, L.: Unbalanced Oil and Vinegar signature schemes. In: Stern, J. (ed.) EUROCRYPT 1999. LNCS, vol. 1592, pp. 206–222. Springer, Heidelberg (1999)

30. Matsumoto, T., Imai, H.: Public quadratic polynomial-tuples for efficient signature verification and message-encryption. In: Günther, C.G. (ed.) EUROCRYPT 1988. LNCS, vol. 330, pp. 419–545. Springer, Heidelberg (1988)
31. Moh, T.: A public key system with signature and master key function. Communications in Algebra 27(5), 2207–2222 (1999), Electronic version, http://citeseer/moh99public.html
32. Seiler, L., Carmean, D., Sprangle, E., Forsyth, T., Abrash, M., Dubey, P., Junkins, S., Lake, A., Sugerman, J., Cavin, R., Espasa, R., Grochowski, E., Juan, T., Hanrahan, P.: Larrabee: a many-core x86 architecture for visual computing. ACM Transactions on Graphics 27(18) (August 2008)
33. Wang, X., Yin, Y.L., Yu, H.: Finding collisions in the full sha-1. In: Shoup, V. (ed.) CRYPTO 2005. LNCS, vol. 3621, pp. 17–36. Springer, Heidelberg (2005)
34. Wolf, C.: Multivariate Quadratic Polynomials in Public Key Cryptography. PhD thesis, Katholieke Universiteit Leuven (2005), http://eprint.iacr.org/2005/393
35. Wolf, C., Preneel, B.: Taxonomy of public key schemes based on the problem of multivariate quadratic equations. Cryptology ePrint Archive, Report 2005/077, 64 pages, May 12 (2005),http://eprint.iacr.org/2005/077/
36. Wulf, W.A., McKee, S.A.: Hitting the memory wall: Implications of the obvious. Computer Architecture News 23(1), 20–24 (1995)
37. Yang, B.-Y., Chen, J.-M.: Building secure tame-like multivariate public-key cryptosystems: The new TTS. In: Boyd, C., González Nieto, J.M. (eds.) ACISP 2005. LNCS, vol. 3574, pp. 518–531. Springer, Heidelberg (2005)

MicroEliece: McEliece for Embedded Devices

Thomas Eisenbarth, Tim Güneysu, Stefan Heyse, and Christof Paar

Horst Görtz Institute for IT Security
Ruhr University Bochum
44780 Bochum, Germany
{eisenbarth,gueneysu,heyse,cpaar}@crypto.rub.de

Abstract. Most advanced security systems rely on public-key schemes based either on the factorization or the discrete logarithm problem. Since both problems are known to be closely related, a major breakthrough in cryptanalysis tackling one of those problems could render a large set of cryptosystems completely useless. The McEliece public-key scheme is based on the alternative security assumption that decoding unknown linear binary codes is NP-complete. In this work, we investigate the efficient implementation of the McEliece scheme on embedded systems what was – up to date – considered a challenge due to the required storage of its large keys. To the best of our knowledge, this is the first time that the McEliece encryption scheme is implemented on a low-cost 8-bit AVR microprocessor and a Xilinx Spartan-3AN FPGA.

1   Introduction

The advanced properties of public-key cryptosystems are required for many cryptographic issues, such as key establishment between parties and digital signatures. In this context, RSA, ElGamal, and later ECC have evolved as most popular choices and build the foundation for virtually all practical security protocols and implementations with requirements for public-key cryptography. However, these cryptosystems rely on two primitive security assumptions, namely the factoring problem (FP) and the discrete logarithm problem (DLP), which are also known to be closely related. With a significant breakthrough in cryptanalysis or a major improvement of the best known attacks on these problems (i.e., the *Number Field Sieve* or *Index Calculus*), a large number of recently employed cryptosystems may turn out to be insecure overnight. Already the existence of a quantum computer that can provide computations on a few thousand qubits would render FP and DLP-based cryptography useless. Though quantum computers of that dimension have not been reported to be built yet, we already want to encourage a larger *diversification* of cryptographic primitives in future public-key systems. However, to be accepted as real alternatives to conventional systems like RSA and ECC, such security primitives need to support efficient implementations with a comparable level of security on recent computing platforms. For example, one promising alternative are public-key schemes based on Multivariate Quadratic (MQ) polynomials for which hardware implementations were proposed on CHES 2008 [11].

C. Clavier and K. Gaj (Eds.): CHES 2009, LNCS 5747, pp. 49–64, 2009.

In this work, we demonstrate the efficient implementation of another public-key cryptosystem proposed by Robert J. McEliece in 1978 that is based on coding theory [22]. The McEliece cryptosystem incorporates a linear error-correcting code (namely a Goppa code) which is hidden as a general linear code. For Goppa codes, fast decoding algorithms exist when the code is known, but decoding codewords without knowledge of the coding scheme is proven NP-complete [5]. Contrary to DLP and FP-based systems, this makes this scheme also suitable for post-quantum era since it will remain unbroken when appropriately chosen security parameters are used [8].

The vast majority[1] of today's computing platforms are embedded systems. Only a few years ago, most of these devices could only provide a few hundred bytes of RAM and ROM which was a tight restriction for application (and security) designers. Thus, the McEliece scheme was regarded impracticable on such small and embedded systems due to the large size of the private and public keys. But nowadays, recent families of microcontrollers provide several hundreds of bytes of Flash-ROM. Moreover, recent off-the-shelf hardware such as FPGAs also contain dedicated memory blocks and Flash memories that support on-chip storage of up to a few megabits of data. In particular, these memories can be used to store the keys of the McEliece cryptosystem.

In this work, we present first implementations of the McEliece cryptosystem on a popular 8-bit AVR microcontroller, namely the ATxMega192, and a Xilinx Spartan-3AN 1400 FPGA which are both suitable for many embedded system applications. To the best of our knowledge, no implementations for the McEliece scheme have been proposed targeting embedded platforms. Fundamental operations for McEliece are based on encoding and decoding binary linear codes in binary extension fields that, in particular, can be implemented very efficiently in dedicated hardware. Unlike FP and DLP-based cryptosystems, operations on binary codes do not require computationally expensive multi-precision integer arithmetic what is beneficial for small computing platforms.

This paper is structured as follows: we start with a brief introduction to McEliece encryption and shortly explain necessary operations on Goppa codes. In Section 4, we discuss requirements and strategies to implement McEliece on memory-constrained embedded devices. Section 5 and Section 6 describe our actual implementations for an AVR 8-bit microprocessor and a Xilinx Spartan-3AN FPGA. Finally, we present our results for these platforms in Section 7.

2   Previous Work

Although invented already more than 30 years ago, the McEliece encryption scheme has never gained much attention due to its large keys and thus has not been implemented in many products. The most recent implementation of the McEliece scheme is due to Biswas and Sendrier [10] and presented a slightly modified version for PCs that achieves about 83 bit security (taken the attack in

---

[1] Already in 2002, 98% of 32-bit microprocessors in world-wide production were integrated in embedded platforms.

Algorithm 1. McEliece Message Encryption

Input: $m, K_{pub} = (\hat{G}, t)$
Output: Ciphertext c
 1: Encode the message m as a binary string of length k
 2: $c' \leftarrow m \cdot \hat{G}$
 3: Generate a random n-bit error vector z containing at most t ones
 4: $c = c' + z$
 5: **return** c

Algorithm 2. McEliece Message Decryption

Input: $c, K_{sec} = (P^{-1}, G, S^{-1})$
Output: Plaintext m
 1: $\hat{c} \leftarrow c \cdot P^{-1}$
 2: Use a decoding algorithm for the code C to decode $\hat{c}$ to $\hat{m} = m \cdot S$
 3: $m \leftarrow \hat{m} \cdot S^{-1}$
 4: **return** m

---

[8] into account). Comparing their implementation to other public key schemes, it turns out that McEliece encryption can even be faster than that of RSA and NTRU [7]. In addition to that, only few further McEliece software implementations have been published up to now and they were all designed for 32 bit architectures [25,26]. The more recent implementation [26] is available only as uncommented C-source code and was nevertheless used for the open-source P2P software Freenet and Entropy [15].

Hardware implementations of the original McEliece cryptosystem do not exist, except for a proof-of-concept McEliece-based signature scheme that was designed for a Xilinx Virtex-E FPGA [9]. Hence, we here present the first FPGA-based hardware and 8-bit software implementation of the McEliece public-key encryption scheme up to date.

3   Background on the McEliece Cryptosystem

The McEliece scheme is a public key cryptosystem based on linear error-correcting codes. The secret key is the generator matrix G of an error-correcting code with dimension k, length n and error correcting capability t. To create a public key, McEliece defined a random $k \times k$-dimensional scrambling matrix S and $n \times n$-dimensional permutation matrix P disguising the structure of the code by computing the product $\hat{G} = S \times G \times P$. Using the public key $K_{pub} = (\hat{G}, t)$ and private key $K_{sec} = (P^{-1}, G, S^{-1})$, encryption and decryption algorithms can be given by Algorithm 1 and Algorithm 2, respectively.

Note that Algorithm 1 only consists of a simple matrix multiplication with the input message and then distributes t random errors on the resulting code word. Thus, the generation of random error vectors requires an appropriate random number generator to be available on the target platform.

Decoding the ciphertext c for decryption as shown in Algorithm 2 is the most time-consuming process and requires several more complex operations in binary extension fields. In Section 3.1 we briefly introduce the required steps for decoding codewords that we need to implement on embedded systems.

As mentioned in the introduction, the main caveat against the McEliece cryptosystem is the significant size of the public and private key. The choice of even a minimal set of security parameters ($m = 10, n = 1024, t = 38, k \geq 644$) according to [23] already translates to a size of 80.5 kByte for the public key and at least 53 kByte for the private key (without any optimizations). However, this setup only provides the comparable security of a 60 bit symmetric cipher. For appropriate 80 bit security, even larger keys, for example the parameters $m = 11, n = 2048, t = 27, k \geq 1751$, are required (more details in Section 3.2).

Many optimizations (cf. Section 4.2) of the original McEliece scheme focus on size reduction of the public key, since the public-key has to be distributed. Hence, a size reduction of K_{pub} is directly beneficial for *all* parties. However, the situation is different when implementing McEliece on embedded platforms: note that the private key must be kept secret at all times and thus should be stored in a protected location on the device (that may be used in a potentially untrustworthy environment). An effective approach for secret key protection is the use of secure on-chip key memories that would require (with appropriate security features such as prohibited memory readback) invasive attacks on the chip to reveal the key. However, secure storage of key bits usually prove costly in hardware so that effective strategies are required to reduce the size of the private key to keep costs low. Addressing this issue, we demonstrate for the first time how to use on-the-fly generation of the large scrambling matrix S^{-1} for the McEliece instead of storing it in memory as in previous implementations. More details on the reduction of the key size are given in Section 4.2.

3.1 Classical Goppa Codes

Theorem 1. *Let $G(z)$ be an irreducible polynomial of degree t over $GF(2^m)$. Then the set*

$$\Gamma(G(z), GF(2^m)) = \{(c_\alpha)_{\alpha \in GF(2^m)} \in \{0,1\}^n \mid \sum_{\alpha \in GF(2^m)} \frac{c_\alpha}{z - \alpha} \equiv 0\} \quad (1)$$

defines a binary Goppa code C of length $n = 2^m$, dimension $k \geq n - mt$ and minimum distance $d \geq 2t+1$. The set of the c_α is called the support of the code. A fast decoding algorithm exists with a runtime of $n \cdot t$.

For each irreducible polynomial $G(z)$ over $GF(2^m)$ of degree t exists a binary Goppa code of length $n = 2^m$ and dimension $k = n - mt$. This code is capable of correcting up to t errors [4] and can be described as a $k \times n$ generator matrix G such that $C = \{mG : m \in F_2^k\}$.

To encode a message m into a codeword c, represent the message m as a binary string of length k and multiply it with the $k \times n$ matrix G.

However, decoding such a codeword r on the receiver's side with a (possibly) additive error vector e is far more complex. For decoding, we use Patterson's algorithm [24] with improvements from [29].

Since $r = c + e \equiv e \mod G(z)$ holds, the syndrome $Syn(z)$ of a received codeword can be obtained from Equation (1) by

$$Syn(z) = \sum_{\alpha \in GF(2^m)} \frac{r_\alpha}{z - \alpha} \equiv \sum_{\alpha \in GF(2^m)} \frac{e_\alpha}{z - \alpha} \mod G(z) \tag{2}$$

To finally recover e, we need to solve the key equation $\sigma(z) \cdot Syn(z) \equiv \omega(z) \mod G(z)$, where $\sigma(z)$ denotes a corresponding error-locator polynomial and $\omega(z)$ denotes an error-weight polynomial. Note that it can be shown that $\omega(z) = \sigma(z)'$ is the formal derivative of the error-locator and by splitting $\sigma(z)$ into even and odd polynomial parts $\sigma(z) = a(z)^2 + z \cdot b(z)^2$, we finally determine the following equation which needs to be solved to determine error positions:

$$Syn(z)(a(z)^2 + z \cdot b(z)^2) \equiv b(z)^2 \mod G(z) \tag{3}$$

To solve Equation (3) for a given codeword r, the following steps have to be performed:

1. From the received codeword r compute the syndrome $Syn(z)$ according to Equation (2). This can also be done using simple table-lookups.
2. Compute an inverse polynomial $T(z)$ with $T(z) \cdot Syn(z) \equiv 1 \mod G(z)$ (or provide a corresponding table). It follows that $(T(z) + z)b(z)^2 \equiv a(z)^2 \mod G(z)$.
3. There is a simple case if $T(z) = z \Rightarrow a(z) = 0$ s.t. $b(z)^2 \equiv z \cdot b(z)^2 \cdot Syn(z) \mod G(z) \Rightarrow 1 \equiv z \cdot Syn(z) \mod G(z)$ what directly leads to $\sigma(z) = z$. Contrary, if $T(z) \neq z$, compute a square root $R(z)$ for the given polynomial $R(z)^2 \equiv T(z) + z \mod G(z)$. Based on a observation by Huber [19] we can then determine solutions $a(z), b(z)$ satisfying

$$a(z) = b(z) \cdot R(z) \mod G(z). \tag{4}$$

Algorithm 3. Decoding Goppa Codes

Input: Received codeword r with up to t errors, inverse generator matrix iG
Output: Recovered message $\hat{m}$
  1: Compute syndrome $Syn(z)$ for codeword r
  2: $T(z) \leftarrow Syn(z)^{-1} \mod G(z)$
  3: **if** $T(z) = z$ **then**
  4:     $\sigma(z) \leftarrow z$
  5: **else**
  6:     $R(z) \leftarrow \sqrt{T(z) + z}$
  7:     Compute $a(z)$ and $b(z)$ with $a(z) \equiv b(z) \cdot R(z) \mod G(z)$
  8:     $\sigma(z) \leftarrow a(z)^2 + z \cdot b(z)^2$
  9: **end if**
 10: Determine roots of $\sigma(z)$ and correct errors in r which results in $\hat{r}$
 11: $\hat{m} \leftarrow \hat{r} \cdot iG$ {Map r_{cor} to $\hat{m}$}
 12: **return** $\hat{m}$

Finally, we use the identified $a(z), b(z)$ to construct the error-locator polynomial $\sigma(z) = a(z)^2 + z \cdot b(z)^2$.

4. The roots of $\sigma(z)$ denote the positions of error bits. If $\sigma(\alpha_i) \equiv 0 \mod G(z)$ with α_i being the corresponding bit of a generator in $GF(2^{11})$, there was an error in the position i in the received codeword that can be corrected by bit-flipping.

This decoding process, as required in Step 2 of Algorithm 2 for message decryption, is finally summarized in Algorithm 3.

3.2 Security Parameters

All security parameters for cryptosystems are chosen in a way to provide sufficient protection against the best known attack (whereas the notion of "sufficient" is determined by the requirements of an application). A recent paper [8] by Bernstein et al. presents a state-of-the-art attack of McEliece making use of a list decoding algorithm [6] for binary Goppa codes.

This attack reduces the binary work factor to break the original McEliece scheme with a $(1024, 524)$ Goppa code and $t = 50$ to $2^{60.55}$ bit operations. According to [8], Table 1 summarizes the security parameters for specific security levels.

4 Design Criteria for Embedded Systems

In this section, we discuss our assumptions, requirements and restrictions which are required when implementing the original McEliece cryptosystem on small, embedded systems. Target platforms for our investigation are 8-bit AVR microprocessors as well as low-cost Xilinx Spartan-3AN FPGAs. Some devices of these platforms come with large integrated Flash-RAMs (e.g., 192 kByte and 2,112 kByte for an AVR ATxMega192 and Spartan-3AN XC3S1400AN, respectively).

4.1 Requirements and Assumptions

For many embedded systems such as prepaid phones or micropayment systems, the short life cycle or comparably low value of the enclosed product often does

Table 1. Security of the McEliece scheme

Security Level	Parameters (n, k, t), errors added	Size K_{pub} in KBits	Size K_{sec} $(G(z), P, S)$ in KBits
Short-term (60 bit)	$(1024, 644, 38), 38$	644	$(0.38, 10, 405)$
Mid-term (80 bit)	$(2048, 1751, 27), 27$	$3,502$	$(0.30, 22, 2994)$
Long-term (256 bit)	$(6624, 5129, 115), 117$	$33,178$	$(1.47, 104, 25690)$

not demand for very long-term security, Hence, mid-term security parameters for public-key cryptosystems providing a comparable security to 64-80 key bits of symmetric ciphers are often regarded sufficient (and help reducing system costs). Hence, our implementations are designed for security parameters that correspond to an 80 bit key size of a symmetric cipher. A second important design requirement is the processing and storage of the private key solely *on-chip* so that all secrets are optimally never used outside the device. With appropriate countermeasures to prevent data extraction from on-chip memories, an attacker can then recover the private key only by sophisticated invasive attacks. For this purpose, AVR μCs provide a lock-bit feature to enable write and read/write protection of the Flash memory [2]. Similar mechanisms are also available for Spartan-3AN FPGAs preventing configuration and Flash readback from chip internals, e.g., using JTAG or ICAP interfaces [27]. Note that larger security parameters of the McEliece scheme are still likely to conflict with this requirement due to the limited amount of permanent on-chip memories of today's embedded platforms.

Analyzing McEliece encryption and decryption algorithms (cf. Section 3.1), the following arithmetic components are required supporting computations in $GF(2^m)$: a multiplier, a squaring unit, calculation of square roots, and an inverter. Furthermore, a binary matrix multiplier for encryption and a permutation element for step 2 in Algorithm 1 are needed. Many arithmetic operations in McEliece can be replaced by table lookups to significantly accelerate computations at the cost of additional memory. For both implementations in this work, our primary goal is area and memory efficiency to fit the large keys and required lookup-tables into the limited on-chip memories of our embedded target platforms.

The susceptibility of the McEliece cryptosystem to side channel attacks has not extensively been studied, yet. However, embedded systems can always be subject to passive attacks such as timing analysis [20] and power/EM analysis [21]. In [28], a successful timing attack on the Patterson algorithm was demonstrated. The attack does not recover the key, but reveals the error vector z and hence allows for efficient decryption of the message c. Our implementations are not susceptible to this attack due to unconditional instruction execution, e.g., our implementation will not terminate after a certain number of errors have been corrected. Differential EM/power attacks and timing attacks are impeded by the permutation and scrambling operations (P and S) obfuscating all internal states, and finally, the large key size. Yet template-like attacks [12] might be feasible if no further protection is applied.

4.2   Reducing Memory Requirements

To make McEliece-based cryptosystems more practical (i.e., to reduce the key sizes), there is an ongoing research to replace the code with one that can be represented in a more compact way.

Using a naïve approach in which the support of the code is the set of all elements in $GF(2^m)$ in lexicographical order and both matrices S, P are totally random, the public key $\hat{G} = S \times G \times P$ becomes a random $n \times k$ matrix. However, since P is a sparse permutation matrix with only a single 1 in each row and column, it is more efficient to store only the positions of the ones, resulting in an array with $n \cdot m$ bits.

Another trick to reduce the public key size is to convert $\hat{G}$ to systematic form $\{I_k \mid Q\}$, where I_k is the $k \times k$ identity matrix. Then, only the $(k \times (n - k))$ matrix Q is published [14].

In the last step of code decoding (Algorithm 3), the k message bits out of the n (corrected) ciphertext bits need to be extracted. Usually, this is done by a mapping matrix iG with $G \times iG = I_k$. But if G is in systematic form, then this step can be omitted, since the first k bits of the corrected ciphertext corresponds to the message bits. Unfortunately, G and $\hat{G}$ cannot both be systematic at the same time, since then $\hat{G} = \{I_k \mid \hat{Q}\} = S \times \{I_k \mid Q\} \times P$ and S would be the identity matrix which is inappropriate for use as the secret key.

For reduction of the secret key size, we chose to generate the large scrambling matrix S^{-1} on-the-fly using a cryptographic pseudo random number generator (CPRNG) and a seed. During key generation, it must be ensured that the seed does not generate a singular matrix S^{-1}. Depending on the target platform and available cryptographic accelerators, there are different options to implement such a CPRNG (e.g. AES in counter mode or a hash-based PRNG) on embedded platforms. However, the secrecy of S^{-1} is not required for hiding the secret polynomial $G(z)$ [14].

5   Implementation on AVR Microprocessors

In this section, we discuss our implementation of the McEliece cryptosystem for 8-bit AVR microcontrollers, a popular family of 8-bit RISC microcontrollers (μC) used in embedded systems. The Atmel AVR processors operate at clock frequencies of up to 32 MHz, provide few kBytes of SRAM, up to hundreds of kBytes of Flash program memory, and additional EEPROM or mask ROM. For our design, we chose an ATxMega192A1 μC due to its 16 kBytes of SRAM and the integrated crypto accelerator engine for DES and AES [2]. The crypto accelerator is particularly useful for a fast implementation of a CPRNG that generates the scrambling matrix S^{-1} on-the-fly. Arithmetic operations in the underlying field $GF(2^{11})$ can be performed efficiently with a combination of polynomial and exponential representation. We store the coefficients of a value $a \in GF(2^{11})$ in memory using a polynomial basis with natural order. Given an $a = a_{10}\alpha^{10} + a_9\alpha^9 + a_8\alpha^8 + \cdots + a_0\alpha^0$, the coefficient $a_i \in GF(2)$ is determined by bit i of an unsigned 16 bit integer where bit 0 denotes the least significant bit. In this representation, addition is fast just by performing an exclusive-or operation on 2×2 registers. For more complex operations, such as multiplication, squaring, inversion and root extraction, an exponential representation is more suitable. Since every element except zero in $GF(2^{11})$ can be written as a power of some primitive element α, all elements in the finite field can also be represented by α^i with $i \in \mathbb{Z}_{2^m-1}$. Multiplication and squaring can then be performed by adding the exponents of the factors over $\mathbb{Z}_{2^m-1}$ such as

$$c = a \cdot b = \alpha^i \cdot \alpha^j = \alpha^{i+j} \mid a, b \in GF(2^{11}), 0 \le i, j \le 2^m - 2. \qquad (5)$$

If one of the elements equals zero, obviously the result is zero. The inverse of a value $d \in GF(2^{11})$ in exponential representation $d = \alpha^i$ can be obtained from a single subtraction in the exponent $d^{-1} = \alpha^{2^{11}-1-i}$ with a subsequent table-lookup. Root extraction, i.e., given a value $a = \alpha^i$ to determine $r = a^{i/2}$ is simple, when i is even and can be performed by a simple right shift on index i. For odd values of i, $m - 1 = 10$ left shifts followed by a reduction with $2^{11} - 1$ determine the square root.

To allow for efficient conversion between the two representations, we employ two precomputed tables (so called *log* and *antilog* tables) that enable fast conversion between polynomial and exponential representation. Each table consists of 2048 11-bit values that are stored as a pair of two bytes in the program memory. Hence, each lookup table consumes 4 kBytes of Flash memory. Due to frequent access, we copy the tables into the faster SRAM at startup time. Accessing the table directly from Flash memory significantly reduces performance, but allows migration to a (slightly) cheaper device with only 4 kBytes of SRAM. For multiplication, squaring, inversion, and root extraction, the operands are transformed on-the-fly to exponential representation and reverted to the polynomial basis after finishing the operation.

5.1   Generation and Storage of Matrices

All matrices as shown in Table 2 are precomputed and stored in Flash memory of the μC. We store the permutation matrix P^{-1} as an array of 2048 16-bit unsigned integers containing 11-bit indices. Matrix G is written in transposed form to simplify multiplications (i.e., all columns are stored as consecutive words in memory for straightforward index calculations). Additionally, arrays for the support of the code, its reverse mapping, and the precomputed inverse polynomials (in the order as they correspond to the ciphertext bits) reside in Flash memory as well. Since the scrambling matrix S^{-1} is too large to be stored in program memory, we opted to generate it on-the-fly from an 80-bit seed, employing the integrated DES-accelerator engine of the ATxMega as a CPRNG.

Encryption is a straightforward binary matrix-vector multiplication and does not require field arithmetic in $GF(2^{11})$. However, the large public-key matrix K_{pub} does not fit into the 192 kByte internal Flash memory. Hence, at least 512 kByte external memory are required for storing the public key $\hat{G}$. Note that the ATxMega can access external memories at the same speed as internal SRAM.

Table 2 shows the requirements of precomputed tables separated by actual size and required size in memory including the necessary 16-bit address alignment and/or padding.

5.2   System and Compiler Limitations

Due to the large demand for memory, we need to take care of some peculiarities in the memory management of the AVR microcontroller. Since originally AVR

Table 2. Sizes of tables and values in memory including overhead for address alignment

Use	Name	Actual Size	Size in Memory
Encryption	Public Key $\hat{G}$	448,256 byte	448,512 byte
Decryption	Private Key S^{-1} (IV only)	10 byte	10 byte
Decryption	Private Key P^{-1} array	2,816 byte	4,096 byte
Decoding	Syndrome table	76,032 byte	110,592 byte
Decoding	Goppa polynomial	309 bits	56 byte
Decoding	ω-polynomial	297 bits	54 byte
Decoding	Log table	22,528 bits	4,096 byte
Decoding	Antilog table	22,528 bits	4,096 byte

microcontrollers supported only a small amount of internal memory, the AVR uses 16 bit pointers to access its Flash memory. Additionally, each Flash cell comprises 16 bit of data, but the μC itself can only handle 8 bit. Hence, one bit of this address pointer must be reserved to select the corresponding byte in the retrieved word, reducing the maximal address range to 64 KByte (or 32K 16 bit words). To address memory segments beyond 64K, additional RAMP-registers need to be used. Additionally, the used avr-gcc compiler internally treats pointers as signed 16 bit integer halving again the addressable memory space. For this reason, all arrays larger than 32 Kbyte need to be split into multiple parts resulting in an additional overhead in the program code.

6   Implementation on Xilinx FPGAs

Since our target device is a low-cost Spartan-3 with moderate logic resources, we only parallelized and unrolled the most time consuming parts of the algorithms such as the polynomial multiplier and inverter. Alike the AVR implementation, we decided to implement less intensive operations of the field arithmetic (i.e., inversion, division, squaring and square roots for single field elements over $GF(2^{11})$) using precomputed log- and antilog tables which are stored in dedicated memory components (BRAM) of the FPGA (cf. Section 5.1). With such precomputed tables being available, the number of computational units in hardware can be reduced what also affects the number of required Lookup-Tables (LUT) in the Configurable Logic Blocks (CLB) of the FPGA. However, note that only 32 BRAMs are available on the Spartan-3AN 1400 FPGA (which is the largest, low-cost device of its class). This limits the option to have more than one instance of each table for allowing parallel access (besides using the dual-port feature of the BRAM). Hence, lookups to these tables need to be serialized in most cases. Since the runtime of polynomial multiplication and polynomial squaring is crucial for the overall system performance (cf. Steps 7 and 8 of Algorithm 3), we opted for a parallel polynomial multiplier instead of using the log and antilog tables as well. The polynomial multiplier consists of 27 coefficient multipliers over $GF(2^{11})$ (the topmost coefficient is treated separately) of which each coefficient multiplication is realized as logic directly in LUTs by linear combination of the input bits and the field polynomial. Hence, the multiplication of a polynomial B with a coefficient a (i.e., $C = a \cdot B \mid a \in GF(2^{11}), B, C \in \frac{F[z]}{G(z)}$) can be performed in a single clock cycle. All field operations, such as root extraction, division, and inversion can be completed in 6 clock cycles using log and antilog tables, of which two clock cycles are for the conversion to exponential representation, two are required for the corresponding operation and additional two cycles for the reverse translation. Note that for several subsequent field computations the conversion can be interleaved with the arithmetic operation so that only 4 cycles for each subsequent operations are required.

The remaining, time-critical component is the polynomial inverter which is used in step 1 and step 2 of Algorithm 3, for example to compute the parity check matrix H on-the-fly. An average of 1024 inverses need to be computed for which we implemented the binary Extended Euclidean Algorithm (EEA) over $GF(2^{11})$ in hardware. Note that each cycle of the polynomial EEA requires exactly one coefficient division (which is realized again using the log and antilog tables). In conclusion, the EEA is the largest component in our design (about 64%) and thus also comprises the critical path of the implementation. For the generation of the inverse scrambling matrix S^{-1} on the FPGA, we implemented a CPRNG based on the 80-bit low-footprint block cipher PRESENT. Note that as an alternative, we store the large static table S^{-1} in the in-system Flash memory of the FPGA. However, due to limitations of the serial SPI-Interface we only can access a single bit of S^{-1} at a maximum frequency of $50\,\mathrm{MHz}$ that significantly degrades our decryption performance.

This limitation also applies to the public-key matrix $\hat{G}$ which is required for the encoding process during encryption. Since this matrix is too large to fit into the 32 18 kBit BRAMs of our Spartan-3AN device, we need to store it in Flash memory. To avoid a performance penalty due to the slow SPI interface to the Flash, we could first load K_{pub} into an external DDR2-333 memory at system startup which then can be accessed via a fast memory controller to retrieve K_{pub} for encryption. With such undamped access to K_{pub}, we could gain a performance speedup for encryption by a factor of 62 (1.15 ms) with respect to loading K_{pub} directly from Flash (71.44 ms). We successfully verified this approach by testing our implementation on a test board providing external SRAM (however, no DDR2 memory). The interface between SRAM and FPGA is realized as 16 bit bus, clocked at $100\,\mathrm{MHz}$ and two clock cycles access time per read.

Due to the limited logic on our FPGA, we thus opted for an individual device configuration for encryption and decryption, of which one can be selected during system startup. Both configurations can be stored within the large, internal Flash memory of the FPGA. Using the multi-boot features of Spartan-3 devices, the corresponding configuration can also be loaded by the FSM (using the internal SPI-interface) during runtime whenever switching between encryption and decryption is necessary. The McEliece implementation (decryption configuration) for the Spartan-3AN FPGA is depicted in Figure 1.

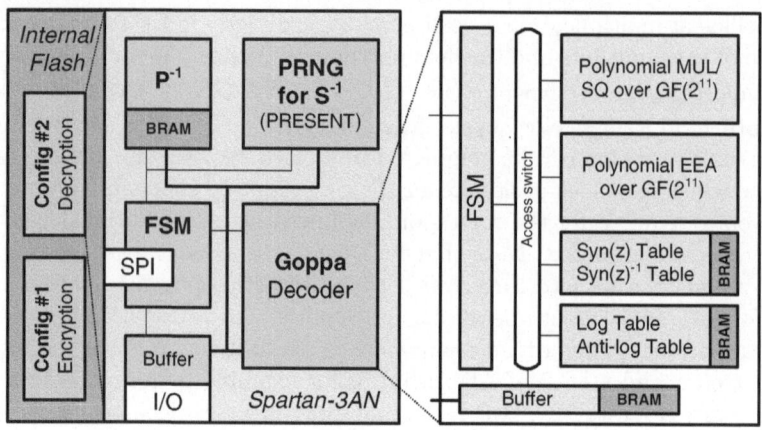

Fig. 1. McEliece implementation on a Spartan-3AN FPGA

7 Results

We now present the results for our two McEliece implementations providing 80 bit security ($n = 2048, k = 1751, t = 27$) for the AVR 8-bit microcontroller and the Xilinx Spartan-3AN FPGA. We report performance figures for the ATxMega192A1 obtained from the avr-gcc compiler v4.3.2 and a Xilinx Spartan-3AN XC3S1400AN-5 FPGA using Xilinx ISE 10.1. The resource requirements for our μC design and FPGA implementation after place-and-route (PAR) are shown in Table 3.

Table 4 summarizes the clock cycles needed for every part of the de- and encryption routines both for the FPGA and the microcontroller implementation.

In our FPGA design, the CPRNG to generate S^{-1} based on the PRESENT block cipher turns out to be a bottleneck of our implementation since the matrix generation does not meet the performance of the matrix multiplication. By replacing the CPRNG by a more efficient solution, we can untap the full

Table 3. Implementation results of the McEliece scheme with $n = 2048, k = 1751, t = 27$ on the AVR ATxMega192 μC and Spartan-3AN XC3S1400AN-5 FPGA after PAR

	Resource	Encryption	Decryption	Available
μC	SRAM	512 Byte	12 kByte	16 kByte
	Flash Memory	684 Byte	130.4 kByte	192 kByte
	External Memory	438 kByte	–	–
$FPGA$	Slices	668 (6%)	11,218 (100%)	11,264
	LUTs	1044 (5%)	22,034(98%)	22,528
	FFs	804 (4%)	8,977 (40%)	22,528
	BRAMs	3 (9%)	20 (63%)	32
	Flash Memory	4,644 KBits	4,644 KBits	16,896 Kbits

Table 4. Performance of McEliece implementations with $n = 2048, k = 1751, t = 27$ on the AVR ATxMega192 μC and Spartan-3AN XC3S1400AN-5 FPGA

	Aspect	ATxMega192 μC	Spartan-3AN 1400
Encrypt.	Maximum frequency	32 MHz	150 MHz
	Encrypt $c^{\prime} = m \cdot \hat{G}$	14,404,944 cycles	(7,889,200)161,480 cycles
	Inject errors $c = c^{\prime} + z$	1,136 cycles	398 cycles
Decryption	Maximum frequency	32 MHz	85 MHz
	Undo permutation $c \cdot P^{-1}$	275,835 cycles	combined with $Syn(z)$
	Determine $Syn(z)$	1,412,514 cycles	360,184 cycles
	Compute $T = Syn(z)^{-1}$	1,164,402 cycles	625 cycles
	Compute $\sqrt{T + z}$	286,573 cycles	487 cycles
	Solve Equation (4) with EEA	318,082 cycles	312 cycles
	Correct errors	15,096,704 cycles	312,328 cycles
	Undo scrambling $\hat{m} \cdot S^{-1}$	1,196,984 cycles	1,035,684/217,800* cycles

* This figure is an estimate assuming that an ideal PRNG for generation of S^{-1} would be available.

Table 5. Comparison of our McEliece designs with single-core ECC and RSA implementations for 80 bit security

	Method	Platform	Time ms/op	Throughput bits/sec
8-bit μC	McEliece encryption	ATxMega192@32MHz	450	3,889
	McEliece decryption	ATxMega192@32MHz	618	2,835
	ECC-P160 (SECG) [17]	ATMega128@8MHz	810/203[1]	197/788[1]
	RSA-1024 $2^{16} + 1$ [17]	ATMega128@8MHz	430/108[1]	2,381/9,524[1]
	RSA-1024 random [17]	ATMega128@8MHz	10,990/2748[1]	93/373[1]
FPGA	McEliece encryption A	Spartan-3AN 1400-5	1.07[2]	1,626,517[2]
	McEliece encryption B	Spartan-3AN 1400-5	2.24[3]	779,948[3]
	McEliece decryption	Spartan-3AN 1400-5	21.61/10.82[4]	81,023/161,829[4]
	ECC-P160 [16]	Spartan-3 1000-4	5.1	31,200
	RSA-1024 random [18]	Spartan-3E 1500-5	51	20,275
	NTRU encryption [3]	Virtex 1000EFG860	0.005	50,876,908

[1] For a fair comparison with our implementations running at 32MHz, timings at lower frequencies were scaled accordingly.

[2] These are estimates are based on the usage of an external DDR-RAM.

[3] These are measurements based on our test setup with external SRAM running at 100MHz.

[4] These are estimates assuming that an ideal PRNG to generate S^{-1} is used.

performance of our implementation. Table 4 also gives estimates for a PRNG that does not incur any wait cycles due to throughput limitations.

The public-key cryptosystems RSA-1024 and ECC-P160 are assumed[2] to roughly achieve a similar margin of 80 bit symmetric security [1]. We finally compare our results to published implementations of these systems that target similar platforms (i.e., AVR ATMega μC and Xilinx Spartan-3 FPGAs). Note that the figures for ECC are obtained from the ECDSA signature scheme.

Embedded implementations of other alternative public key encryption schemes are very rare. The proprietary encryption scheme NTRUEncrypt has received some attention. An encryption-only hardware engine of NTRUEncrypt-251-128-3 for more advanced Xilinx Virtex platform has been presented in [3]. An embedded software implementation of the related NTRUSign performs one signature on an ATMega128L clocked at 7,37 MHz in 619 ms [13]. However, comparable performance figures of NTRU encryption and decryption for the AVR platform are not available.

Note that all throughput figures are based on the number of plaintext bits processed by each system and do not take any message expansion in the ciphertext into account.

8   Conclusions

In this paper, we described the first implementations of the McEliece public-key scheme for embedded systems using an AVR μC and a Xilinx Spartan-3AN FPGA. Our performance results for McEliece providing 80 bit security on these systems exceed the throughput but could not outperform comparable ECC cryptosystems with 160 bit in terms of number of operations per second. However, although our implementations still leave room for further optimizations, our results already show better performance than RSA-1024 on the selected platforms. Thus, we believe with growing memories in embedded systems, ongoing research and further optimizations, McEliece can evolve to a suitable and quantum computer-resistant alternative to RSA and ECC that have been extensively studied for years.

References

1. ECRYPT. Yearly Report on Algorithms and Keysizes (2007-2008). Technical report, D.SPA.28 Rev. 1.1, IST-2002-507932 ECRYPT (July 2008)
2. Atmel Corp. 8-bit XMEGA A Microcontroller. User Guide (February 2009), http://www.atmel.com/dyn/resources/prod_documents/doc8077.pdf
3. Bailey, D.V., Coffin, D., Elbirt, A., Silverman, J.H., Woodbury, A.D.: NTRU in Constrained Devices. In: Koç, Ç.K., Naccache, D., Paar, C. (eds.) CHES 2001. LNCS, vol. 2162, pp. 262–272. Springer, Heidelberg (2001)

---

[2] According to [1], RSA-1248 actually corresponds to 80 bit symmetric security. However, no implementation results for embedded systems are available for this key size.

4. Berlekamp, E.R.: Goppa codes. IEEE Trans. Information Theory IT-19(3), 590–592 (1973)
5. Berlekamp, E.R., McEliece, R.J., van Tilborg, H.C.A.: On the inherent intractability of certain coding problems. IEEE Trans. Information Theory 24(3), 384–386 (1978)
6. Bernstein, D.J.: List Decoding for Binary Goppa Codes. Technical report (2008), http://cr.yp.to/codes/goppalist-20081107.pdf
7. Bernstein, D.J., Lange, T.: eBACS: ECRYPT Benchmarking of Cryptographic Systems, February 17 (2009), http://bench.cr.yp.to
8. Bernstein, D.J., Lange, T., Peters, C.: Attacking and Defending the McEliece Cryptosystem. In: Buchmann, J., Ding, J. (eds.) PQCrypto 2008. LNCS, vol. 5299, pp. 31–46. Springer, Heidelberg (2008)
9. Beuchat, J.-L., Sendrier, N., Tisserand, A., Villard, G.: FPGA Implementation of a Recently Published Signature Scheme. Technical report, INRIA - Institut National de Recherche en Informatique et en Automatique (2004), http://hal.archives-ouvertes.fr/docs/00/07/70/45/PDF/RR-5158.pdf
10. Biswas, B., Sendrier, N.: McEliece crypto-system: A reference implementation, http://www-rocq.inria.fr/secret/CBCrypto/index.php?pg=hymes
11. Bogdanov, A., Eisenbarth, T., Rupp, A., Wolf, C.: Time-Area Optimized Public-Key Engines: MQ-Cryptosystems as Replacement for Elliptic Curves? In: Oswald, E., Rohatgi, P. (eds.) CHES 2008. LNCS, vol. 5154, pp. 45–61. Springer, Heidelberg (2008)
12. Chari, S., Rao, J.R., Rohatgi, P.: Template Attacks. In: Kaliski Jr., B.S., Koç, Ç.K., Paar, C. (eds.) CHES 2002. LNCS, vol. 2523, pp. 13–28. Springer, Heidelberg (2003)
13. Driessen, B., Poschmann, A., Paar, C.: Comparison of Innovative Signature Algorithms for WSNs. In: Proceedings of ACM WiSec 2008. ACM, New York (2008)
14. Engelbert, D., Overbeck, R., Schmidt, A.: A summary of mceliece-type cryptosystems and their security (2007)
15. Freenet and Entropy. Open-Source P2P Network Applications (2009), http://freenetproject.org and http://entropy.stop1984.com
16. Güneysu, T., Paar, C., Pelzl, J.: Special-Purpose Hardware for Solving the Elliptic Curve Discrete Logarithm Problem. ACM Transactions on Reconfigurable Technology and Systems (TRETS) 1(2), 1–21 (2008)
17. Gura, N., Patel, A., Wander, A., Eberle, H., Shantz, S.C.: Comparing Elliptic Curve Cryptography and RSA on 8-bit CPUs. In: Joye, M., Quisquater, J.-J. (eds.) CHES 2004. LNCS, vol. 3156, pp. 925–943. Springer, Heidelberg (2004)
18. Helion Technology Inc. Modular Exponentiation Core Family for Xilinx FPGA. Data Sheet (October 2008), http://www.heliontech.com/downloads/modexp_xilinx_datasheet.pdf
19. Huber, K.: Note on decoding binary Goppa codes. Electronics Letters 32, 102–103 (1996)
20. Kocher, P.C.: Timing Attacks on Implementations of Diffie-Hellman, RSA, DSS, and Other Systems. In: Koblitz, N. (ed.) CRYPTO 1996. LNCS, vol. 1109, pp. 104–113. Springer, Heidelberg (1996)
21. Mangard, S., Oswald, E., Popp, T.: Power Analysis Attacks: Revealing the Secrets of Smartcards. Springer, Heidelberg (2007)
22. McEliece, R.J.: A Public-Key Cryptosystem Based On Algebraic Coding Theory. Deep Space Network Progress Report 44, 114–116 (1978)
23. Menezes, A.J., van Oorschot, P.C., Vanstone, S.A.: Handbook of Applied Cryptography. CRC Press, New York (1996)

24. Patterson, N.: The algebraic decoding of Goppa codes. IEEE Transactions on Information Theory 21, 203–207 (1975)
25. Preneel, B., Bosselaers, A., Govaerts, R., Vandewalle, J.: A Software Implementation of the McEliece Public-Key Cryptosystem. In: Proceedings of the 13th Symposium on Information Theory in the Benelux, Werkgemeenschap voor Informatie en Communicatietheorie, pp. 119–126. Springer, Heidelberg (1992)
26. Prometheus. Implementation of McEliece Cryptosystem for 32-bit microprocessors (c-source) (2009), http://www.eccpage.com/goppacode.c
27. Smerdon, M.: Security Solutions Using Spartan-3 Generation FPGAs. Whitepaper (April 2008),
    http://www.xilinx.com/support/documentation/white_papers/wp266.pdf
28. Strenzke, F., Tews, E., Molter, H., Overbeck, R., Shoufan, A.: Side Channels in the McEliece PKC. In: Buchmann, J., Ding, J. (eds.) PQCrypto 2008. LNCS, vol. 5299, pp. 216–229. Springer, Heidelberg (2008)
29. Sugiyama, Y., Kasahara, M., Hirasawa, S., Namekawa, T.: A Method for Solving Key Equation for Decoding Goppa Codes. IEEE Transactions on Information and Control 27, 87–99 (1975)

Physical Unclonable Functions and Secure Processors

Srini Devadas

Professor and Associate Head
Department of EECS, MIT, Cambridge

Abstract. As computing devices become ever more pervasive, two contradictory trends are appearing. On one hand computing elements are becoming small, disseminated and unsupervised. On the other hand, the cost of security breaches is increasing as we place more responsibility on the devices that surround us. The result of these trends is that physical attacks present an increasing risk that must be dealt with.

Physical Unclonable Functions (PUFs) are a tamper resistant way of establishing shared secrets with a physical device. They rely on the inevitable manufacturing variations between devices to produce an identity for a device. This identity is unclonable, and in some cases is even manufacturer resistant (i.e., it is impossible to produce devices that have the same identity).

We describe a few applications of PUFs, including authentication of individual integrated circuits such as FPGAs and RFIDs, and the production of certificates that guarantee that a particular piece of software was executed on a trusted chip.

We present the design and implementation of two PUF-enabled devices that have been built: a low-cost secure RFID that can be used in anti-counterfeiting and other authentication applications, and a secure processor capable of performing certified execution and higher-level cryptographic functions.

The PUF-enabled RFID uses a simple challenge-response protocol for authentication that shifts complexity to the reader or server and therefore only requires a small number of transistors on the device side. The PUF-enabled processor generates its public/private key pair on power-up so its private key is never left exposed in (on-chip or off-chip) non-volatile storage. It is capable of a broad range of cryptographic functionality, including certified execution of programs.

C. Clavier and K. Gaj (Eds.): CHES 2009, LNCS 5747, p. 65, 2009.

Practical Electromagnetic Template Attack on HMAC

Pierre-Alain Fouque[1], Gaëtan Leurent[1], Denis Réal[2,3], and Frédéric Valette[3]

[1] École normale supérieure/CNRS/INRIA
{Pierre-Alain.Fouque,Gaetan.Leurent}@ens.fr
[2] INSA-IETR, 20 avenue des coesmes, 35043 Rennes, France
Denis.Real@insa-rennes.fr
[3] CELAR, 35 Bruz, France
{Denis.Real,Frederic.Valette}@dga.defense.gouv.fr

Abstract. In this paper, we show a very efficient side channel attack against HMAC. Our attack assumes the presence of a side channel that reveals the Hamming distance of some registers. After a *profiling phase* in which the adversary has access to a device and can configure it, the attack recovers the secret key by monitoring a *single execution* of HMAC-SHA-1. The secret key can be recovered using a "template attack" with a computation of about $2^{32}3^{\kappa}$ compression functions, where κ is the number of 32-bit words of the key. Finally, we show that our attack can also be used to break the secrecy of network protocols usually implemented on embedded devices.

We have performed experiments using a NIOS processor executed on a Field Programmable Gate Array (FPGA) to confirm the leakage model. We hope that our results shed some light on the requirements in term of side channel attack for the future SHA-3 function.

1 Introduction

HMAC is a hash-based message authentication code proposed by Bellare, Canetti and Krawczyk in 1996 [3]. It is very interesting to study for at least three reasons: HMAC is standardized (by ANSI, IETF, ISO and NIST) and widely deployed (*e.g.* SSL, TLS, SSH, IPsec); HMAC has security proofs [2,3]; and it is a rather simple construction. It is used in a lot of Internet standards. For instance embedded devices running IPsec protocols [16] have to implement it. There are many such efficient equipments on the market from router vendors that incorporate security protocols on their systems. It is crucial to study the security of such implementations since Virtual Private Network (VPN) products are widely deployed and used to secure important networks.

Recently, new attacks on HMAC based on Wang *et al.* [31,29,30,28] collision attacks have emerged. However, either their complexity is very high, or they attack a function no more widely used in practice such as HMAC-MD4, or the security model is not really practical such as the related key model [6,24,8]. Here, we focus on more practical attacks on HMAC. We show that when HMAC is

C. Clavier and K. Gaj (Eds.): CHES 2009, LNCS 5747, pp. 66–80, 2009.

implemented on embedded devices, and the attacker has access to the physical device to use a side channel, then there exist devastating attacks. These kind of attacks do not rely on previous collision attacks and can be applied on many hash functions, such as MD5 or SHA-2 for instance. Even though side channel attacks are believed to be very intrusive techniques, we show that our attack can be mounted using little contact with the targeted device.

We choose to illustrate our attack with HMAC since it is used in a lot of Internet standards. Beyond the integrity attacks, the side channel attack we describe, can be used to attack the confidentiality of other Internet standards such as the Layer Two Tunneling Protocol (L2TP [27]) or to attack the key derivation of IPsec in the Internet Key Exchange (IKE [13]) protocol. Our attack can also be applied on a famous side channel countermeasure, proposed by Kocher in [18] which is to derive a specific key for each call to the cryptographic application. This kind of protection is very efficient against classical DPA techniques and makes previous attacks infeasible. However our attack allows to recover the master key after listening to only two derivation processes.

1.1 Description of SHA-1

All the computations in SHA-1 are made on 32-bit words. We use $\boxplus$ to denote the modular addition, and $X^{\lll n}$ to denote the bit-wise rotation of X by n bits. SHA-1 is an iterated hash function following the Merkle-Damgård paradigm. The message is padded and cut into blocks of k bits (with $k = 512$ for SHA-1), and the digest is computed by iterating a compression function, starting with an initial value IV.

The compression function of SHA-1 is an unbalanced Feistel ladder with an internal state of five 32-bit registers A, B, C, D, E. The compression function has two inputs: a chaining value which is used as the initial value of the internal registers A_{-1}, B_{-1}, C_{-1}, D_{-1}; and a message block cut into 16 message words $M_0...M_{15}$. The message is expanded into 80 words $W_0...W_{79}$, such that $W_i = M_i$ for $i < 16$. Then we iterate 80 steps, where each step updates one of the registers. Each step uses one word W_i of the expanded message. If we use A_i, B_i, C_i, D_i, E_i to denote the value of the registers after the step i, the compression function of SHA-1 can be described by:

Step update:	$A_{i+1} = \Phi_i \boxplus A_i^{\lll 5} \boxplus W_i \boxplus E_i \boxplus K_i$
:	$\Phi_i = f_i(B_i, C_i, D_i)$
:	$B_{i+1} = A_i \quad C_{i+1} = B_i^{\lll 30} \quad D_{i+1} = C_i \quad E_{i+1} = D_i$
Input:	$A_{-1} \parallel B_{-1} \parallel C_{-1} \parallel D_{-1} \parallel E_{-1}$
Output:	$A_{-1} \boxplus A_{79} \parallel B_{-1} \boxplus B_{79} \parallel C_{-1} \boxplus C_{79} \parallel D_{-1} \boxplus D_{79} \parallel E_{-1} \boxplus E_{79}$

1.2 Description of HMAC

HMAC is a hash-based message authentication code proposed by Bellare, Canetti and Krawczyk [3]. Let H be an iterated Merkle-Damgård hash function. HMAC is defined by

$$\text{HMAC}_k(M) = H(\bar{k} \oplus \text{opad} \,\|\, H(\bar{k} \oplus \text{ipad} \,\|\, M)),$$

where M is the message, k is the secret key, $\bar{k}$ its completion to a single block of the hash function, ipad and opad are two fixed one-block values.

The security of HMAC is based on that of NMAC. Since H is assumed to be based on the Merkle-Damgård paradigm, we denote by H_k the modification of H where the public IV is replaced by the secret key k. Then NMAC with secret key (k_1, k_2) is defined by:

$$\text{NMAC}_{k_1, k_2}(M) = H_{k_1}(H_{k_2}(M)).$$

We call k_2 the inner key, and k_1 the outer key. Due to the iterative structure of H, HMAC_k is essentially equivalent to $\text{NMAC}_{H(\bar{k} \oplus \text{opad}), H(\bar{k} \oplus \text{ipad})}$.

Any key-recovery attack against NMAC can be used to recover an equivalent inner key $H(\bar{k} \oplus \text{ipad})$ and an equivalent outer key $H(\bar{k} \oplus \text{opad})$ in HMAC. This information is equivalent to the key of HMAC, since it is sufficient to compute any MAC. Most previous attacks against HMAC are of this kind, but our attack is different: we will use a side channel during the computation of $H(\bar{k} \oplus \text{ipad})$ and $H(\bar{k} \oplus \text{opad})$ to recover information about the key k. Thus our attack cannot be used against NMAC.

1.3   Related Work on Side Channel Attacks

Since there is no efficient and practical attacks against HMAC, it is interesting to study the security of this function against side channel attacks. Similarly, Kelsey *et al.* have studied the security of block ciphers using different leakage models [15,14].

A classical side channel attack on HMAC has been proposed without experiments by Lemke *et al.* in 2005 using a Differential Power Analysis [19] in [20]. They show that a forgery attack can be mounted by performing a multi-bit DPA since SHA-1 manipulates 32-bit registers. This attack allows to recover the inner and outer keys of HMAC, but does not allow to retrieve the initial key. Note that other DPA attacks are reported on HMAC based on other hash functions such as [22,10,21]. But none of them allow to retrieve the initial key value as the message is not directly mixed with the initial key but only with the derivated subkeys.

To our knowledge, no Template Attacks (TA) [5,25] have never been applied on HMAC. This is mainly due to the fact that the manipulated registers are 32 bits long which make classical templates attacks infeasible.

1.4   Our Results

The aim of this paper is two-fold. On the one hand, we assume that we have a side channel that leaks the number of flipped bits when a value is loaded from the memory to a register. We show that this side channel is sufficient to recover the secret key used in HMAC. On the other hand, we show that this side channel is available in practice. A similar attack can also be used against other standards or against key derivation process.

Our attack is quite different from previous side-channel attacks: we do not require knowledge of the message being signed, and we only need to measure *one* execution of HMAC. Thus some classical countermeasures against DPA will not affect our attack.

In our attack model, the adversary can profile the device during a first offline stage. This assumption is classical, and is used in template attacks. It seems reasonable since the adversary can buy and test its own device. During this profiling phase, the adversary generates curves when loading data from the memory to a register with all the possible Hamming distances. Then, in a second online stage, he has access to the targeted device during one execution of the implementation. During this phase, according to the recorded values, the Hamming distance can be learned using a single matching between the curves. Finally, all theses Hamming distances give some information about the key, and we can combine them to recover the key, even though the secret-dependent variables are not totally independent. The simulation of the attack has been experimentally tested to show that we are able to recover the secret key in practice and to show that the scale can be generated.

Our attack is based on two important facts. First, the key of HMAC is used as a message in the hash function. The message is processed word by word, as opposed to the IV which is used all at once, so we can extract information that depends on only one word of the key. Second, we have two related computations that leak information on the key. Note that it is important to have two different values of W_0: if HMAC were defined with $k||$ opad and $k||$ ipad, the two series of measures in the first step would be exactly the same, and we would not have enough information for a practical attack.

The main difference between our attack and classical template attacks is that we do not consider the value of a register but its Hamming weight. This presents the advantage to limit the number of profiling (only 33 records are needed instead of 2^{32} for 32-bit registers) even if it gives less information. If the targeted register is manipulated a sufficient number of time, then we can combine the partial information we have recovered to find the entire value of the initial register. To sum up, this attack can be viewed as a combination of template attacks and classical power consumption model. Classical template attacks usually separate keys according to words or substrings of words. To our knowlegde and even if it seems natural, it is the first time that a template attack is based on power consumption models.

1.5 Organization of the Paper

In section 2, we describe SHA-1 and we present how our attack works. In section 3, we give experimental results on a real implementation on the NIOS processor embedded in a FPGA Altera Stratix. Finally, in section 4, we show that our HMAC attack can be applied to other hash functions such as MD5 and SHA2. We also show how a similar attack can be used against other construc tions: it can break the confidentiality of the L2TP protocol and can recover the key materials of the IPsec protocol by attacking the key derivation function.

2   Description of the Attack

2.1   SHA-1 Message Leak

In SHA-1, the message is introduced word by word, as opposed to the IV which is introduced all at once. This means that the values manipulated during the first rounds of SHA-1 depend only on the first message words. For instance, the internal value A_1 depends only on the IV and on m_0. The value A_1 is also used as B_2 and after a rotation as C_3, D_4 and E_5. Each time this value is manipulated, there will be some leakage depending only on m_0. If we can model this leakage, we will be able to recover m_0: for all possible values of m_0, we simulate the leak, and we keep the values that give a good prediction. The full message recovery algorithm is given by Algorithm 1.

Algorithm 1. Recovery of n message words

1: **Profiling Stage**
2:     Study the device and model the leakage

3: **Operational Stage**
4:     Get side-channel information from one run of the hash function

5: **Message recovery**
6:     $\mathcal{S} \leftarrow \{\text{IV}\}$ ▷ $\mathcal{S}$ is the candidate set
7:     **for** $0 \leq i < n$ **do**
8:         $\mathcal{S}' \leftarrow \emptyset$
9:         **for all** $s \in \mathcal{S}$ **do**
10:             **for all** $m_i \in \mathbb{Z}_{2^{32}}$ **do**
11:                 Simulate the leak for (s, m_i)
12:                 **if** it matches the measure **then**
13:                     $\mathcal{S}' \leftarrow \mathcal{S}' \cup (s, m_i)$
14:                 **end if**
15:             **end for**
16:         **end for**
17:         $\mathcal{S} \leftarrow \mathcal{S}'$
18:     **end for**

The complexity of the message recovery depends on the size of the set of good candidates $\mathcal{S}$: each iteration of the loop in line 7 costs $2^{32}|\mathcal{S}|$. If we have γ candidates matching each measure, the set $\mathcal{S}$ will be of size γ^i after iteration i, and the total cost of the algorithm to recover n message words is about $2^{32}\gamma^n$. The value of γ will depend of the number of information leaked through the side channel.

2.2   HMAC Key Leak

If we look at HMAC, we see that the secret key is used as a message for the hash function, so we can use the message leak in SHA-1 to recover the key of

HMAC-SHA-1. Note that this does not apply to NMAC, where the key is used as the initialization vector.

In fact, the secret key is used twice in the HMAC construction: it is used in the inner hash function as $H(\bar{k} \oplus \mathsf{ipad})$ and in the outer hash function as $H(\bar{k} \oplus \mathsf{opad})$. We can collect side-channel information from those two related messages, which gives two sets of measures for the same key. This property is crucial for our attack: with only one set of measures, we would have too many candidates in the set $\mathcal{S}$.

We will now study an implementation of SHA-1 on the NIOS processor to check whether this leakage is sufficient to identify the secret key.

2.3   Modelization of the Attack

The side channel we will use in our practical experiments is a measure of the electromagnetic radiation (see Section 3). We model it as follows: each time the processor loads a data into a register, the electromagnetic signal depends on the number of flipped bits inside this register.

More precisely, we target the `ldw` instruction (*load word*), which loads data from the memory to a register. Since this instruction does not perform any computation, we expect the EM signal to be quite clean, and we should be able to read the number of flipped bits. When a value B is loaded into a register whose previous value was A, we have a transition $A \rightarrow B$, and we can measure the Hamming weight of $A \oplus B$.

2.4   Study of an Implementation of SHA-1

For our experiments, we used the code of XySSL[1] which is an SSL library designed for embedded processor. The compiled code for the round function of SHA-1 is given in Table 1 in Appendix A, together with the values leaked from the `ldw` instructions.

Using this implementation of SHA-1, we have 6 measures at each step of the compression function, which gives 12 measures per key word of HMAC-SHA-1. The Hamming weight of a 32-bit value contains 3.54 bits of entropy, so we expect to have a small number of candidates for each key word. Note that the measures are clearly not independent, which makes it difficult to compute the number of expected candidates. Therefore, we ran some simulations to estimate the complexity of the attack.

2.5   Simulations

To estimate the complexity of the attack, we can run the message recovery step for some initial states. For a given state $(A_i, B_i, C_i, D_i, E_i)$, we consider the 2^{32} values of the message word W_i, and we compute the Hamming weight of the measures given in Table 1. We can count the number of collisions in

---

[1] Available at `http://xyssl.org/code/`

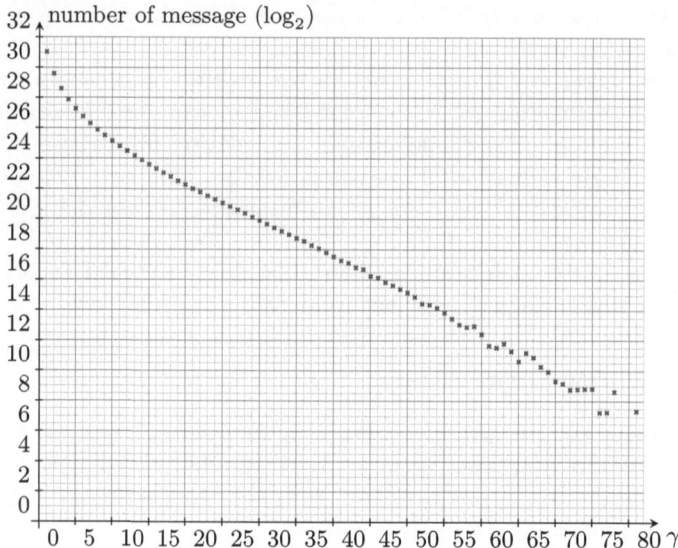

Fig. 1. Distribution of the number of candidates γ, for the first message word W_0

the measures, which will give the distribution of the number of candidates γ matching the measure.

More precisely for the first message word W_0, we know exactly the constants involved in the measures: they are given in Table 1, and we will use them with $W_0 = k_0 \oplus \mathsf{ipad}$ and $W_0 = k_0 \oplus \mathsf{opad}$. The distribution is given in Figure 1. It has an average of 2.79 and a worst case value of 81. Experiments with random initial state instead of the specified IV of SHA-1 give similar results (this simulates the recovery of the other message words). Moreover, if we allow one of the measures to be wrong with a difference of one, then we have 4.60 candidates on average, and 140 in the worst case. This means that we can tolerate some errors in the measures, without affecting too much the computation time.

In the end, we will assume that the measures give about three candidates at each step. We expect that the recovery of a 32κ-bit HMAC key will use a set $\mathcal{S}$ of size about 3^κ, and the complexity of the attack will be about $2^{32} \times 3^\kappa$ which will take a few hours on a PC for a 128-bit key. To identify the correct key among the set $\mathcal{S}$, we can either use a known MAC, or use some leakage in the following steps of the compression function.

3   Experimental Validation on a Known Implementation of HMAC-SHA1

Our side channel attack recovers the message input of the compression function if we can measure the number of bits flipped when loading from the memory to a register with the `ldw` instruction. We experimentally validated this assumption using an implementation of HMAC-SHA1 for a NIOS processor, on an Altera Stratix FPGA. The electromagnetic radiations are measured by near field methods using a loop probe sensitive to the horizontal magnetic field. We assume that the assembly code of is known, and we first study the implementation of the HMAC-SHA1 algorithm. Then, we focus on the leakage of the load instruction.

Electromagnetic radiation signals are now considered as one of the most powerful side channel signals [9,1,23]. One advantage of the EM side channel is that is possible to focus on local emanations.

3.1 Leakage of HMAC-SHA1 Implementation

The Figure 2 shows the radiations measured during the computation of HMAC-SHA1 with a 128-bit key and a 504-bit message M. We can clearly see the 5 executions of the compression function. A sharp analysis of this leakage needs to be done in order to find out some more useful information. During the experiments, we focus on the load instruction (referred as ldw by Stratix Assembly Language) but other instructions could also give information about the Hamming weight of the data used during the execution of SHA-1.

3.2 The Leakage of the ldw Instruction

The analysis in Section 2 shows that information leak from this instruction will be sufficient to recover the secret key. The goal of this section is to validate

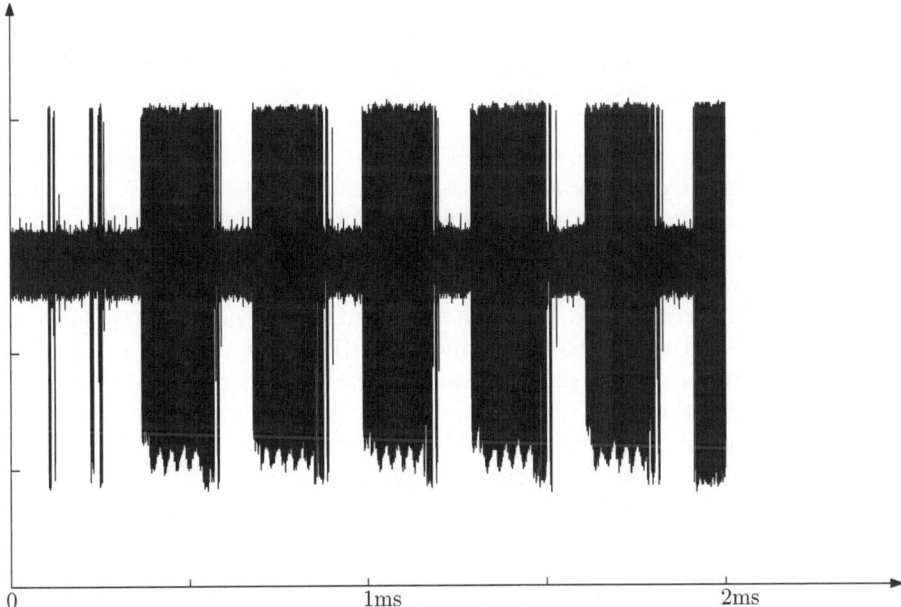

0   1ms   2ms

Fig. 2. HMAC-SHA1 Electromagnetic Execution. We see the three calls of the compression function to compute $h = H(\bar{k} \oplus \mathsf{ipad} \| M)$ and two calls to compute $H(\bar{k} \oplus \mathsf{opad} \| h)$.

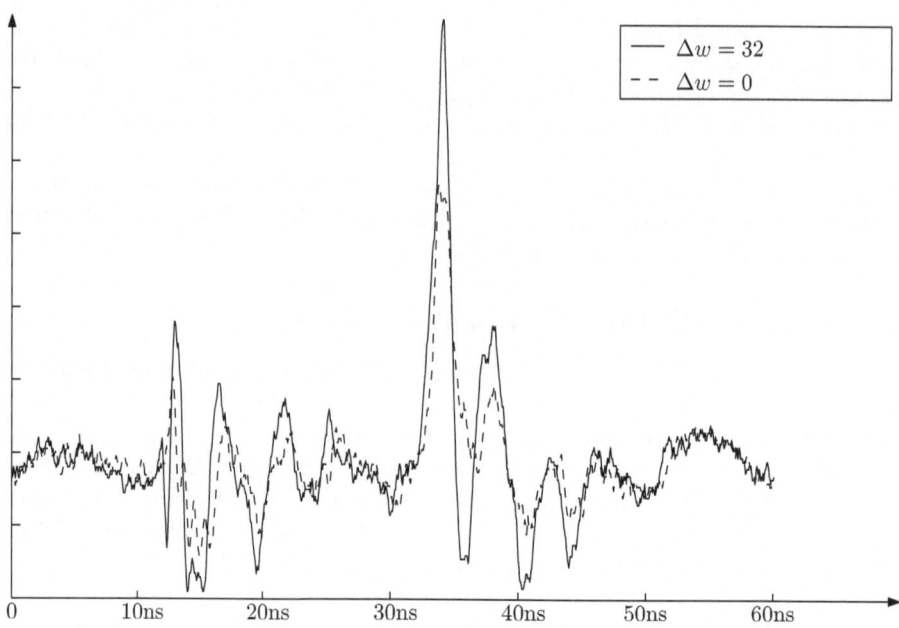

Fig. 3. Extremal Hamming Distances

that the Hamming distance between the mnemonic operands manipulated by the instruction `ldw` leak with electromagnetic radiations. As an example, if the register R, which value was A, is loaded with the new value B, we claim the Hamming distance between A and B leaks with the electromagnetic radiations.

For validation, we precharged the register R to $A = $ `0xae8ceac8`, Altera Stratix being a 32-bit register technology. Then two experiments have been performed: in the first one, we use $B_1 = A = $ `0xae8ceac8` and in the second one, we use $B_2 = \bar{A} = $ `0x51731537`. These two experiments are opposite regarding the Hamming distance: $H(A \oplus B_1) = 0$ while $H(A \oplus B_2) = 32$. Fig. 3 illustrates this link between radiations and Hamming distance.

We must now check if the measures allow to distinguish a small difference in the Hamming weight, and if they are masked by the noise. For a successful attack, we need to be able to distinguish a Hamming distance of 15 from a Hamming distance of 16 since on average the frequency of this Hamming distance is larger than the extremal values. To verify this, we used pairs of values with those Hamming distance. Fig. 4 shows the results with the following pairs: (`0x00000000`, `0xe0f00ff0`), (`0x55555555`, `0x85a55aa5`), (`0x00000000`, `0xffff0000`), (`0xffffffff`, `0x0f0f0ff0`), (`0xaaaaaaaa`, `0x00050000`). We see that the curve depends on the Hamming distance between the two values, and not on the actual value of the register. Moreover, the noise level is sufficiently low in our setting to be able to distinguish a difference of one in the Hamming distance. These curves have been obtained by zooming on figure 3.

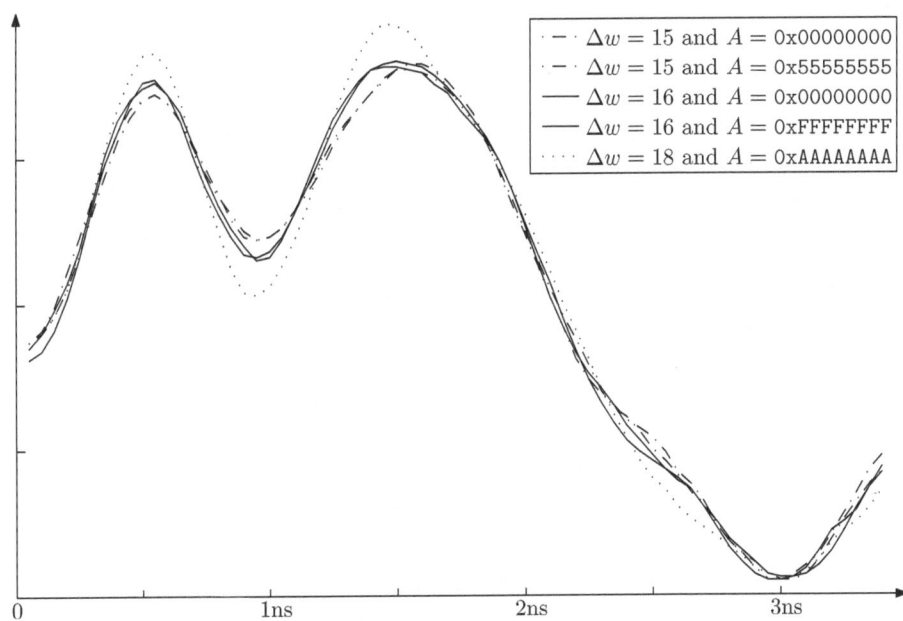

Legend:
- $- \cdot - \Delta w = 15$ and $A = $ 0x00000000
- $\cdot - \cdot \Delta w = 15$ and $A = $ 0x55555555
- —— $\Delta w = 16$ and $A = $ 0x00000000
- —— $\Delta w = 16$ and $A = $ 0xFFFFFFFF
- $\cdots$ $\Delta w = 18$ and $A = $ 0xAAAAAAAA

Fig. 4. Electromagnetic radiations for some Hamming distances

Thus, the Side Channel Analysis procedure can be done in two stages, a profiling stage and an operational stage. 33 measures of load instructions with all the possible Hamming distance are done during the profiling stage. This will allow us to find the Hamming distance of all `ldw` instructions for the operational stage. The profiling stage will also be used to study the timing of the SHA-1 computation, so as to match each instruction with the assembly code. Then, the operational stage consists in a Template Attack [5] on the `ldw` instructions. Following the attack of Section 2, we expect to recover a secret key of κ words with only one HMAC measure and a workload of about $2^{32}3^{\kappa}$ steps of the compressions function.

4   Extension to Other Hash Functions and to Other Usage

In this section, we show that the basic attack we proposed can be extended to other hash functions, works even though the code is unknown and can also be used to recover encryption keys in other protocols.

4.1   Other Hash Function of the MD4 Family

The other functions of the MD4 family (MD5, SHA-2, RIPEMD) have a very similar design and the message words also enter the compression function one by one. The assembly code of a specific implementation should be studied to see how many load instructions are used and what information is leaked, but we expect the attack to be just as efficient. Basically, our attack should work against any hash function based on a Feistel ladder.

4.2   Unknown Implementation

Even if we don't have access to the assembly code of the implementation, our attack is still applicable. The main difference with the previous analysis is that previous value of the targeted register A_{init} is unknown. Anyway, the attacker can improve the profiling stage. Indeed, he can guess the value of A_{init} with Correlation Power Analysis(CPA) [4], making the secret key varying. Let's remark that instead of making a 32 bits CPA, the attacker can do 4 CPA, each one on 8 bits. This procedure permits to limit the computational cost of the profiling stage. Then, thanks to this CPA procedure, the attacker can guess what was a register value before the secret key bits are written on it. Furthermore, the CPA peak permits to localize in time the load instruction.

With all these templates obtain at a very low computational cost, the attacker will be able with only one curve (if the noise is as low as in our setting) to retrieve the full key.

4.3   Other Attack Scenarios

In this section we identify some other construction where our attack can recover a secret key. The basic requirement is that the secret key is used as the message input of the hash function, and we need two related computations where the key enters the hash function in two different states.

Confidentiality: The L2TP Example. The L2TP protocol [27] is a tunneling protocol used to implement VPNs. It uses a shared secret K, and two known values RV and AV. The plaintext is broken into 16-byte chunks, $p_1, p_2, \ldots$ The ciphertext blocks are called $c_1, c_2, \ldots$ and the intermediate values $b_1, b_2, \ldots$ are computed as follows:

$$
\begin{aligned}
b_1 &= MD5(AV\|K\|RV) & c_1 &= p_1 \oplus b_1 \\
b_2 &= MD5(K\|c_1) & c_2 &= p_2 \oplus b_2 \\
b_3 &= MD5(K\|c_2) & c_3 &= p_3 \oplus b_3 \ldots
\end{aligned}
$$

The secret key K enters the hash function in two different states for the computation of b_1 and b_2, so we can apply our attack and recover the key.

Key Derivation. Key derivation is sometimes used as a countermeasure against DPA attacks. The key derivation process described in [17], uses a hash function H, a master key K and a counter ctr, and computes the sessions keys as $SK = H(ctr\|K)$. Using our attack, if we observe only two key derivation process, we have enough information to recover the master key K.

Note About RIPEMD. The RIPEMD family of hash function uses two parallel Feistel ladder, and combines the results in the end of the compression function. This allows us to recover two series of measures, even if the secret key enters the hash function only once. The original function RIPEMD-0 uses two lines with the same permutation and different constants, which gives enough information for our attack. Thus, any construction which uses a secret key as a part of a message to be hashed with RIPEMD-0 is vulnerable. The newer functions RIPEMD-128 and RIPEMD-160 uses different permutations of the message in the two lines; our attack can reduce the key-space but we don't have a practical key recovery with a single hash function call.

4.4 Possible Countermeasure

A possible countermeasure against our attack is to keep the internal state of SHA-1 inside the processor registers. Our attack uses the fact that the internal state is stored in the stack and moved in and out the registers: we measure the radiations during this movement. If all the computations are done inside the registers, we can still measure the radiations during the computations, but the signal will have more noise. Another solution is to load random values between each ldw instruction or to use classical masking methods but this requires a random generator and may downgrade the performance drastically.

5   Conclusion

In this paper, we show that the electromagnetic radiation of a device can leak the number of flipped bits when data is loaded into a register. This information could also be observed using a proper current power analysis. However, EM signal allows to obtain emanations of local instructions and attacks are not intrusive since they can be performed in the near field of the device and do not require to modify the circuitry. Our experimentation studies the ldw instruction since it is easier to characterize during the profiling stage, but the attack could be improved by using other instructions. Our attack targets the message input of the compression function, while previous attacks only considered the IV input. This allows us to attack other standards and to recover crucial information such as the master key in some key-derivation schemes.

Finally, these results give some information about the efficiency of side channel attack on hash functions implemented on embedded processors. It is important to see that the adversary model is very limited: access to a similar device for a profiling stage and then one execution leads to an efficient key recovery.

References

1. Agrawal, D., Archambeault, B., Rao, J.R., Rohatgi, P.: The em side-channel(s). In: Kaliski Jr., B.S., et al [12], pp. 29–45
2. Bellare, M.: New proofs for NMAC and HMAC: Security without collision-resistance. In: Dwork, C. (ed.) CRYPTO 2006. LNCS, vol. 4117, pp. 602–619. Springer, Heidelberg (2006)
3. Bellare, M., Canetti, R., Krawczyk, H.: Keying hash functions for message authentication. In: Koblitz, N. (ed.) CRYPTO 1996. LNCS, vol. 1109, pp. 1–15. Springer, Heidelberg (1996)

4. Brier, E., Clavier, C., Olivier, F.: Correlation power analysis with a leakage model. In: Joye and Quisquater [11], pp. 16–29
5. Chari, S., Rao, J.R., Rohatgi, P.: Template attacks. In: Kaliski Jr., B.S., et al [12], pp. 13–28
6. Contini, S., Yin, Y.L.: Forgery and partial key-recovery attacks on hmac and nmac using hash collisions. In: Lai, X., Chen, K. (eds.) ASIACRYPT 2006. LNCS, vol. 4284, pp. 37–53. Springer, Heidelberg (2006)
7. Cramer, R. (ed.): EUROCRYPT 2005. LNCS, vol. 3494. Springer, Heidelberg (2005)
8. Fouque, P.-A., Leurent, G., Nguyen, P.Q.: Full key-recovery attacks on hmac/nmac-md4 and nmac-md5. In: Menezes, A. (ed.) CRYPTO 2007. LNCS, vol. 4622, pp. 13–30. Springer, Heidelberg (2007)
9. Gandolfi, K., Mourtel, C., Olivier, F.: Electromagnetic analysis: Concrete results. In: Koç, Ç.K., Naccache, D., Paar, C. (eds.) CHES 2001. LNCS, vol. 2162, pp. 251–261. Springer, Heidelberg (2001)
10. Gauravaram, P., Okeya, K.: An update on the side channel cryptanalysis of macs based on cryptographic hash functions. In: Srinathan, K., Rangan, C.P., Yung, M. (eds.) INDOCRYPT 2007. LNCS, vol. 4859, pp. 393–403. Springer, Heidelberg (2007)
11. Joye, M., Quisquater, J.-J. (eds.): CHES 2004. LNCS, vol. 3156. Springer, Heidelberg (2004)
12. Kaliski Jr., B.S., Koç, Ç.K., Paar, C. (eds.): CHES 2002. LNCS, vol. 2523. Springer, Heidelberg (2003)
13. Kaufman, C.: Rfc 4306 - internet key exchange (ike v2) protocol (December 2005), http://www.ietf.org/rfc/rfc4306.txt
14. Kelsey, J., Schneier, B., Wagner, D., Hall, C.: Side channel cryptanalysis of product ciphers. In: Quisquater, J.-J., Deswarte, Y., Meadows, C., Gollmann, D. (eds.) ESORICS 1998. LNCS, vol. 1485, pp. 97–110. Springer, Heidelberg (1998)
15. Kelsey, J., Schneier, B., Wagner, D., Hall, C.: Side channel cryptanalysis of product ciphers. Journal of Computer Security 8(2/3) (2000)
16. Kent, S.: Security architecture for the internet protocol (November 1998), http://www.ietf.org/rfc/rfc2401.txt
17. Kocher, P.: Us patent no. 6,304,658 (2003), http://www.cryptography.com/technology/dpa/Patent6304658.pdf
18. Kocher, P.: Us patent no. 6,539,092 (2003), http://www.cryptography.com/technology/dpa/Patent6539092.pdf
19. Kocher, P.C., Jaffe, J., Jun, B.: Differential power analysis. In: Wiener, M. (ed.) CRYPTO 1999. LNCS, vol. 1666, pp. 388–397. Springer, Heidelberg (1999)
20. Lemke, K., Schramm, K., Paar, C.: Dpa on n-bit sized boolean and arithmetic operations and its application to idea, rc6, and the hmac-construction. In: Joye and Quisquater [11], pp. 205–219
21. McEvoy, R.P., Tunstall, M., Murphy, C.C., Marnane, W.P.: Differential power analysis of hmac based on sha-2, and countermeasures. In: Kim, S., Yung, M., Lee, H.-W. (eds.) WISA 2007. LNCS, vol. 4867, pp. 317–332. Springer, Heidelberg (2008)
22. Okeya, K.: Side channel attacks against hmacs based on block-cipher based hash functions. In: Batten, L.M., Safavi-Naini, R. (eds.) ACISP 2006. LNCS, vol. 4058, pp. 432–443. Springer, Heidelberg (2006)

23. Quisquater, J.-J., Samyde, D.: Electromagnetic analysis (ema): Measures and counter-measures for smart cards. In: Attali, S., Jensen, T. (eds.) E-smart 2001. LNCS, vol. 2140, pp. 200–210. Springer, Heidelberg (2001)

24. Rechberger, C., Rijmen, V.: On authentication with hmac and non-random properties. In: Dietrich, S., Dhamija, R. (eds.) FC 2007 and USEC 2007. LNCS, vol. 4886, pp. 119–133. Springer, Heidelberg (2007)

25. Schindler, W., Lemke, K., Paar, C.: A stochastic model for differential side channel cryptanalysis. In: Rao, J.R., Sunar, B. (eds.) CHES 2005. LNCS, vol. 3659, pp. 30–46. Springer, Heidelberg (2005)

26. Shoup, V. (ed.): CRYPTO 2005. LNCS, vol. 3621. Springer, Heidelberg (2005)

27. Towsley, W., Valencia, A., Rubens, A., Pall, G., Zorn, G., Palter, B.: Rfc 2661 - layer two tunneling protocol "l2tp" (August 1999),
http://www.ietf.org/rfc/rf2661.txt

28. Wang, X., Lai, X., Feng, D., Chen, H., Yu, X.: Cryptanalysis of the hash functions md4 and ripemd. In: Cramer [7], pp. 1–18

29. Wang, X., Yin, Y.L., Yu, H.: Finding collisions in the full sha-1. In: Shoup [26], pp. 17–36

30. Wang, X., Yu, H.: How to break md5 and other hash functions. In: Cramer [7], pp. 19–35

31. Wang, X., Yu, H., Yin, Y.L.: Efficient collision search attacks on sha-0. In: Shoup [26], pp. 1–16

A SHA-1 Code

Table 1. SHA-1 code. The table shows the code for one step of the compression function. The other steps are exactly the same, excepted that the mapping between the internal state values A, B, C, D, E and the stack changes at each round.

Instruction	Stack 76	80	84	88	92	Registers r2	r3	r4	Measure
Begin step 0	A_0	B_0	C_0	D_0	E_0	X_0	Y_0	Z_0	
`ldw r2 76(fp)`						A_0			$X_0 \to A_0$
`roli r4, r2, 5`								$A_0^{\lll 5}$	
`ldw r3, 84(fp)`							C_0		$Y_0 \to C_0$
`ldw r2, 88(fp)`						D_0			$A_0 \to D_0$
`xor r3, r3, r2`							$C_0 \oplus D_0$		
`ldw r2, 80(fp)`						B_0			$D_0 \to B_0$
`and r3, r3, r2`							$(C_0 \oplus D_0) \wedge B_0$		
`ldw r2, 88(fp)`						D_0			$B_0 \to D_0$
`xor r2, r3, r2`						Φ_0			
`add r3, r4, r2`							$\Phi_0 \boxplus A_0^{\lll 5}$		
`ldw r2, 12(fp)`						W_0			$\Phi_0 \to W_0$
`add r3, r3, r2`							$\Phi_0 \boxplus A_0^{\lll 5} \boxplus W_0$		
`ldw r2, 92(fp)`						E_0			$W_0 \to E_0$
`add r3, r3, r2`							$\Phi_0 \boxplus A_0^{\lll 5} \boxplus W_0 \boxplus E_0$		
`movhi r2, 23170`						0x5a820000			
`addi r2, r2, 31129`						0x5a827999			
`add r2, r3, r2`						A_1			
`stw r2, 92(fp)`					A_1				
`ldw r2, 80(fp)`						B_0			$A_1 \to B_0$
`roli r2, r2, 30`						$B_0^{\lll 30}$			
`stw r2, 80(fp)`		C_1							
Begin step 1	B_1	C_1	D_1	E_1	A_1	X_1	Y_1	Z_1	
`ldw r2, 92(fp)`						A_1			$X_1 \to A_1$
`...`									

We have the following relations:

$$\Phi_i = f_i(B_i, C_i, D_i)$$
$$B_{i+1} = A_i$$
$$C_{i+1} = B_i^{\lll 30}$$
$$D_{i+1} = C_i$$
$$E_{i+1} = D_i$$
$$A_{i+1} = \Phi_i \boxplus A_i^{\lll 5} \boxplus W_i \boxplus E_i \boxplus K_i$$
$$X_{i+1} = B_i^{\lll 30}$$
$$Y_{i+1} = \Phi_i \boxplus A_i^{\lll 5} \boxplus W_i \boxplus E_i$$
$$Z_{i+1} = A_i^{\lll 5}$$

We can make 8 measures per step, but this gives only 6 informations leaks, because the transitions $D_{i+1} \to B_{i+1}$ and $B_{i+1} \to D_{i+1}$ leaks the same information as $X_i \to A_i$.

For instance, the leaks related to W_0 are:

$$\Phi_0 \to W_0 \quad W_0 \oplus \text{0x98badcfe}$$
$$W_0 \to E_0 \quad W_0 \oplus \text{0xc3d2e1f0}$$
$$A_1 \to B_0 \quad (W_0 \boxplus \text{0x9fb498b3}) \oplus \text{0xefcdab89}$$
$$X_1 \to A_1 \quad (W_0 \boxplus \text{0x9fb498b3}) \oplus \text{0x7bf36ae2}$$
$$Y_1 \to C_1 \quad (W_0 \boxplus \text{0x45321f1a}) \oplus \text{0x7bf36ae2}$$
$$A_1 \to D_1 \quad (W_0 \boxplus \text{0x9fb498b3}) \oplus \text{0x98badcfe}$$

First-Order Side-Channel Attacks on the Permutation Tables Countermeasure

Emmanuel Prouff[1] and Robert McEvoy[2]

[1] Oberthur Technologies, France
e.prouff@oberthur.com
[2] Claude Shannon Institute for Discrete Mathematics, Coding and Cryptography,
University College Cork, Ireland
robertmce@eleceng.ucc.ie

Abstract. The use of random permutation tables as a side-channel attack countermeasure was recently proposed by Coron [5]. The countermeasure operates by ensuring that during the execution of an algorithm, each intermediate variable that is handled is in a permuted form described by the random permutation tables. In this paper, we examine the application of this countermeasure to the AES algorithm as described in [5], and show that certain operations admit first-order side-channel leakage. New side-channel attacks are developed to exploit these flaws, using correlation-based and mutual information-based methods. The attacks have been verified in simulation, and in practice on a smart card.

Keywords: Side-Channel Attacks, Permutation Tables, CPA, MIA, Masking.

1 Introduction

When a cryptographic algorithm is implemented in hardware or embedded software, information may be leaked about the intermediate variables being processed by the device. The class of implementation attacks called Side-Channel Attacks (SCA) aims to exploit these leakages, and recover secret information [11]. Masking is one of the most popular SCA countermeasures, used to protect sensitive variables (i.e. variables whose statistical distribution is dependent on the secret key) [4]. Masking has been well studied, and has been shown to be effective against a number of types SCA [2,4], but remains ineffective in stronger attack models (*e.g.* Higher-Order SCA [13]).

Recently, Coron presented the permutation tables countermeasure, as an alternative to masking [5]. The new proposal can be viewed as a generalization of the classical approach, where masking is no longer performed through a random translation, but through a random permutation. Like classical masking, the permutation tables countermeasure also requires a random bit string, which is used at the start of the cryptographic algorithm to generate a permutation P. In the case of an encryption algorithm, P is then applied to both the message x to be encrypted and the secret key k, producing $P(x)$ and $P(k)$ respectively. It

C. Clavier and K. Gaj (Eds.): CHES 2009, LNCS 5747, pp. 81–96, 2009.

is these permuted variables that are used by the encryption algorithm. At each stage of the algorithm, the cryptographic operations must be modified so that all of the intermediate variables remain in the permuted form described by P. If the countermeasure is applied correctly, the intermediate variables should all have a uniform distribution independent of sensitive variables, thereby precluding side-channel attacks that rely on statistical dependency of the intermediate variables with the secret key.

In this paper, we examine the application of the permutation tables countermeasure to AES, as described by [5]. We show that certain sensitive intermediate variables in this algorithm are, in fact, not uniformly distributed, and therefore leak side-channel information about the secret key. However, because of the nature of the permutation tables countermeasure, it is not possible to exploit these flaws with classical approaches (such as those used in [6,7,9]). In fact, the main issue is to exhibit a sound *prediction function* to correlate with the leakages in correlation-based SCA (*e.g.* Correlation Power Analysis (CPA) [3]). After modeling the side-channel leakage, we use the method proposed in [17] to exhibit a new prediction function for the permuted sensitive variables. An analytical expression for the optimal prediction function is derived, which, for the correct key hypothesis, maximises the correlation with leakage measurements from the algorithm.

Furthermore, since the flawed intermediate variables do not have a monotonic dependency with the sensitive variables, we consider SCA attacks involving distinguishers able to exploit non-monotonic interdependencies. We investigate how Mutual Information Analysis (MIA) [8,16] can be applied in order to exploit the flaws, and compare it with the correlation-based approach. Both of these new attacks are performed both in simulation and in practice on a smart card, and are successful at breaking the countermeasure described in [5].

2   Preliminaries

2.1   Mathematical Background and Notation

We use calligraphic letters, like $\mathcal{X}$, to denote finite sets (*e.g.* $\mathbb{F}_2^n$). The corresponding capital letter X is used to denote a random variable over $\mathcal{X}$, while the lowercase letter x denotes a particular element from $\mathcal{X}$. The probability of the event $(X = x)$ is denoted $\mathrm{p}\,[X = x]$. The uniform probability distribution over a set $\mathcal{X}$ is denoted by $\mathcal{U}(\mathcal{X})$, and the Gaussian probability distribution with *mean* μ and *standard deviation* σ is denoted by $\mathcal{N}(\mu, \sigma^2)$. The mean of X is denoted by $\mathrm{E}\,[X]$ and its standard deviation by $\sigma[X]$. The correlation coefficient between X and Y is denoted by $\rho\,[X, Y]$. It measures the linear interdependence between X and Y, and is defined by:

$$\rho\,[X, Y] = \frac{\mathrm{Cov}\,[X, Y]}{\sigma\,[X]\,\sigma\,[Y]} \; , \tag{1}$$

where $\mathrm{Cov}\,[X,Y]$, called *covariance of X and Y*, equals $\mathrm{E}\,[XY] - \mathrm{E}\,[X]\,\mathrm{E}\,[Y]$. It can be checked [17] that for every function f measurable on $\mathcal{X}$, the correlation $\rho\,[f(X),Y]$ satisfies:

$$\rho\,[f(X),Y] = \rho\,[f(X), \mathrm{E}\,[Y|X]] \times \rho\,[\mathrm{E}\,[Y|X],Y] \ . \tag{2}$$

This implies (see Proposition 5 in [17]) the following inequality:

$$\rho\,[f(X),Y] \leq \frac{\sigma\,[\mathrm{E}\,[Y|X]]}{\sigma\,[Y]} \ . \tag{3}$$

A sample of a finite number of values taken by X over $\mathcal{X}$ is denoted by $(x_i)_i$ or by (x_i) if there is no ambiguity on the index, and the mean of such a sample is denoted by $\bar{x} = \frac{1}{\#(x_i)} \sum_i x_i$. Given two sample sets (x_i) and (y_i), the empirical version of the correlation coefficient is the *Pearson coefficient*:

$$\widehat{\rho}\,((x_i),(y_i)) = \frac{\sum_i (x_i - \bar{x})(y_i - \bar{y})}{\sqrt{\sum_i (x_i - \bar{x})^2}\sqrt{\sum_i (y_i - \bar{y})^2}} \ , \tag{4}$$

The correlation and Pearson coefficients relate to affine statistical dependencies, and two dependent variables X and Y can be such that $\rho(X,Y) = 0$. To quantify the amount of information that Y reveals about X (whatever the kind of dependency is), the notion of *mutual information* is usually involved. It is the value $\mathrm{I}(X;Y)$ defined by $\mathrm{I}(X;Y) = \mathrm{H}(X) - \mathrm{H}(X|Y)$, where $\mathrm{H}(X)$ is the entropy of X and where $\mathrm{H}(X|Y)$ is the *conditional entropy of X knowing Y* (see [12] for more details).

2.2   Side-Channel Attack Terminology

We shall view an implementation of a cryptographic algorithm as the processing of a sequence of *intermediate variables*, as defined in [2]. We shall say that an intermediate variable is *sensitive* if its distribution is a function of some known data (for example, the plaintext) and the secret key, and is not constant with respect to the secret key. Consequently, the statistical distribution of a sensitive variable depends on both the key and on the distribution of the known data. If a sensitive intermediate variable appears during the execution of a cryptographic algorithm, then that implementation is said to contain a *first-order flaw*. Information arising from a first-order flaw, that can be monitored via a side-channel (such as timing information or power consumption), is termed *first-order leakage*. A *first-order side-channel attack (SCA)* against an implementation is a SCA that exploits a first-order leakage, in order to recover information about the secret key. Similarly, an *r*th-*order SCA (Higher-Order SCA or HO-SCA)* against an implementation is a SCA that exploits leakages at r different times, which are respectively associated with r different intermediate variables.

Remark 1. In [19], an alternative definition for HO-SCA is used, where an *r*th-order SCA is defined with respect to r different *algorithmic* variables (which may

be manipulated simultaneously, or which may correspond to a single intermediate variable). In this paper, we focus on the countermeasure of [5]; therefore,
we adhere to the HO-SCA definition in [5] (which is widely accepted in the
community [2,10,13,14]).

In order to prevent side-channel attacks on cryptographic implementations, many
countermeasures (such as masking and the permutation tables countermeasure)
aim to randomise the leakage caused by each intermediate variable. An implementation of a cryptographic algorithm can be said to possess *first-order SCA
security* if no intermediate variable in the implementation is sensitive. Similarly,
rth-order SCA security requires an implementation to be such that no r-tuple
of its intermediate variables is sensitive.

3   The Permutation Tables Countermeasure

3.1   Generation of Permutation Tables

In order to use permutation tables as a SCA countermeasure, a new permutation table P must be generated at the beginning of each execution of the
cryptographic algorithm. Here, P is described in the context of the AES algorithm, where the intermediate variables are 8-bit words. P comprises two 4-bit
permutations p_1 and p_2, and operates on an 8-bit variable x according to:

$$P(x) = p_2(x_h)||p_1(x_l) \ , \tag{5}$$

where x_h and x_l respectively denote the high and low nibbles of x, and $||$ denotes
concatenation. Upon each invocation of the algorithm, permutations p_1 and p_2
are randomly chosen from a set of permutations $\mathcal{P}$, defined over $\mathbb{F}_2^4$. For efficiency
reasons, the set $\mathcal{P}$ is not defined as the set of all the permutations over $\mathbb{F}_2^4$.
Indeed, in such a case the random generation of an element of $\mathcal{P}$ would be
costly. In [5], Coron defines an algorithm to generate elements of the set $\mathcal{P}$
from a 16-bit random value. Here, we will assume that the random variable P_1
(respectively P_2) associated with the random generation of p_1 (resp. p_2) satisfies
$\mathrm{p}\,[P_1 = p_1] = 1/\#\mathcal{P}$ (resp. $\mathrm{p}\,[P_2 = p_2] = 1/\#\mathcal{P}$) for every $p_1 \in \mathcal{P}$ (resp. $p_2 \in \mathcal{P}$).

3.2   Protecting AES Using Permutation Tables

The Advanced Encryption Standard (AES) is a well-known block cipher, and
details of the algorithm can be found in [5]. Essentially, the AES round function
for encryption operates on a 16-byte state (with each element labelled $a_i, 0 \leq i \leq$
15), and consists of four transformations: AddRoundKey, SubBytes, ShiftRows
and MixColumns.

   In [5], Coron described how to protect the AES encryption algorithm against
side-channel attacks, by using the permutation tables countermeasure. We will
refer to this encryption algorithm as *randomised AES*. Firstly, after the random
permutation P has been generated (as described in [5]), it is applied to each byte

of both the message and the key. For every byte x, we will refer to $u = P(x)$ as the *P-representation* of x. Each permuted value is passed to the AES round function, where it is operated upon by the AES transformations listed above. As noticed by Coron in [5], each of these AES transformations must be carefully implemented, such that: (i) sensitive variables always appear in their P-representation, and (ii) the output of each transformation is in P-representation form. Coron described the following implementations of AddRoundKey and MixColumns (for details of the other transformations, see [5]):

- Randomised AddRoundKey takes two bytes $u = P(x)$ and $v = P(y)$ as inputs, and outputs $P(x \oplus y)$. In order to achieve this, two 8-bit to 4-bit tables are defined (for (u_l, v_l) – resp. (u_h, v_h) – in $(\mathbb{F}_2^4)^2$):

$$\text{XT}_4^1(u_l||v_l) = p_1(p_1^{-1}(u_l) \oplus p_1^{-1}(v_l)) \ , \tag{6}$$
$$\text{XT}_4^2(u_h||v_h) = p_2(p_2^{-1}(u_h) \oplus p_2^{-1}(v_h)) \ . \tag{7}$$

Tables XT_4^1 and XT_4^2 are calculated at the same time as P, and stored in memory. An 8-bit XOR function, denoted by XT_8, is then computed using those table look-ups (for $u, v \in \mathbb{F}_2^8$):

$$\text{XT}_8(u, v) = \text{XT}_4^2(u_h||v_h)||\text{XT}_4^1(u_l||v_l) \ . \tag{8}$$

- Randomised MixColumns is computed as a combination of doubling and XOR operations. To calculate randomised MixColumns from the $P(a_i)$'s (the P-representations of the bytes of the AES state), the XOR operations are computed using the XT_8 function in Eq. (8). For the doubling operations, Coron defined a function D_2, such that when applied to $u = P(x)$, we get $D_2(P(x)) = P(\{02\} \bullet x)$ (where $\{\cdot\}$ denotes hexadecimal notation, and $\bullet$ denotes multiplication modulo $x^8 + x^4 + x^3 + x + 1$). The P-representation of the first byte of the MixColumns output is then calculated using:

$$P(a_0^{new}) = \text{XT}_8(D_2(a_0'), \text{XT}_8(D_2(a_1'), \text{XT}_8(a_1', \text{XT}_8(a_2', a_3')))) \ , \tag{9}$$

where a_i' denotes the P-representation of a_i. The other bytes in the randomised MixColumns output can be similarly calculated.

At the completion of the last encryption round, the inverse permutation P^{-1} is applied to each byte of the AES state, revealing the ciphertext.

4   Security of Randomized AES against First-Order SCA

4.1   Examining the Proof of Security

In [5], the author proposes the following Lemma to argue that the randomised AES implementation is resistant against first-order SCA:

Lemma 1. *For a fixed key and input message, every intermediate byte that is computed in the course of the randomised AES algorithm has the uniform distribution in $\{0, 1\}^8$.*

In [5], the proof of Lemma 1 is based on the fact that any intermediate AES data W is assumed to be represented as $P(W) = P_2(W_h)||P_1(W_l)$. However, this assumption is incorrect for the implementation described in [5] and re-called in Section 3.2. Indeed, when $\text{XT}_8(P(X), P(Y))$ is computed (Eq. (8)), the two functions XT_4^1 and XT_4^2 are parameterized with the intermediate variables $P_1(X_l)||P_1(Y_l)$ and $P_2(X_h)||P_2(Y_h)$ respectively. Namely, the same permutation P_1 (resp. P_2) is applied to the lowest and the highest nibbles of the intermediate data $W = X_l||Y_l$ (resp. $W = X_h||Y_h$). In this case W is not of the form $P(W)$; therefore, the statement made in [5] to prove Lemma 1 is incorrect. Actually, not only the proof but the Lemma itself is incorrect. If two nibbles are equal, e.g. $X_l = Y_l$, then their P_1-representations will also be equal, i.e. $P_1(X_l) = P_1(Y_l)$, irrespective of P_1. Otherwise, if $X_l \neq Y_l$, then $P_1(X_l)$ and $P_1(Y_l)$ behave like two independent random variables, except that they cannot be equal. This implies that the variable $P_1(X_l)||P_1(Y_l)$ will have two different non-uniform distributions depending on whether X_l equals Y_l or not. This gives rise to first-order leakage.

4.2   First-Order Leakage Points

In the randomised AES, the function XT_8 is employed to securely implement every bitwise addition between 8-bit words. To compute the P-representation of $X \oplus Y$ from the P-representations $X' = P(X)$ and $Y' = P(Y)$, the following successive operations are processed:

$$
\begin{aligned}
1. \quad & R_1 \quad \leftarrow \text{XT}_4^1\ (X_l'||Y_l') \\
2. \quad & R_2 \quad \leftarrow \text{XT}_4^2\ (X_h'||Y_h') \\
3. \quad & output \leftarrow \quad R_2||R_1
\end{aligned}
$$

Register $output$ contains $P(X \oplus Y)$ at the end of the processing above. Let us focus on the intermediate result R_1 (the same analysis also holds for R_2). It is computed by accessing the table XT_4^1 at address $Z = X_l'||Y_l'$ which, by construction, satisfies:

$$Z = P_1(X_l)||P_1(Y_l) \ . \tag{10}$$

As discussed in Section 4.1, the manipulation of Z therefore induces a first-order leakage in the AES implementation, whenever (X_l, Y_l) statistically depends on a secret information and a known data. This condition is satisfied when XT_8 is used to process the randomised AddRoundKey and randomised MixColumns operations during the first round of AES:

– **[Randomised AddRoundKey]** During this step, XT_8 takes the pair $(P(A), P(K))$ as operand, where K is a round key byte and A is a known byte of the AES state. In this case, (10) becomes:

$$Z = P_1(A_l)||P_1(K_l) \ . \tag{11}$$

– **[Randomised MixColumns]** During this step, XT_8 takes the pair $(A_1', A_2') = (P(S[A_1 \oplus K_1]), P(S[A_2 \oplus K_2]))$ as operand, with S denoting the AES S-box, with A_1 and A_2 being two known bytes of the AES state and with K_1 and K_2 being two round-key bytes.

In this case, (10) becomes:

$$Z = P_1(S[A_1 \oplus K_1]_l) || P_1(S[A_2 \oplus K_2]_l) \ , \tag{12}$$

where $S[\cdot]_l$ denotes the lowest nibble of $S[\cdot]$.

Both of the leakage points described above are first-order flaws, since they depend on a single intermediate variable Z. In the next sections, we will develop first-order side-channel attacks, that exploit these first order leakages. In both attacks, we will use the notation $Z(k_l)$ (resp. $Z(k_1, k_2)$) for the random variable $Z|(K_l = k_l)$ (resp. $Z|(K_1 = k_1, K_2 = k_2)$), each time we need to specify which key(s) Z is related to. The random variable corresponding to the leakage on Z shall be denoted by L. They are related through the following relationship:

$$L = \varphi(Z) + B \ , \tag{13}$$

where φ denotes a deterministic function called the *leakage function* and B denotes independent noise. We shall use the notation $L(k_l)$ (resp. $L(k_1, k_2)$) when we need to specify the key(s) involved in the leakage measurements.

5   Attacking the Randomised AddRoundKey Operation

There are currently two main ways to perform an attack on the manipulation of a random variable Z. The first method relies on affine statistical dependencies (for example CPA), whereas the second method relies on any kind of statistical dependency (for example MIA). Here, we describe a CPA attack on the first use of randomised AddRoundKey (performing an MIA attack on randomised AddRoundKey is less pertinent, as will be discussed in Section 6).

5.1   CPA Preliminaries

In a CPA [3], the attacker must know a good affine approximation $\hat{\varphi}$ of φ. It is common to choose the Hamming Weight (HW) function for $\hat{\varphi}$, as this is known to be a good leakage model for devices such as 8-bit microcontrollers. The attacker must also know a good affine approximation $\hat{Z}$ of Z. Based on these assumptions, key candidates $k_l^\star$ are discriminated by testing the correlation between $\hat{\varphi}(\hat{Z}(k_l^\star))$ and $L(k_l)$, for a sample of leakage measurements from the target device, and the corresponding known plaintexts.

Here, our attack targets the use of XT_8 when the first randomised AddRoundKey operation is performed. We assume that a sample of N leakages (ℓ_i) has been measured for N known lowest nibbles (a_i) of the AES state. Due to (11) and (13), the ℓ_i's and the a_i's satisfy the following relation:

$$\ell_i = \varphi\left(p_{1,i}(a_i) || p_{1,i}(k_l)\right) + b_i \ , \tag{14}$$

for $1 \leq i \leq N$, where b_i denotes the value of the noise for the ith leakage measurement and where $p_{1,i}$ denotes the permutation used at the time of the ith measurement.

To test a hypothesis $k_l^\star$ on k_l, the following Pearson's coefficient $\widehat{\rho}_{k_l^\star}$ is computed for an appropriate prediction function f:

$$\widehat{\rho}_{k_l^\star} = \widehat{\rho}((\ell_i)_i, (f(a_i, k_l^\star))_i) \ . \tag{15}$$

If f has been well chosen, the expected key will satisfy $k_l = \mathrm{argmax}_{k_l^\star} |\widehat{\rho}_{k_l^\star}|$. This is the case for leakage functions φ in (14) where $\mathrm{E}\left[\varphi[Z(k_l)]\right]$ is not constant on $\mathcal{K}_l$ (recall that $Z(k_l)$ equals $P_1(A)\|P_1(k_l)$). Almost all functions φ satisfy this condition. However, this is not the case for functions φ where $\varphi(X\|Y) = \varphi(X) + \varphi(Y)$ (e.g. $\varphi = \mathrm{HW}$). For those leakage functions, Pearson's coefficient (15) is not a sound key-distinguisher when applied directly to the leakages ℓ_i's. Indeed, in this case, (3) and $\sigma\left[\varphi[Z(K_l)] \mid K_l\right] = 0$ imply that $\rho\left[L(k_l), f(A, k_l)\right]$ is null, regardless of the prediction function f. For such functions φ, it makes sense (see for instance [18]) to focus on higher order moments, and to compute the following Pearson's coefficient for an appropriate function f, which may differ from the case when $o = 1$:

$$\widehat{\rho}_{k_l^\star} = \widehat{\rho}(((\ell_i - \overline{\ell})^o)_i, (f(a_i, k_l^\star))_i) \ . \tag{16}$$

For instance, if $\varphi = \mathrm{HW}$, then the second order centered moments of the $\varphi[Z(k_l)]$'s are different, so (16) must be computed for $o = 2$.

Remark 2. For $o = 1$ (*i.e.* when the CPA focuses on the means), there is no need to center the leakage measurements and the term $\overline{\ell}$ can be omitted. In the other cases, centering the leakage measurements (and thus the predictions) improves the CPA efficiency (see [17]).

When a good approximation $\widehat{\varphi}$ of φ is assumed to be known, the efficiency of the CPA relies on the prediction function f that is chosen. This is especially true in our case where data is not simply masked by the addition of a random value, but by a random permutation, so that removing the effect of the masking (even biased) is difficult. Designing a prediction function f, such that a CPA involving this function in (16) succeeds, is not straightforward. Therefore, to exploit the flaw in (11) using a CPA attack, we need to exhibit a sound prediction function f.

5.2   Designing f_{opt}

The target intermediate variable Z in (11) takes the general form $P_1(X)\|P_1(Y)$, where P_1, X and Y are random variables and where Y depends on k_l. In [17], Prouff *et al.* showed that for every function $\mathcal{C} : L \mapsto \mathcal{C}(L)$, the optimal prediction function f_{opt} is the function $x, y \mapsto \mathrm{E}\left[\mathcal{C}(L(k_l))|X = x, Y = y\right]$. In our case, $\mathcal{C}(L(k_l))$ equals $(L(k_l) - \mathrm{E}\left[L(k_l)\right])^o$ for a given order o. To mount the attack , we need an analytical expression for the function f_{opt} so that it can be estimated even when no information on the noise parameters is known (non-profiling attacks). Therefore, we conducted an analytical study of the function f_{opt} defined by:

$$f_{opt}(x, y) = \mathrm{E}\left[(L(k_l) - \mathrm{E}\left[L(k_l)\right])^o \mid X = x, Y = y\right] , \qquad (17)$$

for $o \in \mathbb{N}$, for L equal to $\hat{\varphi}(Z) + B$, for $B \sim \mathcal{N}(\varepsilon, \sigma^2)$ and for Z equal to $P_1(X) \| P_1(Y)$ with $X, Y \sim \mathcal{U}(\mathbb{F}_2^n)$ and P_1 is a random variable over $\mathcal{P}$.

Below we state the results of our analysis (the derivations of the formulas are given in Appendix A):

- To compute (16), we suggest using the prediction function f defined for every $(a_i, k_l^\star) \in (\mathbb{F}_2^4)^2$ by:

$$f(a_i, k_l^\star) = \sum_{p_1 \in \mathcal{P}} (\hat{\varphi}\left(p_1(a_i) \| p_1(k_l^\star)\right) - \mathrm{E}\left[\hat{\varphi}_{k_l^\star}\right])^o \mathrm{p}\left[P_1 = p_1\right] , \qquad (18)$$

  where:

$$\mathrm{E}\left[\hat{\varphi}_{k_l^\star}\right] = 2^{-4} \sum_{a \in \mathbb{F}_2^4} \sum_{p_1 \in \mathcal{P}} \hat{\varphi}\left(p_1(a) \| p_1(k_l^\star)\right) \mathrm{p}\left[P_1 = p_1\right] . \qquad (19)$$

  For $o \in \{1, 2\}$, it is argued in Appendix A that the functions f above are affine equivalent to f_{opt}.
- Let $\delta_x(y)$ be the function defined by $\delta_x(y) = 1$ if $x = y$, and $\delta_x(y) = 0$ otherwise. If we assume that $o = 2$, $\hat{\varphi} = \mathrm{HW}$, and that P_1 has a uniform distribution over $\mathcal{P}$, we suggest using the following function:

$$f_{opt}(a_i, k_l^\star) = \delta_{a_i}(k_l^\star) , \qquad (20)$$

  which is affine equivalent to f_{opt}.

5.3 Attack Results

In the attack simulations presented in Table 1, we give an estimation of the minimum number of measurements required to achieve the success rate 0.9 for $P_1 \sim \mathcal{U}(\mathcal{P})$ (where $\mathcal{P}$ is designed as proposed in [5]) and $\varphi = \hat{\varphi} = \mathrm{HW}$. In this case, Pearson coefficients have been computed between $((\ell_i - \bar{\ell})^2)_i$ and $(f_{opt}(a_i, k_l^\star))_i$ for the function f_{opt} defined in (18) for $o = 2$. This success rate is defined as the ratio of successful attacks involving N measurements to the number of attacks involving N measurements. We assumed that an attack is successful if the highest correlation is attained for the correct key. The simulations show that for noiseless measurements, key nibbles can be successfully recovered from the randomised AddRoundKey operation using only 100 power traces. We also carried out the CPA attack on a practical smart card implementation of the randomised AES, as described by [5]. We used a Silvercard, which contains a programmable 8-bit PIC16F877 microprocessor, and verified that the power consumption of the card leaks information in the HW model. For each plaintext sent to the card, the encryption operation was performed ten

Table 1. Num. of measurements required in simulated CPA attack on randomised AddRoundKey

Noise standard deviation	Number of measurements
0	100
0.5	1,000
1	1,500
2	4,500
5	60,000
7	230,000
10	900,000

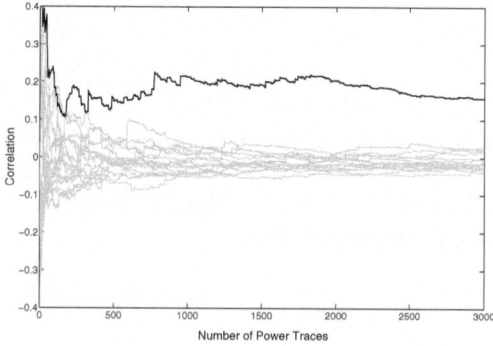

Fig. 1. CPA attack on smart card implementation of randomised AddRoundKey

times (with the same random values used to generate the permutation tables) and an average trace of the power consumption was recorded, in order to reduce the effects of acquisition noise. For the attack, we calculated the correlations between $((\ell_i - \bar{\ell})^2)_i$ and $(f_{opt}(a_i, k_l^\star))_i$ for the simplified function f_{opt} defined in (20). The results of the attack are shown in Fig. 1 for various numbers of power traces. The correlation for the correct key nibble is highlighted, showing that the correct key nibble can be recovered using fewer than 1,000 plaintext/power trace pairs.

6   Attacking the Randomised MixColumns Operation

In this section, we describe CPA and MIA attacks that target the use of XT_8 when the first MixColumns operation is performed. These attacks are of interest, because they allow recovery of two key bytes (*cf.* Eq. (12)), as opposed to a single key nibble when the AddRoundKey operation is targeted. We assume that a sample of N leakages $(\ell_i)_i$ has been measured for N pairs of known AES state values $((a_{1,i}, a_{2,i}))_i$ (where $a_{j,i}$ denotes the known value of byte a_j at the time of the ith measurement ℓ_i). Due to (12) and (13), the ℓ_i's and the $a_{j,i}$'s satisfy the following relation:

$$\ell_i = \varphi(p_{1,i}(S[a_{1,i} \oplus k_1]_l) \| p_{1,i}(S[a_{2,i} \oplus k_2]_l) + b_i \ , \tag{21}$$

where b_i and $p_{1,i}$ are as defined for Eq. (14).

6.1   MIA Preliminaries

In MIA attacks [8], key candidates $k^\star$ are discriminated by estimating the mutual information $I(\hat{\varphi}(\hat{Z}(k^\star)); L(k))$. In an MIA, the attacker is potentially allowed to make weaker assumptions on φ and on Z than in the CPA. Indeed, rather than a good affine approximation of φ and of Z, we only require a pair $(\hat{\varphi}, \hat{Z})$ such that

$I(\hat{\varphi}(\hat{Z}(k)); L(k))$ is non-negligible when the good key k is tested (which may happen even if $\rho(\hat{\varphi}(\hat{Z}(k)), L(k)) = 0$) (see [1,16] for more details). Therefore, we do not require a lengthy derivation for a prediction function f_{opt}, as was required in Section 5.2 for the CPA. After assuming that a good approximation $\hat{\varphi}$ of the leakage function φ is known, an MIA attack can be performed by estimating the mutual information between the random variable L associated with the leakage measurements ℓ_i in (21) and the prediction function $\hat{\varphi}(S[A_1 \oplus k_1^\star]_l || S[A_2 \oplus k_2^\star]_l)$, for various hypotheses $(k_1^\star, k_2^\star)$ on key bytes (k_1, k_2). The mutual information will attain its maximum value for the correct set of key hypotheses.

Remark 3. As noted in Sec. 5, it was less pertinent to use mutual information as a distinguisher when attacking the randomised AddRoundKey operation. The main reason for this is that when φ is the Hamming weight function, the conditional random variable $\varphi(Z(k))$ has the same entropy for each k. As discussed in [8,16], a way to deal with this issue is to focus on the mutual information between $\varphi(Z(k))$ and predictions in the form $\hat{\varphi} \circ \psi(\hat{Z}(k^\star))$, where ψ is any non-injective function. Even if this approach enables recovery of the key, we checked that for various functions ψ (in particular for functions ψ selecting less than 8 bits in $\hat{Z}(k)$), MIA attacks were much less efficient than CPA.

6.2 Attack Results

In simulation, we tested both an MIA attack and a CPA attack, targeting the first call to XT_8 in the first MixColumns operation. For the MIA, we used the Kernel Density and Parametric estimation methods described in [16] to estimate the mutual information. For the same reasons as given in Sec. 5.2 (and Appendix A), the CPA simulations used the pre-processing described in Eq. (16), and the following prediction function:

$$f_{opt}(a_{1,i}, k_1^\star, a_{2,i}, k_2^\star) = \delta_{(S[a_{1,i} \oplus k_1^\star]_l)}(S[a_{2,i} \oplus k_2^\star]_l) \tag{22}$$

In the attack simulations presented in Table 2, we give a rough estimation of the minimum number of measurements required to achieve the success rate 0.9 for the different distinguishers. In these experiments, one key byte was fixed at the correct value, and the distinguishers were calculated for the 2^8 values of the second key byte. Fewer measurements are required for a successful

Table 2. Num. measurements required in MIA and CPA attacks on randomised MixColumns (where '−' implies no successful result with up to 1 million measurements)

Noise standard deviation	0	0.5	1	2	5	7	10	15	20
Nb of measurements [MIA with Kernel]	2,500	20,000	60,000	290,000	−	−	−	−	−
Nb of measurements [Parametric MIA]	na	3,000	4,000	25,000	250,000	500,000	800,000	−	−
Nb of measurements [CPA with f_{opt}]	1,000	1,000	1,500	6,500	120,000	550,000	−	−	−

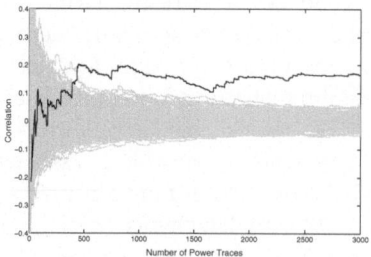
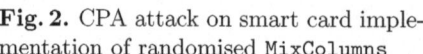
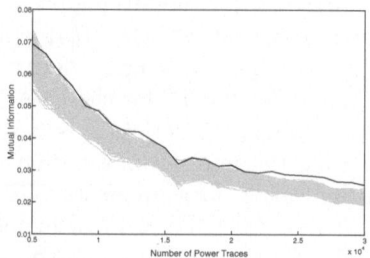

Fig. 2. CPA attack on smart card implementation of randomised `MixColumns`

Fig. 3. MIA attack on smart card implementation of randomised `MixColumns`

attack using CPA than are required when using MIA, for low-noise measurements. This is to be expected, since in the simulations, the HW of the attack variable leaks perfectly, so there is a linear relation between the deterministic part of the leakage and the prediction. MIA is more useful when the relationship between the leakage and the prediction is non-linear. It is interesting to note that when the measurements are noisy, the parametric MIA attack is more efficient than the CPA attack (even in this simulation context that is favourable to CPA).

These attacks were also verified using measurements from the smart card implementation, as shown in Figures 2 and 3 (where the distinguisher for the correct key byte is highlighted). Since the noise in these acquisitions has been reduced due to averaging, the CPA succeeds in fewer measurements ($\sim 2,000$ power traces) than an MIA attack ($\sim 23,000$ traces, using the histogram method to estimate the mutual information [8]).

7 Conclusion

In this paper, we have shown that first-order flaws exist in the permutation tables countermeasure proposed in [5]. In order to exploit this leakage, two attacks have been developed. The first attack applies the recent work of [17] to develop an optimal prediction function for use in a correlation-based attack. The second attack is based on mutual information analysis, and uses estimation methods proposed by [16]. The new attacks were verified in both simulation and practice. In the extended version of this paper [15], we suggest a patch for the permutation tables countermeasure, thereby removing the first-order leakage. It is interesting to note that even if the permutation tables countermeasure is flawed, exploiting this flaw requires more traces than, for instance, an attack on a flawed masking scheme. Therefore, an avenue for further research is to examine the HO-SCA resistance of the (patched) permutation tables countermeasure, as it may also be more HO-SCA resistant than masking.

References

1. Aumonier, S.: Generalized Correlation Power Analysis. Published in the Proceedings of the Ecrypt Workshop Tools For Cryptanalysis 2007(2007)
2. Blömer, J., Merchan, J.G., Krummel, V.: Provably Secure Masking of AES. In: Handschuh, H., Hasan, M.A. (eds.) SAC 2004. LNCS, vol. 3357, pp. 69–83. Springer, Heidelberg (2004)
3. Brier, É., Clavier, C., Olivier, F.: Correlation Power Analysis with a Leakage Model. In: Joye, M., Quisquater, J.-J. (eds.) CHES 2004. LNCS, vol. 3156, pp. 16–29. Springer, Heidelberg (2004)
4. Chari, S., Jutla, C.S., Rao, J.R., Rohatgi, P.: Towards Sound Approaches to Counteract Power-Analysis Attacks. In: Wiener, M. (ed.) CRYPTO 1999. LNCS, vol. 1666, pp. 398–412. Springer, Heidelberg (1999)
5. Coron, J.-S.: A New DPA Countermeasure Based on Permutation Tables. In: Ostrovsky, R., De Prisco, R., Visconti, I. (eds.) SCN 2008. LNCS, vol. 5229, pp. 278–292. Springer, Heidelberg (2008)
6. Coron, J.-S., Giraud, C., Prouff, E., Rivain, M.: Attack and Improvement of a Secure S-Box Calculation Based on the Fourier Transform. In: Oswald, E., Rohatgi, P. (eds.) CHES 2008. LNCS, vol. 5154, pp. 1–14. Springer, Heidelberg (2008)
7. Fumaroli, G., Mayer, E., Dubois, R.: First-Order Differential Power Analysis on the Duplication method. In: Srinathan, K., Rangan, C.P., Yung, M. (eds.) INDOCRYPT 2007. LNCS, vol. 4859, pp. 210–223. Springer, Heidelberg (2007)
8. Gierlichs, B., Batina, L., Tuyls, P., Preneel, B.: Mutual information analysis. In: Oswald, E., Rohatgi, P. (eds.) CHES 2008. LNCS, vol. 5154, pp. 426–442. Springer, Heidelberg (2008)
9. Golić, J., Tymen, C.: Multiplicative Masking and Power Analysis of AES. In: Kaliski Jr., B.S., Koç, Ç.K., Paar, C. (eds.) CHES 2002. LNCS, vol. 2523, pp. 198–212. Springer, Heidelberg (2003)
10. Joye, M., Paillier, P., Schoenmakers, B.: On Second-order Differential Power Analysis. In: Rao, J.R., Sunar, B. (eds.) CHES 2005. LNCS, vol. 3659, pp. 293–308. Springer, Heidelberg (2005)
11. Kocher, P.: Timing attacks on implementations of Diffie-Hellman, RSA, DSS and other systems. In: Koblitz, N. (ed.) CRYPTO 1996. LNCS, vol. 1109, pp. 104–113. Springer, Heidelberg (1996)
12. Menezes, A.J., van Oorschot, P.C., Vanstone, S.A.: Handbook of Applied Cryptography. CRC Press, Boca Raton (1997)
13. Messerges, T.S.: Using Second-order Power Analysis to Attack DPA Resistant software. In: Paar, C., Koç, Ç.K. (eds.) CHES 2000. LNCS, vol. 1965, pp. 238–251. Springer, Heidelberg (2000)
14. Piret, G., Standaert, F.-X.: Security Analysis of Higher-Order Boolean Masking Schemes for Block Ciphers (with Conditions of Perfect Masking). IET Information Security 2(1), March 1–11 (2008)
15. Prouff, E., McEvoy, R.: First-Order Side-Channel Attacks on the Permutation Tables Countermeasure — Extended Version. To appear on the Cryptology ePrint Archive (2009), http://eprint.iacr.org
16. Prouff, E., Rivain, M.: Theoretical and Practical Aspects of Mutual Information Based Side Channel Analysis. In: Abdalla, M., et al. (eds.) ACNS 2009. LNCS, vol. 5536, pp. 502–521. Springer, Heidelberg (2009)

17. Prouff, E., Rivain, M., Bévan, R.: Statistical Analysis of Second Order Differential Power Analysis. IEEE Transactions on Computers 58(6), 799–811 (2009)
18. Soong, T.T.: Fundamentals of Probability and Statistics for Engineers, 3rd edn. John Wiley & Sons, Ltd, Chichester (2004)
19. Waddle, J., Wagner, D.: Toward Efficient Second-order Power Analysis. In: Joye, M., Quisquater, J.-J. (eds.) CHES 2004. LNCS, vol. 3156, pp. 1–15. Springer, Heidelberg (2004)

A   Derivation of f_{opt} for CPA Attacks

This section aims at deriving analytical expressions for the function f_{opt}. We begin with the expression $f_{opt}(x, y) = \mathrm{E}\left[(L - \mathrm{E}\left[L\right])^o \mid X = x, Y = y\right]$ (Eq. (17)), where for clarity reasons and because there is no ambiguity, we use the notation L in place of $L(k_l)$. We recall that the random variable L is assumed to satisfy $L = \hat{\varphi}(Z) + B$ and that Z equals $P_1(X) \| P_1(Y)$ with $P_1 \sim \mathcal{U}(\mathcal{P})$. Since the expectation is linear and the random variables B and (X, Y) are independent, developing $(L - \mathrm{E}\left[L\right])^o$ leads to:

$$f_{opt}(x, y) = \mathrm{E}\left[(\hat{\varphi}(Z) - m)^o \mid (X, Y) = (x, y)\right] + \mu_o$$

$$+ \sum_{i=1}^{o-1} \binom{o}{i} \mu_{o-i} \mathrm{E}\left[(\hat{\varphi}(Z) - m)^i \mid (X, Y) = (x, y)\right] , \quad (23)$$

where m denotes the mean $\mathrm{E}\left[\hat{\varphi}(Z)\right]$ and μ_i denotes the ith order central moment of $B \sim \mathcal{N}(\varepsilon, \sigma^2)$. Let us notice that since μ_1 is zero, the sum in (23) can start from $i = 2$.

Example 1. For o equal to 1 and 2, we respectively have:

$$f_{opt}(x, y) = \mathrm{E}\left[\hat{\varphi}(Z) - m \mid (X, Y) = (x, y)\right]$$

and

$$f_{opt}(x, y) = \mathrm{E}\left[(\hat{\varphi}(Z) - m)^2 \mid (X, Y) = (x, y)\right] + \mu_2 .$$

The prediction function given in (18) corresponds to the development of the terms in (23) that do not depend on noise parameters. It must be noticed that in the cases $o = 1$ and $o = 2$, such an estimation of f_{opt} is perfect since the terms that depend on noise parameters are either null or constant.

Henceforth, we assume that $\mathcal{P}$ is the set of all permutations over $\mathbb{F}_2^n$ and that P_1 is a random variable with uniform distribution over $\mathcal{P}$. This assumption is very favorable to the permutation table countermeasure since it implies that the choice of the masking permutation P_1 is not reduced to a sub-class of the set of permutations over $\mathbb{F}_2^n$.

We now focus on the non-noisy term in (23), namely on the mean $\mathrm{E}\left[(\hat{\varphi}(Z) - m)^o \mid (X, Y) = (x, y)\right]$. Moreover, we denote this conditional mean by $g(x, y)$, and define $\delta_x(y)$ s.t. $\delta_x(y) = 1$ if $y = x$ and $\delta_x(y) = 0$ otherwise (resp. $\overline{\delta_x}(y) = 1 - \delta_x(y)$). We have the following Lemma:

Lemma 2. *Let X and Y be two random variables with uniform distributions over $\mathbb{F}_2^n$ and let P_1 be a random variable uniformly distributed over the set of all permutations over $\mathbb{F}_2^n$. Then for every $\hat{\varphi}$ the function g is 2-valued and satisfies:*

$$g(x, y) = \frac{2^n \delta_x(y) - 1}{2^n - 1} E\left[(\hat{\varphi}(I||I) - m)^o\right] + \frac{2^n \overline{\delta_x}(y)}{2^n - 1} E\left[(\hat{\varphi}(I||J) - m)^o\right] , \quad (24)$$

where I and J are two independent random variables with uniform distribution over $\mathbb{F}_2^n$.

Proof. For every $(x, y) \in \mathbb{F}_2^{2n}$ we have

$$g(x, y) = \sum_{i,j \in \mathbb{F}_2^n} (\hat{\varphi}(i||j) - m)^o \mathrm{p}\left[P_1(x) = i, P_1(y) = j\right] .$$

Since P_1 is assumed to have uniform distribution over the set of permutations over $\mathbb{F}_2^n$, for every $(x, y) \in (\mathbb{F}_2^n)^2$ s.t. $x \neq y$ we have:

$$\mathrm{p}\left[P_1(x) = i, P_1(y) = j\right] = \begin{cases} 1/2^n(2^n - 1) & \text{if } i \neq j \\ 0 & \text{otherwise.} \end{cases} \quad (25)$$

If $x = y$, we have

$$\mathrm{p}\left[P_1(x) = i, P_1(y) = j\right] = \begin{cases} 1/2^n & \text{if } i = j \\ 0 & \text{otherwise.} \end{cases} \quad (26)$$

Combining (25) and (26) gives (24).

When the estimation $\hat{\varphi}$ is the Hamming weight over $\mathbb{F}_2^{2n}$, (24) can be further developed. Indeed, in this case we have:

$$g(x, y) = \frac{2^n \delta_x(y) - 1}{2^n - 1} 2^o E\left[\left(\mathrm{HW}(I) - \frac{n}{2}\right)^o\right] + \frac{2^n \overline{\delta_x}(y)}{2^n - 1} E\left[(\mathrm{HW}(I||J) - n)^o\right] ,$$

since m equals $E\left[\mathrm{HW}\right]$, i.e. n when HW is defined over $\mathbb{F}_2^{2n}$.

As $E\left[\mathrm{HW}(I)\right]$ equals $\frac{n}{2}$ and $E\left[\mathrm{HW}(I||J)\right]$ equals n, the function g is constant equal to 0 when $o = 1$. For $o = 2$, it satisfies:

$$g(x, y) = \delta_x(y) \frac{n2^{n-1}}{2^n - 1} + \frac{n(2^{n-1} - 1)}{2^n - 1} , \quad (27)$$

since we have $E\left[(\mathrm{HW}(I) - \frac{n}{2})^2\right]$ (resp. $E\left[(\mathrm{HW}(I||J) - n)^2\right]$) equal to $\mathrm{Var}\left[\mathrm{HW}(I)\right] = \frac{n}{4}$ (resp. $\mathrm{Var}\left[\mathrm{HW}(I||J)\right] = \frac{n}{2}$).

For $o = 2$, (27) implies that f_{opt} is an affine increasing function of $\delta_x(y)$. Since the correlation coefficient is invariant for any affine transformation of one or both of its parameters, the function $x, y \mapsto \delta_x(y)$ itself (and every affine transformation of it) is actually an optimal prediction function for $o = 2$. Hence, in its simplest form the optimal function for $o = 2$ is defined for every $(x, y) \in \mathbb{F}_2^{n\,2}$ as:

$$f_{opt}(x, y) = \delta_x(y) \ . \tag{28}$$

For $o = 2$, the function f_{opt} in (28) can be applied to conduct CPA attacks in the particular case of Coron's construction of $\mathcal{P}$ (which is not the set of all permutations over $\{0, ..., 15\}$ but a subset of it with cardinality 16^4), without losing a significant factor in attack efficiency (in terms of number of leakage measurements).

Algebraic Side-Channel Attacks on the AES: Why Time also Matters in DPA

Mathieu Renauld[*], François-Xavier Standaert[**],
and Nicolas Veyrat-Charvillon[*]

UCL Crypto Group, Université catholique de Louvain, B-1348 Louvain-la-Neuve
{mathieu.renauld,fstandae,nicolas.veyrat}@uclouvain.be

Abstract. Algebraic side-channel attacks have been recently introduced as a powerful cryptanalysis technique against block ciphers. These attacks represent both a target algorithm and its physical information leakages as an overdefined system of equations that the adversary tries to solve. They were first applied to PRESENT because of its simple algebraic structure. In this paper, we investigate the extent to which they can be exploited against the AES Rijndael and discuss their practical specificities. We show experimentally that most of the intuitions that hold for PRESENT can also be observed for an unprotected implementation of Rijndael in an 8-bit controller. Namely, algebraic side-channel attacks can recover the AES master key with the observation of a single encrypted plaintext and they easily deal with unknown plaintexts/ciphertexts in this context. Because these attacks can take advantage of the physical information corresponding to all the cipher rounds, they imply that one cannot trade speed for code size (or gate count) without affecting the physical security of a leaking device. In other words, more intermediate computations inevitably leads to more exploitable leakages. We analyze the consequences of this observation on two different masking schemes and discuss its impact on other countermeasures. Our results exhibit that algebraic techniques lead to a new understanding of implementation weaknesses that is different than classical side-channel attacks.

1 Introduction

Template attacks [9] are usually considered as the most powerful type of side-channel attacks and can be viewed as divided in two distinct phases. First, an adversary profiles the device that he targets. That is, he builds a probabilistic model for the leakages of this device, usually referred to as templates. Then in a second phase, the adversary uses these templates to compare key-dependent predictions of the leakages with actual measurements. By repeating successive measurements and comparisons, he can eventually identify the secret data that is manipulated by the leaking implementation. This key recovery is generally performed using a divide-and-conquer strategy, recovering small pieces of the

---

[*] Work supported in part by the Walloon Region research project SCEPTIC.
[**] Associate researcher of the Belgian Fund for Scientific Research (FNRS - F.R.S.).

C. Clavier and K. Gaj (Eds.): CHES 2009, LNCS 5747, pp. 97–111, 2009.

key one by one. Following, this description, an important practical question in template attacks is to determine which templates to build. For example, one can decide to profile key bits directly or to profile intermediate operations occurring in a cryptographic implementation. But once templates have been built, it also remains to determine which information to extract from the leaking device. Most side-channel attacks in the literature directly recover key bytes. In this paper, we tackle the question to know whether it is necessary to extract such a precise information or if extracting some function of the intermediate values in a cryptographic computation (that is easier to guess with the side-channels than key bytes) can lead to a successful cryptanalysis with smaller data complexity.

As a matter of fact, this question is not new and several papers already dealt with very similar issues. Most notably, collision attacks such as, *e.g.* [4,5,18,24,25] don't use the side-channels to recover key bytes. They rather detect pairs of plaintexts giving rise to similar leakages for some intermediate values in the target algorithm. Then, in an offline cryptanalysis step, they use these collisions to recover the complete block cipher key. More generally, papers such as [3,7,15] also combine the side-channel leakages with black box (differential, impossible differential, square) cryptanalysis techniques. Still, and as classical side-channel attacks, these attacks mainly exploit the information provided by the first block cipher rounds, where the diffusion is not yet complete. Very recently, a new type of side-channel attacks, denoted as algebraic, has been introduced in order to get rid of this limitation [23]. These attacks (inspired from [10]) first write the target block cipher as a system of quadratic (or cubic, . . .) equations. But since solving such systems of equations is generally too hard, they additionally add the physical information leakages provided by any intermediate computation during the encryption process to the system. In the previously investigated example of the block cipher PRESENT implemented in an 8-bit controller, this allowed to recover the cipher key with the observation of a single encryption. Compared to classical side-channel attacks, algebraic techniques bring two interesting features. First, they can take advantage of the information leakages in all the cipher rounds. Second, they can exploit any type of information leakage. In other words, whatever information about the intermediate computations can be added to the system and the more intermediate computations (*i.e.* clock cycles, generally), the more powerful the key recoveries. Hence, they can be viewed as an extreme version of template attacks in which more intermediate leakage points are targeted but less information need to be recovered from them.

The contributions of this paper are threefold. Since algebraic attacks have first been applied to the block cipher PRESENT, due to its simple algebraic structure, it is interesting to evaluate if a more conservative cipher would lead to significantly different results. Hence, we follow the ideas of [23] and apply them to the AES Rijndael. We show that algebraic attacks are still applicable in this context, but also exhibit an increase in the attacks time complexity. Second, we analyze the security of two different masking schemes and show how algebraic attacks modify previous intuitions (*e.g.* more mask bits do not always imply more security anymore). Eventually, we discuss the influence of different design choices and countermeasures in view of the new algebraic techniques.

2 Algebraic Side-Channel Attacks

A detailed description of algebraic side-channel attacks is given in [23]. In this paper, we aim to evaluate the impact of various parameters on their effectiveness and to apply them to a practical implementation of the AES Rijndael. Hence, this section only provides a high level description of the different attack phases.

2.1 Offline Phase 1: Building the System of Equations

The goal of this first phase is to transform the AES into a big system of low degree boolean equations. In this system, the key bits appear as variables in such a way that solving the system is equivalent to recovering them. In this paper, we exploit the techniques presented in [2] to build our system of equations. In the case of the AES Rijndael with 128-bit plaintext and key, it results in a system of approximately 18 000 equations in 10 000 variables (27 000 monomials).

2.2 Online Phase: Extracting the Physical Information

Since directly solving the system of equations representing the AES Rijndael is generally too hard with present techniques, the idea of algebraic side-channel attacks is to feed this system with additional information. Quite naturally, physical leakages are very good candidates for this additional information. As detailed in the introduction of this paper, the issue is then to decide what to extract from the target implementation (a somewhat arbitrary decision). The more physical information extracted, the easier the solving and the more interesting the algebraic techniques compared to standard DPA. Therefore, it yields two questions:

Which intermediate operations to target? This question mainly depends on the target implementation. For example, our following experiments consider the AES Rijndael with a 128-bit master key in an 8-bit PIC microcontroller. In this context, SubBytes will generally be implemented as a 256-byte table lookup and MixColumn will exploit the description of [13] in which it is implemented with four 256-byte table lookups and 9 XOR operations (giving 13 potential leakage points). It is therefore natural to profile the target device in such a way that the leakages corresponding to all these intermediate computations (additionally considering the key additions) are exploited by the adversary.

Which information to recover from each target operation? Once the target operations have been decided by the adversary, it remains to determine what to learn about them. This again depends on the target device (and on countermeasures that could possibly be included to the implementation). For example, an unprotected implementation of the AES Rijndael in the PIC is such

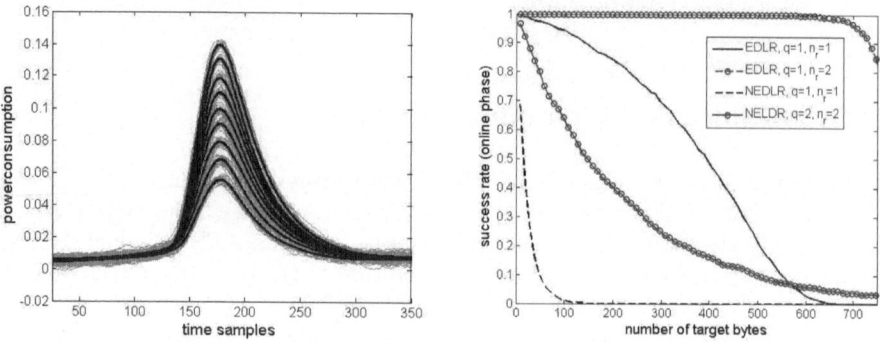

Fig. 1. Leakage traces, mean leakage traces and multiple byte success rate

that for each of the previously mentioned operations, the output data has to commute on an 8-bit bus. During this clock cycle, the leakages will be highly correlated to the Hamming weight of this output data. Hence, it is again natural to perform a template attack such that these Hamming weights will be recovered.

Importantly, algebraic side-channel attacks exploit the leakages of several target operations at once. It implies that one needs to recover a correct information about all these operations at once. Indeed, introducing false information into the system of equations will generally lead to inconsistencies and prevent its proper solving. Fortunately, this can be achieved for our target device. As an illustration, the left part of Figure 1 illustrates the leakage traces of different values commuting on the PIC bus and their mean for different Hamming weights. In the right part of the same figure, we plotted the probability that the Hamming weight of large amounts of target bytes can be recovered with high confidence, possibly exploiting simple error detection (ED) and likelihood rating (LR) techniques (NEDLR means that we do not use these techniques).

- *Error detection* consists in rejecting side-channel information that gives rise to incoherent input and output values for the S-boxes.
- *Likelihood rating* means that we only use a subset of all the Hamming weights values extracted with the templates, starting with the most likely ones.

We stress the fact that all our attacks use a *data complexity* of $q = 1$. That is, we use only the encryption of one single plaintext to perform our measurements. This data complexity is different from the number of traces since we can sometimes repeat the same measurement. In our experiments, we use a repetition umber of $n_r = 1$ or 2. This is significantly different than standard DPA attacks which usually require a much higher data complexity (typically, $10 \leq q \leq 100$). We see that using these techniques, roughly 200 Hamming weights can be recovered with the observation of a single encrypted plaintext and that by repeating the measurement of the same encryption, this number can be up to 700.

As previously mentioned, the implementation of the AES-128 that we attack was profiled in such a way that we recovered the Hamming weights corresponding to AddRoundKey (16 weights), SubBytes (16 weights) and MixColumn ($4 * 13$ weights in our implementation of the AES). For the 10 cipher rounds, it corresponds to a maximum of 788 (not always necessary) correct Hamming weights. It is essential to note that algebraic side-channel attacks in general could work with totally different choices for the target operations and leakages. As a proof of concept and because of their wide use and good connection with our target device, we extracted Hamming weights from our power consumption traces. But one can theoretically exploit any type of information leakage. A particularly interesting open problem is to apply these attacks to advanced circuit technologies with more complex leakage behaviors. It raises the question of how to best extract partial information that is both meaningful (*i.e.* allows to solve the system) and robust (*i.e.* can be easily recovered with high confidence).

2.3 Offline Phase 2: Solving the System

Once the system of equation including the additional side-channel information is written, it remains to attempt solving it. Different solutions have been proposed in the literature for this purpose. The original attack of Courtois and Pieprzyk proposed linearization techniques called XL or XSL [10]. Groebner basis-based techniques have then been suggested as possible alternatives, *e.g.* in [6,12]. Yet another possibility is to use a SAT solver as in [11]. In this paper, we take advantage of this last solution. It implies that the system has to be expressed as a *satisfiability problem*. The satisfiability problem is the reference NP-complete problem and is widely studied (see [14] for a survey). It consists of determining if a boolean formula (a formula combining boolean variables, AND, OR and NOT gates) can be "satisfied", *i.e.* if it exists at least one assignment of the variables such that the whole formula is true. Most SAT solvers require a particular type of formula denoted as a Conjunctive Normal Form (CNF). A CNF is a conjunction (AND) of clauses, each clause being a disjunction (OR) of literals.

In practice, we can write a CNF from our system of equations so that the only valid assignment corresponds to the solution of the system, as detailed in [1]. But this conversion to a boolean formula can be done in a number of ways. For example, the substitution boxes of the AES can either be written as non-linear boolean equations (during the first offline phase of the attack) that are then converted into a boolean formula, or as a set of clauses that are introduced directly into the global CNF. While the first method seems more complex and introduces (much) more intermediate variables, it gave rise to better results in some of our experiments. We conjecture that the SAT solver better handles certain hard instances of the attack with some redundancy and a few additional intermediate variables. As an illustration, the first solution gives about 120 000 clauses of 4 literal per substitution box, versus 2048 clauses of 9 literals for the second solution. The final size of the CNF derived from the system of equations consequently depends on the conversion strategy. To the previous S-boxes, one has to add the linear layers (MixColumn and AddRoundKey) that produce approximately 45 000 clauses of 4 literals per round. Eventually, the additional side-channel information can be defined by approximately 70 000 clauses of up to 8 literals. This gives a formula containing between a minimum of 500 000 and up to several millions of clauses of 1 to 9 literals (the SAT solver used in our experiments [8] was not able to handle more than 5 millions of them).

3   Attacking the AES Key Scheduling Algorithm

Before starting the investigations of algebraic side-channel attacks against the AES, a preliminary remark has to be made about attacks against its key scheduling algorithm. Because of the relatively simple structure of the key expansion in Rijndael, it is possible to write a system of equations only for this part of the algorithm and to solve it successfully with (even a part of) the Hamming weights gathered from its execution. Such attacks then closely connect to the SPA proposed by Mangard in [19]. In the following, we consequently consider the more challenging case in which round keys are pre-computed in a safe environment.

4   Attacking the AES Round Functions

In this section, we assess the practicability of algebraic side-channel attacks against the AES. For this purpose, we considered several situations. First, we attacked an unprotected implementation with a known pair of plaintext/ciphertext. Then, we studied the influence of unknown plaintexts/ciphertexts. Finally, we considered two different masking schemes and analyzed their resistance against algebraic attacks. Each attack exploits the observation of a single encryption trace from which side-channel information is extracted. Additionally, the amount of information recovered by the adversary is used as a parameter in our evaluations. It is measured in "number of rounds of Hamming weights", obtained consecutively (*i.e.* the adversary recovers all the Hamming weights of a few consecutive rounds - the chosen rounds are picked starting in the middle of the block cipher) or randomly (*i.e.* the adversary recovers the same amount of Hamming weights randomly spread over the different intermediate computations).

We first assume that no incorrect side-channel information is used (which would lead the SAT solver to fail). Hence, the experiments in Section 4.1, and 4.2 only focus on the second part of the offline phase described in Section 2.3. Then, in Section 4.4, we discuss the tradeoff between the applicability of this offline phase and the online information extraction described in Section 2.2.

Note that for difficult instances of the attack, the SAT solver is sometimes unable to solve the system in a reasonable time. We consider that an attack has failed whenever the solver has not found a solution within 3600 seconds. Eventually, the SAT solver was running on an Intel-based server with a Xeon E5420 processor cadenced at 2.5GHz running a linux 32-bit 2.6 Kernel.

4.1   Attacking an 8-Bit Device with Known Plaintext/Ciphertext

The results of this first scenario are in Figure 2 (solid lines). Each dot is obtained from averaging on a set of 100 independent experiments. They illustrate that the

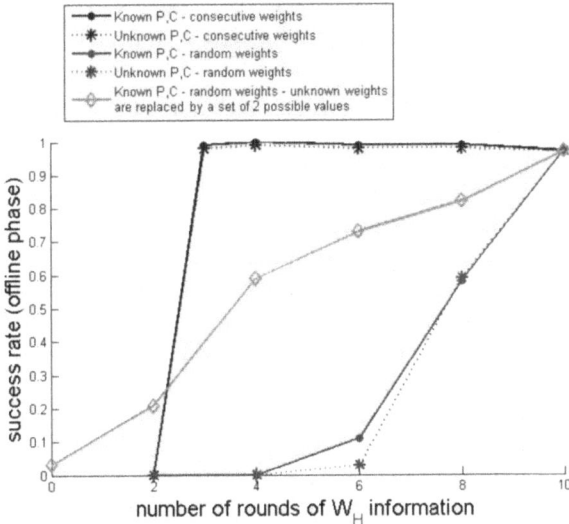

Fig. 2. Success rate of the attacks against an unprotected implementation of the AES Rijndael in function of the amount of exploited leakages. One round of side-channel information is equivalent to 84 known Hamming weights.

success rate of the attacks depends heavily on whether the side-channel leakages correspond to successive operations or not. Indeed, 3 rounds of consecutive leakages (*i.e.* 252 Hamming weights) is enough to solve more than 95% of the cases, whereas 8 rounds of randomly distributed leakages only give a 60% success rate. It is interesting to note that it is in fact the leakage of the MixColumn operation that seems to be the most critical when solving the system. Removing some of the Hamming weights obtained from this operation highly impacts the attack effectiveness. This can be justified by recalling that MixColumn consumes a large amount of clock cycles in a software implementation of the AES.

We finally remark that the SAT solver can also deal with less precise leakages. For example, if one Hamming weight cannot be determined exactly, it is possible to consider a pair of Hamming weights including the correct one and to add this information to the system. This context is investigated in Figure 2 (circled line).

4.2   Attacking an 8-Bit Device with Unknown Plaintext/Ciphertext

The dotted lines in Figure 2 represent the results of an attack with unknown plaintext and ciphertext. It is an interesting scenario to consider since classical DPA attacks such as [17] would generally fail to recover keys in this context. Intuitively, it is also a much harder case, since the plaintext and ciphertext represent 256 unknown variables in our SAT problem. But excepted for a slight reduction of the success rate, our results show that algebraic side-channel attacks against the AES Rijndael are quite strongly immune against unknown inputs.

This can be understood when considering that 3 consecutive rounds of Hamming weight are enough to recover the complete master key.

4.3   Time Complexity and Attack Strategies

The solving times of the SAT problem for different instances of side-channel algebraic attacks (represented in Figure 3) seem to follow an exponential distribution. It means that a small number of instances require much more effort from the SAT solver, which may be due to the intrinsic difficulty of these instances, but also to some poor decisions made by the SAT solver. Therefore, the question to know if good strategies could be applied in order to better cope with this distribution directly arises. This section brings some insights with this respect.

In its basic version, the SAT solver behaves deterministically, always spending the same amount of time on the same boolean formula, for the same result. However, it is possible to introduce randomness in the resolution process, thereby forcing the solver to explore the search space differently. Figure 4 shows how far the solver was able to go towards the resolution of the same SAT problem, when launched 70 times with different random seeds. A successful run has to assign 12577 variables within 3600 seconds. The bold line in the figure shows the maximum number of variables that the solver was able to assign at a given time, averaged over the 70 runs. All the other curves represent single runs. One can observe that the solving time for the same instance varies with random resolutions. As previously, it seems to follow an exponential distribution (*i.e.* the probability that a problem has not been solved after x seconds decreases exponentially in x). Also, most of the time is spent in assigning the first 3000 variables. Once this is done, finding the rest of the solution is almost instantaneous (this is observed with the vertical jump of the single runs in the figure).

These results suggest that performing a serial of short random resolutions can give better results than using a single long resolution. Unfortunately, it is not possible to predict a priori how long the resolution will take. Hence, in order to optimize the attack, we need to adapt the time limit in function of some

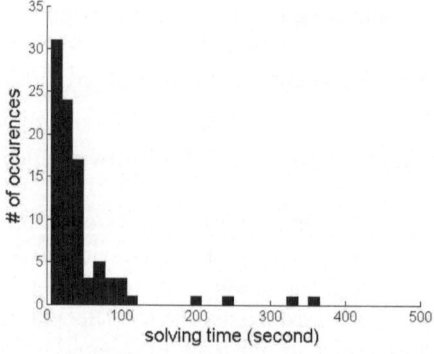

Fig. 3. Distribution of the solving times (8 consecutive rounds of Hamming weights)

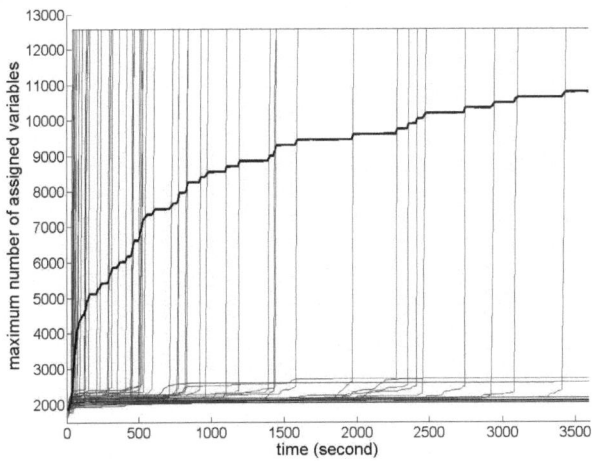

Fig. 4. Evolution of the solving for 70 random runs of the same SAT instance (8 rounds of Hamming weights are known, randomly spread over the block cipher)

preliminary analyzes. For example, in our experiments, 90% of the resolutions succeed in less than 100 seconds. Hence, a reasonable strategy would be to fix the time limit at 100 seconds (rather than 3600, originally). In this setting, most of the attacks will succeed in a single run, and the harder instances will hopefully be solved by another random resolution. Of course, such strategies are mainly interesting for attacks that are successful with high probability within reasonable time constraints, which may not always be the case in practice.

4.4   Global Success Rate

Figure 4.4 finally presents the global success rate of our attack, combining the success rate of the online side-channel phase (given in Figure 1) and the success rate of the offline computation phase (given in Figure 2). It clearly illustrates the tradeoff between these two phases. On the one hand, the success rate of the side-channel information extraction decreases with the quantity of information the attacker tries to recover. On the other hand, the success rate of the algebraic cryptanalysis increases with the quantity of information available. This implies that for a strictly limited amount of measurements (*e.g.* $n_r = 1$ in the figure), the best global success rate does not occur for the maximum amount of Hamming weights inserted into the system. Note that this global success rate could be improved by better dealing with failures in the solving. For example, if an inconsistency is found, it is possible to reject a few Hamming weights and try again to solve the system, hoping that the incorrect leakages have been rejected.

4.5   Attacking Masked Implementations

The previous section suggests that attacking an unprotected implementation of the AES Rijndael in an 8-bit controller is not significantly more difficult

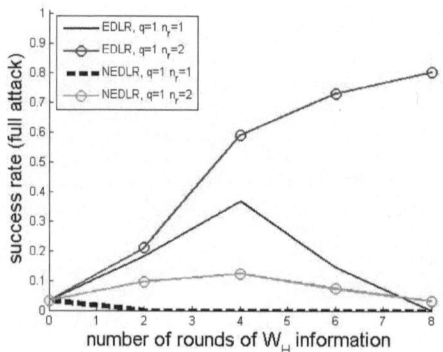

Fig. 5. Global success rate for an unprotected implementation, with known P, C

than attacking the block cipher PRESENT in the same device in terms of data complexity: both attacks succeed with the observation of a single encrypted plaintext. Nevertheless, attacks against the AES exhibit a significantly increased average time complexity (2.5 seconds for 31-round *vs.* 344 seconds for 10-round AES). In this section, we investigate the extent to which the increased algebraic complexity of the AES could lead to more hardly solvable systems if the target implementation is protected with a masking scheme.

Masking is a very popular countermeasure against side-channel attacks. It aims at rendering the power consumption of the device independent of the intermediate values used by the cryptographic algorithm. This is achieved by combining theses values with random "masks" internal to the cryptographic device (hence hidden to the adversary), that vary from execution to execution. The masks are generated at the beginning of the encryption, combined with the plaintext and key, and propagated through the cryptosystem so that every value transiting on the bus is a combination of the original intermediate value and a mask. In terms of security, masking provably prevents certain types of side-channel attacks. But it does not remove the existence of physically exploitable leakages (since the secret data is still manipulated by the device and the correct encryption result has to be produced). In terms of cost overheads, masking affects the performances of an implementation, mainly because the masks have to be propagated through the encryption process in a tractable manner. For this reason, several masking schemes have been proposed in the open literature, trading effectiveness for more random masks (hopefully implying more security).

In the following, we focus on two proposals that we believe illustrative of the state-of-the art countermeasures. Namely, we first focus on the efficient masking scheme proposed by Oswald and Schramm in [20] that uses one different mask for every plaintext byte. Then we consider the AES software implementation resistant to power analysis attacks proposed by Herbst *et al.* in [16].

1. Masking in $GF(2^4)^2$. As noticed in several publications, the most difficult part of the AES to mask is its non-linear layer. This is because for a masked input $x+m$, one needs to generate a masked SubBytes such that: MSubBytes$(x+m)$=SubBytes$(x)+m'$ with m' an output mask that can be propagated through the cipher rounds. The practical issue for designers is to deal with different masks efficiently. If the same mask m is used for all the AES bytes in all the rounds, only one MSubBytes table needs to be computed, which can be done quite easily prior to the encryption. But this may lead to security issues because different intermediate variables are masked with the same value. On the opposite, using a different mask for all the AES bytes implies to recompute MSubBytes anytime an S-box is computed, which is generally too expensive for practical applications. In order to get rid of this limitation, [20] proposes to take advantage of the algebraic structure of the AES S-box to deal with multiple masks efficiently. The main idea is to represent the non-linear part of the AES S-box as a multiplicative inverse in the composite field $GF(2^4)^2$. Doing this, it is possible to compute a masked inverse "on-the-fly" in 14 table lookup operations and 15 XOR operations (this requires to store a total of 1024 bytes of tables in ROM).

Assuming that we can obtain the Hamming weights corresponding to all these intermediate computations, mounting an attack is straightforward. In fact, we can even attack each S-box (anywhere in the cipher) separately and find a solution in less than one second. This simplicity is due to the large quantity of side-channel information leaked (more than 20 Hamming weights for one substitution) and the extreme simplicity of the operations between two leakage points (one XOR or one table look-up operation). As an illustration, Figure 6 presents the results of such an attack mounted with partial information. It turns out that the amount of cycle makes this masked implementation even easier to target than the original (unprotected) implementation, due to the large redundancy of its leakages. We mention that we considered the compact representation described in [20] and only targeted 8-bit intermediate values. But the efficient representation in the same paper would

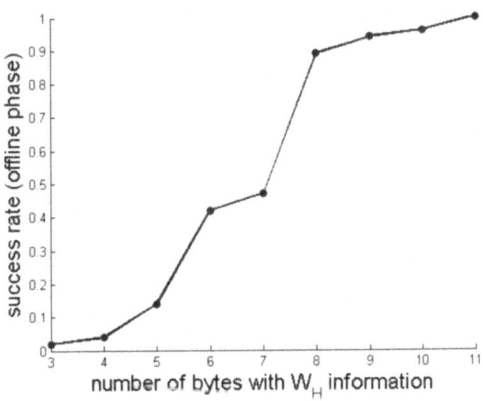

Fig. 6. Attacking one masked AES S-box with partial information leakage

make the situation even worse, allowing 4-bit values to be computed and observed through their Hamming weights.

2. Masking with S-box pre-computation. As previously mentioned, another approach for masking the AES, is to limit the number of 8-bit masks applied to the cipher state so that a reasonable number of masked tables can be pre-computed. The solution proposed by [16] is to use 2 masks m and m' for all the S-boxes inputs and outputs and 4 additional masks (m_1, m_2, m_3, m_4) for the MixColumn operation, resulting in only 48 random bits to generate prior to the encryption. In practice, this scheme turned out to be quite resistant to an algebraic side-channel attack. About 20% of the instances were solved in less than 24 hours of computation, and the fastest solving took about 5 hours. This is surprising, since the masking only adds 48 new variables to the problem while providing 32 more leaked values per round. From an intuitive point of view, this again comes back to the fact that such a masking scheme adds randomness to the computation without too much affecting the number of clock cycles required to encrypt one plaintext. It also relates to the previous observation that Mix-Column is an important source of side-channel information in our attacks and becomes more complex because of the masking. For example, whereas the unprotected AES leaks $W_H(a_0 \oplus a_1)$ during the computation of MixColumn, the masked version leaks $W_H(a_0 \oplus m_1 \oplus a_1 \oplus m_2)$. That is, the masking doubles the number of bytes in each Hamming weight leakage. The quantity of information that one can extract from them is thus significantly less than that of the unmasked implementation. Such a masking scheme in therefore better suited to resist against an algebraic side-channel attack than [20].

5   Countermeasures

The previous section underlines that algebraic side-channel attacks shed a different light on the security of cryptographic implementations. In general, resisting against such attacks should follow different guidelines that we now detail:

1. Use block ciphers with high algebraic complexity.
2. Limit the number of clock cycles in the implementations (*i.e.* have each elementary operation/instruction that has sufficient algebraic complexity).
3. Increase the algebraic complexity of the leakages (*i.e.* use large data buses, add noise and countermeasures to the implementations).

With respect to masking in an 8-bit device, the impact of these observations has been carefully analyzed in the previous section. In addition to the fact that [16] seem to better resist algebraic cryptanalysis than [20], our results obviously confirm that one cannot mask the first/last rounds of a block cipher only. We now discuss how the previous principles apply to practical countermeasures. First, using table-based implementations seems to be the most natural choice to improve resistance against algebraic side-channel attacks. The AES Rijndael has nice features with this respect (*e.g.* the ability to compute four S-boxes and MixColumn in four 256×4-byte table lookups and three XOR operations). Although this solution is not directly applicable to small devices, preliminary experiments suggest that removing all the Hamming weights of the MixColumn operation but the 4 output ones significantly increases the resistance against algebraic side-channel attacks (we could only attack up to 70% of the SAT instances corresponding to an 4-round AES in this context). Hence, implementing table-based implementations in 32-bit (or larger) devices where the leakages would depend on larger amounts of intermediate bits (32 or more) is a challenging scenario for further investigation. Then, increasing the algebraic complexity of the leakages implies to reduce the amount of information provided by any elementary operation. Using large buses is helpful with this respect. But other countermeasures could be helpful: noise addition, dual-rail circuits, ... Eventually, we mention that time randomization would also increase the difficulty of performing the attacks, by preventing the direct profiling of the intermediate operations.

6   Conclusion and Open Problems

This paper shows that algebraic side-channel attacks can be applied to the AES Rijndael implemented in an 8-bit controller. The observation of a single leakage trace can be sufficient to perform a complete key recovery in this context and the attack directly applies to certain masking schemes. Our experiments also suggest interesting scenarios where less information is available (table-based implementations, typically) and for which solving the system of equations describing the target cipher and its leakages becomes challenging. Open problems consequently include the investigations of advanced strategies to attack protected devices (*e.g.* exploiting the leakages in a more flexible manner, dealing with errors, ...). In particular, circuit technologies where the leakage models are significantly different than Hamming weights would be interesting to evaluate.

To properly understand the impact of this result, it is important to relate it with theoretical and practical concerns. From a purely practical point of view, one could argue that algebraic side-channel attacks do not improve previous attacks against leaking ciphers. Indeed, these attacks require to recover the leakages corresponding to a significant amount of operations. In other words, they can be seen as (very) high-order attacks. Hence, a first-order DPA exploiting the leakages corresponding to as many leakage points (but for different plaintexts) would generally succeed as well. But from a theoretical point of view, the data complexity that is required to reach high success rates is significantly decreased. In other words, what eventually matters from a security point of view is the number of observed plaintexts required to recover a key. The attacks in this paper allow reaching a success rate of one with the observation of a single encrypted plaintext (which was never the case for previous side-channel attacks we are aware of). Even if multiple measurements have to be made (we require a maximum of one or two measurements in the present paper), they still correspond to a unique plaintext. By exploiting the leakages in all the cipher rounds, algebraic side-channel attacks also get rid of the computational limitations of classical

DPA in which enumerating key candidates is necessary. Eventually, they have an impact on the assumptions of constructions such as [21]. Hence, we believe these results have both a theoretical and practical relevance.

References

1. Bard, G., Courtois, N., Jefferson, C.: Efficient Methods for Conversion and Solution of Sparse Systems of Low-Degree Multivariate Polynomials over GF(2) via SAT-Solvers, Cryptology ePrint Archive, Report 2007/024
2. Biryukov, A., De Cannière, C.: Block Ciphers and Systems of Quadratic Equations. In: Johansson, T. (ed.) FSE 2003. LNCS, vol. 2887, pp. 274–289. Springer, Heidelberg (2003)
3. Biryukov, A., Khovratovich, D.: Two New Techniques of Side-Channel Cryptanalysis. In: Paillier, P., Verbauwhede, I. (eds.) CHES 2007. LNCS, vol. 4727, pp. 195–208. Springer, Heidelberg (2007)
4. Bogdanov, A.: Improved Side-Channel Collision Attacks on AES. In: Adams, C., Miri, A., Wiener, M. (eds.) SAC 2007. LNCS, vol. 4876, pp. 84–95. Springer, Heidelberg (2007)
5. Bogdanov, A., Kizhvatov, I., Pyshkin, A.: Algebraic Methods in Side-Channel Collision Attacks and Practical Collision Detection. In: Chowdhury, D.R., Rijmen, V., Das, A. (eds.) INDOCRYPT 2008. LNCS, vol. 5365, pp. 251–265. Springer, Heidelberg (2008)
6. Buchmann, J., Pyshkin, A., Weinmann, R.-P.: Block Ciphers Sensitive to Gröbner Basis Attacks. In: Pointcheval, D. (ed.) CT-RSA 2006. LNCS, vol. 3860, pp. 313–331. Springer, Heidelberg (2006)
7. Carlier, V., Chabanne, H., Dottax, E., Pelletier, H.: Generalizing Square Attack using Side-Channels of an AES Implementation on an FPGA. In: The proceedings of FPL 2005, Tampere, Finland, August 2005, pp. 433–437 (2005)
8. http://www.princeton.edu/~chaff/
9. Chari, S., Rao, J., Rohatgi, P.: Template Attacks. In: Kaliski Jr., B.S., Koç, Ç.K., Paar, C. (eds.) CHES 2002. LNCS, vol. 2523, pp. 13–28. Springer, Heidelberg (2003)
10. Courtois, N., Pieprzyk, J.: Cryptanalysis of Block Ciphers with Overdefined Systems of Equations. In: Zheng, Y. (ed.) ASIACRYPT 2002. LNCS, vol. 2501, pp. 267–287. Springer, Heidelberg (2002)
11. Courtois, N., Bard, G.: Algebraic Cryptanalysis of the Data Encryption Standard. In: Galbraith, S.D. (ed.) Cryptography and Coding 2007. LNCS, vol. 4887, pp. 274–289. Springer, Heidelberg (2007)
12. Faugère, J.-C.: Groebner Bases. In: Applications in Cryptology, FSE 2007, Invited Talk (2007), http://fse2007.uni.lu/slides/faugere.pdf
13. FIPS 197, "Advanced Encryption Standard," Federal Information Processing Standard, NIST, U.S. Dept. of Commerce, November 26 (2001)
14. Gu, J., Purdom, P.W., Franco, J., Wah, B.: Algorithms for the Satisfiability problem: a survey. DIMACS Series on Discrete Mathematics and Theoretical Computer Science, vol. 35, pp. 19–151. American Mathematical Society, Providence (1997)
15. Handschuh, H., Preneel, B.: Blind Differential Cryptanalysis for Enhanced Power Attacks. In: Biham, E., Youssef, A.M. (eds.) SAC 2006. LNCS, vol. 4356, pp. 163–173. Springer, Heidelberg (2007)

16. Herbst, C., Oswald, E., Mangard, S.: An AES Smart Card Implementation Resistant to Power Analysis Attacks. In: Zhou, J., Yung, M., Bao, F. (eds.) ACNS 2006. LNCS, vol. 3989, pp. 239–252. Springer, Heidelberg (2006)
17. Kocher, P., Jaffe, J., Jun, B.: Differential Power Analysis. In: Wiener, M. (ed.) CRYPTO 1999. LNCS, vol. 1666, pp. 398–412. Springer, Heidelberg (1999)
18. Ledig, H., Muller, F., Valette, F.: Enhancing Collision Attacks. In: Joye, M., Quisquater, J.-J. (eds.) CHES 2004. LNCS, vol. 3156, pp. 176–190. Springer, Heidelberg (2004)
19. Mangard, S.: A Simple Power-Analysis (SPA) Attackon Implementations of the AES Key Expansion. In: Lee, P.J., Lim, C.H. (eds.) ICISC 2002. LNCS, vol. 2587, pp. 343–358. Springer, Heidelberg (2003)
20. Oswald, E., Schramm, K.: An Efficient Masking Scheme for AES Software Implementations. In: Song, J.-S., Kwon, T., Yung, M. (eds.) WISA 2005. LNCS, vol. 3786, pp. 292–305. Springer, Heidelberg (2006)
21. Petit, C., Standaert, F.-X., Pereira, O., Malkin, T.G., Yung, M.: A Block Cipher based PRNG Secure Against Side-Channel Key Recovery. In: The proceedings of ASIACCS 2008, Tokyo, Japan, March 2008, pp. 56–65 (2008)
22. Pietrzak, K.: A Leakage-Resilient Mode of Operation. In: Joux, A. (ed.) Eurocrypt 2009. LNCS, vol. 5479, pp. 462–482. Springer, Heidelberg (2009)
23. Renauld, M., Standaert, F.-X.: Algebraic Side-Channel Attacks, Cryptology ePrint Archive, report 2009/179, http://eprint.iacr.org/2009/279
24. Schramm, K., Wollinger, T.J., Paar, C.: A New Class of Collision Attacks and Its Application to DES. In: Johansson, T. (ed.) FSE 2003. LNCS, vol. 2887, pp. 206–222. Springer, Heidelberg (2003)
25. Schramm, K., Leander, G., Felke, P., Paar, C.: A Collision-Attack on AES: Combining Side Channel and Differential Attack. In: Joye, M., Quisquater, J.-J. (eds.) CHES 2004. LNCS, vol. 3156, pp. 163–175. Springer, Heidelberg (2004)

Differential Cluster Analysis*

Lejla Batina[1], Benedikt Gierlichs[1], and Kerstin Lemke-Rust[2]

[1] K.U. Leuven, ESAT/SCD-COSIC and IBBT
Kasteelpark Arenberg 10, B-3001 Leuven-Heverlee, Belgium
`firstname.lastname@esat.kuleuven.be`
[2] University of Applied Sciences Bonn-Rhein-Sieg
Grantham-Allee 20, 53757 Sankt Augustin, Germany
`kerstin.lemke-rust@h-brs.de`

Abstract. We propose a new technique called Differential Cluster Analysis for side-channel key recovery attacks. This technique uses cluster analysis to detect internal collisions and it combines features from previously known collision attacks and Differential Power Analysis. It captures more general leakage features and can be applied to algorithmic collisions as well as implementation specific collisions. In addition, the concept is inherently multivariate. Various applications of the approach are possible: with and without power consumption model and single as well as multi-bit leakage can be exploited. Our findings are confirmed by practical results on two platforms: an AVR microcontroller with implemented DES algorithm and an AES hardware module. To our best knowledge, this is the first work demonstrating the feasibility of internal collision attacks on highly parallel hardware platforms. Furthermore, we present a new attack strategy for the targeted AES hardware module.

Keywords: Differential Cluster Analysis, Side-channel Cryptanalysis, Collision Attacks, Differential Power Analysis, AES Hardware.

1   Introduction

Side-channel analysis became a mature area in the past decade with many contributions to new attacks, models and countermeasures since the pioneering results of Differential Power Analysis (DPA) [14]. DPA exploits the fact that information on secret key bits can leak through a side-channel. It typically requires known input to the cryptographic algorithm. Many practical and more recently some theoretical works have been published showing the importance of applying known techniques and ideas from other research communities.

An internal collision attack (CA) is another kind of side-channel analysis. Collision attacks have been introduced by Schramm *et al.* in [22]. A collision in

---

* Work supported in part by the IAP Programme P6/26 BCRYPT of the Belgian State, by FWO project G.0300.07, by the European Commission under contract number ICT-2007-216676 ECRYPT NoE phase II, by K.U. Leuven-BOF (OT/06/40), and by IWT-Tetra STRES (80138).

C. Clavier and K. Gaj (Eds.): CHES 2009, LNCS 5747, pp. 112–127, 2009.

an algorithm occurs if, for at least two different inputs, a function within the algorithm returns the same output. If this happens, the side-channel traces are assumed to be very similar in the time span when the internal collision persists. Collision attacks use the side-channel to detect collisions and afterwards offline computation with or without precomputed tables for key recovery. For both steps there are different approaches proposed in the literature. Considering the assumptions, attacks can be with either chosen or known inputs.

The work of [21] and in particular recent works on collision attacks [3,4,5,6,7] veer away from long sequences of instructions [22,15], e.g. collisions that persist for an entire round, and target short-scale intermediate results. Our approach follows this development and shows that internal collisions can be the source for both DPA and CA leakage.

More precisely, our work introduces Differential Cluster Analysis (DCA) as a new method to detect internal collisions and extract keys from side-channel signals. Our approach is to revisit collision attacks in the unsupervised analysis setting, which can be two-fold e.g. viewed as collision and DPA approach. Our strategy includes key hypothesis testing and the partitioning step similar to those of DPA. Partitioning then yields collisions for the correct key which are detected by cluster analysis. DCA typically requires known input data to the cryptographic algorithm and can be applied to arbitrary algorithmic collisions as well as to implementation specific collisions. Cluster analysis relates to some extent to standard DPA, which is obvious for the univariate case. While DCA is inherently multivariate, the technique inspires a simple extension of standard DPA to multivariate analysis. The most interesting difference is that cluster analysis is sensitive to more general leakage features and does not require a power model for multi-bit collisions.

The idea of clustering for simple side-channel attacks was already used in the work of Walter [25]. Therein, he uses clusters (called buckets) for partitioning sets of similar measurements in order to reveal exponent digits for an attack on sliding windows exponentiation. Our work is also related to Mutual Information Analysis (MIA) [12] in that both approaches can succeed without but benefit from a good power model. Also related to our work is the use of Gaussian mixture models for masked implementations [16]. In this work parameters of different Gaussian components that best fit to the observed mixed multivariate side-channel leakage are estimated without knowing the masks.

Our experiments confirm the findings on two platforms. One platform is an unprotected software implementation of the DES algorithm running on an Atmel AVR microcontroller and the other one is a hardware implementation of AES-128. Collision attacks on platforms like the latter are believed to be unfeasible due to the high parallelism of operations, e.g., [5] states "the collision attacks on AES are mainly constrained to 8-bit software implementations on simple controllers".

The paper is organized as follows: Section 2 describes our new approach to collision detection by use of cluster analysis of measurement samples. Section 3 introduces the new classification of collisions into algorithmic and implementation specific collisions and presents examples for both classes. Experimental results are provided in Section 4 and Section 5 summarizes the results of this work.

2   Differential Cluster Analysis: The General Approach

An internal collision in a cryptographic algorithm occurs if, for at least two inputs $x_i, x_{i'} \in \{0,1\}^u$ with $x_i \neq x_{i'}$ and subkey $k^\circ \in \{0,1\}^v$, values of one particular intermediate state $\Delta \in \{0,1\}^w$ collide. The intermediate state Δ is a specific property of the cryptographic algorithm. Although we provide examples from symmetric schemes the general approach is also valid for public key schemes. In the case of DES for example, the intermediate state is given by a few bits after an S-box access. Let f_k denote the key dependent function $f_k : \{0,1\}^u \mapsto \{0,1\}^w$ that computes Δ for a given input x in a cryptographic algorithm. The function f_k is said to be a *many-to-one collision function* if many inputs are mapped to the same output.

Unlike previous collision attacks that search for similarity between different power traces, the new key recovery attack aims at detecting significantly separated clusters as result of internal collisions. This is an unsupervised learning process. The adversary observes the side-channel response on input patterns, but has incomplete knowledge about the internal state of the system, especially she does not know any key and therefore any true labels of samples. The adversary, however, usually knows the number of different clusters, *i.e.*, the number of possible values for Δ.

DCA assumes that it is feasible to run statistics for all subkey candidates in the algorithm, *i.e.*, v is a small number. For common constructions of ciphering algorithms such as AES and DES this assumption is clearly fulfilled. In the first step of the attack, the adversary classifies measurement samples i_n ($n \in \{1, \dots, N\}$, where N is the total number of samples) with input[1] x_n into 2^w classes according to $f_k(x_n)$ with guessed subkey hypothesis k. As result, the adversary obtains 2^w bins of classified measurements for each key guess.[2] This new attack tests whether clustering statistics, such as good cluster separation or high cluster compactness, indicates a separation of distinct clusters. Note that if $k = k^\circ$ the separation of the samples into the 2^w bins corresponds to the computation of the cryptographic device. If the side-channel leakage of different values of Δ is detectable, this in turn reveals the correct key. If the subkey guess is wrong the measurements are generally classified into bins incorrectly, *i.e.* almost all bins include samples that result from different values of Δ. As a consequence, clusters of measurements resulting from different values of Δ are expected to broaden the statistical distribution of each bin and to smooth out side-channel differences.

DCA classifies objects into classes according to special collisions that relate to known inputs and a key hypothesis. Cluster statistics are used to detect

---

[1] Note that this attack can be applied to both known and chosen input.
[2] Note that not all 2^w states might be possible in a given cryptographic algorithm.

the collisions. Note that this collision and clustering attack is a multivariate approach. Cluster statistics can be applied to measurements samples from a single (univariate) or from multiple (multivariate) time instants. Multivariate DCA benefits if prior knowledge on the relative distances of points in time that contain exploitable side-channel leakage is available, *e.g.* as a result of profiling. Furthermore, additional options exist for combining DCA results. One example is to combine the outcomes of DCA for w 1-bit intermediate states for the analysis of an w-bit intermediate state.

2.1 Cluster Statistics

We provide details about criterion functions for clustering and describe how to measure the clustering quality. In literature, *e.g.* [11,24], cluster statistics use a number of cluster characteristics. In Table 1 characteristics for c clusters $\mathcal{D}_i$, $i \in \{1, \ldots, c\}$ with population n_i of vectors $\boldsymbol{x}$ and total population N are summarized. Note that in case of univariate analysis all vectors have only one element.

The *sum-of-squared-error* is a widely used cluster criterion function. It computes

$$J_{SSE} = \sum_{i=1}^{c} \sum_{\boldsymbol{x} \in \mathcal{D}_i} \| \boldsymbol{x} - \boldsymbol{m}_i \|^2 .$$

This function evaluates the sum of the squared Euclidean distances between the vectors $(\boldsymbol{x} - \boldsymbol{m}_i)$. Informally speaking that is the sum of the scatter over all clusters. The optimal partition minimizes J_{SSE}. An alternative is the *sum-of-squares* criterion. It evaluates the square of the Euclidean distance between the cluster centroids $\boldsymbol{m}_i$ and the total mean vector $\boldsymbol{m}$:

$$J_{SOS} = \sum_{i=1}^{c} n_i \| \boldsymbol{m}_i - \boldsymbol{m} \|^2 .$$

The optimal partition maximizes J_{SOS}. An interesting observation is that the sum of J_{SSE} and J_{SOS} is a constant, thus minimizing J_{SSE} (yielding intra cluster cohesion) is equivalent to maximizing J_{SOS} (yielding inter cluster separation) [24].

In the context of side-channel analysis computing variances can also be useful. In such cases, one can either take variances into account explicitly or normalize

Table 1. Cluster Characteristics

Mean vector for the i-th cluster:	$\boldsymbol{m}_i = \frac{1}{n_i} \sum_{\boldsymbol{x} \in \mathcal{D}_i} \boldsymbol{x}$
Total mean vector:	$\boldsymbol{m} = \frac{1}{N} \sum_{i=1}^{c} n_i \boldsymbol{m}_i$ where $\sum_{i=1}^{c} n_i = N$
Variance vector for the i-th cluster:	$\boldsymbol{v}_i = \frac{1}{n_i} \sum_{\boldsymbol{x} \in \mathcal{D}_i} (\boldsymbol{x} - \boldsymbol{m}_i)^2$
Total variance vector:	$\boldsymbol{v} = \frac{1}{N} \sum_{i=1}^{c} n_i \boldsymbol{v}_i$ where $\sum_{i=1}^{c} n_i = N$
Squared Euclidean norm ($\mathbb{R}^k \to \mathbb{R}$):	$\|(z_1, z_2, \ldots, z_k)\|^2 = \sum_{j=1}^{k} z_j^2$

the measurements before evaluating cluster criteria like J_{SOS} and J_{SSE}. The variance test [23] is a criterion function that evaluates

$$J_{VAR} = \frac{\| \boldsymbol{v} \|^2}{\frac{1}{N} \sum_{i=1}^{c} n_i \| \boldsymbol{v_i} \|^2},$$

i.e. the ratio between the overall variance and the weighted mean of intra cluster variances. The optimal partition maximizes J_{Var}. The student's T-test [13] evaluates the distances between cluster centroids, normalized by intra cluster variances and cluster sizes. We use it in the T-test criterion that evaluates the sum of squared distances for all pairs of clusters

$$J_{STT} = \sum_{i,j=1;i\neq j}^{c} \frac{\| \boldsymbol{m_i} - \boldsymbol{m_j} \|^2}{\sqrt{\frac{\|\boldsymbol{v_i}\|^2}{n_i} + \frac{\|\boldsymbol{v_j}\|^2}{n_j}}}.$$

Again, the optimal partition maximizes J_{STT}.

2.2   The General Approach

Here we summarize our general approach:

1. Measure N samples $\boldsymbol{i_n}$ of power traces while the targeted device computes the cryptographic algorithm with unknown fixed subkey k°. For each sample, store the associated known input x_n for $n = 1, 2, \ldots, N$.
2. For each subkey hypothesis $k \in \{0,1\}^v$ do the following[3]:
   (a) Sort the N measurements according to $\Delta_i = f_k(x_i)$ into 2^w clusters $\mathcal{D}_0, \ldots, \mathcal{D}_{2^w-1}$.
   (b) For each cluster $\mathcal{D}_i$ do the following: Compute mean value $\boldsymbol{m_i}$ and variance $\boldsymbol{v_i}$.
   (c) Compute a cluster criterion J_k (e.g. J_{SSE} or J_{SOS}) to quantitatively assess the quality of the cluster separation.
   (d) Store the pair of k and J_k: (k, J_k).
3. Rank the pairs (k, J_k) according to J_k.
4. Output the key candidate k with the value of J_k that leads to the best clustering quality (min. or max., depending on the criterion function).

2.3   Refinements

Several refinements of the general approach can make the attack more efficient.

(i) Known input x_n is assumed in the general approach. Noise due to non-targeted operations can be highly reduced if the input x_n can be chosen. If the input can be further adaptively chosen this allows to apply an adaptive sieving of key candidates, thereby minimizing the number of samples needed.

---

[3] In practice, steps might be iterated for components of $\boldsymbol{i_n}$, e.g., each iteration might include samples of one or a few clock cycles of the device.

(ii) In the general approach we do not include any assumption on similarity of some clusters. Integration of a side-channel leakage model is possible by merging of clusters, *e.g.*, for a Hamming weight model the number of clusters is reduced from 2^w to $w + 1$. See Sect. 2.4 and 3 for a detailed discussion.

(iii) Depending on the algorithm and the choice of Δ related key hypotheses that lead to so called "ghost peaks" [8] exist. For such key hypotheses the formed clustering is, to a certain degree, identical to the correct clustering. If such related key hypotheses exist this may allow for a special adaption of the analysis.

(iv) Prior normalization of the N samples i_n to zero mean and unit standard deviation is often useful in practice if the spread is due to noise. Transformation to principal components is another useful pre-processing step.

2.4   Detailed Comparison with DPA

Here, we compare DCA with DPA in more detail. For DPA, different variants are known in literature [18]. The original approach for DPA [14] selects one bit of an intermediate state to test whether there is a provable imprint of this bit in the power traces. Kocher's original DPA can be derived from DCA when restricted to two clusters ($w = 1$) and one time instant. The proposed statistics, however, differ to some extent.

Essential differences between DPA and DCA occur for $w > 1$. Multi-bit DPA was first introduced by Messerges *et. al.* [19]. The main idea is that each bit in an intermediate state causes the same amount of leakage and that considering multiple bits at once in an attack may be advantageous. A drawback of this "all-or-nothing" DPA is the inefficient use of measurements: although 2^w clusters exist, only two of them are used. Correlation Power Analysis (CPA) [8] with a Hamming weight or distance model correlates the predicted side-channel leakage with the measured side-channel traces to check for provable correlation. CPA uses all available measurements and can be very efficient in practice. Potential drawbacks are the requirement for a good leakage model, which may be hard to come up with, and the fact that CPA with a Hamming model may not capture all leakage details as Pearson correlation solely tests for equidistance of cluster centroids, *i.e.* linearity.

The general DCA approach uses 2^w clusters and all available measurements. The assumption of a side-channel leakage model is only optional in multi-bit DCA, *e.g.* an adjustment to the Hamming weight or distance model works with $w + 1$ clusters instead. Unlike CPA, DCA can also capture special features of the leakage, *e.g.* unequal densities, non-spherical shapes, and unequal proximities.

Example for $w = 2$ where DCA is advantageous. The following example shows that special cases of side-channel leakage exist that can neither be detected with single-bit DPA nor with Hamming weight based CPA, but still with DCA. We consider a general case with $w = 2$, *i.c.* four equally likely intermediate states "00", "01", "10", "11" that have a mean side-channel leakage of α_0, α_1, α_2, α_3, respectively.

Resistance against single-bit DPA requires that the following conditions must hold: $\alpha_0 + \alpha_1 = \alpha_2 + \alpha_3$ and $\alpha_0 + \alpha_2 = \alpha_1 + \alpha_3$ in order to not reveal any information on either the left or the right bit of the intermediate state. To achieve resistance against Hamming weight CPA the following condition must hold: $\alpha_0 = \alpha_3$.

Resistance against both single-bit DPA and Hamming weight CPA is achieved if the previous three conditions are combined, which leads to $\alpha_0 = \alpha_3$ and $\alpha_1 = \alpha_2$. The trivial solution $\alpha_0 = \alpha_1 = \alpha_2 = \alpha_3$ implies a single joint cluster and no information leakage at all. If $\alpha_0 \neq \alpha_1$ two different clusters arise: a joint cluster of intermediate states "00" and "11" as well as a joint cluster of intermediate states "01" and "10". These two separated clusters can be directly detected with DCA which, assuming that the intermediate state results from a non-injective mapping and depends on a secret subkey value, enables key recovery. We acknowledge that an adapted selection function that tells apart whether the two bits are equal or not would enable multi-bit DPA and CPA. However, the general DCA approach detects such particular features by its own nature while DPA and CPA require a guiding selection function. As mentioned earlier, coming up with such a selection function may not always be easy and can require detailed knowledge of the device's leakage behavior.

3   Differential Cluster Analysis: Applications

In this section we first distinguish between two types of collisions, algorithmic and implementation specific collisions. While an algorithm assumes a more abstract concept, an implementation is a concrete realization of it. There may exist many implementations of an algorithm. On the other hand, having an implementation, one can find just one algorithm corresponding to it.

In general, internal collisions may occur due to the algorithm or may be caused by a particular implementation. In the former case, we speak about algorithmic collisions that are results of non-injective mappings within the algorithm. Algorithmic collisions are therefore typically implementation independent. For algorithmic collisions the adversary guesses the colliding intermediate state as it is computed in the algorithm, *e.g.* a cryptographic standard. Neither a leakage model nor any other information about implementation properties are used here. The question is whether clusters for a complete or partial intermediate state can be distinguished when applying cluster criterion functions.

On the other hand, implementation specific collisions can be observed merely due to the way a certain algorithm is implemented. In other words, there can be ways to implement an algorithm that induce this type of collisions, while the collisions are not obvious in the algorithm. For implementation specific collisions the adversary adds knowledge of confirmed or assumed implementation properties to the algorithmic attack. Examples include targeting particular intermediate states of the implementation or applying a special leakage model. Clusters are built according to such a special implementation specific intermediate state or clusters are merged according to such a special leakage model. Next we present examples for both cases.

3.1   Algorithmic Collisions

Data Encryption Standard. We revisit collisions in isolated S-boxes [22]. The S-box function is $4-to-1$, *i.e.*, it maps four inputs to one output. For each S-box, the collision function $f_k = S_i$ maps a group of four values of $z \in \{0, \ldots, 2^6 - 1\}$ to one cluster of $f_k(z) \in \{0, \ldots, 2^4 - 1\}$. As $z = x \oplus k$ the mapping depends on subkey k given known x. Alternatively, the 4-bit output differential of the first round can be used as a collision function that evolves from tracking the four output bits of the active S-box to the R-register in the next round. For the correct key hypothesis, up to sixteen separated clusters are expected.

Advanced Encryption Standard. An isolated AES S-box is not a collision function because one merely obtains a permutation of cluster labels for each key guess. Targeting only r-bit ($1 \leq r < 8$) of an S-box outcome, however, is an applicable many-to-one collision function, at the cost of loosing predicted bits of the full intermediate result.

Collisions also occur in a single output byte after the MixColumns transformation as described in [21], because this is a $2^{24} - to - 1$ collision function. For the purpose of saving computation time, an attack with two measurement series using two fixed input bytes to each MixColumns transformation is advantageous, if applicable. This leads to a $2^8 - to - 1$ collision function. Let b_{00} be the first output byte of MixColumns, (x_{22}, x_{33}) the two fixed input bytes, S the AES S-box, and (k_{00}, k_{11}) the target for key recovery. Then we have $b_{00} = 02 \cdot S(x_{00} \oplus k_{00}) \oplus 03 \cdot S(x_{11} \oplus k_{11}) \oplus c$, where constant $c = S(x_{22} \oplus k_{22}) \oplus S(x_{33} \oplus k_{33})$. Without loss of generality we assume $c = 0$ and label clusters with the value of b_{00}, as there exists a bijective mapping from b_{00} to the correct cluster label $b_{00} \oplus c$. Similarly, the second measurement series can be used to recover the other two key bytes.

3.2   Implementation Specific Collisions

Examples for this type of collisions include hardware and software implementations at which some internal registers are used multiple times, *e.g.*, in subsequent rounds of an algorithm or during subsequent processing of functional units. Hereby, internal collisions can be caused for intermediate results that are not provided in an algorithmic description.

AES-128 hardware architecture. In Appendix A we describe an AES-128 hardware architecture that leaks such implementation dependent collisions and that we analyze in the experimental part of this paper. The architecture is very compact and suitable for smartcards and other wireless applications, which makes the attacks extremely relevant.

Let $x_i \in \{0, 1\}^8$ ($i \in \{0, 1, \ldots, 15\}$) denote the plaintext byte. Accordingly, let $k_i \in \{0, 1\}^8$ be the corresponding AES key byte. By S we denote the AES S-box. The targeted intermediate result is

$$\Delta_{ii'} = S(x_i \oplus k_i) \oplus S(x_{i'} \oplus k_{i'}) \tag{1}$$

with $i \neq i'$. This intermediate result $\Delta_{ii'}$ is, e.g., given by the differential of two adjacent data cells in the studied AES hardware architecture. $\Delta_{ii'}$ depends on two 8-bit inputs to the AES S-box $(x_i \oplus k_i, x_{i'} \oplus k_{i'})$ and that is crucial for the new attacks. For the key-recovery attack we consider the pairs of known plaintext $(x_i, x_{i'})$ and fixed unknown subkey $(k_i^\circ, k_{i'}^\circ)$. Our attack is enabled because (1) is the result of a 2^8-to-1 collision function.

Using the general approach one evaluates the clustering quality for 2^{16} key hypotheses, but for this particular internal collision a special approach is feasible: for $\Delta_{ii'} = 0$ equation (1) simplifies to

$$S(x_i \oplus k_i) = S(x_{i'} \oplus k_{i'}) \Rightarrow x_i \oplus k_i = x_{i'} \oplus k_{i'} \Rightarrow k_i \oplus k_{i'} = x_i \oplus x_{i'} \,.$$

This leads to the observation that for all key guesses $(k_i, k_{i'})$ satisfying $k_i \oplus k_{i'} = k_i^\circ \oplus k_{i'}^\circ$ the elements assigned to the class $\Delta_{ii'} = 0$ are the same as in the correct assignment (caused by $(k_i^\circ, k_{i'}^\circ)$). This allows for a two-step key recovery attack:

1. Determine the correct xor-difference $k_i^\circ \oplus k_{i'}^\circ$ based on 2^8 hypotheses.
2. Determine the correct pair $(k_i^\circ, k_{i'}^\circ)$ based on 2^8 hypotheses.

We illustrate these steps in Sect. 4.2. As a result of this approach, the complexity of the key search is reduced from 2^{16} to 2^9 hypotheses.

Note that this strategy is not exclusively accredited to DCA. This implementation specific attack strategy can also be applied using standard DPA techniques.

4   Experimental Results

In this section we describe the setups for our experiments, experimental settings and provide results. The empirical evidence shows that the proposed attacks are practical and lead to a successful key recovery.

4.1   DES in Software

Our experimental platform is an unprotected software implementation of the DES running on an Atmel AVR microcontroller. The code executes in constant time and the μC is clocked at about 4MHz. For our first experiments we used a set of $N = 100$ power traces i_n. The samples represent the voltage drop over a 50Ω shunt resistor inserted in the μC's ground line. Power consumption was sampled at 200 MS/s during the first two rounds of DES encryption with a constant key $k^\circ = (k_0^\circ, \ldots, k_7^\circ)$. The plaintexts $x_n = (x_{0n}, \ldots, x_{7n})$ were randomly chosen from a uniform distribution to simulate a known plaintext scenario. The targeted 4-bit intermediate result is $S_1(\tilde{x}_1 \oplus \tilde{k}_1)$ where $\tilde{x}_1$ and $\tilde{k}_1$ denote respectively the six relevant bits of plaintext and roundkey entering the first DES S-box (S_1) in the first round. Figure 1 shows results of our cluster analysis for the criteria J_{SOS}, J_{SSE}, J_{VAR} and J_{STT}. The plots show that all four criteria allow recovery of $\tilde{k}_1$ but also indicate that J_{VAR} and J_{STT} are preferable. Furthermore, Figure 1 nicely illustrates the complementary character of J_{SOS} and J_{SSE}.

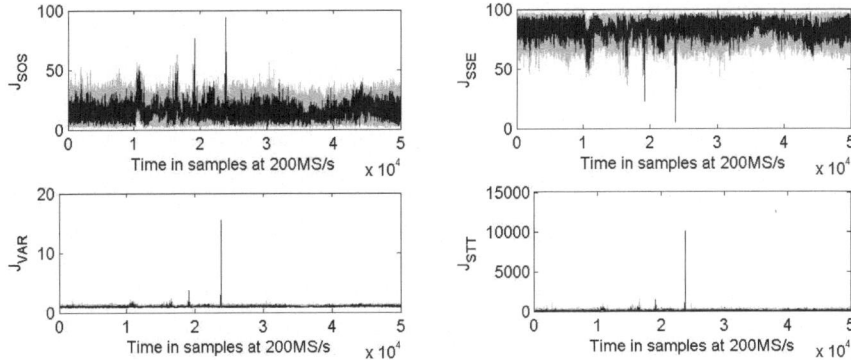

Fig. 1. Results of Cluster Analysis with J_{SOS} (top, left), J_{SSE} (top, right), J_{VAR} (bottom, left) and J_{STT} (bottom,right) when using 2^6 key hypotheses on the subkey $\tilde{k}_1$ of S_1. Results for the correct key guess are plotted in black, all other in gray.

Multivariate DPA. So far, integration of power or differential traces over small time intervals was proposed for the re-construction of a DPA signal in the presence of hardware countermeasures [10] and as a compression technique for measurement traces [18]. More generally, DPA can be extended to capture multiple well-separated leakage signals in two ways. The first approach applies standard DPA statistics and combines the (absolute) scores of multiple leakage signals afterwards. The second approach is to first combine, *e.g.* sum up, the measurement samples from multiple selected time instants before running univariate statistics. Both approaches were tested for CPA and yielded virtually identical results if the second approach takes positive and negative side-channel contributions at the different points in time into account, *e.g.* samples with positive correlation are added and samples with negative correlation are subtracted. As long as the number of instants is small, all possible combinations for combining the leakage at these instants can be tested exhaustively if the right combination is unknown.

Univariate vs. Multivariate Analysis. We study the performance of DCA for univariate and multivariate statistics. To allow a fair comparison we also provide results for univariate and multivariate CPA. Preliminary experiments indicated that the least significant bit (LSB) of $S_1(\tilde{x}_1 \oplus \tilde{k}_1)$ is a preferable target for attacks. We thus focus on attacks using two clusters, *i.e.* $w = 1$, targeting the LSB. Additionally, we identified three points in time (A,B,C) when the LSB leaks most.

For this analysis we used 5000 measurements in one hundred sets of fifty measurements. For each attack we used $N = 15, 20, 25, \ldots, 50$ measurements and repeated the attack one hundred times. We report the percentage of attack scenarios where an attack was successful, *i.e.* where the attack outputs the correct subkey value.

Table 2. Success rates in % for various univariate and multivariate attack scenarios

Test	Model	Point in time	N=15	N=20	N=25	N=30	N=35	N=40	N=45	N=50
CPA	LSB	overall	3	15	37	62	84	95	96	98
CPA	LSB	A	42	64	69	77	89	93	94	96
CPA	LSB	B	64	77	83	93	96	98	98	99
CPA	LSB	C	17	28	29	38	50	55	59	65
J_{SSE}	LSB	overall	3	15	37	62	84	95	96	98
J_{SSE}	LSB	A	42	64	70	77	89	93	94	96
J_{SSE}	LSB	B	64	78	82	93	96	98	98	99
J_{SSE}	LSB	C	18	28	31	38	50	56	59	65
CPA	LSB	AB	70	85	90	96	99	100	100	100
J_{SSE}	LSB	AB	70	83	91	97	99	100	100	100
CPA	LSB	ABC	76	90	96	99	100	100	100	100
J_{SSE}	LSB	ABC	78	94	96	99	100	100	100	100

Table 2 shows the success rates for the various scenarios. Comparing univariate CPA and DCA with the J_{SSE} criterion we observe that the success rates are very similar in all scenarios. The third block of results in the table shows success rates for DCA and CPA still targeting only the LSB but exploiting the multivariate leakage at points A,B, and C. For multivariate CPA, the measurement values were summed up, taking into account the respective signs of the correlation at each point in time, before computing correlation coefficients. Both multivariate attacks perform also very similar. It is clear that multivariate attacks are superior to univariate attacks, but they usually require knowledge of the instants A,B, and C.

4.2 AES in Hardware

Our experimental platform is a prototype chip which implements an 8051-compatible μC with AES-128 co-processor in 0.13 μm sCMOS technology without countermeasures. The architecture of the AES co-processor, which is the target of our attacks, is discussed in detail in Appendix A. The susceptibility of the chip towards templates [9,13], stochastic methods [20,13], DPA and rank correlation has already been analyzed in [2].

For our experiments we obtained a set of $N = 50\,000$ power traces i_n. The samples represent the voltage drop over a 50Ω resistor inserted in the dedicated core VDD supply. Power consumption was sampled at 2 GS/s during the first round of AES-128 encryption with a constant key $k^\circ = (k_0^\circ, \ldots, k_{15}^\circ)$. The plaintexts $x_n = (x_{0n}, \ldots, x_{15n})$ were randomly chosen from a uniform distribution.

In all following examples we focus on the neighboring data cells C0,2 and C0,3 of Figure 4 that represent AES state bytes 8 and 12, respectively. However, we point out that all key bytes can be attacked in the same way since all corresponding state bytes enter the MixColumns circuit at some time.

DCA. We showed in Sect. 3.2 that clustering of the power consumption values caused by the differential $\Delta_{ii'}$ has three interesting properties. First, for the correct hypotheses (k_8, k_{12}) on both involved key bytes *all* power traces are assigned

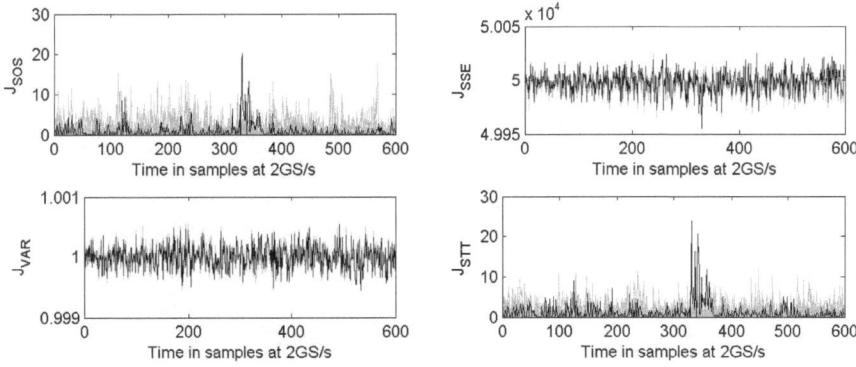

Fig. 2. Results of Cluster Analysis with J_{SOS} (top, left), J_{SSE} (top, right), J_{VAR} (bottom, left) and J_{STT} (bottom,right) when using 2^8 hypotheses on the pair (k_8, k_{12}). Results for the correct xor-difference are plotted in black, all other in gray.

to clusters $\Delta_{ii'}$ correctly which maximizes (minimizes) the cluster statistic J_k over all clusters. Second, for a wrong hypothesis (k_8, k_{12}) where the xor-difference $k_8 \oplus k_{12}$ is correct (*i.e.* equal to $k_8^\circ \oplus k_{12}^\circ$) power traces are assigned to cluster $\Delta_{ii'} = 0$ correctly and uniformly at random to all other clusters $\Delta_{ii'} \neq 0$. Third, for a wrong hypothesis (k_8, k_{12}) power traces are assigned to all clusters $\Delta_{ii'}$ uniformly at random. The second property enables our two step attack.

The results reported in the remainder are based on either $N_1 = 50\,000$ or $N_2 = 10\,000$ measurements. We restrict to univariate analysis and computed J_{SOS}, J_{SSE}, J_{VAR} and J_{STT}.

Step 1: Detecting (k_8, k_{12}) with the correct xor-difference. In order to detect a hypothesis with the correct xor-difference one has to decide whether the cluster $\Delta_{ii'} = 0$ is different from all other clusters. We thus merge the clusters $\Delta_{ii'} \neq 0$ to analyze them as one single cluster. The statistic J_k is then evaluated for the two cluster case, $\Delta_{ii'} = 0$ and the union of the remaining clusters.

We sort N_1 measurements into two clusters $\Delta_{ii'} = 0$ and $\Delta_{ii'} \neq 0$ and evaluate the cluster criterion functions for the two cluster case. Interestingly, for this approach it does not matter whether one applies the Hamming distance model or not, since in both cases the same measurements are assigned to $\Delta_{ii'} = 0$ and $\Delta_{ii'} \neq 0$. Figure 2 shows the resulting cluster criteria. One can observe that the criteria J_{SOS} and J_{SST} yield signals at the points in time when the collision occurs.

Step 2: Detecting the correct key pair. We search all pairs (k_8, k_{12}) with the now known xor-difference $k_8^\circ \oplus k_{12}^\circ$. Detecting the correct pair is easier than detecting a pair with the correct xor-difference, because differences of many clusters yield stronger side-channel signals. We can therefore work with fewer measurements which speeds up the analysis (note that the measurements from step 1 are re-used). For each hypothesis we sort N_2 measurements into 256

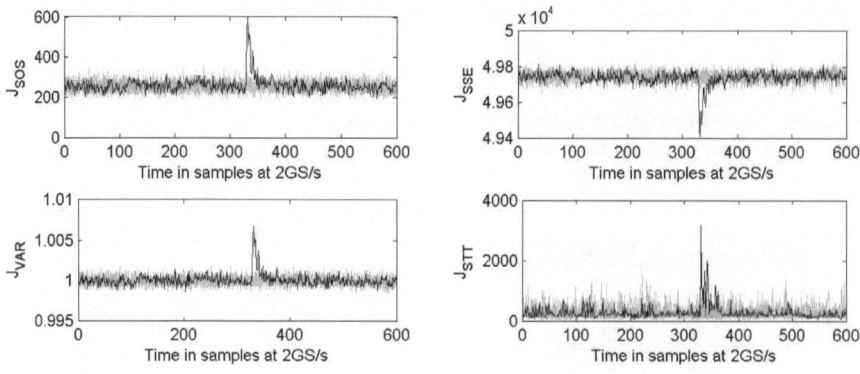

Fig. 3. Results of Cluster Analysis with J_{SOS} (top, left), J_{SSE} (top, right), J_{VAR} (bottom, left) and J_{STT} (bottom,right) when using 2^8 hypotheses on the pair (k_8, k_{12}). Results for the correct key guess are plotted in black, all other in gray.

clusters and evaluate the cluster criteria. Figure 3 shows the cluster criteria for this setting (without any power model).

We observe that all criterion functions are able to identify the correct guess. We also evaluated the cluster criteria under the assumption of the Hamming distance model, where we sorted the measurements into nine clusters according to the Hamming weight of $\Delta_{ii'}$. The results are similar and we do not present them in detail. The results demonstrate the feasibility of our approach on both platforms and show that it works with or without a power model. Among the cluster criteria that we considered J_{STT} gives in general the best results but is particularly error-prone when very few measurements are used.

Complexity. For the two-step attack, approximately 50 000 measurements were needed to reveal the correct two key bytes and $2 \cdot 2^8$ key hypotheses were tested. The general DCA approach tests 2^{16} key hypotheses and the threshold for successful key recovery is approximately at 5000 measurements. As result, our special attack strategy for the AES hardware implementation allows to reduce the number of hypotheses for which cluster criterion functions have to be computed by a factor of 2^7 at the cost of a tenfold increase of measurements.

5   Conclusion

We propose a new technique for side-channel key recovery based on cluster analysis and detection of internal collisions. The technique is broader in applications than previously known DPA attacks. It has obvious advantages when more than two clusters are used. In particular DCA does not require but can benefit from a good leakage model and it is inherently multivariate. DCA inspires a simple extension of standard DPA to multivariate analysis that is also included in this contribution. While previous works focus on internal collisions that are mainly results of algorithmic properties, we additionally consider implementation specific collisions. Our approach is confirmed in practice on two platforms: an AVR microcontroller with implemented DES algorithm and an AES hardware module. This is the first work demonstrating the feasibility of internal collision attacks on highly parallel hardware platforms. Furthermore we present a new attack strategy for the targeted AES hardware module.

Acknowledgements

We sincerely thank the anonymous reviewers for their valuable and insightful comments that helped us to improve the article. We also thank Sören Rinne for providing the DES source code for the AVR microcontroller.

The information in this document reflects only the author's views, is provided as is and no guarantee or warranty is given that the information is fit for any particular purpose. The user thereof uses the information at its sole risk and liability.

References

1. Efficient Implementation of the Rijndael SBox,
   http://www.comms.scitech.susx.ac.uk/fft/crypto/rijndael-sbox.pdf
2. Batina, L., Gierlichs, B., Lemke-Rust, K.: Comparative Evaluation of Rank Correlation Based DPA on an AES Prototype Chip. In: Wu, T.-C., Lei, C.-L., Rijmen, V., Lee, D.-T. (eds.) ISC 2008. LNCS, vol. 5222, pp. 341–354. Springer, Heidelberg (2008)
3. Biryukov, A., Bogdanov, A., Khovratovich, D., Kasper, T.: Collision Attacks on AES-Based MAC: Alpha-MAC. In: Paillier, P., Verbauwhede, I. (eds.) CHES 2007. LNCS, vol. 4727, pp. 166–180. Springer, Heidelberg (2007)
4. Biryukov, A., Khovratovich, D.: Two new techniques of side-channel cryptanalysis. In: Paillier, P., Verbauwhede, I. (eds.) CHES 2007. LNCS, vol. 4727, pp. 195–208. Springer, Heidelberg (2007)
5. Bogdanov, A.: Improved Side-Channel Collision Attacks on AES. In: Adams, C., Miri, A., Wiener, M. (eds.) SAC 2007. LNCS, vol. 4876, pp. 84–95. Springer, Heidelberg (2007)
6. Bogdanov, A.: Multiple-Differential Side-Channel Collision Attacks on AES. In: Oswald, E., Rohatgi, P. (eds.) CHES 2008. LNCS, vol. 5154, pp. 30–44. Springer, Heidelberg (2008)
7. Bogdanov, A., Kizhvatov, I., Pyshkin, A.: Algebraic Methods in Side-Channel Collision Attacks and Practical Collision Detection. In: Chowdhury, D.R., Rijmen, V., Das, A. (eds.) INDOCRYPT 2008. LNCS, vol. 5365, pp. 251–265. Springer, Heidelberg (2008)
8. Brier, E., Clavier, C., Olivier, F.: Correlation Power Analysis with a Leakage Model. In: Joye, M., Quisquater, J.-J. (eds.) CHES 2004. LNCS, vol. 3156, pp. 16–29. Springer, Heidelberg (2004)
9. Chari, S., Rao, J.R., Rohatgi, P.: Template Attacks. In: Kaliski Jr., B.S., Koç, Ç.K., Paar, C. (eds.) CHES 2002. LNCS, vol. 2523, pp. 13–28. Springer, Heidelberg (2003)

10. Clavier, C., Coron, J.-S., Dabbous, N.: Differential Power Analysis in the Presence of Hardware Countermeasures. In: Paar, C., Koç, Ç.K. (eds.) CHES 2000. LNCS, vol. 1965, pp. 252–263. Springer, Heidelberg (2000)
11. Duda, R.O., Hart, P.E., Stork, D.G.: Pattern Classification. John Wiley & Sons, Chichester (2001)
12. Gierlichs, B., Batina, L., Tuyls, P., Preneel, B.: Mutual Information Analysis. In: Oswald, E., Rohatgi, P. (eds.) CHES 2008. LNCS, vol. 5154, pp. 426–442. Springer, Heidelberg (2008)
13. Gierlichs, B., Lemke-Rust, K., Paar, C.: Templates vs. Stochastic Methods.. In: Goubin, L., Matsui, M. (eds.) CHES 2006. LNCS, vol. 4249, pp. 15–29. Springer, Heidelberg (2006)
14. Kocher, P.C., Jaffe, J., Jun, B.: Differential Power Analysis. In: Wiener, M. (ed.) CRYPTO 1999. LNCS, vol. 1666, pp. 388–397. Springer, Heidelberg (1999)
15. Ledig, H., Muller, F., Valette, F.: Enhancing Collision Attacks. In: Joye, M., Quisquater, J.-J. (eds.) CHES 2004. LNCS, vol. 3156, pp. 176–190. Springer, Heidelberg (2004)
16. Lemke-Rust, K., Paar, C.: Gaussian Mixture Models for Higher-Order Side Channel Analysis.. In: Paillier, P., Verbauwhede, I. (eds.) CHES 2007. LNCS, vol. 4727, pp. 14–27. Springer, Heidelberg (2007)
17. Mangard, S., Aigner, M., Dominikus, S.: A Highly Regular and Scalable AES Hardware Architecture. IEEE Trans. Computers 52(4), 483–491 (2003)
18. Mangard, S., Oswald, E., Popp, T.: Power Analysis Attacks. Springer, Heidelberg (2007)
19. Messerges, T.S., Dabbish, E.A., Sloan, R.H.: Investigations of Power Analysis Attacks on Smartcards. In: Proceedings of USENIX Workshop on Smartcard Technology, pp. 151–162 (1999)
20. Schindler, W., Lemke, K., Paar, C.: A Stochastic Model for Differential Side Channel Cryptanalysis.. In: Rao, J.R., Sunar, B. (eds.) CHES 2005. LNCS, vol. 3659, pp. 30–46. Springer, Heidelberg (2005)
21. Schramm, K., Leander, G., Felke, P., Paar, C.: A Collision-Attack on AES Combining Side Channel- and Differential-Attack. In: Joye, M., Quisquater, J.-J. (eds.) CHES 2004. LNCS, vol. 3156, pp. 163–175. Springer, Heidelberg (2004)
22. Schramm, K., Wollinger, T., Paar, C.: A New Class of Collision Attacks and Its Application to DES. In: Johansson, T. (ed.) FSE 2003. LNCS, vol. 2887, pp. 206–222. Springer, Heidelberg (2003)
23. Standaert, F.-X., Gierlichs, B., Verbauwhede, I.: Partition vs. comparison side-channel distinguishers. In: ICISC 2008. LNCS, vol. 5461, pp. 253–267. Springer, Heidelberg (2008)
24. Tan, P.-N., Steinbach, M., Kumar, V.: Introduction to Data Mining. Addison-Wesley, Reading (2006)
25. Walter, C.: Sliding Windows Succumbs to Big Mac Attack. In: Koç, Ç.K., Naccache, D., Paar, C. (eds.) CHES 2001. LNCS, vol. 2162, pp. 286–299. Springer, Heidelberg (2001)
26. Wolkerstorfer, J., Oswald, E., Lamberger, M.: An ASIC Implementation of the AES SBoxes. In: Preneel, B. (ed.) CT-RSA 2002. LNCS, vol. 2271, pp. 67–78. Springer, Heidelberg (2002)

A   The AES Hardware Architecture

A similar implementation is described in [17]. The module includes the following parts: data unit, key unit, and interface. The most important part is the data unit

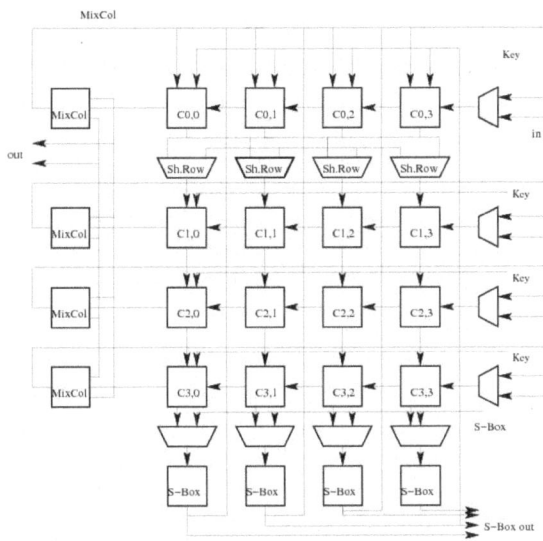

Fig. 4. The architecture of the AES module

(see Figure 4), which includes the AES operation. It is composed of 16 data cells ($C_{i,j}$, where $i, j \in \{0, 1, 2, 3\}$) and four S-Boxes. A data cell consists of flip-flops (able to store 1 byte of data) and some combinational logic (xors gates) in order to perform AddRoundKey operations. It has the ability to shift data vertically and horizontally, which is the feature exploited in our attacks. Load data is done by shifting the input data column by column into the registers of the data cells. The initial AddRoundKey transformation is performed in the fourth clock cycle together with the load of the last column. To calculate one round, the bytes are rotated vertically to perform the S-box and the ShiftRows transformation row by row. In the first clock cycle, the S-Box transformation starts only for the fourth row. Because of pipelining the result is stored after two clock cycles in the first row. S-boxes and the ShiftRows transformations can be applied to all 16 bytes of the state within five clock cycles due to pipelining. The S-Boxes in the AES module are implemented by using composite field arithmetic as proposed by Wolkerstorfer *et al.* [26] following the original idea of Rijmen [1].

Known–Plaintext–Only Attack on RSA–CRT with Montgomery Multiplication

Martin Hlaváč

Department of Algebra, Charles University in Prague,
Sokolovská 83, 186 75 Prague 8, Czech Republic
hlavm1am@artax.karlin.mff.cuni.cz

Abstract. The paper describes a new attack on RSA–CRT employing Montgomery exponentiation. Given the amount of so-called final subtractions during the exponentiation of a known message (not chosen, just known), it creates an instance of the well known Hidden Number Problem (HNP, [2]). Solving the problem reveals the factorization of RSA modulus, i.e. breaks the scheme.

The main advantage of the approach compared to other attacks [14,17] is the lack of the chosen plaintext condition. The existing attacks, for instance, cannot harm so-called Active Authentication (AA) mechanism of the recently deployed electronic passports. Here, the challenge, i.e. the plaintext, is jointly chosen by both parties, the passport and the terminal, thus it can not be conveniently chosen by the attacker. The attack described here deals well with such a situation and it is able to solve the HNP instance with 150 measurements filtered from app. 7000. Once the secret key used by the passport during AA is available to the attacker, he can create a fully functional copy of the RFID chip in the passport he observes.

A possible way to obtain the side information needed for the attack within the electromagnetic traces is sketched in the paper. Having no access to high precision measurement equipment, its existence has not been experimentally verified, yet. The attack, however, should be taken into account by the laboratories testing the resilience of (not only) electronic passports to the side channel attacks.

Keywords: RSA, Chinese Remainder Theorem, Montgomery exponentiation, Hidden Number Problem, side channel attack, electronic passport.

Introduction

Motivated by the recent deployment of the electronic passports, we study the security of it anti-cloning measure called Active Authentication (AA, [5]). As it is an RSA based challenge-response protocol, one can try to attack AA with the well-known Schindler's adaptive chosen plaintext attack [14] or Tomoeda's chosen plaintext attack [17]. It turns out, however, both of these approaches fail in this scenario due to their chosen plaintext condition as the plaintext used in AA is chosen jointly by both parties.

C. Clavier and K. Gaj (Eds.): CHES 2009, LNCS 5747, pp. 128–140, 2009.

In this paper we present a new side channel attack on RSA-CRT with Montgomery multiplication [10]. Being a known plaintext attack, it suits well the AA scenario. The side information that is available to the attacker is the same as in [17], i.e. the amount of the final subtractions during Montgomery exponentiation within one branch of the CRT computation (e.g. exponentiation **mod** p). It is shown such information can be used to obtain modular approximations of one of the factors of the RSA modulus. The side information is stronger variant of the simple timing information used in [14].

The approximations suit perfectly as the input to the well-known Hidden Number Problem [2] which can be efficiently solved using lattice reduction techniques [9,4]. The attack presented using this side information is of independent merit and can be applied in other scenarios where the side information is available.

The existence of the side information in the electronic passport is yet to be proven, however. Our simple measurements show the square-and-multiply-always exponentiation can be identified very well in the electromagnetic trace surrounding the chip. More precise measurements are needed, however, to support the hypothesis that Montgomery multiplication is used and that the amount of the final subtractions is revealed.

As the existence of the side channel implies the insecurity of AA security measure, the attack should be taken into account by the testing laboratories. No further research is needed for this purpose. On the other hand, no theoretical guarantee is given in the paper that the attack always works. Further research is necessary for more theoretical results. The attack validity is supported by the experiments with the emulated side information. As the electronic passports are already deployed, we believe the attack should be made public at this stage already.

The paper is organized as follows. The electronic passport and AA are overviewed together with our simple electromagnetic measurements in Section 1. The RSA-CRT scheme with Montgomery multiplication is described in Section 2. Briefly overviewing the existing attacks, we elaborate the conversion to HNP here, as well. Remarks on HNP relevant to the scenario and the results of the experiments with the emulated observations are given in Section 3. Several possible directions for future research are suggested in Section 4.

1   e-Passport

The electronic passport is a modern travel document equipped with a RFID (Radio Frequency IDentification) chip compatible with ISO 14443 [7] (on the physical layer to the transport layer) and with ISO 7816 [8] (the application layer).

The chip contains digitally signed electronic copy of the data printed on the passport: the machine readable zone (MRZ) including the passport no., the photo of the holder, as well as the public and private key for the Active Authentication (AA) described in the next paragraph.

Algorithm 1. Active authentication

Parties: **T** ... terminal, **P** ... passport

1: **T:** generate random 8-byte value V
2: **T** $\rightarrow$ **P:** V
3: **P:** generate random 106-byte value U
4: **P:** compute $s = m^d \bmod N$, where $m =$ "6A"$||U||w||$"BC", $w = $ SHA-1$(U||V)$ and d
   is the passport's secret AA key securely stored in the protected memory
5: **P** $\rightarrow$ **T:** s, U
6: **T:** verify $m = s^e \bmod N$, where e is the passport's public key stored in publicly
   accessible part of passport memory

1.1 Active Authentication

Besides the required security mechanisms in [6] such as the passive authentication and the basic access control (BAC), the e-passport can optionally employ cryptographically more sophisticated active authentication which aims to make the duplication virtually impossible for the attacker. The challenge-response protocol used in AA is shown in Algorithm 1.

As we can see, the formatted message m being signed by the passport is chosen jointly by the terminal and the passport, thus cannot be conveniently chosen by the attacker on the terminal side.

1.2 Electromagnetic Side Channel Leakage

As previously mentioned, the e-passport is compatible with ISO 14443 on the physical layer. To send the data to the terminal, the so-called near magnetic field is employed. Depending on the data being sent, the passport loads its antenna with a specific impedance circuit. Such an activity propagates in the surrounding magnetic field which is detected by the terminal. The reader is encouraged to see [3] for more details on the physical layer.

The question that is an interesting one to be asked in this scenario is whether the passport can fully control the emanation of the antenna. It is not only the special purpose circuit but also the other parts of the chip that load the antenna with their impedances. Especially, one should ask whether any of the cryptographic operations computed on the chip can be identified in the surrounding field.

During the early stages of the research, we presumed square-and-multiply algorithm with Montgomery exponentiation is employed during AA. This hypothesis is partly supported by the measurements shown on Figure 1. The ratio between the duration of two repetitive patterns corresponds to the execution duration of square and multiply operations and they appear in two series of 512 repetitions. This measurement does not reveal, however, whether the Montgomery multiplication is used. In case it is not, the attack described in the following text can still be employed in other implementations that make use of Montgomery multiplication.

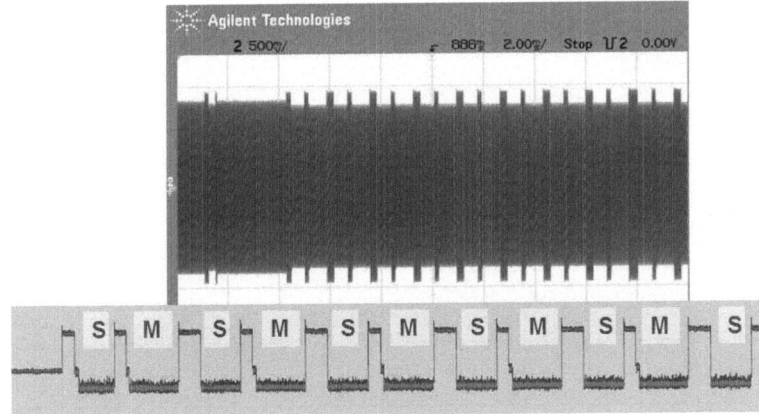

Fig. 1. Electromagnetic side channel measurement on an e-passport during the computation $s = m^d \bmod N$ within AA. The RFID chip on the passport is P5CD072 [13].

Since we presume square-and-multiply-always algorithm (see Algorithm 3) is used for exponentiation, the secret exponent d cannot be directly extracted from these measurements. We suspect however, it is possible to extract some information about the exponentiation if higher precision measurements are available. In fact, we believe the number of so-called final subtractions within the exponentiation $\bmod p$ can be revealed by this side channel. A method that is able to make use of such information and discloses the secret key d is described in the next section.

2   RSA–CRT with Montgomery Multiplication

Let N be the public RSA modulus and e be the public exponent. Let (p, q, d) satisfying $N = pq$, $d = e^{-1} \bmod \phi(N)$ be the corresponding private data.

Being given message m, the private RSA operation $m^d \bmod N$ is computed using Chinese Remainder Theorem as follows

$$s_p = (m_p)^{d_p} \bmod p \tag{1}$$

$$s_q = (m_q)^{d_q} \bmod q \tag{2}$$

$$s = ((s_q - s_p) p_{inv} \bmod q) p + s_p \tag{3}$$

where $d_p = d \bmod (p - 1)$, $d_q = d \bmod (q - 1)$, $m_p = m \bmod p$, $m_q = m \bmod q$ and $p_{inv} p = 1 \,(\bmod q)$. For our attack, we expect the exponentiation in (1) and (2) is computed employing the standard square-and-multiply-always algorithm with Montgomery representation of the integers (see Algorithm 3) with Montgomery constant $R = 2^{\lceil \frac{\log N}{2} \rceil}$.

One of the well-known countermeasures to prevent a simple SPA side channel attack on Algorithm 3 is the execution of the dummy multiplication in step 8.

Algorithm 2. Montgomery multiplication *mont()*

Input: $x, y \in Z_p$
Output: $w = xyR^{-1} \bmod p$

1: $s \leftarrow xy$
2: $t \leftarrow s(-p^{-1}) \bmod R$
3: $g \leftarrow s + tp$
4: $w \leftarrow g/R$
5: **if** $w > p$ **then**
6:        $w \leftarrow w - p$ (final subtraction)
7: **return** w

Algorithm 3. Montgomery exponentiation *expmont()*

Input: $m, p, d \left(= (d_{n-1}e_{d-2} \ldots d_1 d_0)_2\right)$
Output: $x = m^d \bmod p$

1: $u \leftarrow mR \bmod p$
2: $z \leftarrow u$
3: **for** $i \leftarrow n - 2$ **to** 0
4:      $z \leftarrow mont(z, z, p)$
5:      **if** $d_i == 1$ **then**
6:            $z \leftarrow mont(z, u, p)$
7:      **else**
8:            $z' \leftarrow mont(z, u, p)$ (dummy operation)
9: **endfor**
10: $z \leftarrow mont(z, 1, p)$
11: **return** z

This prevents an attacker from distinguishing if the operation $mont(z, u, p)$ was executed or not. We will see, however, this countermeasure has no effect on our attack.

2.1   Schindler's Observation

In [14], Schindler demonstrated an interesting property of the Montgomery multiplication algorithm (Algorithm 1). Let x be a fixed integer in $\mathbb{Z}_p$ and B be randomly chosen from $\mathbb{Z}_p$ with uniform distribution. Then the probability that the final subtraction (step 6 in Algorithm 2) occurs during the computation $mont(x, B)$ is equal to

$$\frac{x \bmod p}{2R} \tag{4}$$

This observation allowed attacking RSA-CRT with an adaptive chosen plaintext timing attack.

2.2   Trick by Tomoeda et al.

In [17], the original Schindler's attack is modified to a chosen plaintext attack. All of the values are chosen in advance by the attacker, i.e. they are not required to be chosen during the attack.

With the probability of the final subtraction computation within one multiplication step given by Schindler (4), Tomoeda gave an estimate on the total number of final subtractions n_i during the whole exponentiation operation $(m_{p,i})^{d_p} \bmod p$, where $m_{p,i} = m_i \bmod p$. In fact, the approximation (5)

$$\frac{m_i R \bmod p}{p} \approx \frac{n_i - n_{min}}{n_{max} - n_{min}} \tag{5}$$

is given for $0 \le i < k$ where $n_{max} = \max_{0 \le i < k} n_i$ and $n_{min} = \min_{0 \le i < k} n_i$ are the maximal and the minimal number of FS during k observations. To justify this approximation, the authors of [17] proposed experimental result similar to the one shown on Figure 2.

Being an approximation, we cannot expect (5) to be as tight as Schindler's high-precision (4). Instead, we can empirically measure minimal precision of (5) in bits. In section 2.4, we will see for 1024 bit modulus we can expect at minimum 4 bits with proper filtering of the measurements.

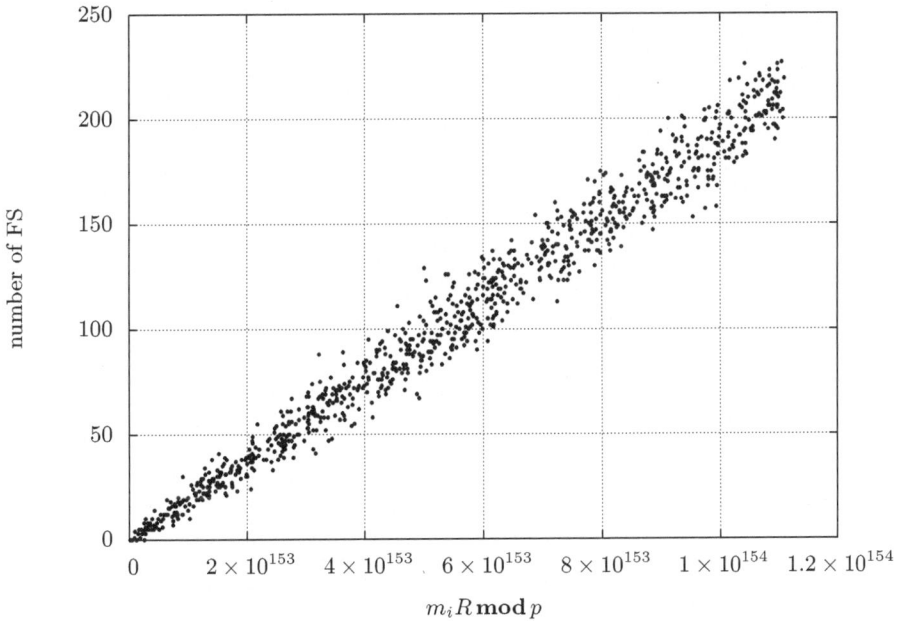

Fig. 2. The relationship between the *known* number of FS during the computation $(m_{p,i})^{d_p} \bmod p$ and the *unknown* value $m_i R \bmod p$. We see it is strongly linear and can be expressed as in (5).

In [17], the attack used 512 measurements (in case without the RSA blinding) to recover 512 bit long prime factor of N, i.e. one bit per measurement was used on average. We will see in section 2.4, however, that the average number of bits extracted per measurement and even their minimum can be much higher.

2.3   Conversion to HNP

Both approaches, Schindler's [14] and Tomoeda's [17], are chosen plaintext attacks on RSA–CRT with Montgomery exponentiation. They cannot be applied on AA in the e-passport scenario, however. As the plaintext (i.e. the formatted challenge) is generated jointly by the terminal and the e-passport, it cannot be conveniently chosen by the attacker.

The main contribution of this paper is the lack of the chosen plaintext condition while recovering the factorization of N. To do this we transform the problem of finding the secret factor of N to the well-known Hidden Number Problem (HNP, see [12]). Being given the approximation (5), we first realize there exists an integer k_i such that $m_i R \bmod p = m_i R - k_i p$. Consequently, we multiply (5) by N obtaining

$$m_i R q - k_i N \approx \frac{n_i - n_{min}}{n_{max} - n_{min}} N \tag{6}$$

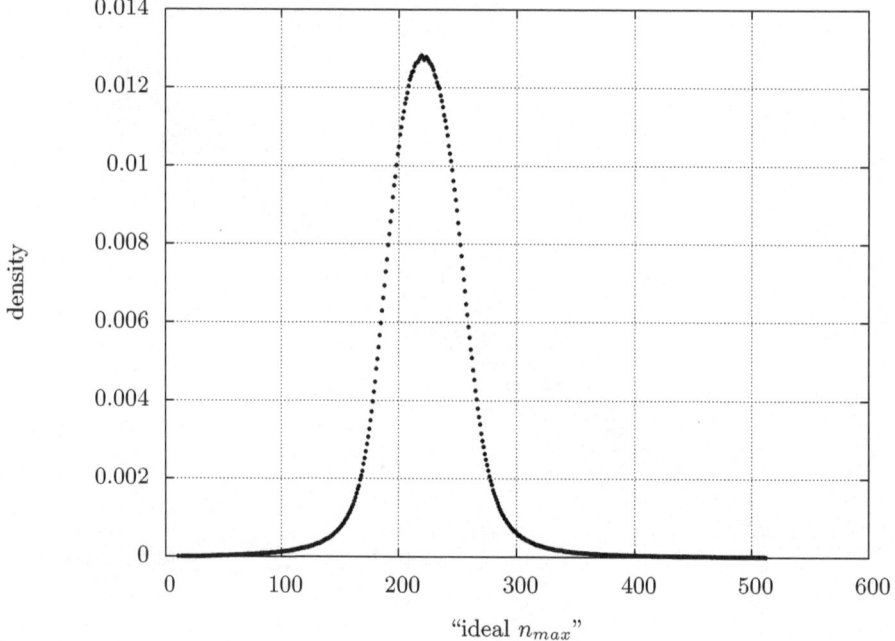

Fig. 3. The distribution of "ideal n_{max}" values computed from (6)

and we substitute $t_i = m_i R \bmod N$ and $u_i = \frac{n_i - n_{min}}{n_{max} - n_{min}} N$ for $0 \le i < k$. We now have a "modular approximation" u_i of a known t_i-multiple of (hidden number) q, i.e.

$$t_i q + k'_i N - u_i \approx 0 \tag{7}$$

for suitable k'_i, $0 \le i < k$.

Even if the values t_i and u_i were taken at random from $\mathbb{Z}_N$, it would hold

$$|t_i q - u_i|_N \le \frac{N}{2} \tag{8}$$

(let us remind $|a|_N = \min_{k \in \mathbb{Z}} (a - kN)$).

However, we expect (7) to be a better approximation than the random one and we can measure its precision in bits and note it as l_i, i.e.

$$|t_i q - u_i|_N \le \frac{N}{2} 2^{-l_i} \tag{9}$$

2.4   Approximation Precision and Filtering

During the one-time precomputation step we simulated the side channel measurements over 2^{12} RSA instances with 1024 bits long mudulus and 2^{12} random

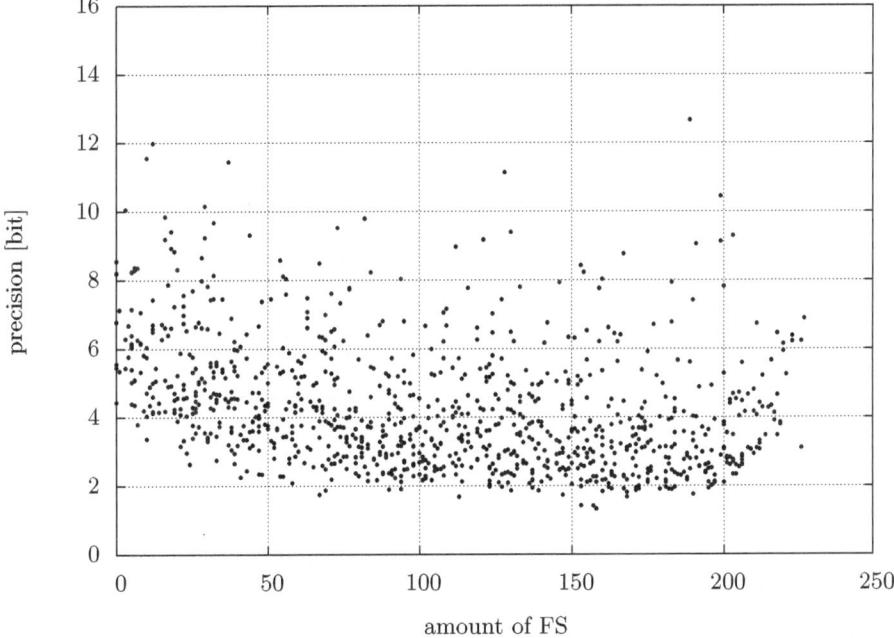

Fig. 4. The precision of the approximation in bits as a function of the amount of FS within the Montgomery exponentiation. During the attack, only the measurements with at most 4 FS are taken into account as their minimal precision is approximately 4 bits.

plaintexts for each instance. The minimal number of FS within the exponentiation **mod** p was 0 while the maximal was 290.

For each measurement we computed so-called "ideal n_{max}", the value for which the approximation (6) becomes equality with $n_{min} = 0$. The value was rounded to the nearest integer. The distribution of these values is shown on Figure 3. The value 224 being the most frequent candidate for "ideal n_{max}" value was used instead of the real value $n_{max} = \max_{0 \le i < k} n_i$ during the following steps. This simple adjustment increased the minimal precision l_{min} by 0.5 bit and even by 1 bit within the filtered measurements described in the next paragraph.

The precision l_i of the i-th approximation u_i (see (9)) was measured as $l_i = -1 + \log N - \log |t_i q - u_i|_N$. The interesting relationship between these values and the number of FS is shown on Figure 4. We see the minimal precision of one single bit is obtained for approximately 150 final subtractions. However, focusing on the experiments with less than 5 final subtractions, the minimal precision jumps to 4 bits. For this reason during the simulated experiment we filter all of the measurements with 5 final subtractions or more resulting in 150 ($2^{7.2}$) suitable measurements from the total of 6797 ($2^{12.7}$) measurements conducted (simulated).

3   Hidden Number Problem

The Hidden Number Problem was first introduced in [2]. Being given k approximations

$$|t_i x - u_i|_N < \frac{N}{2^{l+1}} \tag{10}$$

with $t_i, u_i \in \mathbb{Z}_N$, $l \in \mathbb{N}$ known for $0 \le i < k$, the task is to find the hidden number $x \in \mathbb{Z}_{N^{\frac{1}{2}}}$. In [2], the hidden number is a random unknown value from $\mathbb{Z}_N$, however, this is not the case in our scenario. Here, the hidden number is a factor of N with the expected size in order of $N^{\frac{1}{2}}$. The lattice we use to solve the HNP instance is adjusted for this purpose.

The usual technique to solve HNP is the employment of the lattices. The problem is converted to one of the well studied lattice problem, the Approximate Closest Vector Problem (ACVP). One constructs the lattice $\mathcal{L}$ spanned by the rows of the basis matrix

$$\mathbf{B} = \begin{pmatrix} N & 0 & \cdots & 0 & 0 \\ 0 & N & \ddots & \vdots & \vdots \\ \vdots & \ddots & \ddots & 0 & 0 \\ 0 & \cdots & 0 & N & 0 \\ t_0 & \cdots\cdots & t_{k-1} & N^{\frac{1}{2}}/2^{l+1} \end{pmatrix} \tag{11}$$

and the vector $V = (u_0, \ldots, u_{k-1}, 0)$. The lattice vector

$$H = \left(t_0 x - \alpha_0 N, \ldots, t_{k-1} x - \alpha_{k-1} N, \frac{x N^{\frac{1}{2}}}{2^{l+1}} \right)$$

is the hidden vector for suitable $\alpha_0, \ldots, \alpha_{k-1} \in Z$, as its last coordinate reveals the hidden number x.

The hidden vector H belongs to lattice $\mathcal{L}$. It is unknown, however. The construction of lattice $\mathcal{L}$ and vector V yields existence of such $\alpha_0, \ldots, \alpha_{k-1}$ that $||H - V|| < \frac{N}{2^l}$. The first step in solving ACVP is finding an LLL-reduced basis of $\mathcal{L}$ using the LLL algorithm [9] or its BKZ variant [15] with the time complexity exponential on lattice dimension $k + 1$. Being given the reduced basis, the second step is using Babai's closest plane algorithm [1] to find a vector H' in $\mathcal{L}$ close to V. One can now hope the vector H' reveals the hidden number x in its last coordinate, i.e. H' is equal to hidden vector H or is "similar enough".

It is shown in [12] that the probability of recovering the hidden number using this approach is close to 1 if the precision l of the approximations is in order of $(\log N)^{1/2}$ and reasonable amount of approximations is given.

In our scenario with 1024-bit long modulus N, we would need 32 bit measurement precision in order to have the theoretical guarantee of success. As we have seen previously this would hardly be the case with the electromagnetic side channel which provides us with 4 bits at minimum, 7 bits on average. To overcome this limitation we can lower the imprecision of the approach introduced by Babai's algorithm by heuristically converting the ACVP to Unique-SVP, as shown in Appendix. More importantly, the lattice basis reduction algorithms behave much better in real-life situations than what is guaranteed in theory [4]. Next section shows it is possible in fact to recover the hidden number q in our scenario.

3.1 Experiments with Emulated Observations

We implemented the attack using NTL library [16]. The computing platform was 64-bit GNU/Linux Debian running on Opteron 244 with 2GB RAM.

We first emulated the side channel and extracted the number of final subtractions l_i within the Montgomery exponentiation $s_i = (m_{p,i})^{d_p} \bmod p$. As justified in Figure (4) only the measurements with at most 4 final subtractions were used in order to keep the approximation precision on an acceptable level. In fact, the minimal precision l_{min} within these measurements was 4.2 bits while it was as high as 7.2 bits on average. We have to note however, these values are not known during the attack, thus the lower bound has to be estimated. In order to collect 150 such measurements, the total number of 7000 measurement was emulated. In real life, the physical measuring of such a collection should be feasible in order of hours.

With the side information available, lattice $\mathcal{L}$ was constructed. The dimension of the lattice was 152, since the CVP problem was converted to Unique-SVP adding 1 to the original dimension. The parameter l approximating the minimal number of known bits was chosen from the set $\left\{\frac{7}{2} + \frac{t}{4}, t \in 0, \ldots, 19\right\}$, i.e. 20 lattices were constructed in parallel as the exact precisions l_i of the approximations are not known.

The lattices were first reduced with the basic LLL_XD variant of LLL algorithm implemented in NTL. Following, stronger G_BZK_XD reduction was run with BlockSize initially set to 4 being increased by 2 to up to 20. After each BlockSize increase, the short vector of the reduced lattice was checked. In case it revealed the hidden number q, the attack was successful.

In the experiment with 150 simulated measurements, the attack was successful with parameter l equal to 9 and 9.5. The expensive lattice basis reduction steps took approximately 40 minutes each.

Five different scenarios with random RSA instances were emulated and experimented with. The RSA modulus was successfully factored in each of these instances.

4   Future Research

As mentioned several times, our main hypothesis—that the Montgomery multiplication is used and that the amount of final subtractions leaks—is to be verified. Furthermore, the resilience of other HW modules against this side channel attack in similar scenarios should be verified, as well. The probability of success of the attack under given circumstances is to be elaborated.

5   Conclusion

We presented new known plaintext side channel attack on RSA–CRT with Montgomery exponentiation in this paper. The lack of chosen plaintext condition greatly increases its applicability in the scenarios based on random formatting of the message being signed (probabilistic signature scheme). The existence of the side information we used was questioned. We urge the testing laboratories to verify it in the electronic passport scenario.

Acknowledgment

I would like to thank Dr. Rosa for pointing out [17] and for his guidance and the team of Prof. Lórencz at the department of computer science of FEE CTU in Prague for kindly providing their measurement results. Thanks also goes to the anonymous referees for their helpful comments.

References

1. Babai, L.: On Lovász' lattice reduction and the nearest lattice point problem (shortened version). In: Mehlhorn, K. (ed.) STACS 1985. LNCS, vol. 182, pp. 13–20. Springer, Heidelberg (1984)
2. Boneh, D., Venkatesan, R.: Hardness of computing the most significant bits of secret keys in Diffie-Hellman and related schemes. In: Koblitz, N. (ed.) CRYPTO 1996. LNCS, vol. 1109, pp. 129–142. Springer, Heidelberg (1996)
3. Finkenzeller, K.: RFID Handbook: Fundamentals and Applications in Contactless Smart Cards and Identification, 2nd edn. John Wiley & Sons, Chichester (2003)

4. Gama, N., Nguyen, P.Q.: Predicting lattice reduction. In: Smart, N.P. (ed.) EUROCRYPT 2008. LNCS, vol. 4965, pp. 31–51. Springer, Heidelberg (2008)
5. International Civil Aviation Organization (ICAO). Development of a Logical Data Structure – LDS for Optional Capacity Expansion Technologies,
http://www.iso.org/iso/iso_catalogue/catalogue_tc/
catalogue_detail.htm?csnumber=39693
6. International Civil Aviation Organization (ICAO). Doc 9303, Machine Readable Travel Documents, http://www2.icao.int/en/MRTD/Pages/Doc9393.aspx
7. International Organization for Standardization. ISO/IEC 7816 – Identification cards – Contactless integrated circuit cards – Proximity cards,
http://www.iso.org/iso/iso_catalogue/catalogue_tc/
catalogue_detail.htm?csnumber=39693
8. International Organization for Standardization. ISO/IEC 7816 – Identification cards – Integrated circuit(s) with contacts,
http://www.iso.org/iso/iso_catalogue/catalogue_tc/
catalogue_detail.htm?csnumber=38770
9. Lenstra, A.K., Lenstra Jr., H.W., Lovász, L.: Factoring polynomials with rational coefficients. Mathematische Ann. 261, 513–534 (1982)
10. Montgomery, P.L.: Modular multiplication without trial division. Mathematics of Computation 44, 519–521 (1985)
11. Nguyên, P.Q.: Cryptanalysis of the Goldreich-Goldwasser-Halevi Cryptosystem from Crypto 1997. In: Wiener, M.J. (ed.) CRYPTO 1999. LNCS, vol. 1666, pp. 288–304. Springer, Heidelberg (1999)
12. Nguyen, P.Q., Shparlinski, I.: The insecurity of the Digital Signature Algorithm with partially known nonces. J. Cryptology 15(3), 151–176 (2002)
13. Philips Electronics N.V. P5CD072 – Secure Dual Interface PKI Smart Card Controller,
http://www.nxp.com/acrobat_download/other/identification/sfs095412.pdf
14. Schindler, W.: A timing attack against RSA with the Chinese Remainder Theorem. In: Paar, C., Koç, Ç.K. (eds.) CHES 2000. LNCS, vol. 1965, pp. 109–124. Springer, Heidelberg (2000)
15. Schnorr, C.-P., Euchner, M.: Lattice basis reduction: Improved practical algorithms and solving subset sum problems. Math. Program. 66, 181–199 (1994)
16. Shoup, V.: NTL: A Library for doing Number Theory (2008),
http://www.shoup.net/ntl/
17. Tomoeda, Y., Miyake, H., Shimbo, A., Kawamura, S.-i.: An SPA-based extension of Schindler's timing attack against RSA using CRT. IEICE Transactions 88-A(1), 147–153 (2005)

A   Lattices

We give the definition of a full-rank lattice and overview several basic algorithmic problems associated with it in this section. We point out the state-of-the-art algorithms solving these problems, as well.

Let the set $\mathbf{B} = \{\mathbf{b}_0, \ldots, \mathbf{b}_{k-1}\}$ be a set of linearly independent vectors in $\mathbb{R}^k$. The lattice $\mathcal{L}$ spanned by the vectors in $\mathbf{B}$ is defined as $\mathcal{L} = \sum x_i \mathbf{b}_i, x_i \in \mathbb{Z}$. In such case, B is a basis of lattice L. A $k \times k$-type matrix over $\mathbb{R}$ whose rows are the vectors $\mathbf{b}_0, \ldots, \mathbf{b}_{k-1}$ is the called basis matrix of $\mathcal{L}$ and we will note it as $\mathbf{B}$, as well. The volume of a lattice $\mathcal{L}$ is defined as det $\mathbf{B}$, where $\mathbf{B}$ is any basis of $\mathcal{L}$.

i-th successive Minkowski minimum $\lambda_i(\mathcal{L})$ of lattice $\mathcal{L}$ is the radius of the smallest sphere containing at least i linearly independent (non-zero) vectors of $\mathcal{L}$. Especially, we see the first Minkowski minimum is the length of the shortest non-zero lattice vector and we denote it as $\lambda(L)$. The ratio $\frac{\lambda_2(\mathcal{L})}{\lambda_1(\mathcal{L})}$ is called the gap of the lattice.

A.1   Problems

Two lattice problems that are interesting in scope of this paper are the Unique shortest vector problem (Unique-SVP) and the Closest vector problem (CVP). Being given the lattice and its gap, Unique-SVP problem is to find the shortest vector of the lattice. Analogically, CVP problem is to find closest lattice vector to a given non-lattice vector. Sometimes, CVP is viewed as a non-homogenic variant of SVP.

A.2   Solutions

The usual approach to solve Unique-SVP is the LLL algorithm [9] or one of its variants [15]. In [4], it is experimentally shown it is possible to solve Unique-SVP if the gap $\frac{\lambda_2}{\lambda_1}$ is at least 1.021^k with BKZ-20 variant of LLL algorithm.

One can try to solve CVP with Babais closest plane algorithm [Ba85], the experience shows, however, the heuristic conversion to Unique-SVP provides better results. We use the same technique as in [11], i.e. we construct lattice $\mathcal{L}'$ with the basis matrix $\mathbf{B}' = \begin{pmatrix} \mathbf{B} & 0 \\ V & 1 \end{pmatrix}$. As the lattices $\mathcal{L}$ and $\mathcal{L}'$ have the same determinant and approximately the same dimension, we can expect their respective shortest vectors to be approximately of the same size. Given the fact that the hidden vector H is in $\mathcal{L}$ and close to V (section 3), we see the vector $V-H$ is short and belongs to $\mathcal{L}'$. In fact, we can expect is to be the shortest vector of $\mathcal{L}'$. If the gap $\frac{\lambda_2}{\lambda_1}$ is sufficiently large, we can use the lattice basis reduction techniques and check if the short vector found reveals the hidden number x in $(k+1)$-st coordinate (follows from the construction of lattice L in section 3).

A New Side-Channel Attack
on RSA Prime Generation

Thomas Finke, Max Gebhardt, and Werner Schindler

Bundesamt für Sicherheit in der Informationstechnik (BSI)
Godesberger Allee 185–189
53175 Bonn, Germany
{Thomas.Finke,Maximilian.Gebhardt,Werner.Schindler}@bsi.bund.de

Abstract. We introduce and analyze a side-channel attack on a straight-forward implementation of the RSA key generation step. The attack exploits power information that allows to determine the number of the trial divisions for each prime candidate. Practical experiments are conducted, and countermeasures are proposed. For realistic parameters the success probability of our attack is in the order of 10–15 %.

Keywords: Side-channel attack, RSA prime generation, key generation.

1 Introduction

Side-channel attacks on RSA implementations have a long tradition (e.g. [8,9,12,13]). These attacks aim at the RSA exponentiation with the private key d (digital signature, key exchange etc.). On the other hand only a few papers on side-channel attacks, resp. on side-channel resistant implementations, exist that focus on the generation of the primes p and q, the private key d and the public key $(e, n = pq)$ (cf., e.g. [4,1]). If the key generation process is performed in a secure environment (e.g., as part of the smart card personalisation) it is infeasible to mount any side-channel attack. However, devices may generate an RSA key pair before the computation of the first signature or when applying for a certificate. In these scenarios the primes may be generated in insecure environments.

Compared to side-channel attacks on RSA exponentiation with the secret key d the situation for a potential attacker seems to be less comfortable since the primes, resp. the key pair, are generated only once. Moreover, the generation process does not use any (known or chosen) external input.

We introduce and analyse a power attack on a straight-forward implementation of the prime generation step where the prime candidates are iteratively incremented by 2. The primality of each prime candidate v is checked by trial divisions with small primes until v is shown to be composite or it has passed all trial divisions. To the 'surviving' prime candidates the Miller-Rabin primality test is applied several times. We assume that the power information discovers the number of trial divisions for each prime candidate, which yields information on p and q, namely $p(\mathrm{mod}\,s)$ and $q(\mathrm{mod}\,s)$ for some modulus s, which is a product

C. Clavier and K. Gaj (Eds.): CHES 2009, LNCS 5747, pp. 141–155, 2009.

of small primes. The attack will be successful if s is sufficiently large. Simulations and experimental results show that for realistic parameters (number of small primes for the trial divisions under consideration of the magnitude of the RSA primes) the success probability is in the order of 10–15%, and that our assumptions on the side-channel leakage are realistic.

Reference [4] considers a (theoretical) side channel attack on a careless implementation of a special case of a prime generation algorithm proposed in [7] that is successful in about 0.1% of the trials. Reference [3] applies techniques from [1], which were originally designated for the shared generation of RSA keys. We will briefly discuss these aspects in Section 6.

The intention of this paper is two-fold. First of all it presents a side-channel attack which gets by with weak assumptions on the implementation. Secondly, the authors want to sensibilise the community that RSA key generation in potentially insecure environments may bear risks. The authors want to encourage the community to spend more attention on the side-channel analysis of the RSA key generation process.

The paper is organized as follows: In Section 2 we have a closer look at the RSA prime generation step. In Section 3 we explain our attack and its theoretical background. Section 4 and Section 5 provide results from simulations and conclusions from the power analysis of an exemplary implementation on a standard microcontroller. The paper ends with possible countermeasures and final conclusions.

2   Prime Generation

In this section we have a closer look at the prime generation step. Moreover, we formulate assumptions on the side-channel leakage that are relevant for the next sections.

Definition 1. *For any $k \in N$ a k-bit integer denotes an integer that is contained in the interval $[2^{k-1}, 2^k)$. For a positive integer $m \geq 2$ as usually $Z_m := \{0, 1, \ldots, m-1\}$ and $Z_m^* := \{x \in Z_m \mid \gcd(x, m) = 1\}$. Further, $b \pmod m$ denotes that element in Z_m that has the same m-remainder as b.*

Pseudoalgorithm 1 (prime generation)
  1) Generate a (pseudo-)random odd integer $v \in [2^{k-1}, 2^k)$
  2) Check whether v is prime. If v is composite then goto Step 1
  3) $p := v$ (resp., $q := v$)

Pseudoalgorithm 1 represents the most straight-forward approach to generate a random k-bit prime. In Step 2 trial divisions by small odd primes from a particular set $\mathcal{T} := \{r_2, \ldots, r_N\}$ are performed, and to the 'surviving' prime candidates the Miller-Rabin primality test (or, alternatively, any other probabilistic primality test) is applied several times. The 'trial base' $\mathcal{T} := \{r_2, \ldots, r_N\}$ (containing all odd primes $\leq$ some bound B) should be selected to minimize the average run-time of Step 2.

By the prime number theorem

$$\# \text{ primes } \in [2^{k-1}, 2^k) \approx \frac{2^k}{\log_e(2^k)} - \frac{2^{k-1}}{\log_e(2^{k-1})} = \frac{2^{k-1}}{\log_e(2)} \left(\frac{2}{k} - \frac{1}{k-1} \right). \quad (1)$$

Consequently, for a randomly selected odd integer $v \in [2^{k-1}, 2^k)$ we obtain

$$\text{Prob}\left(v \in [2^{k-1}, 2^k) \text{ is prime} \right) \approx \frac{2}{\log_e(2)} \left(\frac{2}{k} - \frac{1}{k-1} \right) \approx \frac{2}{k \log_e(2)}. \quad (2)$$

For $k = 512$ and $k = 1024$ this probability is $\approx 1/177$ and $\approx 1/355$, respectively. This means that in average 177, resp. 355, prime candidates have to be checked to obtain a k-bit prime. The optimal size $|\mathcal{T}|$ depends on k and on the ratio between the run-times of the trial divisions and the Miller-Rabin tests. This ratio clearly is device-dependent.

Hence Pseudoalgorithm 1 requires hundreds of calls of the RNG (random number generator), which may be too time-consuming for many applications. Pseudoalgorithm 2 below overcomes this problem as it only requires one k-bit random number per generated prime. References [2,11], for example, thus recommend the successive incrementation of the prime candidates or at least mention this as a reasonable option. Note that the relevant part of Pseudoalgorithm 2 matches with Algorithm 2 in [2] (cf. also the second paragraph on p. 444). The parameter t in Step 2d depends on the tolerated error probability.

Pseudoalgorithm 2 (prime generation)
   1) Generate a (pseudo-)random odd integer $v_0 \in [2^{k-1}, 2^k)$
          $v := v_0$;
   2) a) $i := 2$;
      b)   while $(i \leq N)$ do {

```
            if (r_i divides v) then {
                v := v + 2; GOTO Step 2a; }
            i++;

        }
```

      c) $m := 1$;
      d)   while $(m \leq t)$ do {

```
            apply the Miller-Rabin primality test to v;
            if the primality test fails then {
                v := v + 2;  GOTO Step 2a; }
            else m++;

        }
```

   3) $p := v$ (resp., $q := v$)

Pseudoalgorithm 2 obviously 'prefers' primes that follow long prime gaps but until now no algebraic attack is known that exploits this property. However, the situation may change if side-channel analysis is taken into consideration. We formulate two assumptions that will be relevant for the following.

Assumptions 1. a) Pseudoalgorithm 2 is implemented on the target device.
b) Power analysis allows a potential attacker to identify for each prime candidate v after which trial division the while-loop in Step 2b terminates. Moreover, he is able to realize whether Miller-Rabin primality test(s) have been performed.

Remark 1. We may assume that
(i) a strong RNG is applied to generate the odd number v_0 in Step 1 of Pseudoalgorithm 2.
(ii) the trial division algorithm itself and the Miller-Rabin test procedure are effectively protected against side-channel attack. This means that the side-channel leakage does not reveal any information on the dividend of the trial divisions, i.e. on the prime candidates v.

Remark 2. (i) If any of the security assumptions from Remark 1 are violated it may be possible to improve our attack or to mount a different, even more efficient side-channel attack. This is yet outside the scope of this paper. In the following we merely exploit Assumption b)
(ii) Assumption b) is clearly fulfilled if the attacker is able to determine the beginning or the end of each trial divisions. If all trial divisions require essentially the same run-time (maybe depending on the prime candidates v) it suffices to identify the beginning of the while-loop or the incrementation by 2 in Step 2b. The run-time also reveals whether Miller-Rabin tests have been performed.
(iii) It may be feasible to apply our attack also against software implementations on PCs although power analysis is not applicable there. Instead, the attacker may try to mount microarchitectural attacks (cache attacks etc.).
(iv) We point out that more efficient (and more sophisticated) prime generation algorithms than Pseudoalgorithm 2 exist (cf. Section 6 and [1,7,11], Note 4.51(ii), for instance).

3   The Attack

In Section 3 we describe and analyze the theoretical background of our attack. Empirical and experimental results are presented in Section 4 and Section 5.

3.1   Basic Attack

We assume that the candidate $v_m := v_0 + 2m$ in Pseudoalgorithm 2 is prime, i.e. $p = v_m$. If for $v_j = v_0 + 2j$ Pseudoalgorithm 2 returned to Step 2a after the trial division by r_i then v_j is divisible by r_i. This gives

$$\left.\begin{array}{l} v_j \equiv 0 \pmod{r_i} \\ v_j = \quad v_0 + 2j \\ p = v_m = v_0 + 2m \end{array}\right\} \Rightarrow p = v_j + 2(m-j) \equiv 2(m-j) \pmod{r_i}. \quad (3)$$

Let

$$\mathcal{S}_p := \{2\} \cup \{r \in \mathcal{T} \mid \text{division by } r \text{ caused a return to 2a for at least one } v_j\}. \quad (4)$$

We point out that 'caused a return ...' is not equivalent to 'divides at least one v_j'. (Note that it may happen that $r \in \mathcal{T} \setminus \mathcal{S}_p$ divides some v_j but loop 2b terminates earlier due to a smaller divisor r' of v_j.) We combine all equations of type (3) via the Chinese Remainder Theorem (CRT). This yields a congruence

$$a_p \equiv p \pmod{s_p} \quad \text{for} \quad s_p := \prod_{r \in \mathcal{S}_p} r \tag{5}$$

with known a_p. As $pq = n$ we have

$$a_q := q \equiv a_p^{-1} n \pmod{s_p}. \tag{6}$$

By observing the generation of q we obtain

$$b_q \equiv q \pmod{s_q} \quad \text{and} \quad b_p \equiv p \equiv b_q^{-1} n \pmod{s_q} \tag{7}$$

where $\mathcal{S}_q$ and s_q are defined analogously to $\mathcal{S}_p$ and s_p. Equations (5), (6) and (7) give via the CRT integers c_p, c_q and s with

$$s := \mathrm{lcm}\,(s_p, s_q), \quad c_p \equiv p \pmod{s}, \quad c_q \equiv q \pmod{s} \quad \text{and} \quad 0 \leq c_p, c_q < s. \tag{8}$$

By (8)

$$p = sx_p + c_p \quad \text{and} \quad q = sy_q + c_q \quad \text{with unknown integers} \quad x_p, y_q \in \mathbb{N} \tag{9}$$

while c_p, c_q and s are known. Lemma 1 transforms the problem of finding p and q into a zero set problem for a bivariate polynomial over Z.

Lemma 1. *(i) The pair (x_p, y_q) is a zero of the polynomial*

$$f : Z \times Z \to Z, \quad f(x,y) := sxy + c_p y + c_q x - t \quad \text{with} \quad t := (n - c_p c_q)/s. \tag{10}$$

(ii) In particular

$$t \in \mathbb{N}, \quad f \text{ is irreducible over } Z, \quad \text{and} \tag{11}$$

$$0 < x_p, y_q < \max\left\{\frac{p}{s}, \frac{q}{s}\right\} < \frac{2^k}{s}. \tag{12}$$

Proof. Obviously,

$$0 = pq - n = (sx_p + c_p)(sy_q + c_q) - n = s^2 x_p y_q + sc_p y_q + sc_q x_p - (n - c_p c_q),$$

which verifies (i). Since $n \equiv c_p c_q \pmod{s}$ the last bracket is a multiple of s, and hence $t \in Z$. Since $c_p \equiv p \not\equiv 0 \pmod{r_j}$ and $c_q \equiv q \not\equiv 0 \pmod{r_j}$ for all prime divisors r_j of s we conclude $\gcd(s, c_p) = \gcd(s, c_q) = 1$, and in particular $\gcd(s, c_p, c_q, t) = 1$. Assume that $f(x,y) = (ax + by + c)(dx + ey + f)$ for suitably selected integers a, b, c, d, e and f. Comparing coefficients immediately restricts to $(a = e = 0)$ or $(b = d = 0)$. The gcd-properties yield $\gcd(bd, bf) = 1 = \gcd(bd, cd)$, resp. $\gcd(ae, af) = 1 = \gcd(ae, ce)$, and thus $b = d = 1$, resp. $a = e = 1$, leading to a contradiction. Assertion (12) is obvious.

In general finding zeroes of bivariate polynomials over Z is difficult. It is well-known that 'small' integer solutions yet can be found efficiently with the LLL-algorithm, which transforms the zero set problem to finding short vectors in lattices.

Theorem 1. *(i) Let $p(x,y)$ be an irreducible polynomial in two variables over Z, of maximum degree δ in each variable separately. Let X, Y be upper bounds for the absolute value of the searched solutions x_0, y_0. Define $\tilde{p}(x,y) := p(xX, yY)$ and let W be the absolute value of the largest coefficient of $\tilde{p}$. If*

$$XY < W^{2/(3\delta)}$$

then in time polynomial in $(\log W, \delta)$, one can find all integer pairs (x_0, y_0) with $p(x_0, y_0) = 0$, $|x_0| < X$, $|y_0| < Y$.
(ii) Let p and q be k-bit primes and $n = p \cdot q$. If integers s and c_p are given with $s \geq 2^{\frac{k}{2}}$ and $c_p \equiv p \pmod s$ then one can factorize n in time polnomial in k.

Proof. (i) [5], Corollary 2
(ii) We apply assertion (i) to the polynomial $f(x,y)$ from Lemma 1. By (12) we have $0 < x_p < X := 2^k/s$ and $0 < y_q < Y := 2^k/s$. Let $\tilde{f}(x,y) = f(xX, yY)$ and let W denote the maximum of the absolute values of the coefficients of $\tilde{f}(x,y)$. Then $W \geq sXY = \frac{2^{2k}}{s}$, and for $s > 2^{\frac{k}{2}}$ we get

$$XY = \left(\frac{2^k}{s}\right)^2 < \left(\frac{2^{2k}}{s}\right)^{\frac{2}{3}} \leq W^{\frac{2}{3}}$$

where the first inequality follows from $2^k < s^2$ by some equivalence transformations. Since the degree δ in each variable is one by (i) we can find (x_p, y_q) in time polynomial in k.

3.2   Gaining Additional Information

Theorem 1 demands $\log_2(s) > 0.5k$. If $\log_2(s)$ is only slightly larger than $0.5k$ the dimension of the lattice ($\rightarrow$ LLL-algorithm) has to be very large which affords much computation time. For concrete computations thus $\log_2(s) \geq C > 0.5k$ is desirable for some bound C that is reasonably larger than $0.5k$.

If $\log_2(s) \geq C$ Theorem 1 can be applied and then the work is done. If $\log_2(s) < C$ one may multiply s by some relatively prime integer s_1 (e.g. the product of some primes in $\mathcal{T} \setminus (\mathcal{S}_p \cup \mathcal{S}q)$) with $\log_2(s) + \log_2(s_1) > C$. Of course, the adversary has to apply Theorem 1 to any admissible pair of remainders $(p(\mathrm{mod}\,(s \cdot s_1)), q(\mathrm{mod}\,(s \cdot s_1)))$. Theorem 1 clearly yields the factorization of n only for the correct pair $(p(\mathrm{mod}\,(s \cdot s_1)), q(\mathrm{mod}\,(s \cdot s_1)))$, which increases the workload by factor 2^{s_1}. Note that $p(\mathrm{mod}\,(s \cdot s_1))$ determines $q(\mathrm{mod}\,(s \cdot s_1))$ since $n(\mathrm{mod}\,(s \cdot s_1))$ is known.

The basic attack explained in Subsection 3.1 yet does not exploit all information. Assume that the prime $r_u \in \mathcal{T}$ does not divide s, which means that

$r_u \notin \mathcal{S}_p \cup \mathcal{S}_q$ or, equivalently, that the trial division loop 2b in Algorithm 2 has never terminated directly after a division by r_u. Assume that during the search for p the prime candidates $v_{j_1}, \ldots, v_{j_\tau}$ have been divided by r_u. Then

$$
\left.
\begin{aligned}
v_{j_1} &= p - 2(m - j_1) \not\equiv 0 \pmod{r_u} \\
&\quad\vdots \\
v_{j_\tau} &= p - 2(m - j_\tau) \not\equiv 0 \pmod{r_u} \\
v_m &= p \qquad\qquad\quad \not\equiv 0 \pmod{r_u}
\end{aligned}
\right\} \implies
\tag{13}
$$

$$
p \not\equiv 0, 2(m - j_1), \ldots, 2(m - j_\tau) \pmod{r_u}.
\tag{14}
$$

This yields a 'positive list'

$$
L'_p(r_u) = \{0, 1, \ldots, r_u - 1\} \setminus \{0, 2(m - j_1)(\bmod\ r_u), \ldots, 2(m - j_\tau)(\bmod\ r_u)\}
\tag{15}
$$

of possible r_u-remainders of p. Analogously, one obtains a positive list $L'_q(r_u)$ for possible r_u-remainders of q. The relation $p \equiv nq^{-1}(\bmod\ r_u)$ reduces the set of possible r_u-remainders of p further to

$$
L_p(r_u) := L'_p(r_u) \cap \left(nL'_q(r_u)^{-1}(\bmod\ r_u) \right), \quad \text{and finally}
\tag{16}
$$

$$
(p(\bmod\ r_u), q(\bmod\ r_u)) \in \{(a, na^{-1}(\bmod\ r_u)) \mid a \in L_p(r_u)\}.
\tag{17}
$$

For prime r_u equations (16) and (17) provide

$$
I(r_u) := \log_2 \left(\frac{r_u}{|L_p(r_u)|} \right) = \log_2 (r_u) - \log_2 (|L_p(r_u)|)
\tag{18}
$$

bit of information. From the attacker's point of view the most favourable case clearly is $|L_p(r_u)| = 1$, i.e. $I(r_u) = \log_2(r_u)$, which means that $p(\bmod\ r_u)$ is known. The attacker may select some primes $r_{u_1}, \ldots, r_{u_w} \in \mathcal{T} \setminus (\mathcal{S}_p \cup \mathcal{S}_q)$ that provide much information $I(r_{u_1}), \ldots, I(r_{u_w})$ (or, maybe more effectively, selecting primes with large ratios $I(r_{u_1})/\log_2(r_{u_1}), \ldots, I(r_{j_w})/\log_2(r_{j_w})$) where w clearly depends on the gap $C - \log_2(s)$. Then he applies Theorem 1 to $(s \cdot s_1)$ with $s_1 = r_{u_1} \cdots r_{u_w}$ for all $|L_p(r_{u_1})| \cdots |L_p(r_{u_w})|$ admissible pairs of remainders $(p(\bmod\ (s \cdot s_1)), q(\bmod\ (s \cdot s_1)))$. Compared to a 'blind' exhaustive search without any information $p(\bmod\ s_1)$ (17) reduces the workload by factor $2^{I(r_{u_1}) + \cdots + I(r_{u_w})}$ or, loosely speaking, reduces the search space by $I(r_{u_1}) + \cdots + I(r_{u_w})$ bit.

After the primes p and q have been generated the secret exponents $d_p \equiv e^{-1}(\bmod\ (p-1))$ and $d_q \equiv e^{-1}(\bmod\ (q-1))$ are computed. These computations may provide further information. Analogously to Assumptions 1 we formulate

Assumptions 2. The exponents d_p and d_q are computed with the extended Euclidean algorithm, and the power consumption allows the attacker to determine the number of steps that are needed by the Euclidean algorithm.

We may clearly assume $p > e$, and thus the first step in the Euclidean algorithm reads $p - 1 = \alpha_2 e + x_3$ with $\alpha_2 \geq 1$ and $x_3 = p - 1(\bmod\ e)$. The following steps depend only on the remainder $p - 1 (\mathrm{mod}\, e)$, resp. on $p(\mathrm{mod}\, e)$. As $p \not\equiv 0(\mathrm{mod}\, e)$ for $j \in \mathbb{N}_0$ we define the sets

$$M'(j) = \{x \in \mathbb{Z}_e^* \mid \text{for } (e, x{-}1) \text{ the Euclidean alg. terminates after j steps}\}. \quad (19)$$

Assume that the attacker has observed that the computation of d_p and d_q require v_p and v_q steps, respectively. By definition, $p(\mathrm{mod}\, e) \in M'(v_p - 1)$ and $q(\mathrm{mod}\, e) \in M'(v_q - 1)$. Similarly as above

$$M_p := M'(v_p - 1) \cap \left(n \left(M'(v_q - 1) \right)^{-1} (\mathrm{mod}\, e) \right), \quad \text{and finally} \quad (20)$$

$$(p(\mathrm{mod}\, e), q(\mathrm{mod}\, e)) \in \{(a, na^{-1}(\mathrm{mod}\, e)) \mid a \in M_p\}. \quad (21)$$

If e is relatively prime to s, resp. to $s \cdot s_1$ (e.g. for $e = 2^{16} + 1$), the attacker may apply Theorem 1 to $s \cdot e$, resp. to $s \cdot s_1 \cdot e$, and the gain of information is $I(e) = \log_2(e) - \log_2(|M_p|)$. If $\gcd(s, e) > 1$, resp. if $\gcd(s \cdot s_1, e) > 1$, one uses $e' := e/\gcd(s, e)$, resp. $e' := e/\gcd(s \cdot s_1, e)$, in place of e.

Remark 3. In Section 2 we assumed $p, q \in [2^{k-1}, 2^k)$. We note that our attack merely exploits $p, q < 2^k$, and it also applies to unbalanced primes p and q.

4   Empirical Results

The basic basic attack reveals congruences $p(\mathrm{mod}\, s)$ and $q(\mathrm{mod}\, s)$ for some modulus s. If s is sufficiently large Theorem 1 allows a successful attack. The term $\log_2(s)$ quantifies the information we get from side-channel analysis. The product s can at most equal $\prod_{r \in \mathcal{T}} r$ but usually it is much smaller. Experiments show that the bitsize of s may vary considerably for different k-bit starting candidates v_0 for Pseudoalgorithm 2. Theorem 1 demands $\log_2(s) > \frac{k}{2}$, or (from a practical point of view) even better $\log_2(s) \geq C$ for some bound C which allows to apply the LLL-algorithm with moderate lattice dimension. We investigated the distribution of the bitlength of s for $k = 512$ and $k = 1024$ bit primes and for different sizes of the trial division base $\mathcal{T} = \{r_2 = 3, 5, 7, \ldots, r_N\}$ by a large number of simulations. Note that two runs of Pseudoalgorithm 2 generate an RSA modulus n of bitsize $2k - 1$ or $2k$.

We implemented Pseudoalgorithm 2 (with $t = 20$ Miller-Rabin-tests) in MAGMA [10] and ran the RSA-generation process 10000 times for each of several pairs (k, N). For $k = 512$ and $N = 54$ we obtained the empirical cumulative distribution shown in Figure 1. The choice $N = 54$ is natural since $r_{54} = 251$ is the largest prime smaller than 256, and thus each prime of the trial division base can be represented by one byte. Further results for $k = 512$ and $k = 1024$ are given in Table 1, resp. in Table 2. The ideas from Subsection 3.2 are considered below.

Further on, we analysed how the run-time of the LLL-algorithm and thus the run-time of the factorization of the RSA modulus n by Theorem 1 depends on the bitsize of s. We did not use the original algorithm of Coppersmith from

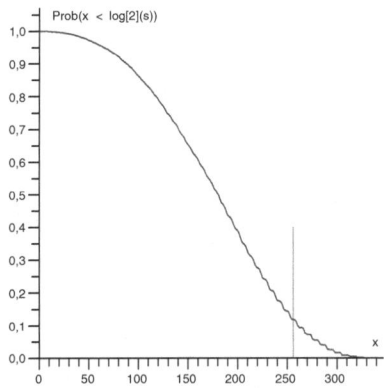

Fig. 1. Basic attack: Cumulative distribution for $r_N = 251$ and $k = 512$ (1024-bit RSA moduli)

Table 1. Basic attack

N	r_N	$Prob(\log_2(s) > 256)$	$Prob(\log_2(s) > 277)$	$\log_2(\prod_{r \leq r_N} r)$
		$k = 512$		
54	251	0.118	0.055	334.8
60	281	0.188	0.120	388.3
70	349	0.283	0.208	466.5

Table 2. Basic attack

N	r_N	$Prob(\log_2(s) > 512)$	$Prob(\log_2(s) > 553)$	$\log_2(\prod_{r < r_N} r)$
		$k = 1024$		
100	541	0.125	0.065	729.7
110	601	0.178	0.113	821.2
120	659	0.217	0.150	914.5

[5]. Instead we implemented Coron's algorithm from [6] in the computer algebra system MAGMA [10].

The LLL-reduction in Coron's algorithm uses a lattice of dimension $\omega = (\tilde{k} + \delta)^2 - \tilde{k}^2$ where δ denotes the degree of the polynomial and $\tilde{k}$ an adjustable parameter of the algorithm. In our case $\delta = 1$, so the lattice dimension is $\omega = 2\tilde{k} + 1$. For $\omega = 15$ our implementation (i.e. the LLL-substep) never terminated in less than one hour; we stopped the process in these cases. Table 3 provides empirical results. (More sophisticated implementations may allow to get by with smaller s ($\rightarrow$ larger ω) but this is irrelevant for the scope of this paper.)

If the basic attack yields $\log_2(s) \leq k/2$ or $\log_2(s) < C$ the attacker may apply the techniques from Subsection 3.2. Since the information $I(e)$ is deduced from the computation of $\gcd(e, p - 1)$ it is independent of the outcome of the basic attack while $I(r_{u_1}) + \cdots + I(r_{u_w})$ depends on the size of s. If $\log_2(s)$ is contained

Table 3. Empirical run-times for different lattice dimensions and moduli

$Bitsize(n = pq)$	ω	$Min\{Bitsize(s)\mid^{factorization}_{succesfull}\}$	$\approx run-time(sec)$
512	5	156	0.01
512	7	148	0.07
512	9	144	0.24
512	11	141	1.1
512	13	139	4.6
1024	5	308	0.02
1024	7	294	0.17
1024	9	287	0.66
1024	11	281	3.1
1024	13	277	13.2
1536	5	462	0.05
1536	7	440	0.36
1536	9	428	1.6
1536	11	420	8.1
1536	13	415	41.5
2048	5	616	0.06
2048	7	587	0.76
2048	9	571	3.3
2048	11	560	18.5
2048	13	553	87.4

in $[230, 260]$, a relevant range for $k = 512$, for $r_N = 251$ the mean value of $I(r_{u_1}) + \cdots + I(r_{u_w})$ is nearly constant.

Simulations for $k = 512$, $e = 2^{16} + 1$, and $\mathcal{T} = \{3, \ldots, 251\}$ show that $(I(2^{16} + 1), \log_2(2^{16} + 1)) = (6.40, 16)$, while $(I(r_{u_1}), \log_2(r_{u_1})) = (2.81, 6.39)$, resp. $(I(r_{u_1}) + I(r_{u_2}), \log_2(r_{u_1}) + \log_2(r_{u_2})) = (4.80, 13.11)$, resp. $(I(r_{u_1}) + I(r_{u_2}) + I(r_{u_3}), \log_2(r_{u_1}) + \log_2(r_{u_2}) + \log_2(r_{u_3})) = (6.42, 20.04)$, where r_{u_1}, r_{u_2} and r_{u_3} (in this order) denote those primes in $\mathcal{T} \setminus (\mathcal{S}_p \cup \mathcal{S}_q)$ that provide maximum information.

Multiplying the modulus s from the basic attack by e, resp. by $e \cdot r_{u_1}$, resp. by $e \cdot r_{u_1} \cdot r_{u_2}$, resp. by $e \cdot r_{u_1} \cdot r_{u_2} \cdot r_{u_3}$, increases the bitlength of the modulus by 16 bits, resp. by $16 + 6.39 = 22.39$, resp. by 29.11, resp. by 36.04 in average although the average workload increases only by factor $2^{16-6.40} = 2^{9.6}$, resp. by $2^{9.6} \cdot 2^{6.39-2.81} = 2^{13.18}$, resp. by $2^{17.91}$, resp. by $2^{23.22}$.

Combining this with the run-time of 13.2 seconds given in Table 3 with our implementation we can factorize a 1024-Bit RSA modulus in at most $2^{13.18} \cdot 13.2 \, sec \approx 34 \, hours$ (in about 17 hours in average) if the modulus s gained by the basic attack only consists of $277 - 22 = 255$ bits. According to our experiments this happens with probability ≈ 0.119. Table 1 shows that the methods of Subsection 3.2 double (for our LLL-implementation) the success probability. Further on we want to emphasize that by Theorem 1 the attack becomes principally feasible if the basic attack yields $\log_2(s) > 256$. So, at cost of increasing the run-time by factor $2^{13.18}$ the modulus n can be factored with the LLL-algorithm if the basic attack yields a modulus s with $\log_2(s) > 256 - 22 = 234$. This means that the success probability would increase from 11.8% to 21.2% (cf. Table 1).

5   Experimental Results

The central assumption of our attack is that the power consumption reveals the exact number of trial divisions for the prime candidates $v_0 = v, v_1 = v + 2, \ldots$. To verify that this assumption is realistic we implemented the relevant part of Pseudoalgorithm 2 (Step 1 to Step 2b) on a standard microcontroller (Atmel ATmega) and conducted measurements.

The power consumption was simply measured as a voltage drop over a resistor that was inserted into the GND line of this chip. An active probe was used. As the controller is clocked by its internal oscillator running at only 1MHz a sampling-rate of 25 MHz was sufficient. The acquired waveforms were high-pass-filtered and reduced to one peak value per clock cycle.

Figure 2 shows the empirical distribution of the number of clock cycles per trial division. We considered 2000 trial divisions $testdiv(v, r)$ with randomly selected 512 bit numbers v and primes $r < 2^{16}$. The number of clock cycles are contained in the interval $[24600, 24900]$, which means that they differ not more than about 0.006μ cycles from their arithmetic mean μ. In our standard case $\mathcal{T} = \{3, \ldots, 251\}$ a maximum sequence of 53 consecutive trial divisions may occur. We point out that it is hence not necessary to identify the particular trial divisions, it suffices to identify those positions of the power trace that correspond to the incrementation of the prime candidates by 2. Since short and long run-times of the individual trial divisions should compensate to some extent, this conclusion should remain valid also for larger trial bases and for other implementations of the trial divisions with (somewhat) larger variance of the run-times.

The crucial task is to find characteristic parts of the power trace that allow to identify the incrementation operations or even the individual trial divisions. The trial division algorithm and the incrementation routine were implemented in a straight-forward manner in an 8-bit arithmetic. Since the incrementation operations leave the most significant parts of the prime candidates $v_0, v_1, \ldots$ unchanged and since all divisors are smaller than 255 it is reasonable to expect that the power consumption curve reveals similar parts. Observing the following sequence of operations confirmed this conjecture.

Prime generation and trial divisions

```
rnd2r ();       // generates an odd 512 bit random number v
testdiv512 (v,3);   // trial division by 3
testdiv512 (v,5);
testdiv512 (v,7);
incrnd (v);         // increments v by 2
testdiv512 (v,3);
incrnd (v);
```

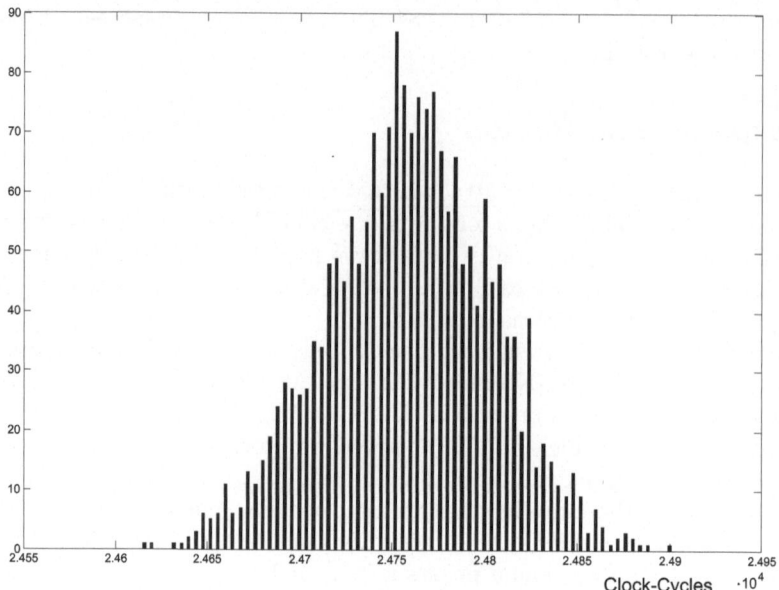

Fig. 2. Empirical run-times of trial divisions

```
testdiv512 (v,3);
testdiv512 (v,5);
```

We measured the power-consumption x_i for each clock cycle i. We selected short sample sequences $\{y_1 = x_t, \ldots, y_M = x_{t+M-1}\} \subset \{x_1, x_2, \ldots\}$ that correspond to 10 to 20 consecutive cycles, and searched for similar patterns in the power consumption curve. For fixed sample pattern $(y_1, \ldots, y_M)$ we used the 'similarity function',

$$a_j = \frac{1}{M} \sum_{i=1}^{M} |x_{i+j} - y_i| \quad \text{for shift parameter } j = 1, \ldots, N - M, \qquad (22)$$

which compares the sample sequence $(y_1, \ldots, y_M)$ with a subsequence of power values of the same length that is shifted by j positions. A small value a_j indicates that $(x_j, \ldots, x_{j+M-1})$ is 'similar' to the sample sequence $(y_1, \ldots, y_M)$. It turned out that it is even more favourable to consider the minimum within 'neighbourhoods' rather than local minima. More precisely, we applied the values

$$b_j = \min\{a_j, \ldots, a_{j+F-1}\} \qquad (23)$$

with $F \approx 100$. Figure 3 shows three graphs of b_j-values. The vertical grey bars mark the position of the selected sample sequence $(y_1, \ldots, y_M)$. For Curve (1) the sample sequence was part of the random number generation process. Obviously, this sample sequence does not help to identify any trial division or the

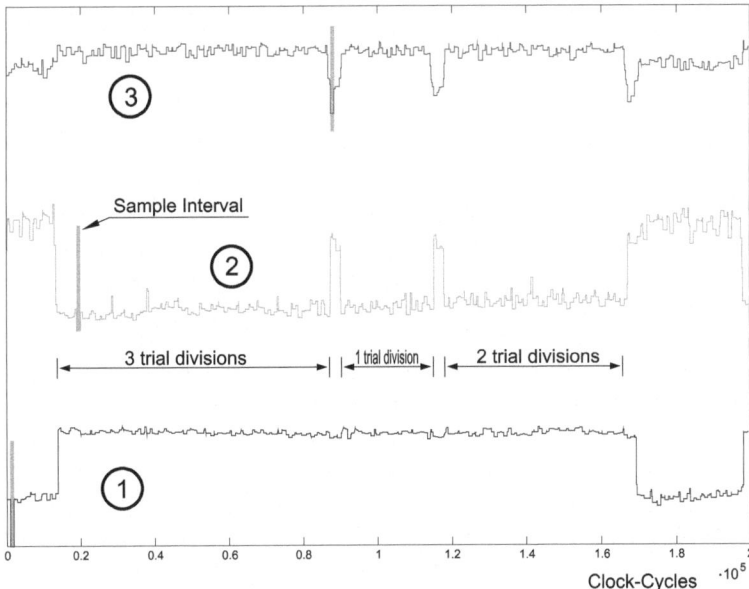

Fig. 3. Similarity curves (b_j-values)

incrementation steps. For Curve (2) we selected a sample sequence within a trial division. The high peaks of Curve (2) stand for 'large dissimilarity' and identify the incrementation steps. Curve (3) shows the b_j-values for a sample sequence from the incrementation step, and low peaks show the positions of the incrementation steps. Our experiments showed that the procedure is tolerant against moderate deviations of M, F and the sample pattern $(y_1, \ldots, y_M)$ from the optimal values.

6   Countermeasures and Alternative Implementations

Our attack can be prevented by various countermeasures. The most rigorous variants are surely to divide each prime candidate by all elements of the trial base or to generate each prime candidate independent from its predecessors ($\rightarrow$ Pseudoalgorithm 1). However, both solutions are very time-consuming and thus may be inappropriate for many applications due to performance requirements. Clearly more efficient is to XOR some fresh random bits to every τ^{th} prime candidate v in order to compensate the side-channel leakage of the trial divisions of the previous τ prime candidates. These random bits should at least compensate the average information gained from the leakage or, even better, compensate the maximum information leakage that is possible (worst-case scenario). In analogy to (5) let s_τ denote the product of all primes of $\mathcal{T}$, after which the while loop in Step 2b of Pseudoalgorithm 2 has terminated for at least one of the last τ prime candidates v_j. For $\tau = 10$, for instance, the while-loop must have terminated at least three times after the trial division by 3 and at least once after a trial division by 5. In the worst case, the remaining six loops terminated after the division by one of the six largest primes of $\mathcal{T}$, which gives the (pessimistic) inequality $\log_2(s_\tau) \leq \log_2(3 \cdot 5 \cdot r_{N-5} \cdots r_N)$. For $k = 512$ and $\mathcal{T} = \{3, 5, \ldots, 251\}$, for instance, $\log_2(s_{10}) < \log_2(3 \cdot 5 \cdot 227 \cdot \ldots \cdot 251) \approx 51.2$. Simulations showed that the average value of $\log_2(s_{10})$ is much smaller (≈ 18.6). By applying the ideas from Subsection 3.2 the attacker may gain some additional information, in the worst case yet not more than $\sum_{j=4}^{48} \log_2(r_j/(r_j - 6)) \approx 10.66$ bit. The designer should be on the safe side if he selects randomly at least 8 bytes of each 10th prime candidate v and XORs 64 random bits to these positions. (Simulations indicate that the average overall gain of information is less than 24 bit.)

We mention that several other, more sophisticated prime generation algorithms have been proposed in literature. For instance, the remainders of the first prime candidate v_0 with respect to all primes in $\mathcal{T}$ may be stored in a table, saving the trial divisions for all the following prime candidates in favour of modular additions of all table entries by 2 ([11], Note 4.51 (ii)). This prevents our attack but, of course, a careless implementation of the modular additions may also reveal side-channel information. A principal disadvantage is that a large table has to be built up, stored and managed, which may cause problems in devices with little resources. Alternatively, in a first step one may determine an integer v that is relatively prime to all primes of a trial base $\mathcal{T}$. The integer v then is increased by a multiple of the product of all primes from the trial base until v is a prime. Reference [3] applies techniques from [1], which were originally designated for the shared generation of RSA keys. The authors of [3] yet point out that they do not aim at good performance, and in fact, for many applications performance aspects are crucial.

Reference [7] proposes a prime generation algorithm that uses four integer parameters P (large odd number, e.g. the product of the first $N-1$ odd primes), w, and $b_{min} \leq b_{max}$. The algorithm starts with a randomly selected integer $y_0 \in \mathbb{Z}_P^*$ and generates prime candidates $v_j = v_j(y_j) = (v+b)P+y_j$ or $v_j(y_j) = (v+b+1)P - y_j$, respectively, for some integer $b \in [b_{min}, b_{max}]$ and $y_j = 2^j y_0 (\mathrm{mod}\ P)$ until a prime is found, or more precisely, until some v_j passes the primality tests. Reference [4] describes a (theoretical) side channel attack on a special case of this scheme (with $b_{min} = b_{max} = 0$ and known (P, w)) on a careless implementation that reveals the parity of the prime candidates v_j. This attack is successful in about 0.1% of the trials, and [4] also suggests countermeasures. For further information on primality testing we refer the interested reader to the relevant literature.

7   Conclusion

This paper presents an elementary side-channel attack which focuses on the RSA key generation. The attack works under weak assumptions on the side-channel leakage, and practical experiments show that these assumption may be realistic. If the attack is known it can be prevented effectively.

Reference [4] and the above results demonstatrate that the RSA key generation process may be vulnerable to side-channel attacks. It appears to be reasonable to analyse implementations of various key generation algorithms in this regard. New attacks (possibly in combination with weaker assumptions than Remark 1) and in particular effective countermeasures may be detected.

References

1. Boneh, D., Franklin, J.: Efficient Generation of Shared RSA keys. In: Kaliski Jr., B.S. (ed.) CRYPTO 1997. LNCS, vol. 1294, pp. 425–439. Springer, Heidelberg (1997)
2. Brandt, J., Damgard, I., Landrock, P.: Speeding up Prime Number Generation. In: Matsumoto, T., Imai, H., Rivest, R.L. (eds.) ASIACRYPT 1991. LNCS, vol. 739, pp. 440–449. Springer, Heidelberg (1993)
3. Chabanne, H., Dottax, E., Ramsamy, L.: Masked Prime Number Generation, http://www.cosic.esat.kuleuven.be/wissec2006/papers/29.pdf
4. Clavier, C., Coron, J.-S.: On the Implementation of a Fast Prime Generation Algorithm. In: Paillier, P., Verbauwhede, I. (eds.) CHES 2007. LNCS, vol. 4727, pp. 443–449. Springer, Heidelberg (2007)
5. Coppersmith, D.: Small Solutions to Polynomial Equations, and Low Exponent Vulnerabilities. J. Crypt. 10(4), 233–260 (1997)
6. Coron, J.S.: Finding Small Roots of Bivariate Integer Polynomial Equations: A Direct Approach. In: Menezes, A. (ed.) CRYPTO 2007. LNCS, vol. 4622, pp. 379–394. Springer, Heidelberg (2007)
7. Joye, M., Paillier, P.: Fast Generation of Prime Numbers on Portable Devices. An Update. In: Goubin, L., Matsui, M. (eds.) CHES 2006. LNCS, vol. 4249, pp. 160–173. Springer, Heidelberg (2006)
8. Kocher, P.C.: Timing Attacks on Implementations of Diffie-Hellman, RSA, DSS and Other Systems. In: Koblitz, N. (ed.) CRYPTO 1996. LNCS, vol. 1109, pp. 104–113. Springer, Heidelberg (1996)
9. Kocher, P.C., Jaffe, J., Jun, B.: Differential Power Analysis. In: Wiener, M. (ed.) CRYPTO 1999. LNCS, vol. 1666, pp. 388–397. Springer, Heidelberg (1999)
10. Magma, Computational Algebra Group, School of Mathematics and Statistics, University of Sydney
11. Menezes, A., van Oorschot, P.C., Vanstone, S.A.: Handbook of Applied Cryptography. CRC Press, Boca Raton (1997)
12. Schindler, W.: A Timing Attack against RSA with the Chinese Remainder Theorem. In: Paar, C., Koç, Ç.K. (eds.) CHES 2000. LNCS, vol. 1965, pp. 110–125. Springer, Heidelberg (2000)
13. Schindler, W.: A Combined Timing and Power Attack. In: Paillier, P., Naccache, D. (eds.) PKC 2002. LNCS, vol. 2274, pp. 263–279. Springer, Heidelberg (2002)

An Efficient Method for Random Delay Generation in Embedded Software

Jean-Sébastien Coron and Ilya Kizhvatov

Université du Luxembourg
6, rue Richard Coudenhove-Kalergi
L-1359 Luxembourg
{jean-sebastien.coron,ilya.kizhvatov}@uni.lu

Abstract. Random delays are a countermeasure against a range of side channel and fault attacks that is often implemented in embedded software. We propose a new method for generation of random delays and a criterion for measuring the efficiency of a random delay countermeasure. We implement this new method along with the existing ones on an 8-bit platform and mount practical side-channel attacks against the implementations. We show that the new method is significantly more secure in practice than the previously published solutions and also more lightweight.

Keywords: Side channel attacks, countermeasures, random delays.

1 Introduction

Insertion of random delays in the execution flow of a cryptographic algorithm is a simple yet rather effective countermeasure against side-channel and fault attacks. To our knowledge, random delays are widely used for protection of cryptographic implementations in embedded devices, mainly smart cards. It belongs to a group of *hiding* countermeasures, that introduce additional noise (either in time, amplitude or frequency domain) to the side channel leakage while not eliminating the informative signal itself. This is in contrary to *masking* countermeasures, that eliminate correlation between the side channel leakage and the sensitive data processed by an implementation.

Hiding countermeasures increase complexity of attacks while not rendering them completely impossible. They are not treated in academia as extensively as masking but are of great importance in industry. A mixture of multiple hiding and masking countermeasures would often be used in a real-life protected implementation to raise the complexity of attacks above the foreseen capabilities of an adversary.

There are two connected problems that arise in this field. The first one is to develop efficient countermeasures, and the second one is how to measure the efficiency of the countermeasures. In this paper we tackle both tasks for the case of the random delays.

C. Clavier and K. Gaj (Eds.): CHES 2009, LNCS 5747, pp. 156–170, 2009.

Random delays. Most side-channel and fault attacks require an adversary to know precisely when the target operations occur in the execution flow. This enables her to synchronize multiple traces at the event of interest as in the case of Differential Power Analysis (DPA) and to inject some disturbance into the computations at the right time as in the case of fault attacks. By introducing random delays into the execution flow the synchronization is broken, which increases the attack complexity. This can be done in hardware with the so called *Random Process Interrupts* (RPI) as well as in software by placing "dummy" cycles at some points of the program. We give preliminary information on software random delays in Sect. 2.

Related work. First detailed treatment of the countermeasure was done by Clavier *et al.* in [1]. They showed that the number of traces for a successful DPA attack against RPI grows quadratically or linearly with the variance of the delay (when integration is used). Mangard presented statistical analysis of random disarrangement effectiveness in [2]. Amiel *et al.* [3] performed practical evaluation of random delays as a protection against fault attacks.

To date, the only effort to *improve* the random delays countermeasure in software was published by Benoit and Tunstall in [4]. They suggested to modify the distribution of an individual independently generated random delay so that the variance of the sum increases and the mean, in turn, decreases. As a result, they achieve some improvement. We outline their method briefly here in Sect. 3.

Our Contribution. In this work, we propose a significantly more efficient algorithm for generating random delays in software (see Sect. 4). Our main idea is to generate random delays non-independently in order to obtain a much greater variance of the cumulative delay for the same mean.

We also introduce a method for estimating the efficiency of random delays based on the coefficient of variation (see Sects. 2 and 5). This method shows how much variance is introduced by the sum of the delays for a given performance overhead. We show that the plain uniform delays and the Benoit-Tunstall method [4] both have efficiency in $\Theta\left(1/\sqrt{N}\right)$ only, where N is the number of delays in the sum, whereas our method achieves $\Theta(1)$ efficiency with the growth of N. For example, compared to the plain uniform delays and to the Benoit-Tunstall method, for the sum of 10 delays our method is more than twice as efficient, and for the sum of 100 delays – over 6 times more efficient.

Finally, we implement our new method along with the previously known methods on an 8-bit Atmel AVR microcontroller and demonstrate by mounting practical side-channel attacks that it is indeed more efficient and secure (see Sect. 6). It is also more lightweight in terms of implementation.

2   Software Random Delays and Their Efficiency

A common way of implementing random delays in software is placing loops of "dummy" operations (like NOP instructions) at some points of the program. The number of loop iterations varies depending on the delay value.

A straightforward method is to generate individual delays independently with durations uniformly distributed in the interval $[0, a]$ for some $a \in \mathbb{N}$. We refer to this method as *plain uniform delays*. It is easily implementable in cryptographic devices as most of them have a hardware random number generator (RNG) on board.

In [1] and [2] it was shown that the complexity of a DPA attack (expressed as the number of power consumption traces required) grows quadratically or linearly (in case integration techniques are used) with the standard deviation of the trace displacement in the attacked point. That is why we are interested in making the variance of random delays as large as possible.

Here are our preliminary assumptions about the attacker's capabilities.

1. An attacker knows the times when the cryptographic algorithm execution starts and ends. This is commonly possible by monitoring I/O operations of a device, or operations like EEPROM access.
2. It is harder for an attacker to eliminate multiple random delays than a few ones.
3. The method of delay generation and its parameters are known to an attacker.

Note that it could be possible to place two sufficiently large and uniformly distributed delays in the beginning and in the end of the execution. That would make each point in the trace uniformly distributed over time when looking from the start of from the end, which is actually the worst case for an attacker. Unfortunately, in this case it would be relatively easy to synchronize the traces with the the the help of cross-correlation (see [5] for an example). So we assume that in this case resynchronization of traces *can* be performed by an attacker. Therefore, we want to break the trace with relatively short (to keep performance) random delays in multiple places.

It can be still possible to detect delays produced by means of "dummy" loops in a side-channel trace because of a regular instruction pattern. To partially hinder this, "dummy" random data may be processed within a loop. We do not address this issue in this paper, just following the simple (but natural) assumption 2.

So an attacker will typically face the sum of several random delays. Following the Central Limit Theorem, the distribution of the sum of N *independent* (and *not necessarily uniform*) delays converges to normal with mean $N\mu_d$ and variance $N\sigma_d^2$, where μ_d and σ_d^2 are correspondingly the mean and the variance of the duration of an individual random delay. In other words, the distribution of the sum of independent delays depends only on the mean and the variance of individual delays but *not* on their particular distribution.

With all the above in mind, we adhere to the following criteria for random delay generation.

1. The sum of random delays from start or end to some point within the execution should have the greatest possible variance.
2. The performance overhead should be possibly minimal.

When estimating efficiency of random delay generation, one might be interested what performance overhead is required to achieve the given variation of the sum of N delays. Performance overhead can be naturally measured as the mean μ of this sum. We suggest to estimate efficiency of random delay generation methods in terms of the coefficient of variation σ/μ, where σ is the standard deviation for the sum of N random delays. The greater this efficiency ratio σ/μ, the more efficient the method is.

3   Method of Benoit and Tunstall

In [4], Benoit and Tunstall propose a way to improve the efficiency of the random delays countermeasure. Their aim is to increase the variance and decrease the mean of the sum of random delays while not spoiling the distribution of an individual random delay. To achieve this aim, the authors modify the distribution of an independently generated individual delay from the uniform to a pit-shaped one (see Figure 1). This increases the variance of the individual delay. Furthermore, some asymmetry is introduced to the pit in order to decrease the mean of an individual delay.

The delays are generated independently, so if an individual delay has mean μ_{BT} and variance σ^2_{BT}, the distribution of the sum of N delays converges to normal (as in the case of plain uniform delays) with mean $N\mu_{\mathrm{BT}}$ and variance $N\sigma^2_{\mathrm{BT}}$.

The authors estimate efficiency of their method by comparing it to plain uniform random delays. In an example, they report an increase of the variance by 33% along with a decrease of the mean by 20%. Distributions for a single delay and for the sum of 10 delays (for the parameters from the example mentioned above, see [4]) are shown in Figure 1 in comparison to plain uniform delays.

We note that the authors also pursued an additional criterion for the difficulty of deriving the distribution of the random delay. But it seems reasonable to consider this distribution to be known to an adversary, at least if the method is published.

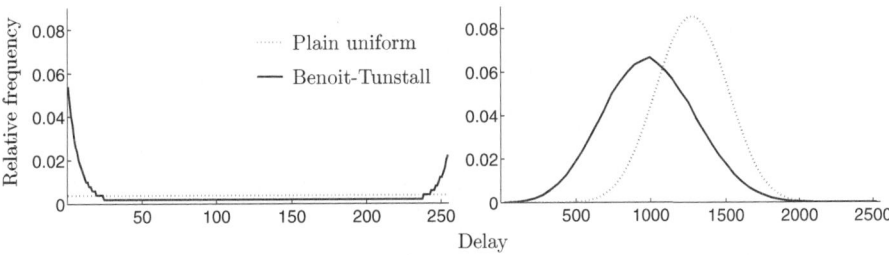

Fig. 1. Distribution for the method of Benoit and Tunstall [4] compared to plain uniform delays: 1 delay (left) and sum of 10 delays (right)

4   Our New Method: Floating Mean

In this section we present our new method for random delay generation in software. The main idea of the method is to generate random delays non-independently. This significantly improves the variance of the cumulative delay and the method is also more efficient compared to [4] and to plain uniform random delays.

By $x \sim \mathcal{DU}[y, z]$ we will denote a random variable x following discrete uniform distribution on $[y, z]$, $y, z \in \mathbb{Z}$, $y < z$.

Our method is as follows. First, we fix some $a \in \mathbb{N}$ which is the maximum delay length.[1] Additionally, we fix another parameter $b \in \mathbb{N}$, $b \leq a$. These implementation parameters a and b are fixed in an implementation and do not change between different executions of an algorithm under protection.

Now, in each execution, we first produce a value $m \in \mathbb{N}$ randomly uniformly on $[0, a - b]$, and then generate individual delays independently and uniformly on $[m, m + b]$. In other words, within any given execution individual random delays have a fixed mean $m + b/2$. But this mean varies from execution to execution, hence our naming of the method.

The resulting histograms in comparison to plain uniform delays are depicted in Figure 2. This figure also shows how the properties of the method vary dependent on the ratio b/a of the parameters of the method, that can take possible values between 0 and 1.

In fact, Floating mean is a pure trade-off between the quality of the distribution of single delay within a trace and that of the sum of the delays. When b/a is small (like the case $b = 50$, $a = 255$, $b/a \approx 0.2$ in Figure 2), the distribution of an individual delay within a trace has a comparatively small variance, but the variance of a single delay across traces and of the sum of the delays is large. When b/a is large (like the case $b = 200$, $a = 255$, $b/a \approx 0.8$ in Figure 2), the distribution of an individual delay within a trace has large variance, but the distribution of the sum of the delays converges to normal. The extreme case

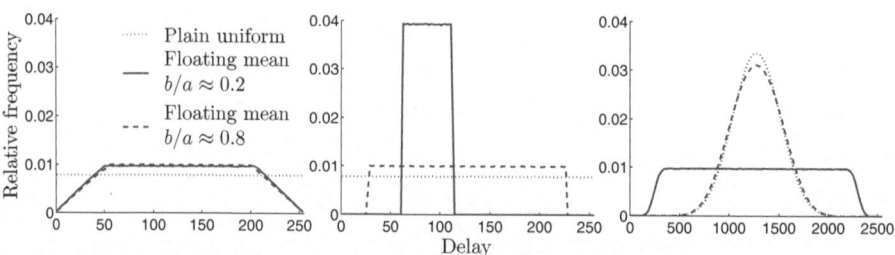

Fig. 2. Distribution for the Floating mean method with different b/a ratio compared to plain uniform delays: histogram for 1 delay (left), for 1 delay within a single trace, *i.e.* for some fixed m (center) and for the sum of 10 delays (right), $a = 255$

---

[1] We consider a and other parameters below to be integers as in an embedded device integer arithmetic would be the only option when generating delays.

$b/a = 0$ just means that within an execution all delays have same length m, while the distribution of the sum of N delays is uniform on the N-multiples in $[0, aN]$. In the other extreme case, $b/a = 1$, the methods simply converges to plain uniform delays with each delay generated uniformly on $[0, a]$.

To calculate the parameters of the distribution of the sum S_N of N delays, we represent an individual delay as a random variable $d_i = m + v_i$, where $m \sim \mathcal{DU}[0, a - b]$ and $v_i \sim \mathcal{DU}[0, b]$ for $i = 1, 2, \ldots N$ are independent random variables. The sum is then expressed as

$$S_N = \sum_{i=1}^{N} d_i = Nm + \sum_{i=1}^{N} v_i .$$

For the mean, we have

$$E(S_N) = E(Nm) + E\left(\sum_{i=1}^{N} v_i\right) = N \cdot \frac{a - b}{2} + N \cdot \frac{b}{2} = \frac{Na}{2} .$$

For the variance, since m and v_i are independent, all v_i are identically distributed and

$$\mathrm{Var}(m) = \frac{(a - b + 1)^2 - 1}{12} , \quad \mathrm{Var}(v_i) = \frac{(b + 1)^2 - 1}{12} , \quad i = 1, 2, \ldots, N$$

we have

$$\mathrm{Var}(S_N) = \mathrm{Var}\left(Nm + \sum_{i=1}^{N} v_i\right) = N^2 \cdot \mathrm{Var}(m) + N \cdot \mathrm{Var}(v_1)$$

$$= N^2 \cdot \frac{(a - b + 1)^2 - 1}{12} + N \cdot \frac{b^2 + 2b}{12} .$$

So, the variance of the sum of N delays is in $\Theta\left(N^2\right)$, in comparison to plain uniform delays and the method of [4] that both have variances in $\Theta\left(N\right)$. This is because we generate random delays non-independently; namely in our solution the lengths of the individual random delays are correlated: they are short if m is small, or they are longer if m is larger. This enables us to get a much larger variance than if the delays were generated independently, as in the plain uniform method and the method of [4].

At the same time, if we look at the delays within a single execution and thus under fixed m, the mean for the sum of N delays becomes $N(m + b/2)$. This implies that the cumulative delay for a given execution and therefore the length of the execution depends on m. An adversary can thus accept only the short traces, as they have short individual delays, and reject the long ones; this can lower the complexity of the attack.

In order to relieve an adversary of such a benefit, we can generate the first half of random delays (in the first half of the execution) uniformly on $[m, m+b]$ (that is, with mean $m + b/2$), and the second half of delays – uniformly on

$[a - m - b, a - m]$ (that is, with mean $a - m - b/2$). In this way, the distribution of the sum of all the $N = 2M$ delays for a given execution is independent of m (the mean is $aN/2$ and the variance is $N(b^2 + 2b)/12$). So an adversary cannot gain any additional information about the distribution of the delays within an execution by observing its length. Still, the variance of the sum of $L < M$ delays from start or end to some point up to the middle of the execution is in $\Theta(L^2)$.

Floating mean method is described in Algorithm 1.1. It is easily implementable in software on a constrained platform that has a built-in RNG producing uniformly distributed bytes since parameters a and b can be naturally chosen so that $a - b = 2^s - 1$ and $b = 2^t - 1$, where $s, t \in \mathbb{N}$ and $2^s + 2^t < 2^n + 2$ for an n-bit target microcontroller. Random integers in the range $[0, 2^s - 1]$ and $[0, 2^t - 1]$ can be obtained by a simple bit-wise AND with bit masks $2^s - 1$ and $2^t - 1$ correspondingly. The method requires no additional memory, as opposed to [4]. We are describing our implementation of Floating mean in Sect. 6 and Appendix B.

Algorithm 1.1 Floating mean method for generation of random delays

---

Input: $a, b \in \mathbb{N}, b \le a, N = 2M \in \mathbb{N}$
   $m \leftarrow \mathcal{DU}[0, a - b]$
   for $i = 1$ to $N/2$ **do**
      $d_i \leftarrow m + \mathcal{DU}[0, b]$
   end for
   for $i = N/2 + 1$ to N **do**
      $d_i \leftarrow a - m - \mathcal{DU}[0, b]$
   end for
Output: $d_1, d_2, \ldots, d_N$

4.1   A Method That Does Not Quite Work: Floating Ceiling

In this section we present another method that is based on the same principle as the previous method: generate random delays non-independently to improve the variance of the cumulative sum. However we explain below why this method does not quite work.

The method is as follows. First, we fix some implementation parameter $a \in \mathbb{N}$ which determines the maximum length of an individual random delay. Now, prior to generation of the first delay in each execution of the algorithm we produce a value $c \in \mathbb{N}$ randomly uniformly on $[1, a - 1]$. After that, within the execution we generate individual delays randomly uniformly on $[0, c]$. Loosely speaking, c is the "ceiling" for the length of the random delays that varies from execution to execution. The resulting distributions are shown in Figure 6 in Appendix A. For the sum S_N of N delays we obtain the following mean and variance (see Appendix A):

$$E(S_N) = N \cdot \frac{a}{4}, \quad \mathrm{Var}(S_N) = N^2 \cdot \frac{a^2 - 2a}{48} + N \cdot \frac{2a^2 + 5a}{72}.$$

As in the Floating mean method, here the variance of the sum of the delays is also in N^2 since we generate delays non-independently. However we have the same undesired property as in the Floating mean method without the two halves. Namely the mean length of the cumulative delay within a single trace (*i.e.* with c fixed) is $Nc/2$. So an adversary can judge the mean length of the delays within an execution by the total length of the execution that he can definitely measure.

If we try to fix this in the same manner by generating the first half of random delays uniformly on $[0, c]$ and the second half – uniformly on $[0, a - c]$, the mean of the sum of all $N = 2M$ random delays within an execution becomes constant and equal to $Na/4$. However, one can see that for a given execution the distribution of the sum (and in particular its variance) still depends on c; therefore an adversary could still derive information from c in a given execution by measuring its length. For example, since the variance of the sum is maximal when $c = 0$ or $c = a$, an adversary could select those executions in which a large deviation from the mean is observed; this would likely correspond to small c or large c; then the adversary would concentrate his attack on those executions only.

The complete Floating ceiling method is defined by Algorithm 1.2. It does not require any tables to be stored in memory, as opposed to [4]. However, its implementation requires random integers on $[0, c]$ for arbitrary positive integer c. This can be inconvenient on constrained platforms as this requires to omit RNG outputs larger than c, thus leading to a performance decrease.

Algorithm 1.2 Floating ceiling method for generation of random delays

---

Input: $a \in \mathbb{N}, N = 2M \in \mathbb{N}$
  $c \leftarrow \mathcal{DU}[1, a - 1]$
  for $i = 1$ to $N/2$ **do**
    $d_i \leftarrow \mathcal{DU}[0, c]$
  end for
  for $i = N/2 + 1$ to N **do**
    $d_i \leftarrow \mathcal{DU}[0, a - c]$
  end for
Output: $d_1, d_2, \ldots, d_N$

---

5 Comparing Efficiency

In this section we compare our new method with the existing ones based on the efficiency metrics σ/μ suggested in Sect. 2.

Efficiency ratios σ/μ for the sum of N delays for the new method and for the existing ones are given in Table 1. Note that we are mostly interested in the coefficient of variation somewhere around the middle of the trace.

In Figure 3, the efficiency ratio σ/μ for the sum of N delays for different methods is depicted against N. For all methods, we have considered the maximum delay length $a = 255$. The mean $\mu_{\mathrm{BT}} = 99$ and the variance $\sigma^2_{\mathrm{BT}} = 9281$

Table 1. Efficiency ratios σ/μ for different random delay generation methods

Plain uniform	Benoit-Tunstall	Floating mean
$\frac{1}{\sqrt{3N}} = \Theta\left(\frac{1}{\sqrt{N}}\right)$	$\frac{\sigma_{\mathrm{BT}}}{\mu_{\mathrm{BT}}} \cdot \frac{1}{\sqrt{N}} = \Theta\left(\frac{1}{\sqrt{N}}\right)$	$\frac{\sqrt{N((a-b+1)^2-1)+b^2+2b}}{a\sqrt{3N}} = \Theta\left(1\right)$

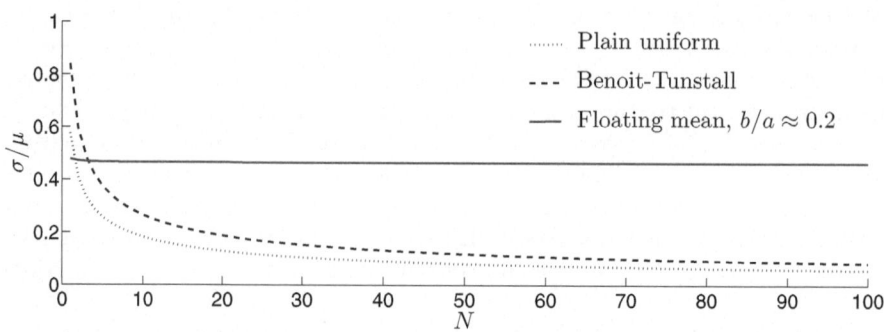

Fig. 3. Efficiency of the random delay generation algorithms in terms of the efficiency ratio σ/μ against the number of delays N in a sum

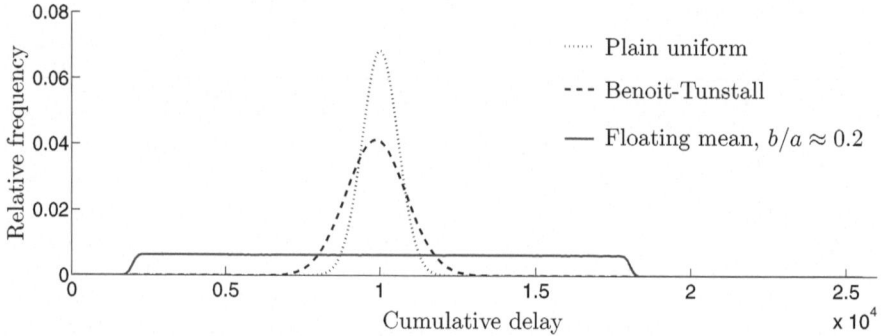

Fig. 4. Distributions of the sum of 100 delays for random delay generation algorithms, for the case of equal means

of an individual delay in the Benoit-Tunstall method was estimated empirically for the parameters used as an example in [4].

It can be seen that our new Floating mean method presented in Sect. 4 is more efficient compared to the previously published ones. Figure 4 further illustrates the difference, presenting the distributions of the sum of 100 random delays for different methods with the parameters that yield the same performance penalty, *i.e.* the same mean of the sum. We see that for the same average performance penalty, our method has a much larger variance.

In the case of independently generated individual delays the efficiency ratio σ/μ for the sum of any N delays is $\sigma_d/\mu_d \cdot 1/\sqrt{N}$, where σ_d and μ_d are the standard deviation and the mean of an individual delay. One can increase σ_d/μ_d ratio to improve efficiency, which was done in [4], but with an increase of the number of delays in the sum the efficiency of such methods decreases like $\Theta\left(1/\sqrt{N}\right)$, asymptotically tending to 0. Whereas for our method the efficiency is in $\Theta(1)$, so with an increase of the number of delays it tends to a nonzero constant value. This can be seen in Figure 3.

Thus, when implementing our method, one can benefit from using shorter but more frequent delays, as this does not cause the decrease in efficiency. This is an advantage as frequent short delays may be harder to eliminate than the long but less frequent ones.

6   Implementation and Resistance to Practical Attacks

Here we present comparison between the practical implementations of plain uniform delays, table method of Benoit and Tunstall [4] and the new Floating mean method by mounting Correlation Power Analysis (CPA) attack [6] against them.

We have implemented the methods on an 8-bit Atmel AVR microcontroller. Each delay is a multiple of 3 processor cycles (this granularity cannot be further reduced for this platform). Further details on our implementation are presented in Appendix B.

Random delays were introduced into AES-128 encryption. We put 10 delays per each round: before AddRoundKey, 4 per SubBytes+ShiftRows, before each MixColumn and after MixColumns. 3 "dummy" AES rounds that also incorporated random delays were added in the beginning and in the end of the encryption. Thus, the first SubByte operation of the first encryption round, which is the target for our attacks, is separated from the start of the execution, which is in turn our synchronization point, by 32 random delays.

The parameters of the methods were chosen to ensure (nearly) the same performance overhead across the methods. They were made sufficiently small to enable attacks with a reasonable number of traces. For the Floating mean method we used parameters $a = 18$ and $b = 3$. For the table method of Benoit and Tunstall, the p.d.f. of the pit-shaped distribution was generated using the formula $y = \lceil ak^x + bk^{N-x} \rceil$ from [4] with the parameters $N = 19$, $a = 40$, $b = 34$ and $k = 0.7$. These parameters were chosen so that they lead to the table of 256 entries with the inverse c.d.f. of the distribution. We use this table to produce delay values on $[0, 19]$ by indexing it with a random byte. For the plain uniform delays, the individual delay values were generated on $[0, 16]$. On our 8-bit platform we can efficiently produce random integers only on $[0, 2^i - 1]$ for $i = 1, 2, ..., 8$ (see Sect. 4), so we could not make the performance overhead for this method to be exactly the same as for the other methods.

We mounted CPA attack [6] in the Hamming weight power consumption model against the first AES key byte for each of the methods, first SubByte operation being the attack target. As a reference benchmark for our measurement conditions we performed CPA attack against the implementation without random delays. For implementations with random delays, we used power consumption

Table 2. Practical effect of the sum of 32 delays for different methods

	No delays	Plain uniform	Benoit-Tunstall [4]	Floating mean
μ, cycles	0	720	860	862
σ, cycles	0	79	129	442
σ/μ	–	0.11	0.15	0.51
CPA, traces	50	2500	7000	45000

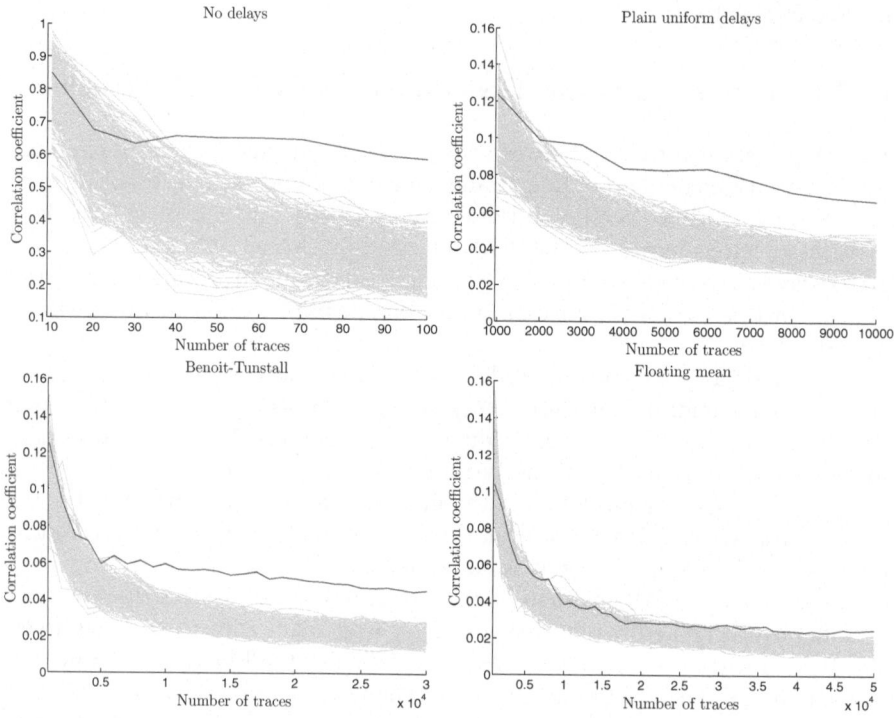

Fig. 5. CPA against random delays: correlation coefficient for all key byte guesses against the number of power consumption traces. The trace for the correct guess is highlighted.

traces *as is* without any alignment or integration to make a consistent comparison. Table 2 presents the number of traces required for a successful (with the 1st-order success rate close to 1) key byte recovery along with estimated mean μ, standard deviation σ and efficiency ratio σ/μ of the sum of 32 delays for each of the methods. Figure 5 presents the CPA attack results.

It can be seen that the Floating mean method is more secure in practice already for small delay durations and for a small number of delays. To break our implementation, we require 45000 traces for Floating mean and 7000 traces for

Benoit-Tunstall. That is, for the *same performance penalty* the Floating mean method requires 6 times more curves to be broken. This ratio will increase with the number of delays. However, our method is more efficient already for less than 10 delays in the sum, as can be seen from Figure 4. This is important for symmetric algorithm implementations that are relatively short. For inherently long implementations of public key algorithms the number of delays in the sum will be naturally large.

7   Conclusion

We proposed a new method for random delay generation in embedded software – the Floating mean method – and introduced a way to estimate efficiency of the random delays countermeasure. We presented the lightweight implementation of our method for protection of AES encryption on an 8-bit platform. We mounted practical CPA attacks showing that for the same level of performance the implementation of the new method requires 6 times more curves to be broken compared to the method of Benoit and Tunstall [4]. Thus, our method is significantly more efficient and secure.

References

1. Clavier, C., Coron, J.-S., Dabbous, N.: Differential power analysis in the presence of hardware countermeasures. In: Paar, C., Koç, Ç.K. (eds.) CHES 2000. LNCS, vol. 1965, pp. 252–263. Springer, Heidelberg (2000)
2. Mangard, S.: Hardware countermeasures against DPA—a statistical analysis of their effectiveness. In: Okamoto, T. (ed.) CT-RSA 2004. LNCS, vol. 2964, pp. 222–235. Springer, Heidelberg (2004)
3. Amiel, F., Clavier, C., Tunstall, M.: Fault analysis of DPA-resistant algorithms. In: Breveglieri, L., Koren, I., Naccache, D., Seifert, J.-P. (eds.) FDTC 2006. LNCS, vol. 4236, pp. 223–236. Springer, Heidelberg (2006)
4. Tunstall, M., Benoit, O.: Efficient use of random delays in embedded software. In: Sauveron, D., Markantonakis, K., Bilas, A., Quisquater, J.-J. (eds.) WISTP 2007. LNCS, vol. 4462, pp. 27–38. Springer, Heidelberg (2007)
5. Nagashima, S., Homma, N., Imai, Y., Aoki, T., Satoh, A.: DPA using phase-based waveform matching against random-delay countermeasure. In: IEEE International Symposium on Circuits and Systems—ISCAS 2007, May 2007, pp. 1807–1810 (2007)
6. Brier, E., Clavier, C., Benoit, O.: Correlation power analysis with a leakage model. In: Joye, M., Quisquater, J.-J. (eds.) CHES 2004. LNCS, vol. 3156, pp. 135–152. Springer, Heidelberg (2004)

A   Distribution for the Floating Ceiling Method

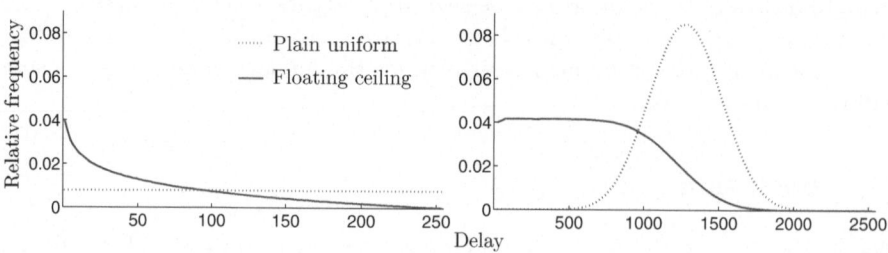

Fig. 6. Distribution for the Floating ceiling compared to plain uniform delays: 1 delay (left) and sum of 10 delays (right) for $a = 255$

To calculate the mean and the variance for the Floating ceiling method, we represent an i-th individual delay as a random variable $d_i \sim \mathcal{DU}[0, c]$, $i = 1, 2, \ldots N$, where in turn $c \sim \mathcal{DU}[1, a - 1]$. The sum of N delays is expressed as

$$S_N = \sum_{i=1}^{N} d_i \ .$$

For the mean, since d_i are identically distributed, we have

$$E(S_N) = NE(d_1) = N \sum_{c=1}^{a-1} \frac{1}{a-1} E\left(d_1 | c\right) = N \cdot \frac{1}{a-1} \sum_{c=1}^{a-1} \frac{c}{2} = N \cdot \frac{a}{4} \ .$$

For the variance, in turn,

$$\mathrm{Var}(S_N) = E\left(S_N^2\right) - \left(E\left(S_N\right)\right)^2 \ .$$

Again, since d_i are identically distributed, we have

$$E\left(S_N^2\right) = E\left(\left(\sum_{i=1}^{N} d_i\right)^2\right) = E\left(\sum_{i=1}^{N} d_i^2\right) + 2E\left(d_1 d_2 + d_1 d_3 + \ldots + d_{N-1} d_N\right)$$

$$= NE\left(d_1^2\right) + \binom{N}{2} \cdot 2E\left(d_1 d_2\right) \ .$$

Now, having

$$E\left(d_1^2\right) = \sum_{c=1}^{a-1} \frac{1}{a-1} E\left(d_i^2 \mid c\right) = \frac{1}{a-1} \sum_{c=1}^{a-1} \frac{1}{c+1} \sum_{j=0}^{c} j^2 = \frac{4a^2 + a}{36}$$

and (since $d_i | c$ and $d_j | c$ are independent for $i \neq j$ and identically distributed)

$$E\left(d_1 d_2\right) = \sum_{c=1}^{a-1} \frac{1}{a-1} E\left(d_1 d_2 \mid c\right) = \frac{1}{a-1} \sum_{c=1}^{a-1}\left(E(d_1 \mid c)\right)^2 = \frac{2a^2 - a}{24},$$

we finally obtain

$$\mathrm{Var}(S_N) = N^2 \cdot \frac{a^2 - 2a}{48} + N \cdot \frac{2a^2 + 5a}{72}.$$

B   Implementation of Random Delays for an 8-bit AVR Platform

Here we present the reference implementation of several delay generation methods in the 8-bit AVR assembly language. Throughout the code, the following registers are reserved: RND for obtaining the random delay duration, FM for storing the value of m used in Floating mean during the execution, MASK for the bit mask that truncates random values to the desired length.

Common ATmega16 microcontroller that we used does not have a built-in RNG. Hence, we have simulated the RNG by pre-loading a pool of pseudorandom numbers to microcontroller's SRAM from the host PC prior to each execution and pointing the X register at the beginning of the pool. Random numbers are then loaded successively from SRAM to RND register by calling the randombyte function:

```
randombyte:
    ld   RND, X+   ; X is the dedicated address register
    ret            ;   that is used only in this function
```

First, here is the basic delay generation routine. It produces delays of length $3 \cdot \mathrm{RND} + C$ cycles, where C is the constant overhead per delay. To reduce this overhead, the delay generation can be implemented in-line to avoid the cost of entering and leaving the function. The part of the code specific for delay generation methods is omitted and will be given below.

```
randomdelay:
    rcall randombyte   ; obtain a random byte in RND
    ;
    ; <place for method-specific code>
    ;
    tst   RND     ; mind balancing between zero and
    breq  zero    ;   non-zero delay values!
    nop
    nop
dummyloop:
    dec   RND
    brne  dummyloop
zero:
    ret
```

Here are specific code parts for delay value generation. For plain uniform delays, the code is just:

```
and   RND, MASK     ; truncate random value to the desired length
```

The code for the floating mean is one instruction longer (namely, addition of the value m).

```
and   RND, MASK     ; truncate random value to the desired length
add   RND, FM       ; add 'floating mean'
```

Floating mean also requires initialization (namely, generation of m) in the beginning of each execution:

```
rcall randombyte    ; obtain a random byte in RND
mov FM, RND
ldi MASK, 0x0f
and FM, MASK        ; trucate mean to the desired length
ldi MASK, 0x03      ; set mask for future use in individual delays
```

and "flipping" FM in the middle of the execution to make the total execution length independent of the value of m.

```
ldi MASK, 0x0f
sub MASK, FM
mov FM, MASK
ldi MASK, 0x03
```

Finally, for the method of Benoit and Tunstall, the delay value is generated as follows.

```
ldi ZH, high(bttable)
mov ZL, RND
ld RND, Z
```

Here bttable is the table of 256 byte entries with the c.d.f of the pit-shaped distribution that is pre-loaded into SRAM.

It can be seen that Floating mean is more "lightweight" in terms of both memory and code than the table method of Benoit and Tunstall.

Higher-Order Masking and Shuffling for Software Implementations of Block Ciphers

Matthieu Rivain[1,2], Emmanuel Prouff[1], and Julien Doget[1,3,4]

[1] Oberthur Technologies, France
[2] University of Luxembourg, Luxembourg
[3] Université catholique de Louvain, Belgium
[4] University of Paris 8, France
{m.rivain,e.prouff,j.doget}@oberthur.com

Abstract. Differential Power Analysis (DPA) is a powerful side channel key recovery attack that efficiently breaks block ciphers implementations. In software, two main techniques are usually applied to thwart them: masking and operations shuffling. To benefit from the advantages of the two techniques, recent works have proposed to combine them. However, the schemes which have been designed until now only provide limited resistance levels and some advanced DPA attacks have turned out to break them. In this paper, we investigate the combination of masking and shuffling. We moreover extend the approach with the use of higher-order masking and we show that it enables to significantly improve the security level of such a scheme. We first conduct a theoretical analysis in which the efficiency of advanced DPA attacks targeting masking and shuffling is quantified. Based on this analysis, we design a generic scheme combining higher-order masking and shuffling. This scheme is scalable and its security parameters can be chosen according to any desired resistance level. As an illustration, we apply it to protect a software implementation of AES for which we give several security/efficiency trade-offs.

1 Introduction

Side Channel Analysis (SCA in short) exploits information that leaks from physical implementations of cryptographic algorithms. This leakage (*e.g.* the power consumption or the electro-magnetic emanations) may indeed reveal information on the secret data manipulated by the implementation. Among SCA attacks, two classes may be distinguished. The set of so-called *Profiling SCA* corresponds to a powerful adversary who has a copy of the attacked device under control and who uses it to evaluate the distribution of the leakage according to the processed values. Once such an evaluation is obtained, a maximum likelihood approach is followed to recover the secret data manipulated by the attacked device. The second set of attacks is the set of so-called *Differential Power Analysis* (DPA) [11]. It corresponds to a more realistic (and much weaker) adversary than the one considered in Profiling SCA, since the adversary is only able to observe the device behavior and has no *a priori* knowledge of the implementation details.

C. Clavier and K. Gaj (Eds.): CHES 2009, LNCS 5747, pp. 171–188, 2009.

This paper only deals with the set of DPA as it includes a great majority of the attacks encountered *e.g.* by the smart card industry.

Block ciphers implementations are especially vulnerable to DPA attacks and research efforts have been stepped up to specify implementation schemes counteracting them. For software implementations, one identifies two main approaches: masking [2, 7] and shuffling [8]. However, some advanced DPA techniques exist that defeat these countermeasures [3, 15]. A natural approach to improve the DPA resistance is to mix masking and shuffling [8, 25, 26]. This approach seems promising since it enables to get the best of the two techniques. However, the schemes that have been proposed so far [8, 26] only focus on first-order masking which prevents them from reaching high resistance levels. This is all the more serious that advanced DPA attacks have turned out to be quite efficient in breaking them [25, 26].

In this paper, we conduct an analysis to quantify the efficiency of an attack that targets either a masked implementation or a shuffled implementation or a masked-and-shuffled implementation. Based on this analysis, we design a new scheme combining higher-order masking and shuffling to protect software implementations of block ciphers. This scheme is scalable and its parameters can be specified to achieve any desired resistance level. We apply it to protect a software implementation of AES and we show how to choose the scheme parameters to achieve a given security level with the minimum overhead.

2   Preliminaries

2.1   Masking and Shuffling Countermeasures

To protect cryptographic implementations against DPA, one must reduce the amount of information that leaks on sensitive intermediate variables during the processing. A variable is said to be *sensitive* if it is a function of the plaintext and a guessable part of the secret key (that is not constant with respect to the latter).

To thwart DPA attacks, countermeasures try to make leakages as independent as possible of sensitive variables. Nowadays, two main approaches are followed to achieve such a purpose in software: the *masking* and the *shuffling*. We briefly recall hereafter the two techniques.

The core idea behind masking is to randomly split every sensitive variable X into $d+1$ shares $M_0,..., M_d$ in such a way that the relation $M_0 \star ... \star M_d = X$ is satisfied for a group operation $\star$ (*e.g.* the x-or or the modular addition). Usually, $M_1,..., M_d$ (called *the masks*) are randomly picked up and M_0 (called *the masked variable*) is processed to satisfy $M_0 \star ... \star M_d = X$. The parameter d is usually called *the masking order*. When carefully implemented (namely when all the shares are processed at different times), d^{th}-order masking perfectly withstands any DPA exploiting less than $d+1$ leakage signals simultaneously. Although attacks exploiting $d+1$ leakages are always theoretically possible, in practical settings their complexity grows exponentially with d [2]. The design of efficient higher-order masking schemes for block ciphers is therefore of great interest.

However, even for small d, dealing with the propagation of the masks through the underlying scheme is an issue. For linear operations, efficient and simple solutions exist that induce an acceptable overhead irrespective of d. Actually, the issue is to protect the non-linear S-boxes computations. In the particular case $d = 1$, a straightforward solution called the *table re-computation* exists (see for instance [1, 14]). Straightforward generalizations of the method to higher orders d do not provide security *versus* higher-order DPA. Indeed, whatever the number of masks, an attack targeting two different masked input/output is always possible (see for instance [17]). To bypass this flaw, [23] suggests to re-compute a new table before every S-box computation. This solution is very costly in terms of timings and [5] shows the feasibility of third-order attacks, so the scheme is only secure for $d < 3$. An alternative solution for $d = 2$ has been proposed in [21] but the timing overhead is of the same order.

Shuffling consists in spreading the signal containing information about a sensitive variable X over t different signals $S_1, \ldots, S_t$ leaking at different times. This way, if the spread is uniform, then for every i the probability that S_i corresponds to the manipulation of X is $\frac{1}{t}$. As a consequence, the signal-to-noise ratio of the instantaneous leakage on X is reduced by a factor of t (see Sect. 3.2 for details). Applying shuffling is straightforward and does not relate to the nature (linear or non-linear) of the layer to protect. Moreover, shuffling is usually significantly less costly than higher-order masking when applied to non-linear layers.

Since higher-order masking is expensive and since first-order masking can be defeated with quite reasonable efforts [17], a natural idea is to use shuffling together with first-order masking. A few schemes have already been proposed in the literature [8, 26]. In [8], an 8-bit implementation of AES is protected using first-order masking and shuffling. The work in [26] extends this scheme to a 32-bit implementation with the possible use of instructions set extension. Furthermore, [26] proposes some advanced DPA attacks on such schemes whose practicability is demonstrated in [25]. These works show that combining first-order masking with shuffling is definitely not enough to provide a strong security level. A possible improvement is to involve higher-order masking. This raises two issues. First, a way to combine higher-order masking with shuffling must be defined (especially for S-boxes computations). Secondly, the security of such a scheme should be quantifiable. It would indeed be of particular interest to have a lower bound on the resistance of the overall implementation by choosing *a priori* the appropriate trade-off between masking and shuffling orders. In the rest of the paper, we address those two issues.

2.2   Notations and Leakage Model

We use the calligraphic letters, like $\mathcal{X}$, to denote finite sets (*e.g.* $\mathbb{F}_2^n$). The corresponding capital letter X is used to denote a random variable over $\mathcal{X}$, while the lowercase letter x - a particular element from $\mathcal{X}$. The expectation of X is denoted by $\mathrm{E}[X]$, its variance by $\mathrm{Var}[X]$ and its standard deviation by $\sigma[X]$. The correlation coefficient [27] between X and Y is denoted by $\rho[X, Y]$. It measures the linear interdependence between X and Y and is defined by:

$$\rho\,[X,Y] = \frac{\mathrm{Cov}\,[X,Y]}{\sigma\,[X]\,\sigma\,[Y]}\ , \tag{1}$$

where $\mathrm{Cov}\,[X,Y]$, called *covariance of X and Y*, equals $\mathrm{E}\,[(X-\mathrm{E}\,[X])(Y-\mathrm{E}\,[Y])]$ or $\mathrm{E}\,[XY] - \mathrm{E}\,[X]\,\mathrm{E}\,[Y]$ equivalently.

In the next sections, we investigate the security of the combination of masking and shuffling towards DPA. Our analysis is conducted in the Hamming weight leakage model that we formally define hereafter. This model is very common for the analysis of DPA attacks [9, 20, 26] and it has been practically validated several times [15, 17].

Definition 1 (Hamming weight model). *The leakage signal S_i produced by the processing of a variable D_i satisfies:*

$$S_i = \delta_i + \beta_i \cdot \mathrm{H}(D_i) + N_i\ , \tag{2}$$

where δ_i denotes a constant offset, β_i is a real value, $\mathrm{H}(\cdot)$ denotes the Hamming weight function and N_i denotes a noise with mean 0 and standard deviation σ.

When several leakage signals S_i's are jointly considered, we shall make three additional assumptions: (1) the constant β_i is the same for the different S_i's (without loss of generality, we consider $\beta_i = 1$), (2) noises N_i's are mutually independent and (3) the noise standard deviation σ is the same for the different N_i's.

3   Analysis of Advanced DPA Attacks against Masking and Shuffling

Higher-order DPA attacks aim at recovering information on a sensitive variable X by considering several non-simultaneous leakage signals. Let us denote by **S** the multivariate random variable corresponding to those signals. The attack starts by converting **S** into an univariate random variable by applying it a function g. Then, a *prediction function f* is defined according to some assumptions on the device leakage model. Eventually, every guess $\hat{X}$ on X is checked by estimating the correlation coefficient between the combined leakage signal $g(\mathbf{S})$ and the so-called *prediction* $f(\hat{X})$.

As argued in several works (see for instance [20, 12, 13, 23]), the absolute value of the correlation coefficient $\rho\,[f(X), g(\mathbf{S})]$ (corresponding to the correct key guess) is a sound estimator of the efficiency of a correlation based DPA characterized by the pair of functions (f, g). In [13, 24], it is even shown that the number of leakage measurements required for the attack to succeed can be approximated by $c \cdot \rho\,[f(X), g(\mathbf{S})]^{-2}$ where c is a constant depending on the number of key guesses and the required success rate. In the following, we exhibit in the Hamming weight model (see Sect. 2.2) explicit formulae of this coefficient for advanced DPA attacks where the sensitive variable is either (1) protected by (higher-order) masking, or (2) protected by shuffling or (3) protected with a combination of the two techniques.

3.1 Defeating Masking: Higher-Order DPA

When d^{th}-order masking is used, any sensitive variable X is split into $d+1$ shares $X \oplus \mathbf{M}$, M_1, ..., M_d, where $\mathbf{M}$ denotes the sum $\bigoplus_i M_i$. In the following, we shall denote $X \oplus \mathbf{M}$ by M_0. The processing of each share M_i respectively results in a leakage signal S_i. Since the M_i's are assumed to be mutually independent, every tuple of d signals or less among the S_i's is independent of X. Thus, to recover information about X, the joint distribution of all the $d+1$ signals must be considered. Higher-order DPA consists in combining the $d+1$ leakage signals by the mean of a so-called *combining function* $\mathsf{C}(\cdot, \cdots, \cdot)$. This enables the construction of a signal that is correlated to the sensitive variable X.

Several combining functions have been proposed in the literature. Two of them are commonly used: the *product combining* [2] which consists in multiplying the different signals and the *absolute difference combining* [15] which computes the absolute value of the difference between two signals. As noted in [5, Sect. 1], the latter can be extended to higher orders by induction. Other combining functions have been proposed in [9, 16]. In a recent paper [20], the different combining functions are compared for second-order DPA in the Hamming weight model. An improvement of the product combining called *normalized product combining* is proposed and it is shown to be more efficient than the other combining functions[1]. In this paper, we therefore consider the normalized product combining generalized to higher orders:

$$\mathsf{C}\left(S_0, S_1 \cdots, S_d\right) = \prod_{i=0}^{d} \left(S_i - \mathrm{E}\left[S_i\right]\right) \ . \tag{3}$$

We shall denote by $C_d(X)$ the combined leakage signal $\mathsf{C}\left(S_0, S_1 \cdots, S_d\right)$ where the S_i's correspond to the processing of the shares $X \oplus \mathbf{M}$, M_1, ..., M_d in the Hamming weight model. The following lemma gives the expectation of $C_d(X)$ given $X = x$ for every $x \in \mathbb{F}_2^n$. The proof is given in the extended version of this paper [22].

Lemma 1. *Let $x \in \mathbb{F}_2^n$, then the expectation of $C_d(x)$ satisfies:*

$$\mathrm{E}\left[C_d(x)\right] = \left(-\frac{1}{2}\right)^d \left(\mathrm{H}(x) - \frac{n}{2}\right) \ . \tag{4}$$

Lemma 1 shows that the expectation of $C_d(x)$ is an affine function of the Hamming weight of x. According to the analysis in [20], this implies that the Hamming weight of X maximizes the correlation. For the reasons given in [20], this function can therefore be considered as an optimal prediction for $C_d(X)$. Hence, the HO-DPA we focus here consists in estimating the correlation between the Hamming weight of the target variable $\mathrm{H}(X)$ and the combined leakage $C_d(X)$. The next proposition provides the exact value of this correlation. The proof is given in the extended version of this paper [22].

---

[1] This assertion is true while considering a noisy model. In a fully idealized model, other combining may provide better results (see [20]).

Proposition 1. *Let X be a random variable uniformly distributed over $\mathbb{F}_2^n$. The correlation between $\mathrm{H}(X)$ and $C_d(X)$ satisfies:*

$$\rho\left[\mathrm{H}(X), C_d(X)\right] = (-1)^d \frac{\sqrt{n}}{(n + 4\sigma^2)^{\frac{d+1}{2}}} \ . \tag{5}$$

Notation. The correlation coefficient in (5) shall be referred as $\rho(n, d, \sigma)$.

3.2 Defeating Shuffling: Integrated DPA

When shuffling is used, the signal containing information about the sensitive variable X is randomly spread over t different signals S_1, ..., S_t. As a result, the correlation between the prediction and one of these signals is reduced by a factor t compared to the correlation without shuffling. In [3], an *integrated DPA attack* (also called *windowing attack*) is proposed for this issue. The principle is to add the t signals all together to obtain an *integrated signal*. The correlation is then computed between the prediction and the integrated signal. The resulting correlation is reduced by a factor $\sqrt{t}$ instead of t without integration. This is formalized in the next proposition.

Proposition 2. *Let $(S_i)_{1 \leqslant i \leqslant t}$ be t random variables identically distributed and mutually independent. Let Y denote a signal S_j's whose index j is a random variable uniformly distributed over $\{1, \cdots, t\}$. Let X be a random variable that is correlated to Y and that is independent of the remaining S_i's. For every measurable function f, the correlation between $f(X)$ and $S_1 + \cdots + S_t$ satisfies:*

$$\rho\left[f(X), S_1 + \cdots + S_t\right] = \frac{1}{\sqrt{t}} \rho\left[f(X), Y\right] \ . \tag{6}$$

Proof. On one hand we have $\mathrm{Cov}\left[f(X), S_1 + \cdots + S_t\right] = \mathrm{Cov}\left[f(X), Y\right]$ and on the other hand we have $\sigma\left[S_1 + \cdots + S_t\right] = \sqrt{t}\ \sigma\left[Y\right]$. Relation (6) straightforwardly follows.   ◇

3.3 Defeating Combined Masking and Shuffling: Combined Higher-Order and Integrated DPA

When masking is combined with shuffling, any sensitive variable X is split into $d + 1$ shares $X \oplus \mathbf{M}$, M_1, ..., M_d whose manipulations are randomly spread over t different times yielding t different signals S_i. The $(d + 1)$-tuple of signals indices corresponding to the shares hence ranges over a subset I of the set of $(d + 1)$-combinations from $\{1, \cdots, t\}$. This subset depends on how the shuffling is performed (*e.g.* the shares may be independently shuffled or shuffled all together).

To bypass such a countermeasure, an adversary may combine integrated and higher-order DPA techniques. The most pertinent way to perform such a combined attack is to design a so-called *combined-and-integrated* signal by summing all the possible combinations of $d + 1$ signals among S_1, ..., S_t [25, 26]. That is, the combined-and-integrated signal, denoted $IC_{d,I}(X)$, is defined by:

$$IC_{d,I}(X) = \sum_{(i_0,\dots,i_d) \in I} \mathsf{C}(S_{i_0}, \cdots, S_{i_d}) . \tag{7}$$

We show in the following proposition that the correlation coefficients of the combined attacks relate on $\rho(n, d, \sigma)$. The proof is given in the extended version of this paper [22].

Proposition 3. *Let X, M_1, ..., M_d be a family of $d + 1$ n-bit random variables uniformly distributed and mutually independent. Let I be a set of $(d + 1)$-combinations from $\{1, \cdots, t\}$ and let $(i_0, \cdots, i_d)$ be a random vector uniformly distributed over I. Let $(D_i)_i$ be a family of random variables such that $(D_{i_0}, D_{i_i}, \cdots, D_{i_d}) = (X \oplus \bigoplus_i M_i, M_1, \cdots, M_d)$ and, for every $j \neq i_0, ..., i_d$, D_j is uniformly distributed and mutually independent of $(D_i)_{i \neq j}$. Let $(S_i)_i$ be a family of t signals following the Hamming weight model corresponding to the processing of the D_i's. Then we have:*

$$\rho\left[\mathrm{H}(X), IC_{d,I}(X)\right] = \frac{1}{\sqrt{\#I}} \rho(n, d, \sigma) .$$

4   A Generic Scheme Combining Higher-Order Masking and Shuffling

In this section, we propose a generic scheme to protect block cipher implementations by combining higher-order masking and shuffling. First we introduce the general block cipher model and then we describe the proposed scheme. Afterward, we investigate the possible attack paths and we deduce a strategy for choosing the scheme parameters (*i.e.* the masking and shuffling orders, see Sect. 4.2).

4.1   Block Cipher Model

A block cipher is parameterized by a *master key* and it transforms a plaintext block into a ciphertext block through the repetition of key-dependent *round transformations*. We denote by p, and we call *state*, the temporary value taken by the ciphertext during the algorithm. In practice, the cipher is *iterative*, which means that it applies several times the same round transformation φ to the state. This round transformation is parameterized by a *round key k* that is derived from the master key.

In our model, φ is composed of different operations: a key addition layer (by xor), a non-linear layer γ and a linear layer λ:

$$\varphi[k](p) = [\lambda \circ \gamma](p \oplus k) .$$

We assume that the non-linear layer applies the same non-linear transformation S, called *S-box*, on N independent n-bit parts p_i of the state:

$\gamma(p) = \big(S(p_1), \cdots, S(p_N)\big)$. For efficiency reasons, the S-box is usually implemented by a look-up table. The linear layer λ is composed of L linear operations λ_i that operate on L independent l-bit parts $p_{i(l)}$ of the state: $\lambda(p) = \big(\lambda_1(p_{1(l)}), \cdots, \lambda_L(p_{L(l)})\big)$. We also denote by $l' \leqslant l$ the minimum number of bits of a variable manipulated during the processing of λ_i. For instance, the Mix-Columns layer of AES applies to columns of $l = 32$ bits but it manipulates some elements of $l' = 8$ bits. We further assume that the λ_i's are sufficiently similar to be implemented by one *atomic operation* that is an operation which has the same execution flow whatever the index i.

Remark 1. Linear and non-linear layers may involve different state indexing. In AES for instance, the state is usually represented as a 4×4 matrix of bytes and the non-linear layer usually operates on its elements $p_1,...,p_{16}$ vertically (starting at the top) and from left to right. In this case, the operation λ_1 corresponding to the AES linear layer (that is composed of ShiftRows followed by MixColumns [6]) operates on $p_{1(32)} = (p_1, p_6, p_{11}, p_{16})$.

In the sequel, we shall consider that the key addition and the non-linear layer are merged in a *keyed substitution layer* that adds each key part k_i to the corresponding state part p_i before applying the S-box S.

4.2   Our Scheme

In this section, we describe a generic scheme to protect a round φ by combining higher-order masking and operations shuffling. Our scheme involves a dth-order masking for an arbitrarily chosen d. Namely, the state p is split into $d+1$ shares $m_0, ..., m_d$ satisfying:

$$m_0 \oplus \cdots \oplus m_d = p . \tag{8}$$

In practice, $m_1, ..., m_d$ are random masks and m_0 is the masked state defined according to (8). In the sequel, we shall denote by $(m_j)_i$ (resp. $(m_j)_{i(l)}$) the ith n-bit part (resp. the ith l-bit part) of a share m_j. At the beginning of the ciphering the masks are initialized to zero. Then, each time a part of a mask is used during the keyed substitution layer computation, it is refreshed with a new random value (see below). For the reasons given in Sect. 2.1, our scheme uses two different approaches to protect the keyed substitution layer and the linear layer. These are described hereafter.

Protecting the keyed substitution layer. To protect the keyed substitution layer, we use a single d'^{th}-order masked S-box (for some $d' \leqslant d$) to perform all the S-box computations. As explained in Sect. 2.1, such a method is vulnerable to a second-order DPA attack targeting two masked inputs/outputs. To deal with this issue, we make use of a high level of shuffling in order to render such an attack difficult and to keep an homogeneous security level (see Sect. 4.4).

The input of S is masked with d' masks $r_1, ..., r_{d'}$ and its output is masked with d' masks $s_1, ..., s_{d'}$. Namely, a masked S-box S^* is computed that is defined for every $x \in \{0, 1\}^n$ by:

$$S^*(x) = S\left(x \oplus \bigoplus_{j=1}^{d'} r_j\right) \oplus \bigoplus_{j=1}^{d'} s_j . \tag{9}$$

This masked S-box is then involved to perform all the S-box computations. Namely, when the S-box must be applied to a masked variable $(m_0)_i$, the d masks $(m_j)_i$ of this latter are replaced by the d' masks r_j which enables the application of S^*. The d' masks s_j of the obtained masked output are then switched for d new random masks $(m_j)_i$.

The high level shuffling is ensured by the addition of dummy operations. Namely, the S-box computation is performed t times: N times on a relevant part of the state and $t - N$ times on dummy data. For such a purpose, each share m_j is extended by a dummy part $(m_j)_{N+1}$ that is initialized by a random value at the beginning of the ciphering. The round key k is also extended by such a dummy part k_{N+1}. For each of the t S-box computations, the index i of the parts $(m_j)_i$ to process is read in a table T. This table of size t contains all the indices from 1 to N stored at random positions and its $t - N$ other elements equal $N + 1$. Thanks to this table, the S-box computation is performed once on every of the N relevant parts and $t - N$ times on the dummy parts. The following algorithm describes the whole protected keyed substitution layer computation.

Algorithm 1. Protected keyed substitution layer

INPUT: the shares $m_0, ..., m_d$ s.t. $\bigoplus m_i = p$, the round key $k = (k_1, \cdots, k_{N+1})$
OUTPUT: the shares $m_0, ..., m_d$ s.t. $\bigoplus m_i = \gamma(p \oplus k)$

1. **for** $i_T = 1$ **to** t

// Random index pick-up
2. $i \leftarrow T[i_T]$

// Masks conversion : $(m_0)_i \Leftarrow p_i \bigoplus_j r_j$
3.   **for** $j = 1$ **to** d' **do** $(m_0)_i \leftarrow ((m_0)_i \oplus r_j) \oplus (m_j)_i$
4.   **for** $j = d' + 1$ **to** d **do** $(m_0)_i \leftarrow (m_0)_i \oplus (m_j)_i$

// key addition and S-box computation: $(m_0)_i \Leftarrow S(p_i \oplus k_i) \oplus \bigoplus_j s_j$
5. $(m_0)_i \leftarrow S^*((m_0)_i \oplus k_i)$

// Masks generation and conversion: $(m_0)_i \Leftarrow S(p_i \oplus k_i) \oplus \bigoplus_j (m_j)_i$
6.   **for** $j = 1$ **to** d'
7. $(m_j)_i \leftarrow \mathsf{rand}()$
8. $(m_0)_i \leftarrow ((m_0)_i \oplus (m_j)_i) \oplus s_j$
9.   **for** $j = d' + 1$ **to** d
10. $(m_j)_i \leftarrow \mathsf{rand}()$
11. $(m_0)_i \leftarrow (m_0)_i \oplus (m_j)_i$

12. **return** $(m_0, \cdots, m_d)$

Remark 2. In Steps 3 and 8, we used round brackets to underline the order in which the masks are introduced. A new mask is always introduced before removing an old mask. Respecting this order is mandatory for the scheme security.

Masked S-box computation. The look-up table for S^* is computed dynamically at the beginning of the ciphering by performing d' table re-computations such as proposed in [23]. This method has been shown to be insecure for $d' > 2$, or for $d' > 3$ depending on the table re-computation algorithm [5, App. A]. We will therefore consider that one can compute a masked S-box S^* with $d' \leqslant 3$ only. The secure computation of a masked S-box with $d' > 3$ is left to further investigations.

Indices table computation. Several solutions exist in the literature to randomly generate indices permutation over a finite set [10, 18, 19]. Most of them can be slightly transformed to design tables T of size $t \geqslant N$ containing all the indices 1 to N in a random order and whose remaining cells are filled with $N+1$. However, few of those solutions are efficient when implemented in low resources devices. In our case, since t is likely to be much greater than N, we have a straightforward algorithm which tends to be very efficient for $t \gg N$. This algorithm is given in Appendix A (Algorithm 3).

Protecting the linear layer. The atomic operations λ_i are applied on each part $(m_j)_{i(l)}$ of each share m_j in a random order. For such a purpose a table T' is constructed at the beginning of the ciphering that is randomly filled with all the pairs of indices $(j, i) \in \{0, \cdots, d\} \times \{1, \cdots, L\}$. The linear layer is then implemented such as described by the following algorithm.

Algorithm 2. Protected linear layer

INPUT: the shares m_0, ..., m_d s.t. $\bigoplus m_i = p$
OUTPUT: the shares m_0, ..., m_d s.t. $\bigoplus m_i = \lambda(p)$

1. **for** $i_{T'} = 1$ **to** $(d+1) \cdot L$
2. $(j, i) \leftarrow T'[i_{T'}]$ // Random index look-up
3. $(m_j)_{i(l)} \leftarrow \lambda_i((m_j)_{i(l)})$ // Linear operation
4. **return** $(m_0, \cdots, m_d)$

Indices table computation. To implement the random generation of a permutation T' on $\{0, \cdots, d\} \times \{1, \cdots, L\}$, we followed the outlines of the method proposed in [4]. However, since this method can only be applied to generate permutations on sets with cardinality a power of 2 (which is not *a priori* the case for T'), we slightly modified it. The new version can be found in Appendix A (Algorithm 4).

4.3   Time Complexity

In the following we express the time complexity of each step of our scheme in terms of the parameters (t, d, d', N, L) and of constants a_i that depend on the implementation and the device architecture. Moreover, we provide practical values of these constants (in number of clock cycles) for an AES implementation protected with our scheme and running on a 8051-architecture.

Generation of T (see Appendix A). Its complexity $\mathcal{C}_T$ satisfies:

$$\mathcal{C}_T = t \times a_0 + N \times a_1 + f(N, t) \times a_2 \, ,$$

where $f(N, t) = t \sum_{i=0}^{N-1} \frac{1}{t-i}$. As argued in Appendix A, $f(N, t)$ can be approximated by $t \ln \left(\frac{t}{t-N} \right)$ for $t \gg N$.

Example 1. For our AES implementation, we got $a_0 = 6$, $a_1 = 7$ and $a_2 = 9$.

Generation of T'. Let q denote $\lceil \log_2((d+1)L) \rceil$. The complexity $\mathcal{C}_{T'}$ satisfies:

$$\mathcal{C}_{T'} = \begin{cases} q \times a_0 + 2^q \times (a_1 + q \times a_2) \text{ if } q = \log_2((d+1)L), \\ q \times a_0 + 2^q \times (a_1 + q \times a_2) + 2^q \times a_3 \text{ otherwise.} \end{cases}$$

Example 2. For our AES implementation, we got $a_0 = 3$, $a_1 = 15$ and $a_2 = 14$, $a_3 = 17$.

Generation the Masked S-box. Its complexity $\mathcal{C}_{MS}$ satisfies:

$$\mathcal{C}_{MS} = d' \times a_0 \, .$$

Example 3. For our AES implementation, we got $a_0 = 4352$.

Protected keyed Substitution Layer. Its complexity $\mathcal{C}_{SL}$ satisfies:

$$\mathcal{C}_{SL} = t \times (a_0 + d \times a_1 + d' \times a_2) \, .$$

Example 4. For our AES implementation, we got $a_0 = 55$, $a_1 = 37$ and $a_2 = 18$.

Protected Linear Layer. Its complexity $\mathcal{C}_{LL}$ satisfies:

$$\mathcal{C}_{LL} = (d+1)L \times a_0 \, .$$

Example 5. For our AES implementation, we got $a_0 = 169$.

4.4   Attack Paths

In this section, we list attacks combining higher-order and integrated DPA that may be attempted against our scheme. Section 3 is then involved to associate each attack with a correlation coefficient that depends on the leakage noise deviation σ, the block cipher parameters (n, N, l', L) and the security parameters (d, d', t). As argued, these coefficients characterize the attacks efficiencies and hence the overall resistance of the scheme.

Remark 3. In this paper, we only consider known plaintext attack *i.e.* we assume the different sensitive variables uniformly distributed. In a chosen plaintext attack, the adversary would be able to fix the value of some sensitive variables which could yield better attack paths. We do not take such attacks into account and let them for further investigations.

Every sensitive variable in the scheme is (1) either masked with d unique masks or (2) masked with d' masks shared with other sensitive variables (during the keyed substitution layer).

(1). In the first case, the $d+1$ shares appear during the keyed substitution layer computation and the linear layer computation. In both cases, their manipulation is shuffled.

(1.1). For the keyed substitution layer (see Algorithm 1), the $d+1$ shares all appear during a single iteration of the loop among t. The attack consists in combining the $d+1$ corresponding signals for each loop iteration and to sum the t obtained combined signals. Proposition 2 implies that this attack can be associated with the following correlation coefficient ρ_1:

$$\rho_1(t, d) = \frac{1}{\sqrt{t}}\rho(n, d, \sigma) \ . \tag{10}$$

(1.2). For the linear layer (see Algorithm 2), the $d+1$ shares appear among $(d+1)\cdot L$ possible operations. The attack consists in summing all the combinations of $d+1$ signals among the $(d+1)\cdot L$ corresponding signals. According to Proposition 3, this attack can be associated with the following correlation coefficient ρ_2:

$$\rho_2(L, d) = \frac{1}{\sqrt{\binom{(d+1)\cdot L}{d+1}}}\rho(l', d, \sigma) \ . \tag{11}$$

Remark 4. In the analysis above, we chose to not consider attacks combining shares processed in the linear layers together with shares processed in the keyed substitution layer. Actually, such an attack would yield to a correlation coefficient upper bounded by the maximum of the two correlations in (10) and (11).

(2). In the second case, the attack targets a d'^{th}-order masked variable occurring during the keyed substitution layer. Two alternatives are possible.

(2.1). The first one is to simultaneously target the masked variable (that appears in one loop iteration among t) and the d' masks that appear at fixed times (*e.g.* in every loop iteration of Algorithm 1 or during the masked S-box computation). The attack hence consists in summing the t possible combined signals obtained by combining the masked variable signal (t possible times) and the d' masks signals (at fixed times). According to Proposition 3, this leads to a correlation coefficient ρ_3 that satisfies:

$$\rho_3(t, d') = \frac{1}{\sqrt{t}}\rho(n, d', \sigma) \ . \tag{12}$$

(2.2). The second alternative is to target two different variables both masked with the same sum of d' masks (for instance two masked S-box inputs or outputs).

These variables are shuffled among t variables. The attack hence consists in summing all the possible combinations of the two signals among the t corresponding signals. According to Proposition 3, this leads to a correlation coefficient ρ_4 that satisfies:

$$\rho_4(t) = \frac{1}{\sqrt{t \cdot (t-1)}} \rho(n, 2, \sigma) \ . \tag{13}$$

4.5   Parameters Setting

The security parameters (d, d', t) can be chosen to satisfy an arbitrary resistance level characterized by an upper bound ρ^* on the correlation coefficients corresponding to the different attack paths exhibited in the previous section. That is, the parameters are chosen to satisfy the following inequality:

$$\max(|\rho_1|, |\rho_2|, |\rho_3|, |\rho_4|) \leqslant \rho^* \ . \tag{14}$$

Among the 3-tuples (d, d', t) satisfying the relation above, we select one among those that minimize the timing complexity (see Sect. 4.3).

5   Application to AES

We implemented our scheme for AES on a 8051-architecture. According to Remark 1, the ShiftRows and the MixColumns were merged in a single linear layer applying four times the same operation (but with different state indexings). The block cipher parameters hence satisfy: $n = 8$, $N = 16$, $l = 32$, $l' = 8$ and $L = 4$.

Remark 5. In [8], it is claimed that the manipulations of the different bytes in the MixColumns can be shuffled. However it is not clear how to perform such a shuffling in practice since the processing differs according to the byte index.

Table 1 summarizes the timings obtained for the different steps of the scheme for our implementation.

Remark 6. The unprotected round implementation has been optimized, in particular by only using variables stored in DATA memory. Because of memory constraints and due to the scalability of the code corresponding to the protected round, many variables have been in stored in XDATA memory which made the implementation more complex. This explains that, even for $d = d' = 0$ and $t = 16$ (*i.e.* when there is no security), the protected round is more time consuming than the unprotected round.

We give hereafter the optimal security parameters (t, d, d') for our AES implementation according to some illustrative values of the device noise deviation σ and of correlation bound ρ^*. We consider three noise deviation values: 0, $\sqrt{2}$ and $4\sqrt{2}$. In the Hamming weight model, these values respectively correspond to a signal-to-noise ratio (SNR) to $+\infty$, 1 and $\frac{1}{4}$. We consider four correlation

Table 1. Timings for the different steps of the scheme for an AES implementation on a 8051-architecture

T Generation	$\mathcal{C}_T = 112 + t\left(6 + 9\sum_{i=0}^{15}\frac{1}{t-i}\right)$
T' Generation	$\mathcal{C}_{T'} = 3q + 2^q(15 + 14q) \quad [+17 \times 2^q]$
Masked S-box Generation	$\mathcal{C}_{MS} = 4352d'$
Pre-computations	$\mathcal{C}_T + \mathcal{C}_{T'} + \mathcal{C}_{MS}$
Substitution Layer	$\mathcal{C}_{SL} = t(55 + 37d + 18d')$
Linear Layer	$\mathcal{C}_{LL} = 676(d + 1)$
Protected Round	$\mathcal{C}_{SL} + \mathcal{C}_{LL} = 676(d + 1) + t(55 + 37d + 18d')$
Unprotected Round	432

Table 2. Optimal parameters and timings according to SNR and ρ^*

ρ^*	SNR = $+\infty$			SNR = 1			SNR = $\frac{1}{4}$		
	t	d d'	timings	t	d d'	timings	t	d d'	timings
10^{-1}	16	1 1	3.66×10^4	16	1 1	3.66×10^4	16	1 0	2.94×10^4
10^{-2}	20	3 3	8.57×10^4	20	2 2	6.39×10^4	16	1 1	3.66×10^4
10^{-3}	1954	4 3	5.08×10^6	123	3 3	3.13×10^5	16	2 2	5.75×10^4
10^{-4}	195313	5 3	5.75×10^8	12208	4 3	3.15×10^7	19	3 3	8.35×10^4

bounds: 10^{-1}, 10^{-2}, 10^{-3}, and 10^{-4}. The security parameters and the corresponding timings for the protected AES implementation are given in Table 5. Note that all the rounds have been protected.

When SNR = $+\infty$, the bound $d' \leqslant 3$ implies an intensive use of shuffling in the keyed substitution layer. The resulting parameters for correlation bounds 10^{-3} and 10^{-4} imply timings that quickly become prohibitive. A solution to overcome this drawback would be to design secure table re-computation algorithms for $d' \geqslant 3$. Besides, these timings underline the difficulty of securing block ciphers implementations with pure software countermeasures. When the leakage signals are not very noisy (SNR = 1), timings clearly decrease (by a factor from 10 to 20). This illustrates, once again, the soundness of combining masking with noise addition. This is even clearer when the noise is stronger (SNR = $\frac{1}{4}$), where it can be noticed that the addition of dummy operations is almost not required to achieve the desired security level.

6   Conclusion

In this paper, we have conducted an analysis that quantifies the efficiency of advanced DPA attacks targeting masking and shuffling. Based on this analysis, we have designed a generic scheme combining higher-order masking and shuffling.

This scheme generalizes to higher orders the solutions previously proposed in the literature. It is moreover scalable and its security parameters can be chosen according to any desired resistance level. As an illustration, we applied it to protect a software implementation of AES for which we gave several security/efficiency trade-offs.

References

1. Akkar, M.-L., Giraud, C.: An Implementation of DES and AES, Secure against Some Attacks. In: Koç, Ç.K., Naccache, D., Paar, C. (eds.) CHES 2001. LNCS, vol. 2162, pp. 309–318. Springer, Heidelberg (2001)
2. Chari, S., Jutla, C., Rao, J., Rohatgi, P.: Towards Sound Approaches to Counteract Power-Analysis Attacks. In: Wiener, M. (ed.) CRYPTO 1999. LNCS, vol. 1666, pp. 398–412. Springer, Heidelberg (1999)
3. Clavier, C., Coron, J.-S., Dabbous, N.: Differential Power Analysis in the Presence of Hardware Countermeasures. In: Paar, C., Koç, Ç.K. (eds.) CHES 2000. LNCS, vol. 1965, pp. 252–263. Springer, Heidelberg (2000)
4. Coron, J.-S.: A New DPA Countermeasure Based on Permutation Tables. In: Ostrovsky, R., De Prisco, R., Visconti, I. (eds.) SCN 2008. LNCS, vol. 5229, pp. 278–292. Springer, Heidelberg (2008)
5. Coron, J.-S., Prouff, E., Rivain, M.: Side Channel Cryptanalysis of a Higher Order Masking Scheme. In: Paillier, P., Verbauwhede, I. (eds.) CHES 2007. LNCS, vol. 4727, pp. 28–44. Springer, Heidelberg (2007)
6. FIPS PUB 197. Advanced Encryption Standard. National Institute of Standards and Technology (November 2001)
7. Goubin, L., Patarin, J.: DES and Differential Power Analysis – The Duplication Method. In: Koç, Ç.K., Paar, C. (eds.) CHES 1999. LNCS, vol. 1717, pp. 158–172. Springer, Heidelberg (1999)
8. Herbst, P., Oswald, E., Mangard, S.: An AES Smart Card Implementation Resistant to Power Analysis Attacks. In: Zhou, J., Yung, M., Bao, F. (eds.) ACNS 2006. LNCS, vol. 3989, pp. 239–252. Springer, Heidelberg (2006)
9. Joye, M., Paillier, P., Schoenmakers, B.: On Second-order Differential Power Analysis. In: Rao, J.R., Sunar, B. (eds.) CHES 2005. LNCS, vol. 3659, pp. 293–308. Springer, Heidelberg (2005)
10. Knuth, D.: The Art of Computer Programming, 3rd edn., vol. 2. Addison-Wesley, Reading (1988)
11. Kocher, P., Jaffe, J., Jun, B.: Differential Power Analysis. In: Wiener, M. (ed.) CRYPTO 1999. LNCS, vol. 1666, pp. 388–397. Springer, Heidelberg (1999)
12. Mangard, S.: Hardware Countermeasures against DPA – A Statistical Analysis of Their Effectiveness. In: Okamoto, T. (ed.) CT-RSA 2004. LNCS, vol. 2964, pp. 222–235. Springer, Heidelberg (2004)
13. Mangard, S., Oswald, E., Popp, T.: Power Analysis Attacks – Revealing the Secrets of Smartcards. Springer, Heidelberg (2007)
14. Messerges, T.: Securing the AES Finalists against Power Analysis Attacks. In: Schneier, B. (ed.) FSE 2000. LNCS, vol. 1978, pp. 150–164. Springer, Heidelberg (2001)

15. Messerges, T.: Using Second-order Power Analysis to Attack DPA Resistant Software. In: Paar, C., Koç, Ç.K. (eds.) CHES 2000. LNCS, vol. 1965, pp. 238–251. Springer, Heidelberg (2000)
16. Oswald, E., Mangard, S.: Template Attacks on Masking—Resistance is Futile. In: Abe, M. (ed.) CT-RSA 2007. LNCS, vol. 4377, pp. 243–256. Springer, Heidelberg (2006)
17. Oswald, E., Mangard, S., Herbst, C., Tillich, S.: Practical Second-order DPA Attacks for Masked Smart Card Implementations of Block Ciphers. In: Pointcheval, D. (ed.) CT-RSA 2006. LNCS, vol. 3860, pp. 192–207. Springer, Heidelberg (2006)
18. Patarin, J.: How to Construct Pseudorandom and Super Pseudorandom Permutation from one Single Pseudorandom Function. In: Rueppel, R.A. (ed.) EUROCRYPT 1992. LNCS, vol. 658, pp. 256–266. Springer, Heidelberg (1993)
19. Pieprzyk, J.: How to Construct Pseudorandom Permutations from Single Pseudorandom Functions Advances. In: Damgård, I.B. (ed.) EUROCRYPT 1990. LNCS, vol. 473, pp. 140–150. Springer, Heidelberg (1991)
20. Prouff, E., Rivain, M., Bévan, R.: Statistical Analysis of Second Order Differential Power Analysis. IEEE Trans. Comput. 58(6), 799–811 (2009)
21. Rivain, M., Dottax, E., Prouff, E.: Block Ciphers Implementations Provably Secure Against Second Order Side Channel Analysis. In: Nyberg, K. (ed.) FSE 2008. LNCS, vol. 5086, pp. 127–143. Springer, Heidelberg (2008)
22. Rivain, M., Prouff, E., Doget, J.: Higher-order Masking and Shuffling for Software Implementations of Block Ciphers. Cryptology ePrint Archive (2009), http://eprint.iacr.org/
23. Schramm, K., Paar, C.: Higher Order Masking of the AES. In: Pointcheval, D. (ed.) CT-RSA 2006. LNCS, vol. 3860, pp. 208–225. Springer, Heidelberg (2006)
24. Standaert, F.-X., Peeters, E., Rouvroy, G., Quisquater, J.-J.: An Overview of Power Analysis Attacks Against Field Programmable Gate Arrays. IEEE 94(2), 383–394 (2006)
25. Tillich, S., Herbst, C.: Attacking State-of-the-Art Software Countermeasures-A Case Study for AES. In: Oswald, E., Rohatgi, P. (eds.) CHES 2008. LNCS, vol. 5154, pp. 228–243. Springer, Heidelberg (2008)
26. Tillich, S., Herbst, C., Mangard, S.: Protecting AES Software Implementations on 32-Bit Processors Against Power Analysis. In: Katz, J., Yung, M. (eds.) ACNS 2007. LNCS, vol. 4521, pp. 141–157. Springer, Heidelberg (2007)
27. Wasserman, L.: All of Statistics: A Concise Course in Statistical Inference. Springer Texts in Statistics (2005)

A   Algorithms for Index Tables Generations

Generation of T. To generate T, we start by initializing all the cells of T to the value $N + 1$. Then, for every $j \leqslant N$, we randomly generate an index $i < t$ until $T[i] = N + 1$ and we move j into $T[i]$. The process is detailed hereafter.

Algorithm 3. Generation of T

INPUT: state's length N and shuffling order t
OUTPUT: indices permutation table T

---

1. **for** $i \leftarrow 0$ **to** $t - 1$
2.   **do** $T[i] \leftarrow N + 1$ // Initialization of T
3. $j \leftarrow 1$
4. **for** $j \leftarrow 1$ **to** N
5.   **do** $i \leftarrow rand(t)$ **while** $T[i] = N+1$ // Generate random index $i < t$
6. $T[i] = j$ and $j \leftarrow j + 1$
7. **return** T

Complexity Anlysis of loop 4-to-6. The expected number $f(N,t)$ of iterations of the loop 4-to-7 in Algorithm 3 satisfies:

$$f(N,t) = t \cdot (H_t - H_{t-N}) \ , \tag{15}$$

where for every r, H_r denotes the rth *Harmonic number* defined by $H_r = \sum_{i=1}^{r} \frac{1}{i}$.

Let us argue about (15). For every $j \leqslant N$, the probability that the loop **do-while** ends up after i iterations is $\left(\frac{t-j}{t}\right) \cdot \left(\frac{j}{t}\right)^{i-1}$: at the jth iteration of the **for** loop, the test $T[i] = N+1$ succeeds with probability $p_j = \left(\frac{j}{t}\right)$ and fails with probability $1 - p_j = \left(\frac{t-j}{t}\right)$. One deduces that for every $j \leqslant N$, the expected number of iterations of the loop **do-while** is $\sum_{i \in \mathbb{N}} i \cdot p_j^{i-1} \cdot (1 - p_j)$. We eventually get that the number of iterations $f(N,t)$ satisfies $f(N,t) = \sum_{j=0}^{N-1} \sum_{i \in \mathbb{N}} i \cdot \left(p_j^{i-1} - p_j^{i}\right)$, that is $f(N,t) = \sum_{j=0}^{N-1} \sum_{i \in \mathbb{N}} i \cdot p_j^{i-1} - \sum_{j=0}^{N-1} \sum_{i \in \mathbb{N}} (i+1) \cdot p_j^{i} + \sum_{j=0}^{N-1} \sum_{i \in \mathbb{N}} p_j^{i}$. As the two first sums in the right-hand side of the previous equation are equal, one deduces that $f(N,t)$ equals $\sum_{j=0}^{N-1} \sum_{i \in \mathbb{N}} p_j^{i}$ that is $\sum_{j=0}^{N-1} \frac{1}{1-p_j}$. Eventually, since p_j equals $\frac{j}{t}$, we get $f(N,t) = \sum_{j=0}^{N-1} \frac{t}{t-j}$ which is equivalent with (15).

Since H_r tends towards $\ln(r) + \gamma$, where γ is the Euler-Mascheroni constant, we can approximate $H_t - H_{t-N}$ by $\ln(t) - \ln(t - N)$. We eventually get the following relation for $t \gg N$:

$$f(N,t) \approx t \cdot \ln\left(\frac{t}{t-N}\right).$$

Generation of T'. In view of the previous complexity, generating a permutation with the same implementation as for T is not pertinent (in this case $t = N$).

To generate the permutation T', we follow the outlines of the method proposed in [4]. However, since this method can only be applied to generate permutations on sets with cardinality a power of 2 (which is not a priori the case for T'), we slightly modified it. Let 2^q be the smallest power of 2 which is greater than $(d+1)L$. Our algorithm essentially consists in designing a q-bit random permutation T' from a fixed q-bit permutation π and a family of q random values in $\mathbb{F}_2^q$ (Steps 1 to 6 in Algorithm 4). Then, if $(d+1)L$ is not a power of 2, the table T' is transformed into a permutation over $\{0, \cdots, d\} \times \{1, \cdots, L\}$ by deleting the elements which are strictly greater than $(d+1)L - 1$. The process is detailled in pseudo-code hereafter.

Algorithm 4. Generation of T'

INPUT: parameters (d, L) and a n'-bit permutation π with $q = \lceil \log_2((d+1)L) \rceil$
OUTPUT: indices permutation table T'

---

1. **for** $i \leftarrow 0$ **to** $q - 1$
2.   **do** $alea_i \leftarrow rand(q)$ // Initialization of aleas
3. **for** $j \leftarrow 0$ **to** $2^q - 1$
4.   **do** $T'[j] \leftarrow \pi[j]$
5.   **for** $i \leftarrow 0$ **to** $q - 1$
6.   **do** $T'[j] \leftarrow \pi[T'[j] \oplus alea_i]$ // Process the ith index
7. **if** $q \neq (d+1)L$
8.   **then for** $j \leftarrow 0$ **to** $(d+1)L - 1$
9.   **do** $i \leftarrow j$
10.   **while** $T'[i] \geq (d+1)L$
11.   **do** $i \leftarrow i + 1$
12. $T'[j] \leftarrow T'[i]$
13. **return** T'

With Algorithm 4, it is not possible to generate all the permutations over $\{0, \cdots, d\} \times \{1, \cdots, L\}$. In our context, we assume that this does not introduce any weakness in the scheme.

Complexity Anlysis of loop 8-to-12. The number of iterations of loop 8-to-12 in Algorithm 4 in the worst case is 2^q.

A Design Methodology for a DPA-Resistant Cryptographic LSI with RSL Techniques

Minoru Saeki[1], Daisuke Suzuki[1,2], Koichi Shimizu[1], and Akashi Satoh[3]

[1] Information Technology R&D Center, Mitsubishi Electric Corporation
Kamakura Kanagawa, Japan
{Saeki.Minoru@db,Suzuki.Daisuke@bx,
Shimizu.Koichi@ea}.MitsubishiElectric.co.jp
[2] Graduate School of Environmental and Information Sciences,
Yokohama National University
Yokohama Kanagawa, Japan
[3] Research Center for Information Security,
National Institute of Advanced Industrial Science and Technology (AIST)
Chiyoda Tokyo, Japan
akashi.satoh@aist.go.jp

Abstract. A design methodology of Random Switching Logic (RSL) using CMOS standard cell libraries is proposed to counter power analysis attacks against cryptographic hardware modules. The original RSL proposed in 2004 requires a unique RSL-gate for random data masking and glitch suppression to prevent secret information leakage through power traces. However, our new methodology enables to use general logic gates supported by standard cell libraries. In order to evaluate its practical performance in hardware size and speed as well as resistance against power analysis attacks, an AES circuit with the RSL technique was implemented as a cryptographic LSI using a 130-nm CMOS standard cell library. From the results of attack experiments that used a million traces, we confirmed that the RSL-AES circuit has very high DPA and CPA resistance thanks to the contributions of both the masking function and the glitch suppressing function. This is the first result demonstrating reduction of the side-channel leakage by glitch suppression quantitatively on real ASIC.

1 Introduction

Since Kocher et al. proposed side-channel attacks [1,2], many countermeasures such as data masking and power equalization techniques have been studied and implemented as FPGA and ASIC circuits. However, most of them require custom logic gates and/or design tools specialized to their methods, or implementations using standard libraries can be compromised by glitch signals, unbalanced signal delays and power consumptions. A DPA countermeasure called Random Switching Logic (hereafter RSL) [3] is a strong countermeasure that combines a masking function using random numbers and a function that suppresses glitch signals. The original authors further improve RSL in [4] by adding re-mask processing to

C. Clavier and K. Gaj (Eds.): CHES 2009, LNCS 5747, pp. 189–204, 2009.

each RSL gate to counter high-order DPA [5,6] and such DPAs as identify random numbers for each waveform with the use of filtering [7,8]. RSL is, however, difficult to develop under a general environment compared with other countermeasures like Wave Dynamic Differential Logic (WDDL) [9] since it uses a logic gate that does not exist in standard CMOS cell libraries. In contract, we propose a design methodology of RSL that enables implementation in a standard CMOS library, and show the design flow using a standard design tool. An AES circuit with the RSL countermeasure is designed and implemented on a cryptographic LSI using a 130-nm CMOS standard cell library [10], and its hardware performances in gate counts and operating speed are evaluated. DPA and CPA experiments are also applied on the LSI, and the effectiveness of our methodology as DPA and CPA countermeasure is demonstrated. In addition, the effects of the RSL function that suppresses glitches is detailed in the evaluation of the power analysis resistance.

2   RSL Using Standard Cell and Security Evaluation

RSL is a DPA countermeasure at the transistor level using a custom RSL gate that consists of a majority logic gate with output-control signal shown in Fig. 1. Table 1 shows a pseudo code of NAND operation by RSL. Random numbers are used to mask data and then signal transitions of the RSL gate lose correlation with selection functions of DPA. However, a simple random masking without signal delay control may cause glitches that do have correlation with the selection functions [3,11]. In order to prevent the glitches, the RSL gate manages the delay relationship of the output-control signal (en), input signals (x_z, y_z), and random mask signals (r_z). Re-mask operation is independently performed in each RSL gate so that high-order DPAs [5,6] cannot identify the random numbers in each waveform even using filtering techniques [7,8]. For more details about RSL, refer to [4].

This section proposes a new method to realize RSL using only a standard CMOS cell library, while the original RSL scheme using the custom gate requires high development cost and long design period. Fig. 2 shows an RSL NAND logic using standard CMOS gates, a Majority-Inverter (MAJI) or OR-AND-Inverter (OAI222), and a NOR for output control, which as a whole provide compatible operation with the RSL-NAND gate in Fig. 1. We call this compatible logic "pseudo RSL". The majority operation needs to be performed at the front step, and the output-control signal is enabled at the final step to suppress glitch signals on the output of the pseudo RSL gate.

According to [12], side-channel leakage occurs in non-linear data operations and is amplified in linear/non-linear operations. We hence estimate side-channel leakage of the MAJI logic in Fig. 2, which operates non-linearly on data inputs, using the leakage models mentioned in [12]. The change of the MAJI gate inputs (x, y, r) at the i-th cycle is denoted as

$$(x_{i-1}, y_{i-1}, r_{i-1}) \rightarrow (x_i, y_i, r_i).$$

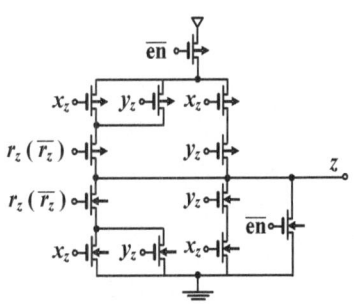

Fig. 1. An RSL-NAND(NOR) gate

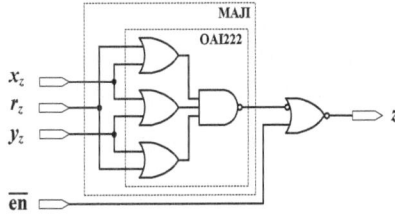

Fig. 2. NAND operation by pseudo RSL

Table 1. Pseudo-code of NAND operation by RSL

NAND operation by RSL
input en, $x = a \oplus r_x, y = b \oplus r_y,$
$\qquad r_{xz} = r_x \oplus r_z, r_{yz} = r_y \oplus r_z, r_z$
output $z = \overline{a \cdot b} \oplus r_z$
begin
/*Operation 1: Suppress glitches */
en <= 0;
/*Operation 2(a): Re-mask x */
$x_z <= x \oplus r_{xz} \ (= a \oplus r_z)$
/*Operation 2(b): Re-mask y */
$y_z <= y \oplus r_{yz} \ (= b \oplus r_z)$
/*Operation 3: Input data
to the RSL gate */
$z <=$ RSL-NAND$(x_z, y_z, r_z,$ en$)$
/*Operation 4: The enable signal en
rise after all other
input signals are fixed */
en <= 1 after max_delay(x_z, y_z, r_z);
end

As the selection functions of DPA, we use $a_i (= x_i \oplus r_i), b_i (= y_i \oplus r_i)$, which are unmasked signals of x_i, y_i at the i-th cycle. The average number of signal transitions, $N(i)$ and $N(i + 1)$, including glitches caused by the input signals (x, y, r) at MAJI gate, are evaluated. The transitions for two (the i-th and $(i+1)$-th) cycles are used to assess the bias of the numbers based on the cycle when an operation including the DPA selection function takes place. Table 2 shows the transition counts of the MAJI gate for each input pattern in each delay condition. For example, in the delay condition 1, when $(a_i, b_i, r_i) = (1, 0, 0)$, the average number of signal transitions during $(x_{i-1}, y_{i-1}, r_{i-1}) \rightarrow (x_i, y_i, r_i)$ is $N(i) = 1$, and the number during $(x_i, y_i, r_i) \rightarrow (x_{i+1}, y_{i+1}, r_{i+1})$ is $N(i + 1) = 1/2$. Next, the amount of leakage is calculated from Table 2 for each delay condition and selection function. The amount of leakage at the i-th cycle, $N_{\text{diff}}(i)$, is defined according to [12] as follows.

$$N_{\text{diff}}(i) = N_{\alpha=1}(i) - N_{\alpha=0}(i)$$

where $N_{\alpha=\delta}$ is the average number with the selection function being α and its value δ.

Table 3 shows the amount of leakage for each delay condition calculated from Table 2. Each condition includes cases where $N_{\text{diff}} \neq 0$, and thus pseudo RSL does not meet the security conditions described in [12] when it is strictly evaluated as a single gate. However, leakage model in [12] is inapplicable to pseudo RSL, which assumes the leakage is amplified by following gates. As stated before, pseudo RSL prevents the spread of the glitches after MAJI by the output-control

Table 2. Signal transitions of a MAJI gate

Delay condition 1			
delay(x) < delay(y) < delay(r)			
a_i, b_i, r_i	x_i, y_i, r_i	$N(i)$	$N(i+1)$
0 0 0	0 0 0	1/2	1/2
0 0 1	1 1 1	1/2	1/2
0 1 0	0 1 0	1	1
0 1 1	1 0 1	1	1
1 0 0	1 0 0	1	1/2
1 0 1	0 1 1	1	1/2
1 1 0	1 1 0	1/2	1
1 1 1	0 0 1	1/2	1
Delay condition 2			
delay(x) < delay(r) < delay(y)			
a_i, b_i, r_i	x_i, y_i, r_i	$N(i)$	$N(i+1)$
0 0 0	0 0 0	1/2	1/2
0 0 1	1 1 1	1/2	1/2
0 1 0	0 1 0	1/2	1
0 1 1	1 0 1	1/2	1
1 0 0	1 0 0	1	1/2
1 0 1	0 1 1	1	1/2
1 1 0	1 1 0	1	1
1 1 1	0 0 1	1	1
Delay condition 3			
delay(r) < delay(x) < delay(y)			
a_i, b_i, r_i	x_i, y_i, r_i	$N(i)$	$N(i+1)$
0 0 0	0 0 0	1/2	1/2
0 0 1	1 1 1	1/2	1/2
0 1 0	0 1 0	1/2	1
0 1 1	1 0 1	1/2	1
1 0 0	1 0 0	1	1
1 0 1	0 1 1	1	1
1 1 0	1 1 0	1	1/2
1 1 1	0 0 1	1	1/2

Table 3. Leakage amount of a MAJI gate

Delay condition 1		
delay(x) < delay(y) < delay(r)		
Cycle	Selection function	N_{diff}
i	a_i	0
i	b_i	0
$i+1$	a_i	0
$i+1$	b_i	1/2
Delay condition 2		
delay(x) < delay(r) < delay(y)		
Cycle	Selection function	N_{diff}
i	a_i	1/2
i	b_i	0
$i+1$	a_i	0
$i+1$	b_i	1/2
Delay condition 3		
delay(r) < delay(x) < delay(y)		
Cycle	Selection function	N_{diff}
i	a_i	1/2
i	b_i	0
$i+1$	a_i	0
$i+1$	b_i	0

signal, and hence the 1/2 transition bias of the MAJI gate in Table 3 does not propagate. Therefore, the amount of leakage in the entire circuit is at most $k/2$, where k is the the number of MAJI gates whose input bits are used as selection functions.

In short, pseudo RSL provides the same level of security as RSL if there exists a lower limit of $|N_{\text{diff}}|$ detectable by DPA, denoted as ϵ, such that $k/2 < \epsilon$. It is hard to give the threshold ϵ value, because it depends on the side-channel evaluation environment and the device characteristics, but it is easy to calculate the maximum of k. For instance, k is 2 for the AES circuit in Section 5. The circuit size is about 30 Kgates and the average number of signal transitions per cycle using virtual delay is approximately 15,000, which implies that if DPA fails to detect the bias of 1/15,000 transitions per DPA trace, pseudo RSL can make the circuit sufficiently secure.

3   Design Flow of Cryptographic Circuits Using RSL

Fig. 3 shows the design flow of the RSL and pseudo RSL circuits, and Fig. 4 is the abstract of the hardware architecture. It is assumed that the circuit is

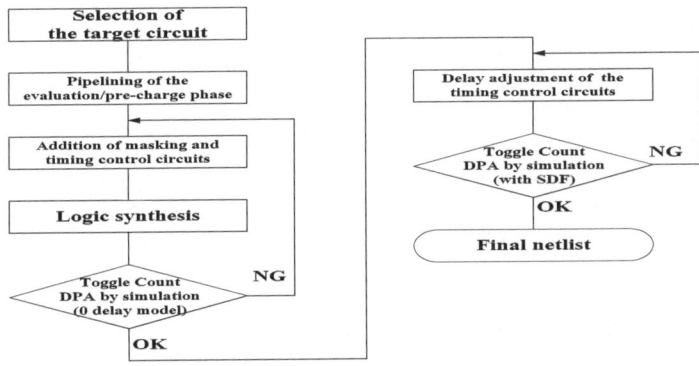

Fig. 3. Overview chart of design flow

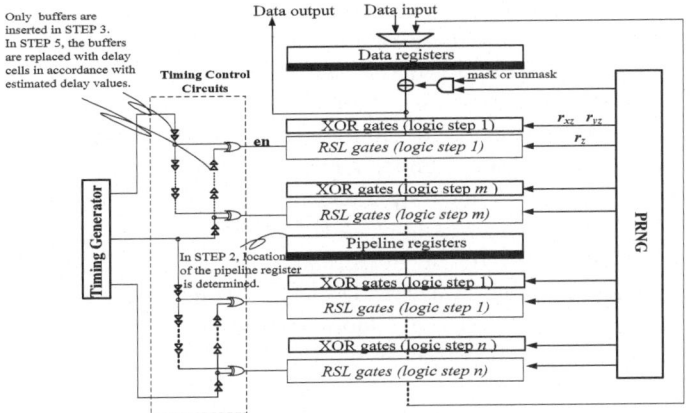

Fig. 4. Cryptographic circuit using RSL

designed in RTL and netlist is generated by a logic synthesis tool. In this flow, RSL or pseudo RSL circuit is generated from an existing cryptographic design without DPA countermeasure. The flow is largely made up of the following five steps.

STEP 1 Selection of the target circuit.
STEP 2 Pipelining of the evaluation/pre-charge phase.
STEP 3 Addition of masking and timing control circuits.
STEP 4 Logic synthesis.
STEP 5 Delay adjustment of the timing control circuit.

In STEP 1, a cryptographic circuit is selected to convert into RSL logics, but circuits that fall into the following two cases cannot be used. The first case is that the circuit data paths contain successive non-linear operations. For example, if SubBytes circuit of AES is implemented by AND-XOR-type 2-step logic such as

Positive Polarity Reed-Muller (PPRM), it contains high-order product terms. Then the precise delay control and random number re-masking for each 2-input NAND is extremely difficult. Integer arithmetic that includes successive non-linear operations such as addition and multiplication is not suitable either. In contrast, SubBytes implementation that employs composite field arithmetic [13] is very suitable because it has a nested structure of AND-XOR operations with a few non-linear steps. The second case is that the circuit is designed as behavior model or using a large look-up table description, because its logic structure depends on synthesis tools.

In STEP 2, a tentative logic synthesis is performed to evaluate delays in the whole circuit chosen in STEP 1. Afterward, a pipeline register is inserted to divide the circuit into two parts and thereby to let the RSL evaluation and pre-charge phases run in parallel. The register location is determined in consideration of the tradeoff between the operating frequency and the circuit area. A high frequency is expected by dividing the critical path at the center, but there may be a large number of intermediate signals that require a large register. In our prototypic LSI described in Section 5, the pipeline register is located at the output of SubBytes to minimize the number.

STEP 3 replaces logic gates for non-linear operations with RSL or pseudo RSL gates, and puts random number generators for data masking. Note that RSL need not be applied to linear operations such as XOR because they do not generate side-channel leakage as long as the inputs to them are masked. The replacement is applied only to the gates related to secret key operations. Fig. 5 shows a conversion example of a NAND gate, in which three random number input ports are added. In the case of a loop architecture, the initial data masking and the final data unmasking need an additional XOR circuit shown in Fig. 4. The above operations determine the number of random number generator and unmasking/re-masking circuits to be added. The timing control logic for output-control signals of each RSL gate is also designed. At this time, timing adjustment to prevent glitches is not required. In the implementation example of AES in Section 5, the number of RSL stages from the data register to the pipeline register is 4, but no RSL stage exists between the pipeline register to data register because there is only a linear transformation. Pseudo-random number generators using two 33-bit LFSR are designed in the AES circuit. Quality and characteristics of random numbers required by a masking method is an open question.

In STEP 4, ordinary logic synthesis is performed while the structures of RSL gates and the timing control circuits are protected not to be modified. In consideration of the increase of the critical path delay for timing adjustment in the next step, logic synthesis must be performed with some margin for performance in speed.

In STEP 5, the delay is adjusted by modifying the timing control logic part of the netlist created in STEP 4. First, the maximum delay before the RSL gates (logic step 1) located at the upper-most of Fig. 4 is extracted. The path of the output-control signal is not included in the delay. Next, temporary buffers in the timing control circuit are replaced to have delays longer than the extracted

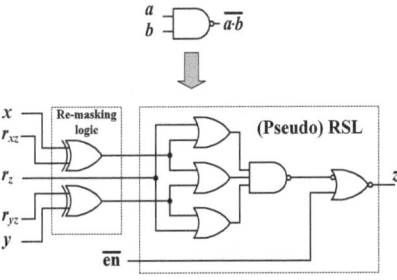

Fig. 5. RSL conversion for NAND Gate

Table 4. Our implementation environment

Process	TSMC 130nm CL013G [10]
Logic synthesis	Design Complier version 2004.12.SP4
Simulator	NC-Verilog version 05.40-p004

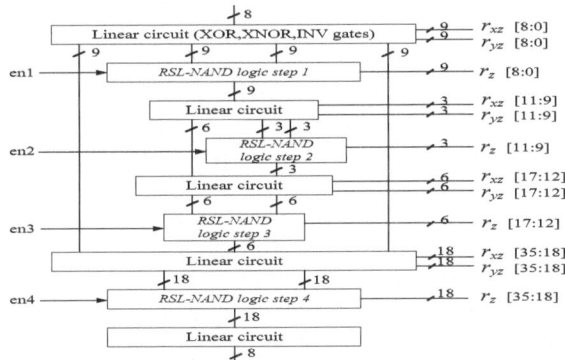

Fig. 6. SubBytes logic with composite field arithmetic

delay. This modification is repeated until timing simulation using the netlist and Standard Delay File (SDF) created in STEP 4 and 5, meets the security conditions of RSL.

4 Performance Evaluation of the Prototype Circuit

Table 4 summarizes our implementation environment applying the pseudo RSL scheme to the Verilog-HDL source code [14] designed with composite field arithmetic for the SubBytes function. It uses one round function block with a 16-byte SubBytes block for the loop architecture, and key expansion is processed on-the-fly using a 4-byte SubBytes block. The pseudo RSL is only applied to the 16-byte SubBytes block, which is the only non-linear function in the round operations. The composite gate OAI222 in the TSMC's 130-nm CMOS standard cell library CL013G is used as the majority logic. The pipeline register is placed at the output of SubBytes and is implemented as negative edge triggered flip-flops to have the same number of clock cycles as that of the original circuit without pseudo RSL. As described in the previous section, the number of RSL stages containing SubBytes before the pipeline register is 4, and numbers of the pseudo RSL gates in each stage are 9, 3, 6, and 18. Therefore three 36-bit random numbers

Table 5. Performance evaluation result

Evaluation item	Before applying [14]	After applying
Gate counts	14.5 Kgate	30.5 Kgate
Maximum delay	16.77 ns	14.77 ns
Maximum operation frequency	59.6 MHz	33.8 MHz
Processing performance (at f_{max})	763 Mbps	432 Mbps

Table 6. Our experimental environment

Parameters	Explanation
Target device	TSMC 130-nm cryptographic LSI on SASEBO-R
Operating frequency	24 MHz (standard setting on the board)
Measuring point	Resistance (2.2 Ω) between power supply and ASIC
Oscilloscope	Agilent DSO8104A
Sampling frequency	2 GHz
Number of power traces	1,000,000 traces

(r_z, r_{xz}, and r_{yz}) are needed for the mask operations. Fig. 6 summarizes this configuration.

Table 5 shows the performances of the AES circuits with and without the pseudo RSL. The experimental cryptographic LSI that contains these AES circuits is operated at a low frequency of 24 MHz, and hence speed constraints for the circuits were not specified to a logic synthesis tool. The throughput of 432 Mbps at 33.8 MHz of the pseudo RSL circuit can be applied to many embedded applications including smart cards. The gate counts are doubled to 30.5 Kgates, including all the components of RSL-AES such as the encryption block, key scheduler and pseudo-random number generator, but this number is small enough for any practical use. A simple performance comparison cannot be made between an AES circuit using pseudo RSL and the one using WDDL in [15] because their design environments are different. However, we believe that our pseudo RSL logic has high advantages over WDDL in both hardware size and speed because the AES circuit with WDDL in [15] requires about three times as many gates as and its operating speed is degraded down to 1/4 in comparison to the original AES circuit without WDDL.

5   Power Analysis Attack against Prototype LSI

Power analysis attacks against the pseudo RSL-AES circuit implemented on the experimental LSI is performed by using a circuit board SASEBO-R (Side-channel Attack Standard Evaluation Board, see Fig. 16 in Appendix A), which is specially designed for the side-channel attack experiments [16]. The evaluation environment is detailed in Table 6. The random masking and glitch suppressing functions in the pseudo RSL-AES circuit can be enabled separately by setting a mode register. The DPA and CPA attacks are performed on four possible combinations of the functions to reveal the final round key.

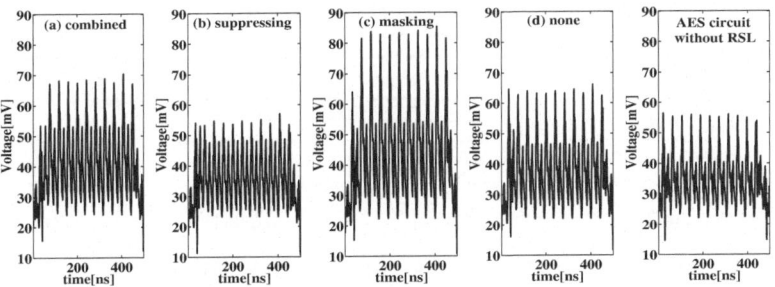

Fig. 7. Comparison of power traces

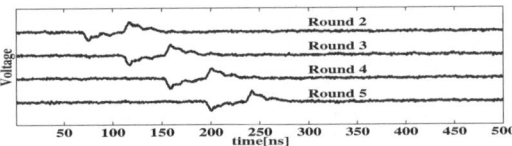

Fig. 8. Example of leakage analysis results

5.1   Comparison of the Power Traces

Fig. 7 shows the mean power traces of 1,000 encryptions in each operation mode: (a) **"combined"** enabling the both functions, (b) **"suppressing"** with only the glitch suppressing, (c) **"masking"** with only the random masking, and (d) **"none"** disabling the both functions. The original AES circuit without the pseudo RSL is also measured as the power trace in Fig. 7.

In all figures, large and relatively small peaks can be observed synchronizing rising and falling edges of system clock, respectively. In the pseudo RSL-AES circuit, by comparing the traces of **none** and **masking**, or those of **suppressing** and **combined**, it is observed that the peaks (power consumption) are increased by the mask operation. The glitch suppression works to reduce the peaks as can be seen in the same way by comparison of **none** and **suppressing**, or **masking** and **combined**.

The glitch suppression enables to reduce the increase of the peak current that the data masking causes. The decrease in peak current of circuits is an important issue for some devices such as contactless smart cards. We thus believe that RSL is suitable for those devices.

5.2   Leakage Analysis

Fig. 8 shows DPA traces in **none** with correct predictions for one input bit to SubBytes circuit in successive four rounds. The spikes in DPA traces appear at appropriate time frame corresponding to the target rounds, and thus they are information leakage but not noise signals. We call this evaluation using correct internal information known by evaluators "leakage analysis". Even if information leakage is confirmed by the leakage analysis, it does not mean the implementation

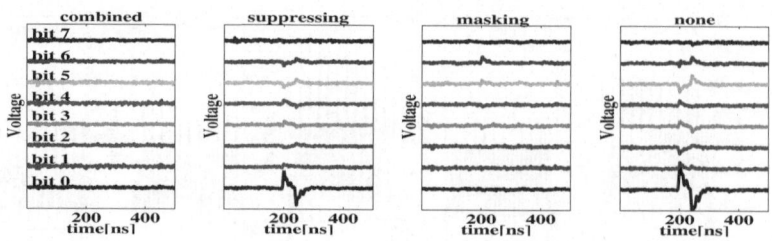

Fig. 9. Leakage analysis results with SubBytes input as selection function

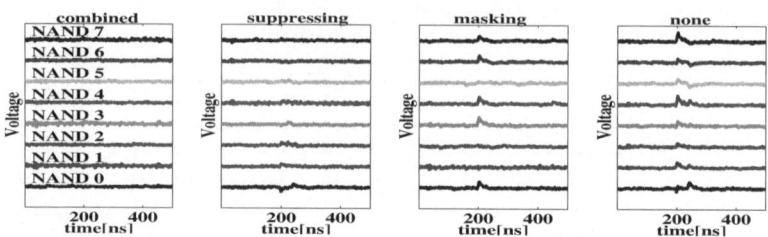

Fig. 10. Leakage analysis results with NAND gate input as selection function

is weak against power analysis attacks because the spikes for wrong keys (called ghost peaks) would be larger than correct ones [17], or key search space would be too large to attack while we know the correct key in the leakage analysis. Therefore, the analysis is the most powerful scheme to evaluate security of cryptographic modules against power analysis attacks.

Figs. 9 and 10 show the DPA traces by the leakage analysis of the RSL-AES circuit in four operation modes. Each input bit of one 8-bit S-box and each input bit of eight NAND gates in SubBytes circuit are used as selection functions. The effects of the random mask are not taken into account for the selection functions. In other words, random numbers are assumed to be 0 in the functions. The traces for **suppressing**, **masking** and **none** show spikes of information leakage while no leakage is observed in **combined**. The peak values of **suppressing** and **none** in Fig. 9 are larger than those in Fig. 10, but the values of **masking** in Fig. 9 is smaller than that of Fig. 10. Therefore, a selection function suitable for attacks varies depending on the circuit implementation. Both in Figs. 9 and 10, the peak values for **suppressing** and **masking** are smaller than those for **none** while their shapes are similar. These results clearly show that each countermeasure has some independent effect to reduce the leakage, but the pseudo RSL scheme that is a combination of the two countermeasures provides very high security.

5.3   CPA and DPA Attacks

CPA [17] and DPA attacks are performed using 1,000,000 power traces to reveal the final round key of the pseudo RSL-AES circuit in each mode. In the following discussion, the results for two out of sixteen 8-bit S-box circuits are displayed.

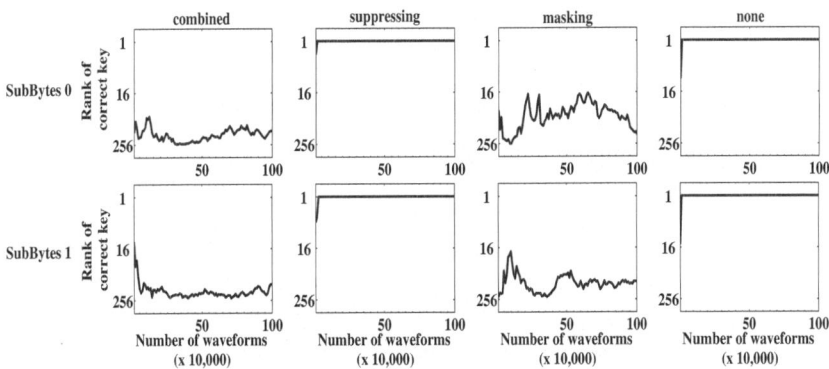

Fig. 11. CPA results of SubBytes 0-1

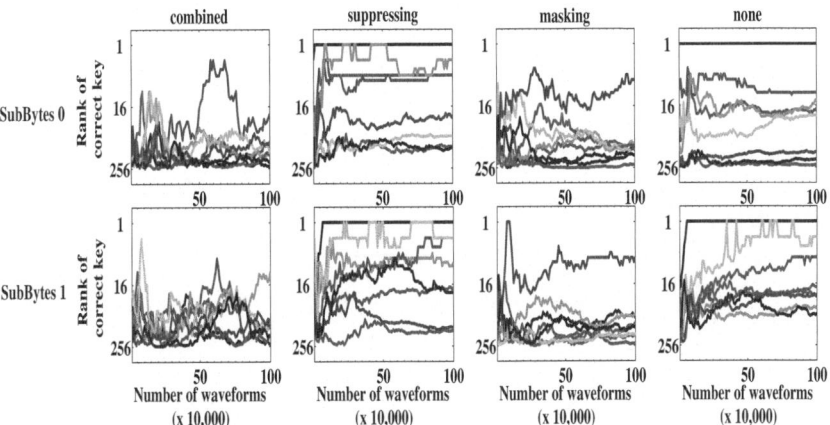

Fig. 12. 1-bit DPA results of SubBytes 0-1 with SubBytes input as selection function

CPA results are shown in Fig. 11, where Hamming distance of a data register is used for the selection function. The vertical axis shows the logarithm plots representing the ranks of a correct 8-bit partial key and the horizontal axis shows the number of traces. The correct keys for the two S-boxes were found with very few number of traces in **suppressing** and **none**, but the attack failed in **combined** and **masking** using the random mask. This is because the Hamming distance of the masked data cannot be estimated from the partial key and a ciphertext.

Single-bit DPA results are shown in Fig. 12, where each input bit of the 8-bit S-box is used as the selection function. Therefore, eight estimations for one correct partial key are shown in each graph. The correct key was identified by some selection functions in **suppressing** and **none** but the attack was completely failed in **combined**. There are some selection functions that lead relatively high rank for the correct key, but it would be difficult to identify the answer. In contrast, The DPA using NAND gate input as the selection function successfully

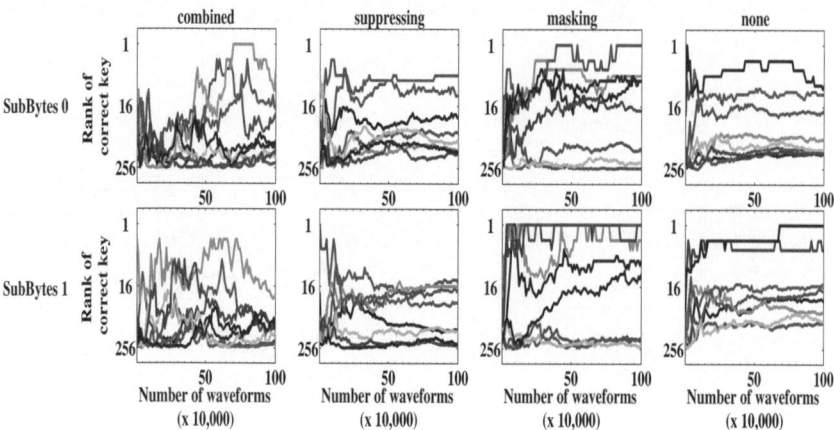

Fig. 13. 1-bit DPA results of SubBytes 0-1 with NAND gate input as selection function

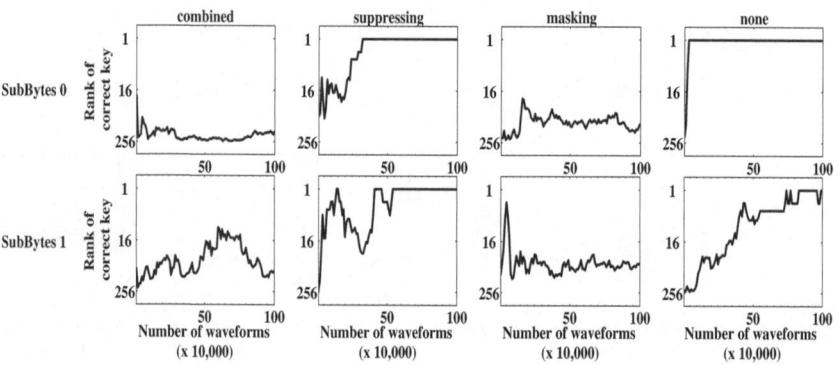

Fig. 14. 8-bit DPA results of SubBytes 0-1 with with SubBytes input as selection function

revealed the correct key as shown in Fig. 13. These results are consistent with the leakage analysis in Figs. 9 and 10.

Multi-bit DPAs, extended version of DPA, are more powerful attacks [18,19]. We explain the attack results by the multi-bit DPA in [19], which worked fine as an attack to our circuit. This is an attack method that uses the sum of absolute values of differential power of multi-bits. Fig. 14 shows the results of 8-bit DPA using the 8-bit input to the S-box circuit in the final round. In the same way as the single-bit DPA, the correct key was identified in **suppressing** and **none**, but not in **combined** and **masking**. The multi-bit DPA that generates only one DPA trace for each partial key makes the key estimation easier than the single-bit DPA that creates eight DPA traces that contain wrong keys. In Fig. 15, each input to eight NAND gates of the S-box circuit is used as a selection function. The results also show the correct key estimation more clearly than the single-bit DPA. However, the attack in **combined** still failed. It would be difficult

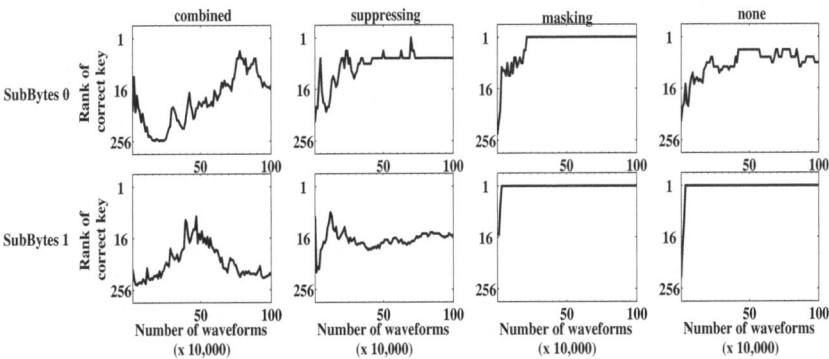

Fig. 15. 8-bit DPA results of SubBytes 0-1 with NAND gate input as selection function

to form a special selection function such as the input to the specific NAND gate without design information about the target device. However, the method presented in [20] enables us to seek an efficient selection function. Therefore, the glitch suppressing function in the pseudo RSL is important for pursuing sufficient security for power analysis attacks.

6   Conclusion

In this paper, we proposed design methodology of RSL using CMOS standard libraries, which we call "pseudo RSL". Conditions to guarantee security of the pseudo RSL are also discussed. AES circuits with (and without) the pseudo RSL gates are implemented on an experimental LSI using a 130-nm CMOS standard cell library according to the proposed design flow. The pseudo RSL scheme doubles the hardware resource and halves the operating speed in comparison with original AES circuit. However, this result shows high advantages in hardware performance over the WDDL scheme that needs three times as many gates and reduces the operating frequency down to 1/4. Various CPA and DPA attacks on the LSI controlling data masking and glitch suppressing function demonstrated that the pseudo RSL-AES circuit has a very high security against the attacks. We also confirmed that the AES circuit with the pseudo RSL has a very high power analysis resistance thanks to the contributions of the masking and the glitch suppressing functions. This is the first result demonstrating reduction of the side-channel leakage by glitch suppression quantitatively on real ASIC.

However, in the pseudo RSL AES circuit, a specific 1-bit intermediate signal was found to be used as a selection function to identify a specific 1-byte partial key (See Appendix A). We have confirmed that some signal paths do not satisfy the delay conditions for pseudo RSL by static timing analysis using design tools. Therefore, new LSIs have been developed, where the detailed delay information to satisfy the security conditions were fed back to their layouts. We have just started evaluation of the new LSIs, and no information leakage has been found so far. Detailed experimental results will soon be reported.

Acknowledgments

SASEBO boards were developed by AIST and Tohoku University in undertaking projects sponsored by METI (Ministry of Economy, Trade and Industry, Japan).

References

1. Kocher, P.C.: Timing Attacks on Implementations of Diffe-Hellmann, RSA, DSS, and Other Systems. In: Koblitz, N. (ed.) CRYPTO 1996. LNCS, vol. 1109, pp. 104–113. Springer, Heidelberg (1996)
2. Kocher, P.C., Jaffe, J., Jun, B.: Differential Power Analysis. In: Wiener, M. (ed.) CRYPTO 1999. LNCS, vol. 1666, pp. 388–397. Springer, Heidelberg (1999)
3. Suzuki, D., Saeki, M., Ichikawa, T.: Random Switching Logic: A Countermeasure against DPA based on Transition Probability. Cryptology ePrint Archive, Report 2004/346 (2004)
4. Suzuki, D., Saeki, M., Ichikawa, T.: Random Switching Logic: A New Countermeasure against DPA and Second-Order DPA at the Logic Level. IEICE Trans. Fundamentals E90-A(1), 160–168 (2007)
5. Messerges, T.S.: Using Second-Order Power Analysis to Attack DPA Resistant Software. In: Paar, C., Koç, Ç.K. (eds.) CHES 2000. LNCS, vol. 1965, pp. 238–251. Springer, Heidelberg (2000)
6. Waddle, J., Wagner, D.: Towards Efficient Second-Order Power Analysis. In: Joye, M., Quisquater, J.-J. (eds.) CHES 2004. LNCS, vol. 3156, pp. 1–15. Springer, Heidelberg (2004)
7. Tiri, K., Schaumont, P.: Changing the Odds against Masked Logic. In: Biham, E., Youssef, A.M. (eds.) SAC 2006. LNCS, vol. 4356, pp. 134–146. Springer, Heidelberg (2007)
8. Tiri, K., Schaumont, P.: Masking and Dual-Rail Logic Don't Add Up. In: Paillier, P., Verbauwhede, I. (eds.) CHES 2007. LNCS, vol. 4727, pp. 95–106. Springer, Heidelberg (2007)
9. Tiri, K., Verbauwhede, I.: A Logic Level Design Methodology for a Secure DPA Resistant ASIC or FPGA Implementation. In: Design, Automation and Test in Europe Conference (DATE 2004), pp. 246–251 (2004)
10. TSMC Webpage, http://www.tsmc.com/english/default.htm
11. Mangard, S., Popp, T., Gammel, B.M.: Side-Channel Leakage of Masked CMOS Gates. In: Menezes, A. (ed.) CT-RSA 2005. LNCS, vol. 3376, pp. 351–365. Springer, Heidelberg (2005)
12. Suzuki, D., Saeki, M., Ichikawa, T.: DPA Lekage Models for CMOS Logic Circuits. In: Rao, J.R., Sunar, B. (eds.) CHES 2005. LNCS, vol. 3659, pp. 366–382. Springer, Heidelberg (2005)
13. Satoh, A., Morioka, S., Takano, K., Munetoh, S.: A Compact Rijndael Hardware Architecture with S-Box Optimization. In: Boyd, C. (ed.) ASIACRYPT 2001. LNCS, vol. 2248, pp. 239–254. Springer, Heidelberg (2001)
14. Cryptographic Hardware Project: Project Webpage, http://www.aoki.ecei.tohoku.ac.jp/crypto/web/cores.html
15. Tiri, K., Hwang, D., Hojat, A., Lai, B., Yang, S., Schaumont, P., Verbauwhede, I.: Prototype IC with WDDL and Differential Routing - DPA Resistance Assessment. In: Rao, J.R., Sunar, B. (eds.) CHES 2005. LNCS, vol. 3659, pp. 354–365. Springer, Heidelberg (2005)

16. Side-channel Attack Standard Evaluation Board (SASEBO) Webpage,
    http://www.rcis.aist.go.jp/special/SASEBO/index-en.html
17. Brier, E., Clavier, C., Olivier, F.: Correlation Power Analysis with a Leakage Model.
    In: Joye, M., Quisquater, J.-J. (eds.) CHES 2004. LNCS, vol. 3156, pp. 16–29.
    Springer, Heidelberg (2004)
18. Messerges, T.S., Dabbish, E.A., Sloan, R.H.: Investigations of Power Analysis At-
    tacks on Smartcards. In: USENIX 1999 (1999), http://www.usenix.org/
19. Bevan, R., Knudsen, R.: Ways to enhance differential power analysis. In: Lee, P.J.,
    Lim, C.H. (eds.) ICISC 2002. LNCS, vol. 2587, pp. 327–342. Springer, Heidelberg
    (2003)
20. Suzuki, D., Saeki, M., Matsumoto T.: Self-Contained Template Attack: How to
    Detect Weak Bits for Power Analysis without Reference Devices. In: SCIS 2009,
    1A1-2 (2009) (in Japanese)

Appendix A: Weakness Detected in Our Prototype LSI

A photograph of an evaluation board used in our experiment is shown in Fig. 16.
Our pseudo RSL-AES co-processor can toggle on/off independently the two functions to prevent leakage, the random masking and glitch suppressing functions
by setting the values of the mode register implemented in the prototype LSI
shown in Fig. 17.

As for **combined**, Fig. 18 shows the results of attacks by 1 bit DPA that uses
all input data to the SubBytes circuit as the selection function. The figure shows
that there is a bit where the correct answer key ranks 1 with a million traces.
Similar tendency is also observed in the experimental result shown in Fig. 19
changing selection functions.

This could be caused by some input signals not satisfying the delay conditions
of the pseudo RSL. We have already confirmed some timing violations in a report
of static timing analysis. Therefore, new LSIs have been developed, where the detailed delay information to satisfy the security conditions were fed back to their
layouts. We have just started evaluation of the new LSIs, and no information leakage has been found so far. Detailed experimental results will soon be reported.

Also, in Fig. 18, as the number of trace increases, the correct key goes higher in
rank and stays there for a while, and then goes down. This is caused by the bias

Fig. 16. Experimental board SASEBO-R

Fig. 17. Our prototype LSI

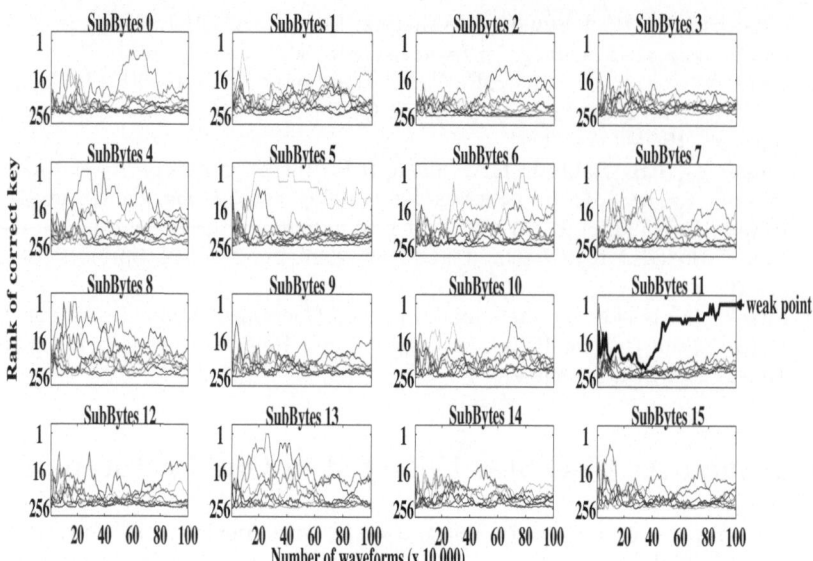

Fig. 18. 1-bit DPA result of all on **combined** with SubBytes input as selection function

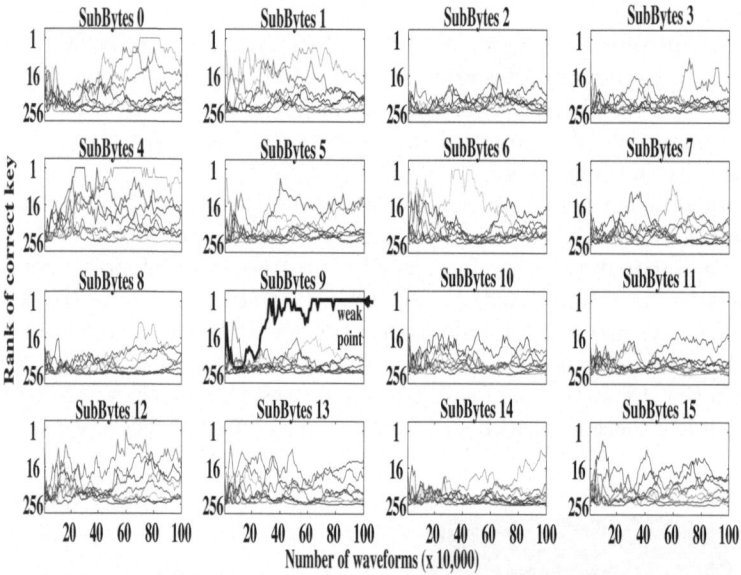

Fig. 19. 1-bit DPA result of all on **combined** with NAND gate input as selection function

of pseudo-random number for the masking. This trend is changed if the order of selecting trace data changes in the statistical processing. This case shows that it is difficult to know how many waveforms justify the end of attacks.

A Design Flow and Evaluation Framework for DPA-Resistant Instruction Set Extensions

Francesco Regazzoni[1,4], Alessandro Cevrero[2,3], François-Xavier Standaert[1],
Stephane Badel[3], Theo Kluter[2], Philip Brisk[2], Yusuf Leblebici[3],
and Paolo Ienne[2]

[1] UCL Crypto Group, Université catholique de Louvain, Louvain-la-Neuve, Belgium
{francesco.regazzoni,fstandae}@uclouvain.be
[2] School of Computer and Communication Sciences - EPFL, Lausanne, Switzerland
first_name.last_name@epfl.ch
[3] School of Engineering - EPFL, Lausanne, Switzerland
first_name.last_name@epfl.ch
[4] ALaRI - University of Lugano, Lugano, Switzerland
regazzoni@alari.ch

Abstract. Power-based side channel attacks are a significant security risk, especially for embedded applications. To improve the security of such devices, protected logic styles have been proposed as an alternative to CMOS. However, they should only be used sparingly, since their area and power consumption are both significantly larger than for CMOS. We propose to augment a processor, realized in CMOS, with custom instruction set extensions, designed with security and performance as the primary objectives, that are realized in a protected logic. We have developed a design flow based on standard CAD tools that can automatically synthesize and place-and-route such hybrid designs. The flow is integrated into a simulation and evaluation environment to quantify the security achieved on a sound basis. Using MCML logic as a case study, we have explored different partitions of the PRESENT block cipher between protected and unprotected logic. This experiment illustrates the tradeoff between the type and amount of application-level functionality implemented in protected logic and the level of security achieved by the design. Our design approach and evaluation tools are generic and could be used to partition any algorithm using any protected logic style.

1 Introduction

Security is a fundamental requirement for modern embedded systems. Mathematically strong cryptographic algorithms are insufficient due to the advent of side channel attacks, which exploit weaknesses in the underlying hardware platform rather than directly attacking the algorithm itself. At present, there is no perfect protection against side channel attacks. Hence, combining countermeasures implemented at different abstraction levels is necessary to reach a significant security level. In this context, solutions exploiting a dedicated technology such as protected logic styles are interesting because they directly tackle

C. Clavier and K. Gaj (Eds.): CHES 2009, LNCS 5747, pp. 205–219, 2009.

the problem of information leakage at their source. These logic styles can also be combined with software countermeasures to increase the difficulty of performing an attack.

The main drawback of protected logic styles proposed so far is that their area and power consumption are both significantly greater than that of traditional CMOS. They can also significantly increase the design time compared to circuits built from standard cells. Hence, complete processors and ASICs implemented in protected logic styles are generally too costly for practical applications, and would likely have low production volumes.

To overcome this issue without compromising security, protected logic styles must be used sparingly. With this respect, an interesting alternative is to build processors and ASICs that are realized primarily in CMOS logic, with a small and security-critical portion of the design realized in a protected logic. This creates a new and challenging partitioning problem that designers must be able to solve. But doing so will be quite difficult unless there is a suitable metric to evaluate and compare the security of a computation performed in either CMOS or a protected logic. Additionally, CAD tools must be able to support complex design flows that mix different logic styles. Finally, there is a distinct need for a comprehensive evaluation framework that combines a simulation environment with suitable metrics and provides a fair and accurate comparison of designs in respect to different criteria (e.g., power consumption, area, throughput, security).

To this end, this paper proposes a complete design flow for implementing and verifying circuits realized combining protected and non protected logic styles. Our design flow is built from standard CAD tools and is integrated with a methodology to evaluate the security of the designs that have been partitioned, following a theoretical framework for analyzing the information leakage provided by side-channel attacks. Focusing on a processors augmented with custom Instruction Set Extensions (ISEs) realized in protected logic styles, we explore the tradeoffs between the type and amount of application-level functionality implemented in protected logic and the level of security that can be achieved by the design. In our experiments, in particular, we vary the portions of the cryptographic algorithm that are realized in protected logic and CMOS, which gives us a better understanding of the tradeoffs between the usage of protected logic and security.

Starting from an RTL description of the target processor and a software implementation of a cryptographic algorithm, our tool allows the user to manually select the sensitive parts of the algorithm to be protected. Our design flow then automatically generates the new software, the ISEs and their interface to the processor, synthesizing a complete system as described above. The power consumption of the full system is simulated at the SPICE level while running the application. These power traces are then used to compute an information theoretic metric in order to evaluate the information leakage of the protected core. We have selected MOS Current Mode Logic (MCML) as the protected logic for use in this study. However, the ideas presented in this paper are generally amenable to any type of protected logic that is compatible with a CMOS process.

The remainder of the paper is organized as follows. Section 2 summarizes previous works in the area of side channel attacks. Section 3 recalls the metric used to evaluate the side-channel leakage and presents several extensions that were necessary to make it usable in the context of our design flow. Section 4 describes our hybrid design and evaluation methodology. Section 5 presents the results we obtained applying our methodology to the PRESENT block cipher and discusses the security vs. cost tradeoffs. Section 6 concludes the paper.

2   Background and Related Work

Side channel attacks are a powerful and easy to perform class of physical attacks. Consequently they have received much attention from the scientific community. The most frequently considered sources of side-channel information are power consumption and timing characteristics [13,12]. To perform Differential Power Analysis (DPA), the attacker executes the cryptographic algorithm on a target device multiple times, and then uses statistical methods to evaluate the information observed from the executions. Countermeasures, such as algorithmic techniques [7,23], architectural approaches [11,18,19], and hardware-related methods [20,27] can help to protect against DPA. Even if none of them are perfect, these countermeasures increase the efforts required to mount a successful attack. In this paper, we are mainly concerned with technological solutions, usually denoted as *DPA-resistant logic styles* in the literature.

Many DPA-resistant logic styles have been proposed in the past. *Sense Amplified Base Logic (SABL)* [27], for example, combines dual-rail and pre-charged logic [17]. SABL cells consume constant power, provided that they are designed and implemented in a carefully balanced way. Other proposed DPA-resistant logic styles include: *Wave Dynamic Differential Logic (WDDL)* [28], which balances circuit activity with complementary logic gates; *Dynamic Current Mode Logic (DyCML)* [2,15], a dual-rail pre-charge logic, similar to SABL, but with a reduced power-delay product; *Low-swing current mode logic (LSCML)* [9], which is similar to DyCML, but is independent of transistor sizes and load capacitances; *Masked Dual-Rail Pre-charge Logic (MDPL)* [21], which attempts to eliminate routing constraints that plague other dual-rail style gates; *Gammel-Fischer Logic (GF)* [8], a form of masked logic that protects against information leakage in the presence of glitches; finally, MCML that will be our running example [22], a MOS transistor-based current mode logic.

One of the key challenges when implementing protected logic styles is to analyze the DPA-resistance of the different operators in an application. This task is even more critical when partitioning a design between CMOS and a protected logic. To address possible shortcomings, Standaert et al. [24] introduced a combination of metrics that can be used to describe the amount of information leaked by a cryptographic device and the effectiveness of a side-channel adversary to exploit this information. When analyzing new countermeasures, it is primarily the information theoretic metric that is most useful, since it quantifies the reduction in information leakage resulting from the countermeasure using a sound criteria.

In theory, this metric yields an adversary-independent image of the asymptotic security of the device. This metric was first applied to DPA-resistant logic styles by Mace et al. [16], who describe in detail how to compute the entropy of a secret key conditionally with respect to the physical leakage in different scenarios. In this paper, we follow and extend this application of the metric.

Various studies on partitioning designs between CMOS and protected logic have been published in the literature. The most relevant work related to our concerns is probably the one of Tillich and Großschädl [26]. These authors analyze the resistance against side channel attacks of a processor extended with custom ISE for AES. They consider the possibility of implementing the most security-critical portions of the processor datapath in a DPA-resistant logic style. Our paper extends these initial ideas, providing different novel contributions. We present a fully automated design flow that allows realizing and simulating a complete environment (core + protected ISE). This proves the feasibility of combining CMOS and protected logic styles on the same chip and provides realistic measurements for area and power consumption. We also provide a more precise evaluation of the resistance against power analysis attacks for each design, due to the integration of an objective metric that quantifies the information leaked by different protected implementation. Lastly, our quantitative analysis applies jointly to security and performance issues and drives the process of ISE identification and synthesis; to the best of our knowledge, prior ISE identification methods have been driven primarily by performance [10,4,25].

3   Security Evaluation

The evaluation of the power consumption leakage provided by our simulation environment follows the principles of [24]. The goal of this methodology (that we don't detail here) is to provide fair evaluation metrics for side-channel leakage and attacks. In particular and as far as evaluating countermeasures or protected logic styles is concerned, the information theoretic metric that we now summarize allows being independent of a particular DPA attack scenario. It intuitively measures the resistance against the strongest possible type of side-channel attacks. In summary, let K be a random variable representing the key that is to be recovered in the side-channel attack; let X be a random variable representing the known plaintexts entering the target (leaking) operations; and let L be a random variable representing the power consumption traces generated by a computation with input x and key k. In our design environment, L is the output of a SPICE simulation trace T to which we add a certain amount of normally distributed random noise R, i.e. $L = T + R$. We compute the conditional entropy (or mutual information) between the key K and its corresponding leakage L, that is,

$$H[K|L] = -\sum_k \Pr[k] \cdot \sum_x \Pr[x] \int \Pr[l|k, x] \cdot \log_2 \Pr[k|l, x] \, dl.$$

There are different factors that influence this conditional entropy. The first is the shape of the simulated traces T. The second is the standard deviation of the noise in the leakage L. The number of dimensions in the traces is also important. Simulated traces contain several thousands of samples. Hence, directly applying multivariate statistics on these large dimensionality variables is hardly tractable. Mace et al. [16] reduce the dimensionality using Principal Component Analysis (PCA), and then evaluate the conditional entropy. Thus, the number of dimensions remaining after PCA is a third parameter that we consider.

Our usage of the metric builds on Mace et al.'s in two respects. First, we move from a 1-dimensional analysis to a multi-dimensional analysis, and we discuss the extent to which more samples increase the estimated value of $H[K|L]$. Additionally, we analyze complete designs, rather than 2-input logic gates, thereby establishing the scalability of the aforementioned metric and evaluation tools.

4   The Proposed Hybrid Design Flow

This section describes the entire design flow, from RTL to the integration with the information theoretic metric discussed in the preceding section. For any application, there exists a range of possible architectural and electrical implementations, such as ASIC vs. processor, or standard cell vs. full custom design. The choice of the platform has historically been dictated by performance, area, and/or power consumption, each of which can be measured accurately. Once the initial design point is fixed, designers consider a fine-grained space of possible solutions, and only at this point is security typically considered, often based mainly on empirical evaluation. The aim of this work is to bring security to the forefront of design variables for embedded systems by associating it with a clear quantitative metric. To achieve our goal, we propose a flexible and fully automated design flow based on standard CAD tools. The flow supports partitioning of a design between CMOS and protected logic, and includes a sound metric to measure resistance against side-channel attacks.

Figure 1 depicts the design flow, which is an extension of prior simulation-based methodologies to evaluate resistance to power analysis attacks [6,22]. The key idea is to use commodity from EDA tools and to leverage on transistor level simulation to provide a high degree of accuracy. The design flow that is used in this study targets an embedded processor that is augmented with instruction set extensions. The ISE are designed using information leakage as part of the objectives to optimize, and are implemented in a protected logic style. This point represents a major innovation, since in recent years, ISEs have been used primarily to improve performance, rather than to enhance security.

The flow has two inputs: the RTL description of an embedded processor that supports ISEs, and a software implementation of a cryptographic algorithm. The software code is passed to a tool that automatically extracts its data flow graph. The user manually select from the aforementioned graph the portion of the algorithm to be implemented in protected ISEs. Once this selection is done, the remainder of the flow is fully automated: the rewritten software, including an explicit call to the ISE is generated, along with an RTL description of the ISE and its interface to the processor. The output of the flow is a place-and-routed

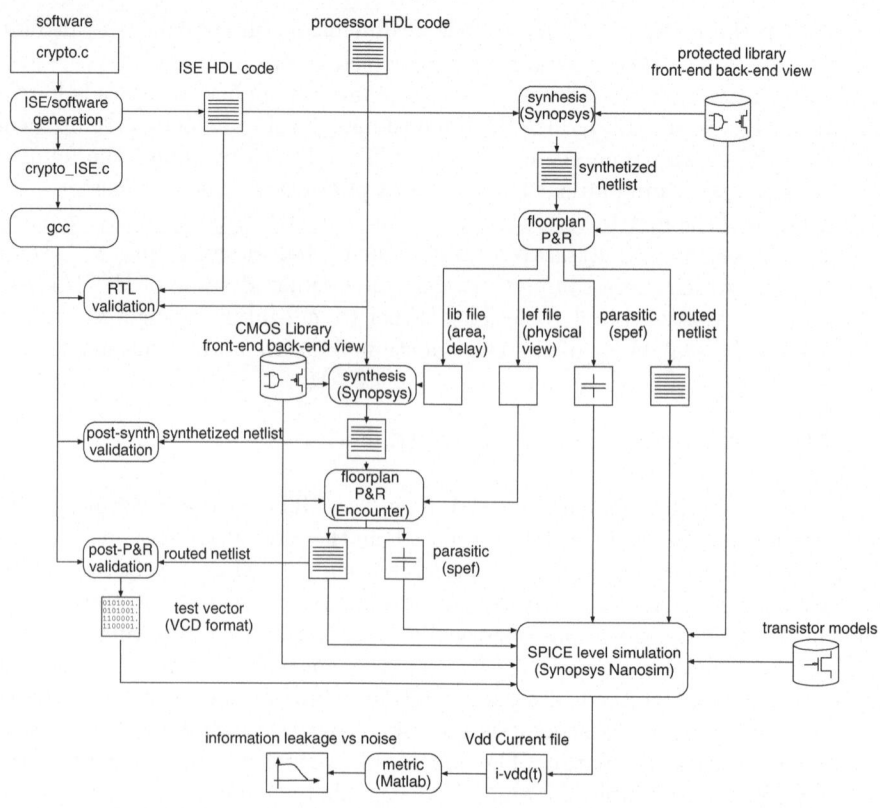

Fig. 1. Full view of the design flow

hardware design that is evaluated using a sound metric that measures information leakage. By iterating through different ISE implementations, an architect could generate a set of a design among which select the most suitable one.

In the newly generated software, calls to the ISE are automatically generated using the correct syntax, consistent with the RTL code, and thus supported by the compiler. The corresponding binary file is generated and simulated using an interpreter that mimics full system behavior, including the boot loader, and generates the corresponding values of the processor pins for each cycle of software execution. The pin values are then used in every validation step, including the generation of test vectors for use in SPICE level simulation. The SPICE level simulation of the full core (core + protected ISE) generates power traces which are used to measure the leakage of the processor.

To generate the customized processor, we begin with an HDL description of the processor core and a software implementation of a cryptographic algorithm. The first step is to select from an automatically generated data flow graph one or more ISEs to realize in protected logic; at present, this selection is the only step still performed manually, although we intend to automate it in the future. Once the HDL description of the ISE(s) has been generated, the circuit is synthesized using protected logic based on a standard cell library using Synopsys Design Compiler. The circuit is then placed and routed with Cadence Design Systems SoC Encounter. A parasitics file (in spef format) is produced, along with the Verilog netlist of the circuit and an sdf file for back annotation of the delays. The library file describing the ISE (i.e., timing information, power and dimension) and the layout exchange format file (the abstract view for placement and routing) are generated to integrate the ISE as a black box during synthesis and placement and routing of the complete design. Next, the complete design (i.e., the processor augmented with ISEs as a black-box) is synthesized and placed and routed using a standard CMOS flow. For the front end, the ISE library file is loaded by Design Compiler; the unit is recognized by the synthesis engine and the ISE's timing information is used during the synthesis process. During the physical design phase, the ISE is treated as a macro just like typical IP blocks and is pre-placed into the core. The flow produces the spef and sdf files and the Verilog netlists of the whole design.

Post-place-and-route simulation is now performed using ModelSim, with the previously generated sdf files (CMOS and protected logic) under the considered cryptographic benchmark. This simulation verifies the functionality of the processor and generates the test vectors for transistor-level simulation that will be used to generate power traces that will be input to the security evaluation. Synposys Nanosim performs transistor-level simulation, using the spef files for the protected ISE and CMOS core, with the relative Verilog netlists, SPICE models of the technology cells and the transistor models. This simulation generates vector-based time-varying power profiles which are stored in a simple text format. This data typically corresponds to the simulated traces represented by the variable T in Section 3 which concludes the treatment of the flow.

5   Case Study and Results

In this section we present the results of the evaluation of our design flow evaluated with different metrics of performances and security.

5.1   PRESENT Algorithm and the Considered Versions Overview

PRESENT [5] is a block cipher based on an SP-network that consists of 31 rounds. The block length is 64 bits and two key lengths of 80 and 128 bits are supported. During the encryption process, three different transformations are iterated 31 times. The three basic transformations are: *addRoundKey, sBoxLayer*, and *pLayer*: the first is function of the state and the secret key, while the final two are only functions of the state. At the completion of the last round an extra addRoundKey transformation is performed. The added key is different in each round and these round keys are generated by a key schedule routine that takes the secret key and executes an expansion as specified in algorithm description. The evaluation performed in this work, is done on a reduced version of the PRESENT algorithm, composed of just addRoundKey and sBoxLayer of 4 by 4

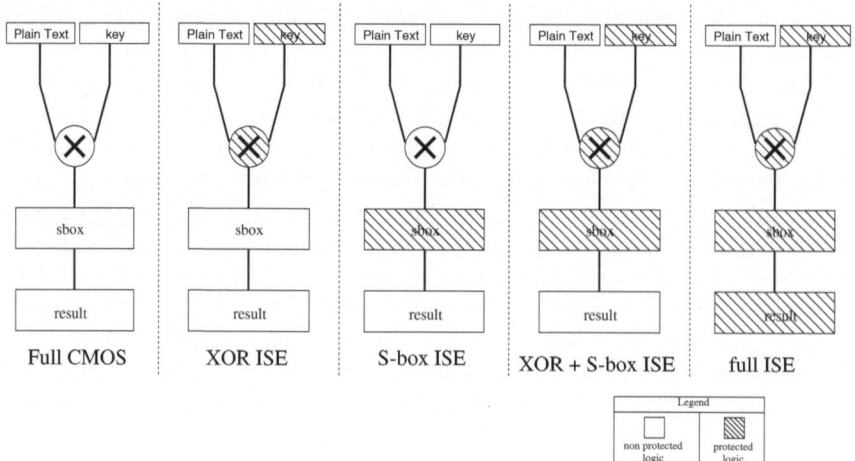

Fig. 2. Considered implementations for the algorithm

bit. We selected PRESENT as case of study for this work because the size of its S-box permits an exhaustive search of the design space without compromising the generality of the proposed methodology and results.

For our experiments, we considered the five possible implementations of the algorithm, that are depicted in Figure 2. Each implementation has a different section realized in protected logic. The first version, *Full CMOS*, is the reference version, in which the core is completely implemented in CMOS and the software does not leverage on any kind of ISE. In the second, *XOR ISE*, the full program is executed on the CMOS core, except for the secret key, that is stored into a protected register and the keyAddition, that is implemented using protected ISE. The third version, *S-box ISE*, implements only the sBoxLayer in a protected ISE, while the rest of the algorithm executes on the CMOS core. The fourth, *XOR + S-box ISE*, stores the secret key in a protected register and executes both addRoundKey and sBoxLayer using a protected ISE, but writes the result back to the processor register file, which is unprotected. Lastly, *full ISE* implements addRoundKey and sBoxLayer in protected logic, and stores the secret key and the result in a protected register that is part of the ISE as well.

5.2   Experimental Setup

The processor used in this work is an OpenRISC 1000 [14], a five stage pipelined in-order embedded processor. The processor provides a 32-bit datapath and a 32-entry single write-port, dual read port register file. The processor includes extra opcode space to support ISEs and is provided with a gcc cross-compiler.

We have selected MOS Current-Mode Logic (MCML) as secure logic style to implement the protected ISEs. MCML cells are low-swing, fully-differential circuits built with networks of differential pairs biased with a constant current source [1]. The constant DC current and the differential nature of the cells

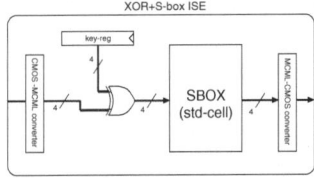

```
// Calculate S-box (plaintext XOR key)
int PRESENT_XOR+S-box-ISE(int plaintex) {
1 int result = 0; // initialize the result
  // call the new instruction to calculate s-box(pt ^key)
2 Instr_1(plaintext, result);
3 return result; }; // return the result
```

Fig. 3. Example of an ISE and its source in C: the XOR + S-box

provide an almost constant current consumption profile, which is independent of the switching activity. In theory, this results in a dramatic decrease of side-channel leakage and thus increased resistance against DPA attacks [22,29]. The increased DPA-resistance in differential logic circuits is obtained by the simultaneous and opposite switching of differential signal pairs resulting in almost perfect cancellation of current transients. In order to obtain consistently robust circuits, it is therefore critical to match the time constants in the two wires of each pair. This implies that each pair of wires must be physically routed along the exact same path, in order to equalize the length and parasitics of individual routes. To achieve this, the design flow proposed in [3] is used in this work. This entirely automated methodology enables the implementation of standard cell based-differential circuits from RTL with true differential routing, using a classical timing-driven design flow without human intervention. With this approach, the sensitive parts of the processor are implemented with secure logic and converter circuits are inserted at the boundary to interface between the two different logic styles. This increases security in a transparent way, without additional effort from the designers.

One example of an automatically generated ISE, reported in Figure 3, depicts the hardware view corresponding to the XOR + S-box ISE. The figure includes the converters between CMOS and MCML, which are necessary to interface the protected logic with the processor. These conversion circuits are automatically added at the inputs and outputs of the ISE.

We performed our experiments using the following versions of the design tools: Mentor Graphics Modelsim 6.2d for logic simulation, Synopsys Design Compiler 2007.12 for synthesis, and Cadence Design System SoC Encounter 7.1, for placementing and routing. Our CMOS target library was a 0.18μm commercial standard cell library. SPICE level simulation was carried on using Synopsys Nanosim 2007.03, and the transistor model is BSIM 3.3.

5.3   Results

Each of the five implementations presented in this paper has been synthesized to run at a clock frequency of 100 MHz, under worst-case process conditions. The same clock frequency is used for each of the ISEs.

Table 1 reports the area and average power consumption of the base Open-RISC1000 processor, as well as all the four versions augmented with protected

Table 1. Area occupation and average power consumption of each implementation

Version	Power Consumption (mW)	Full Die Size (mm²)	ISE Size (mm²)	Gate Count (GE)
full CMOS	87.77	1.8603	-	139071
XOR ISE	129.24	1.9810	0.1207	140787
S-box ISE	129.42	1.9838	0.1235	140843
XOR + S-box ISE	129.81	1.9844	0.1241	140853
full ISE	129.83	1.9849	0.1246	140865

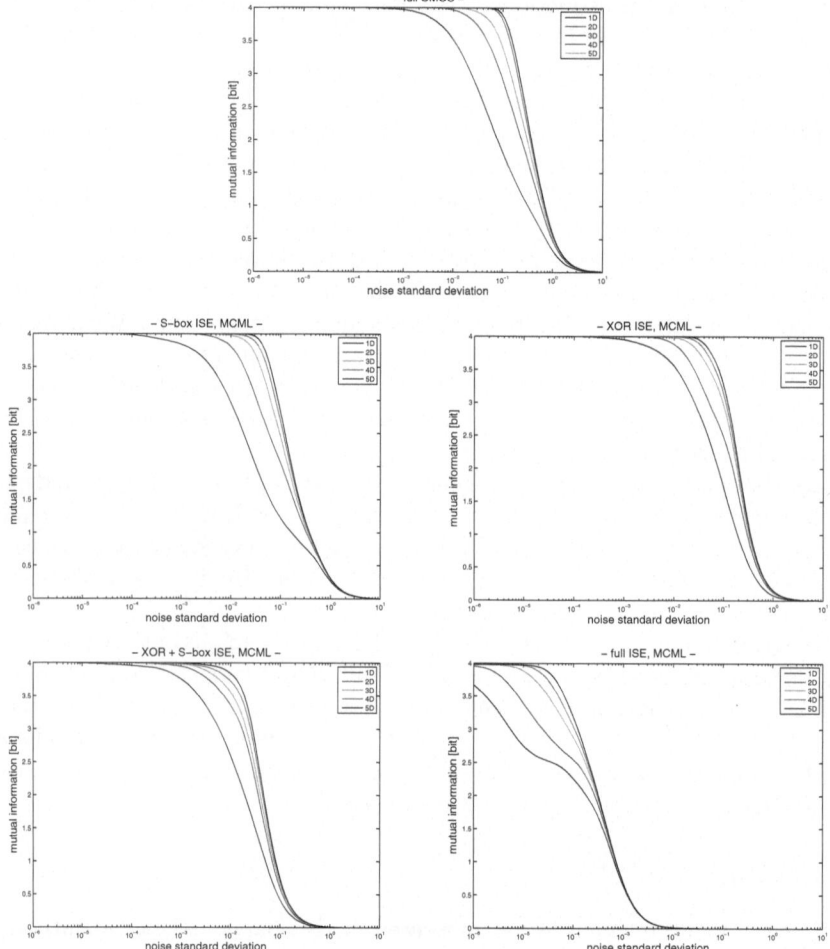

Fig. 4. Mutual information leaked by different implementations in function of a noise standard deviation, for different dimensions kept after application of the PCA

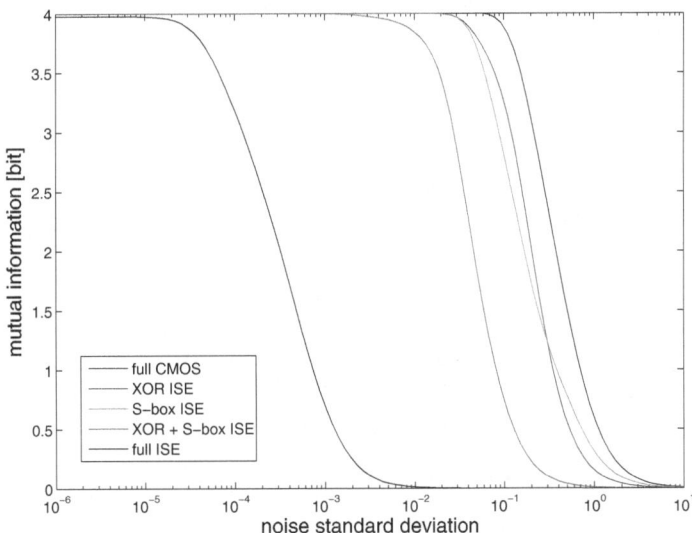

Fig. 5. Mutual information leaked by different implementations in function of a noise standard deviation, with 5 dimensions kept after application of the PCA

ISEs. The average power consumption is calculated for each core during the execution of the PRESENT algorithm, including calls to the ISEs. We report the silicon area occupation and the gate count. The absolute silicon area provides a clear measure for the physical cost of each implementation, while the number of equivalent gates highlights the complexity of the circuit. We calculated the number of equivalent gates for each implementation, with appropriate weights to account for the disparity in sizes between reference gates for CMOS and MCML. In our experiments, the difference in area penalty between the largest and smallest ISE is 0.2%. This is primarily due to the small size of the PRESENT algorithm, which tends to be overshadowed by the size of the conversion circuitry at the CMOS-MCML boundary.

The full ISE implementation, which is the most resistant to DPA, increases the power consumption by 47.9% with respect to the full CMOS design, while the area overhead is to 6.7%. A similar level of leakage using the same protected logic would be possible by implementing a full processor in MCML. Our results show that this would increase the power consumption by a factor of approximately 40 time higher compared to a CMOS implementation, while increasing the total area 2.65 times. The large power difference is due to the static current consumption of MCML gates, whose power consumption becomes close to CMOS gates for high switching activity and operating frequancy. The MCML library, which has been developed internally, has not been tuned for battery operated devices; an MCML library tagetting these devices would significantly improve the results.

Preliminary results of our security evaluations using the information theoretic metric of Section 3 are plotted in Figure 4 for different implementations. They

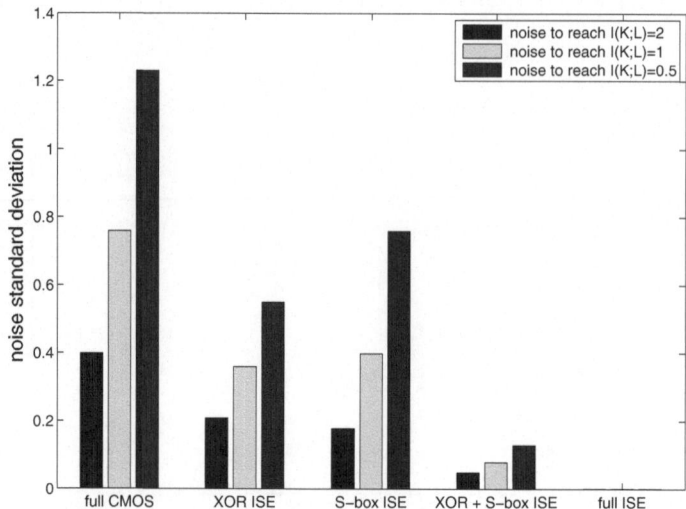

Fig. 6. Noise standard deviation required to reach a certain level of information leakage, with 5 dimensions kept after application of the PCA

show that increasing the number of dimensions to evaluate the mutual information $I(K, L)$ improves the quality of the evaluations up to a certain extent, where the noise variance is sufficiently large for hiding the small information leakage in the higher dimensions of the transformed traces.

Figures 5 and 6 compare the results for the five different processor and ISE combinations. Figure 5 plots the mutual information leaked by the different implementations. Figure 6 takes the opposite point of view, and illustrates the amount of noise that is required to reduce the leakage up to a threshold. The goal of a robust countermeasure is to reduce the information leakage.

These two figures concisely illustrate and confirm our intuition, namely, that protecting a part of the algorithm reduces the leakage; however, the overall security of a system depends on its weakest point. Consequently, there is a significant improvement when considering a fully protected ISE. The analysis shows that the fully protected ISE has no obvious logical weaknesses; however, it remains to be determined the extent to which a fabricated piece of silicon can be built to confirm the results of the simulations. Anyway, these results confirm the applicability of our proposed design flow up to the analysis of the side channel leakage. The computation of the evaluation metrics, including the selection of the points of interest in the leakage traces with a PCA, is fully automated.

6   Conclusions

With the increased use of embedded systems in security applications, protection against side channel attacks has become increasingly important. This paper summarizes the first attempt to integrate a meaningful information leakage metric into an industrial design flow for secure systems. Our results establish the feasibility of the proposed flow, and show that the use of ISEs in protected logic styles is a reasonable and low-cost method to provide the desired security. Future work will focus on manufacturing the most promising implementations. It is in fact well known that the decisive proof of robustness is obtained only when the actual fabricated microchip is attacked using high frequency probes and an oscilloscope. Nonetheless, our design flow is fundamental to perform a deeper design space exploration before the fabrication.

Acknowledgements

The authors would like to thank Federico Ferrari for fruitful discussions as well as the anonymous referees for their constructive comments. This work was partially supported by HIPEAC Network of Excellence (collaboration grant) and by the Walloon Region project E-USER (WIST program). François-Xavier Standaert is an associate researched of the Belgian Fund for Scientific Research (FNRS-F.R.S.).

References

1. Alioto, M., Palumbo, G.: Model and Design of Bipolar and MOS Current-Mode Logic: CML, ECL and SCL Digital Circuits. Springer, Dordrecht (2005)
2. Allam, M.W., Elmasry, M.I.: Dynamic current mode logic (DyCML): A new low-power high-performance logic style. IEEE Journal of Solid-State Circuits 36(3), 550–558 (2001)
3. Badel, S., Guleyupoglu, E., Inac, O., Martinez, A.P., Vietti, P., Gürkaynak, F.K., Leblebici, Y.: A generic standard cell design methodology for differential circuit styles. In: Proceedings of the Design, Automation and Test in Europe Conference and Exhibition, Munich, March 2008, pp. 843–848 (2008)
4. Bartolini, S., Branovic, I., Giorgi, R., Martinelli, E.: A performance evaluation of ARM ISA extension for elliptic curve cryptography over binary finite fields. In: Proceedings of the 16th Symposium on Computer Architecture and High Performance Computing, Foz do Iguaçu, Brazil, October 2004, pp. 238–245 (2004)
5. Bogdanov, A., Knudsen, L.R., Leander, G., Paar, C., Poschmann, A., Robshaw, M.J.B., Seurin, Y., Vikkelsoe, C.: PRESENT: An ultra-lightweight block cipher. In: Paillier, P., Verbauwhede, I. (eds.) CHES 2007. LNCS, vol. 4727, pp. 450–466. Springer, Heidelberg (2007)
6. Bucci, M., Guglielmo, M., Luzzi, R., Trifiletti, A.: A power consumption randomization countermeasure for DPA-resistant cryptographic processors. In: Macii, E., Paliouras, V., Koufopavlou, O. (eds.) PATMOS 2004. LNCS, vol. 3254, pp. 481–490. Springer, Heidelberg (2004)
7. Coron, J.-S., Goubin, L.: On boolean and arithmetic masking against differential power analysis. In: Paar, C., Koç, Ç.K. (eds.) CHES 2000. LNCS, vol. 1965, pp. 231–237. Springer, Heidelberg (2000)
8. Fischer, W., Gammel, B.M.: Masking at gate level in the presence of glitches. In: Rao, J.R., Sunar, B. (eds.) CHES 2005. LNCS, vol. 3659, pp. 187–200. Springer, Heidelberg (2005)

9. Hassoune, I., Macé, F., Flandre, D., Legat, J.-D.: Low-swing current mode logic (LSCML): A new logic style for secure and robust smart cards against power analysis attacks. Microelectronics Journal 37(9), 997–1006 (2006)

10. Intel's advanced encryption standard (AES) instructions set (white paper) (April 2009)

11. Irwin, J., Page, D., Smart, N.P.: Instruction Stream Mutation for Non-Deterministic Processors. In: Proceedings of the 13th International Conference on Application-specific Systems, Architectures and Processors, San Jose, Calif., July 2002, pp. 286–295 (2002)

12. Kocher, P.C., Jaffe, J., Jun, B.: Differential power analysis. In: Wiener, M. (ed.) CRYPTO 1999. LNCS, vol. 1666, pp. 398–412. Springer, Heidelberg (1999)

13. Kocher, P.C.: Timing attacks on implementations of diffie-hellman, rsa, dss, and other systems. In: Koblitz, N.I. (ed.) CRYPTO 1996. LNCS, vol. 1109, pp. 104–113. Springer, Heidelberg (1996)

14. Lampret, D.: OpenRISC 1000 Architecture Manual (April 2006)

15. Macé, F., Standaert, F.-X., Hassoune, I., Legat, J.-D., Quisquater, J.-J.: A dynamic current mode logic to counteract power analysis attacks. In: Proceedings of the XIX Conference on Design of Circuits and Integrated Systems, Bordeaux, France (November 2004)

16. Macé, F., Standaert, F.-X., Quisquater, J.-J.: Information theoretic evaluation of side-channel resistant logic styles. In: Paillier, P., Verbauwhede, I. (eds.) CHES 2007. LNCS, vol. 4727, pp. 427–442. Springer, Heidelberg (2007)

17. Mangard, S., Oswald, E., Popp, T.: Power Analysis Attacks: Revealing the Secrets of Smart Cards. Advances in Information Security. Springer, New York (2007)

18. May, D., Muller, H.L., Smart, N.P.: Non-deterministic processors. In: Varadharajan, V., Mu, Y. (eds.) ACISP 2001. LNCS, vol. 2119, pp. 115–129. Springer, Heidelberg (2001)

19. May, D., Muller, H.L., Smart, N.P.: Random register renaming to foil DPA. In: Koç, Ç.K., Naccache, D., Paar, C. (eds.) CHES 2001. LNCS, vol. 2162, pp. 28–38. Springer, Heidelberg (2001)

20. Moore, S., Anderson, R., Cunningham, P., Mullins, R., Taylor, G.: Improving Smart Card security using self-timed circuits. In: Proceedings of the 8th International Symposium on Advanced Research in Asynchronous Circuits and Systems, Manchester, April 2002, pp. 211–218 (2002)

21. Popp, T., Mangard, S.: Masked dual-rail pre-charge logic: DPA-resistance without routing constraints. In: Rao, J.R., Sunar, B. (eds.) CHES 2005. LNCS, vol. 3659, pp. 172–186. Springer, Heidelberg (2005)

22. Regazzoni, F., Eisenbarth, T., Poschmann, A., Großschädl, J., Gurkaynak, F., Macchetti, M., Toprak, Z., Pozzi, L., Paar, C., Leblebici, Y., Ienne, P.: Evaluating resistance of MCML technology to power analysis attacks using a simulation-based methodology. In: Gavrilova, M.L., Tan, C.J.K., Moreno, E.D. (eds.) Transactions on Computational Science IV. LNCS, vol. 5430, pp. 230–243. Springer, Heidelberg (2009)

23. Rostovtsev, A.G., Shemyakina, O.V.: AES side channel attack protection using random isomorphisms. Cryptology e-print archive (March 2005), http://eprint.iacr.org/

24. Standaert, F.-X., Malkin, T., Yung, M.: A unified framework for the analysis of side-channel key recovery attacks. In: Joux, A. (ed.) EUROCRYPT 2009. LNCS, vol. 5479, pp. 443–461. Springer, Heidelberg (2009)

25. Tillich, S., Großschädl, J.: Instruction set extensions for efficient AES implementation on 32-bit processors. In: Goubin, L., Matsui, M. (eds.) CHES 2006. LNCS, vol. 4249, pp. 270–284. Springer, Heidelberg (2006)
26. Tillich, S., Großschädl, J.: Power analysis resistant AES implementation with instruction set extensions. In: Paillier, P., Verbauwhede, I. (eds.) CHES 2007. LNCS, vol. 4727, pp. 303–319. Springer, Heidelberg (2007)
27. Tiri, K., Akmal, M., Verbauwhede, I.: A dynamic and differential CMOS logic with signal independent power consumption to withstand differential power analysis on Smart Cards. In: Proceedings of the 28th European Solid-State Circuits Conference, Florence, September 2002, pp. 403–406 (2002)
28. Tiri, K., Verbauwhede, I.: A logic level design methodology for a secure DPA resistant ASIC or FPGA implementation. In: Proceedings of the Design, Automation and Test in Europe Conference and Exhibition, Paris, February 2004, pp. 246–251 (2004)
29. Toprak, Z., Leblebici, Y.: Low-power current mode logic for improved DPA-resistance in embedded systems. In: Proceedings of the IEEE International Symposium on Circuits and Systems, Kobe, Japan, May 2005, pp. 1059–1062 (2005)

Crypto Engineering: Some History and Some Case Studies

(Extended Abstract)

Christof Paar

Chair for Embedded Security
Electrical Engineering and Information Sciences Dept.
Ruhr-Universität Bochum
www.crypto.rub.de

Abstract of the Extended Abstract. In this extended abstract, I will first try to describe briefly the developments in the cryptographic engineering community over the last decade. After this, some hopefully instructive case studies about cryptographic implementations in the real world will be given.

1 Timing the Market or: Why Did CHES Grow So Rapidly?

Exactly 10 years have passed since Cetin Koc and myself started the CHES (Cryptographic Hardware and Embedded Systems) workshop series. When the idea for the workshop was conceived in 1998 we had planned for a small, highly focused workshop in our favorite research area of cryptographic hardware, and we expected 50, perhaps 60, attendees. When almost 160 people showed up, Cetin and I knew that there was something special about this field. Even though this was a pleasant surprise, I had no idea how broad the area of cryptographic engineering would evolve in the years to come. In the following I would like to take the chance and speculate a bit about the reasons why CHES has grown to one of the most important events in applied cryptography.

At that time of the first CHES, my main interest and expertise was in hardware and software algorithms for asymmetric cryptography. Since public-key algorithms were very arithmetic intensive and both hardware and software performance was a far cry from what it is today, it was clear that much research was needed. Implementing RSA with an astronomically large modulus of 512 bit on 1990s PCs with acceptable run times was a major undertaking. Thus, at the time we started to plan CHES, the main focus in cryptographic engineering was on fast public-key implementation techniques, such as, for example [8,5]. Even though there was certainly some work done on fast block cipher implementations (especially DES, but also IDEA and other ciphers), most of the scientifically challenging work targeted asymmetric implementations.

Thus, my view on the field where research in crypto engineering should take place was roughly described by the rather compact Table 1.

C. Clavier and K. Gaj (Eds.): CHES 2009, LNCS 5747, pp. 220–224, 2009.

Table 1. My world view on crypto engineering, ca. 1998

	HW impl.	SW impl.
asymmetric alg.	x	x

In the late 1990s, several developments took place which lead to an almost explosive growth of the area of cryptographic engineering. I see (at least) four main driving forces:

Side-Channel Attacks. The notion of

$$Crypto\ Engineering = Efficient\ Implementation$$

started to change in the second half of the 1990s. In 1996, the Bell Core attack as one of the first fault injection attack was published [3]. In the next three years timing attacks, simple power analysis and differential power analysis were presented [9]. Not only the smart card industry was under shell shock, but the crypto implementation community realized very quickly that its formula had to be extended to:

$$Crypto\ Engineering = Efficient\ Implementation + Secure\ Implementation$$

AES. In 1997 the AES selection process had started. For the community of implementers, the AES competition became interesting in earnest in 1998/99, in other words, after the algorithms had been submitted and the first ciphers were excluded. This sparked an increase interest in the implementation aspects of symmetric-key algorithms.

Cryptology Research Matured. Until the early 1990s, there were relatively few researchers working in cryptography outside government agencies. The field of cryptographic implementations was only a niche discipline with even fewer active people doing research. Publications were scattered over the cryptographic and engineering literature. The cryptographic community was well served by two flagship conferences, namely CRYPTO and EUROCRYPT, which were sufficient for reporting the most important developments in cryptology every year. However, the increasing number of researchers together with the increased understanding of many theoretical and practical issues in cryptology triggered a specialization and extension of the crypto conference landscape. Starting with FSE (Fast Software Encryption) in 1993, PKC (Public-Key Cryptography) in 1998 and CHES in 1999, several new workshops in sub-areas of cryptography served the need of a larger and more specialized scientific community. I believe this is the natural and healthy evolution of a discipline which is maturing.

Internet Boom. The last development which helped to push CHES and the field of crypto engineering was the dot-com boom in the late 1990s. There was both a perceived and a real need for everything that was related to information technology. Applied cryptography was considered part of the whole brave new world of the Internet area, and many companies started or enlarged their security groups. As part of that development, crypto engineering was also receiving more attention.

Table 2. The field of crypto engineering in 2009

	HW impl.		SW impl.		Secure impl.		TRNG	cryptanal. HW	PUF
	lightweight	high speed	lightweight	high speed	passive	fault inj.			
asymmetric	x	x	x	x	x	x	x	x	x
symmetric	x	x	x	x	x	x			

All of these factors contributed to extend the scope of CHES considerably. Within three years, there were more than 200 attendees and more than 100 submissions. Hence in hindsight, the reason why CHES has become such an important conference was there was almost a perfect *market timing* for starting CHES in 1999: the time was simply ripe for such an event.

In the years since then, new topics such as lightweight crypto, true random number generators (TRNG), cryptanalytical hardware and physical unclonable functions (PUF) were also added to the repertoire of topics treated at CHES. Thus, a current listing of the sub-areas of modern crypto engineering is more accurately described by this table:

The table should certainly not be taken as the final verdict on the spectrum of topics within crypto engineering. For instance, new topics like Trojan hardware (as evident by the Hot Topics Session of this year's CHES), are emerging and should certainly be included.

2 Embedded Cryptography in the Wild: Some Case Studies

Cryptography has sneaked into everything, from web browsers and email programs to cell phones, bank cards, cars and even into medical devices. In the near future we will find many new exciting applications for cryptography such as RFID tags for anti-counterfeiting or car-to-car communications. I want to briefly mention research projects we have been involved in which cryptography was instrumental for securing new embedded applications.

Lightweight Cryptography for RFID Tags and such. With the advent of pervasive computing, an increasing demand to integrate cryptographic functions in applications has risen. Different from the past, it is often desirable to have cryptographic primitives that are as small as possible. There are two main reasons for this. First, there are applications constrained by a hard limit with respect to gate count or power. The prime example are RFID tags on which it is simply physically impossible to implement RSA-1024. The second reason are applications which are heavily cost constrained, e.g., high-volume consumer devices. Here it would be possible to integrate non-optimized crypto engines but it is highly desirable to use implementations which cause the smallest possible cost increase for the product.

With this goal in mind, a team of researchers developed the PRESENT block cipher [2]. It can be implemented with as little as 1000 gate equivalences [11] which is close to the theoretical limit if one has to store 64 state bits and 80 key bits. For the asymmetric case, we developed an extremely small elliptic curve engine which requires between roughly 10,000 and 15,000 gate equivalences, depending on the security level [10].

High Performance Elliptic Curve Engine for Car-to-Car Communication. Air pollution is not the only health hazard posed by cars. They are also quite deadly when it comes to accidents. In the developed world, traffic fatalities are, by a far margin, the most common cause of death caused by accidents. Both in the European Union and in the USA there are more than 40,000 traffic fatalities annually, and world-wide they are the leading cause of death for people in the age range of 10–24. Given that many mechanically safety measure such as seat belts, air bags and anti-block brake (ABS) systems are very far advanced, there has been a push in the last few years to develop electronic driver assistant systems. One major motivation is to reduce the number of accidents. Some driver assistant systems are based on car-to-car (C2C) and car-to-infrastructure (C2I) communication. If such systems were in place, many collisions between vehicles could be avoided. One requirement of C2C and C2I systems is that the communication should be secure. It does not take much fantasy to imagine how an attacker could cause quite serious trouble if, for instance, faked collision warning messages are issued to cars driving on the German autobahn with 200 km/h.

The IEEE Standard 1609 calls for a digital signature over every position message sent out by every car. In a high-density traffic environment that could translate in 1000 or more digital signatures which have to be verified per second. The challenge here is to develop an ECC engine that can support thousands of verifications per second at affordable costs. We developed new ECC engines that make use of the DSP cores available on modern FPGAs. Our engine can verify more than 2000 ECC signatures (224 bit NIST curve) with a mid-size commercial FPGA [6]. Previously such speeds were only achievable with expensive and power-consuming parallel CPUs or with ASICs. Our design scales theoretically to more than 30,000 signatures per second on high-end FPGAs.

Side-Channel Attacks against Remote Keyless Entry Systems. Ever since side-channel analysis (SCA) were proposed (cf. the first section of this abstract) it was recognized that they pose a major risk for real-world systems. There had been many anecdotal reports, especially from the smart card industry, about the vulnerability against SCA. However, despite hundreds of research papers in this area, there had been hardly any descriptions of an SCA against an actual system.

Last year we attacked the KeeLoq remote keyless entry system using SCA. KeeLoq was an instructive target. It is a 1980s symmetric cipher against which several analytical attack had been proposed in short sequence [1,4,7]. However, due to the mode of operation of keyless entry systems, the required plaintext-ciphertext pairs are almost impossible to obtain in the real world. In contrast, using a DPA-like attack, we showed that both the individual transmitter keys (which are typically embedded in garage door openers or car keys) as well as system-wide manufacturer keys can be extracted. Once the manufacturer key has been recovered after a few thousands measurements, transmitters can be cloned after simply eavesdropping on one or two communications.

References

1. Bogdanov, A.: Attacks on the KeeLoq Block Cipher and Authentication Systems. In: 3rd Conference on RFID Security 2007, RFIDSec 2007 (2007), http://rfidsec07.etsit.uma.es/slides/papers/paper-22.pdf
2. Bogdanov, A., Leander, G., Knudsen, L.R., Paar, C., Poschmann, A., Robshaw, M.J.B., Seurin, Y., Vikkelsoe, C.: PRESENT - An Ultra-Lightweight Block Cipher. In: Paillier, P., Verbauwhede, I. (eds.) CHES 2007. LNCS, vol. 4727, pp. 450–466. Springer, Heidelberg (2007)
3. Boneh, D., DeMillo, R.A., Lipton, R.J.: On the importance of checking computations (1996), http://citeseer.ist.psu.edu/491209.html
4. Courtois, N.T., Bard, G.V., Wagner, D.: Algebraic and Slide Attacks on KeeLoq. In: Nyberg, K. (ed.) FSE 2008. LNCS, vol. 5086, Springer, Heidelberg (2008)
5. Eldridge, S.E., Walter, C.D.: Hardware implementation of Montgomery's modular multiplication algorithm. IEEE Transactions on Computers 42(6), 693–699 (1993)
6. Güneysu, T., Paar, C.: Ultra High Performance ECC over NIST Primes on Commercial FPGAs. In: Oswald, E., Rohatgi, P. (eds.) CHES 2008. LNCS, vol. 5154, pp. 62–78. Springer, Heidelberg (2008)
7. Indesteege, S., Keller, N., Dunkelman, O., Biham, E., Preneel, B.: A Practical Attack on KeeLoq. In: Smart, N.P. (ed.) EUROCRYPT 2008. LNCS, vol. 4965, Springer, Heidelberg (2008)
8. Koc, C.K., Acar, T., Burton, J., Kaliski, S.: Analyzing and comparing montgomery multiplication algorithms. IEEE Micro 16(3), 26–33 (1996)
9. Kocher, P.C., Jaffe, J., Jun, B.: Differential power analysis. In: Wiener, M. (ed.) CRYPTO 1999. LNCS, vol. 1666, p. 388. Springer, Heidelberg (1999)
10. Kumar, S.: Elliptic Curve Cryptography for Constrained Devices. PhD thesis, Electrical Engineering and Information Sciences Department, Ruhr-University of Bochum (2006)
11. Rolfes, C., Poschmann, A., Leander, G., Paar, C.: Ultra-Lightweight Implementations for Smart Devices-Security for 1000 Gate Equivalents. In: Grimaud, G., Standaert, F.-X. (eds.) CARDIS 2008. LNCS, vol. 5189, pp. 89–103. Springer, Heidelberg (2008)

Hardware Accelerator for the Tate Pairing in Characteristic Three Based on Karatsuba-Ofman Multipliers

Jean-Luc Beuchat[1], Jérémie Detrey[2], Nicolas Estibals[2], Eiji Okamoto[1],
and Francisco Rodríguez-Henríquez[3]

[1] Graduate School of Systems and Information Engineering, University of Tsukuba,
1-1-1 Tennodai, Tsukuba, Ibaraki, 305-8573, Japan
[2] CACAO project-team, LORIA, INRIA Nancy - Grand Est, Bâtiment A, 615,
rue du Jardin Botanique, 54602 Villers-les-Nancy Cédex, France
[3] Computer Science Department, Centro de Investigación y de Estudios Avanzados
del IPN, Av. Instituto Politécnico Nacional No. 2508, 07300 México City, México

Abstract. This paper is devoted to the design of fast parallel accelerators for the cryptographic Tate pairing in characteristic three over supersingular elliptic curves. We propose here a novel hardware implementation of Miller's loop based on a pipelined Karatsuba-Ofman multiplier. Thanks to a careful selection of algorithms for computing the tower field arithmetic associated to the Tate pairing, we manage to keep the pipeline busy. We also describe the strategies we considered to design our parallel multiplier. They are included in a VHDL code generator allowing for the exploration of a wide range of operators. Then, we outline the architecture of a coprocessor for the Tate pairing over $\mathbb{F}_{3^m}$. However, a final exponentiation is still needed to obtain a unique value, which is desirable in most of the cryptographic protocols. We supplement our pairing accelerator with a coprocessor responsible for this task. An improved exponentiation algorithm allows us to save hardware resources.

According to our place-and-route results on Xilinx FPGAs, our design improves both the computation time and the area-time trade-off compared to previoulsy published coprocessors.

Keywords: Tate pairing, η_T pairing, elliptic curve, finite field arithmetic, Karatsuba-Ofman multiplier, hardware accelerator, FPGA.

1 Introduction

The Weil and Tate pairings were independently introduced in cryptography by Menezes, Okamoto & Vanstone [34] and Frey & Rück [16] as a tool to attack the discrete logarithm problem on some classes of elliptic curves defined over finite fields. The discovery of constructive properties by Mitsunari, Sakai & Kasahara [38], Sakai, Oghishi & Kasahara [42], and Joux [26] initiated the proposal of an ever increasing number of protocols based on bilinear pairings: identity-based encryption [11], short signature [13], and efficient broadcast encryption [12] to mention but a few.

C. Clavier and K. Gaj (Eds.): CHES 2009, LNCS 5747, pp. 225–239, 2009.

Miller described the first iterative algorithm to compute the Weil and Tate pairings back in 1986 [35,36]. In practice, the Tate pairing seems to be more efficient for computation (see for instance [20,31]) and has attracted a lot of interest from the research community. Supersingular curves received considerable attention since significant improvements of Miller's algorithm were independently proposed by Barreto *et al.* [4] and Galbraith *et al.* [17] in 2002. One year later, Duursma & Lee gave a closed formula in the case of characteristic three [14]. In 2004, Barreto *et al.* [3] introduced the η_T approach, which further shortens the loop of Miller's algorithm. Recall that the modified Tate pairing can be computed from the reduced η_T pairing at almost no extra cost [7]. More recently, Hess, Smart, and Vercauteren generalized these results to ordinary curves [24,45,23].

This paper is devoted to the design of a coprocessor for the Tate pairing on supersingular elliptic curves in characteristic three. We propose here a novel architecture based on a pipelined Karatsuba-Ofman multiplier over $\mathbb{F}_{3^m}$ to implement Miller's algorithm. Thanks to a judicious choice of algorithms for tower field arithmetic and a careful analysis of the scheduling, we manage to keep the pipeline busy and compute one iteration of Miller's algorithm in only 17 clock cycles (Section 2). We describe the strategies we considered to design our parallel multiplier in Section 3. They are included in a VHDL code generator allowing for the exploration of a wide range of operators. Section 4 describes the architecture of a coprocessor for the Tate pairing over $\mathbb{F}_{3^m}$. We summarize our implementation results on FPGA and provide the reader with a thorough comparison against previously published coprocessors in Section 5.

For the sake of concision, we are forced to skip the description of many important concepts of elliptic curve theory. We suggest the interested reader to review [46] for an in-depth coverage of this topic.

2   Reduced η_T Pairing in Characteristic Three Revisited

In the following, we consider the computation of the reduced η_T pairing in characteristic three. Table 1 summarizes the parameters of the algorithm and the supersingular curve. We refer the reader to [3, 8] for more details about the computation of the η_T pairing.

2.1   Computation of Miller's Algorithm

We rewrote the reversed-loop algorithm in characteristic three described in [8], denoting each iteration with parenthesized indices in superscript, in order to emphasize the intrinsic parallelism of the η_T pairing (Algorithm 1). At each iteration of Miller's algorithm, two tasks are performed in parallel, namely: a sparse multiplication over $\mathbb{F}_{3^{6m}}$, and the computation of the coefficients for the next sparse operation. We say that an operand in $\mathbb{F}_{3^{6m}}$ is sparse when some of its coefficients are either zero or one.

Table 1. Supersingular curves over $\mathbb{F}_{3^m}$

Underlying field	$\mathbb{F}_{3^m}$, where m is coprime to 6
Curve	$E : y^2 = x^3 - x + b$, with $b \in \{-1, 1\}$
Number of rational points	$N = \#E(\mathbb{F}_{3^m}) = 3^m + 1 + \mu b 3^{(m+1)/2}$, with $$\mu = \begin{cases} +1 & \text{if } m \equiv 1,\ 11 \pmod{12}, \text{ or} \\ -1 & \text{if } m \equiv 5,\ 7 \pmod{12} \end{cases}$$
Embedding degree	$k = 6$
Distortion map	$\psi : E(\mathbb{F}_{3^m})[\ell] \longrightarrow E(\mathbb{F}_{3^{6m}})[\ell] \setminus E(\mathbb{F}_{3^m})[\ell]$ $(x, y) \longmapsto (\rho - x, y\sigma)$ with $\rho \in \mathbb{F}_{3^3}$ and $\sigma \in \mathbb{F}_{3^2}$ satisfying $\rho^3 = \rho + b$ and $\sigma^2 = -1$
Tower field	$\mathbb{F}_{3^{6m}} = \mathbb{F}_{3^m}[\rho, \sigma] \cong \mathbb{F}_{3^m}[X, Y]/(X^3 - X - b, Y^2 + 1)$
Final exponentiation	$M = \left(3^{3m} - 1\right) \cdot \left(3^m + 1\right) \cdot \left(3^m + 1 - \mu b 3^{(m+1)/2}\right)$
Parameters of Algorithm 1	$$\lambda = \begin{cases} +1 & \text{if } m \equiv 7,\ 11 \pmod{12}, \text{ or} \\ -1 & \text{if } m \equiv 1,\ 5 \pmod{12}, \text{ and} \end{cases}$$ $$\nu = \begin{cases} +1 & \text{if } m \equiv 5,\ 11 \pmod{12}, \text{ or} \\ -1 & \text{if } m \equiv 1,\ 7 \pmod{12} \end{cases}$$

Sparse multiplication over $\mathbb{F}_{3^{6m}}$ (lines 6 and 7). The intermediate result $R^{(i-1)}$ is multiplied by the sparse operand $S^{(i)}$. This operation is easier than a standard multiplication over $\mathbb{F}_{3^{6m}}$.

The choice of a sparse multiplication algorithm over $\mathbb{F}_{3^{6m}}$ requires careful attention. Bertoni *et al.* [6] and Gorla *et al.* [18] took advantage of Karatsuba-Ofman multiplication and Lagrange interpolation, respectively, to reduce the number of multiplications over $\mathbb{F}_{3^m}$ at the expense of several additions (note that Gorla *et al.* study standard multiplication over $\mathbb{F}_{3^{6m}}$ in [18], but extending their approach to sparse multiplication is straightforward). In order to keep the pipeline of a Karatsuba-Ofman multiplier busy, we would have to embed in our processor a large multioperand adder (up to twelve operands for the scheme proposed by Gorla *et al.*) and several multiplexers to deal with the irregular datapath. This would negatively impact the area and the clock frequency, and we prefer considering the algorithm discussed by Beuchat *et al.* in [10] which gives a better trade-off between the number of multiplications and additions over the underlying field when $b = 1$. We give here a more general version of this scheme which also works when $b = -1$ (Algorithm 2). It involves 17 multiplications and 29 additions over $\mathbb{F}_{3^m}$, and reduces the number of intermediate variables compared to the algorithms mentioned above. Another nice feature of this scheme is that it requires the addition of at most four operands.

We suggest to take advantage of a Karatsuba-Ofman multiplier with seven pipeline stages to compute $S^{(i)}$ and $R^{(i-1)} \cdot S^{(i)}$. We managed to find a scheduling that allows us to perform a multiplication over $\mathbb{F}_{3^m}$ at each clock cycle, thus keeping the pipeline busy. Therefore, we compute lines 6 and 7 of Algorithm 1 in 17 clock cycles.

Algorithm 1. Computation of the reduced η_T pairing in characteristic three. Intermediate variables in uppercase belong to $\mathbb{F}_{3^{6m}}$, those in lowercase to $\mathbb{F}_{3^m}$.

Input: $P = (x_P, y_P)$ and $Q = (x_Q, y_Q) \in E(\mathbb{F}_{3^m})[\ell]$.
Output: $\eta_T(P, Q)^M \in \mathbb{F}^*_{3^{6m}}$.

1: $x_P^{(0)} \leftarrow x_p - \nu b;\ y_P^{(0)} \leftarrow -\mu b y_P$;
2: $x_Q^{(0)} \leftarrow x_Q;\ y_Q^{(0)} \leftarrow -\lambda y_Q$;

3: $t^{(0)} \leftarrow x_P^{(0)} + x_Q^{(0)}$;
4: $R^{(-1)} \leftarrow \lambda y_P^{(0)} \cdot t^{(0)} - \lambda y_Q^{(0)} \sigma - \lambda y_P^{(0)} \rho$;

5: **for** $i = 0$ to $(m-1)/2$ **do**
6:     $S^{(i)} \leftarrow - \left(t^{(i)}\right)^2 + y_P^{(i)} y_Q^{(i)} \sigma - t^{(i)} \rho - \rho^2$;
7:     $R^{(i)} \leftarrow R^{(i-1)} \cdot S^{(i)}$;

8:     $x_P^{(i+1)} \leftarrow \sqrt[3]{x_P^{(i)}};\ y_P^{(i+1)} \leftarrow \sqrt[3]{y_P^{(i)}}$;
9:     $x_Q^{(i+1)} \leftarrow \left(x_Q^{(i)}\right)^3;\ y_Q^{(i+1)} \leftarrow \left(y_Q^{(i)}\right)^3$;
10:    $t^{(i+1)} \leftarrow x_P^{(i)} + x_Q^{(i)}$;
11: **end for**

12: **return** $\left(R^{((m-1)/2)}\right)^M$;

Computation of the coefficients of the next sparse operand $S^{(i+1)}$ (lines 8 to 10). The second task consists of computing the sparse operand $S^{(i+1)}$ required for the next iteration of Miller's algorithm. Two cubings and an addition over $\mathbb{F}_{3^m}$ allow us to update the coordinates of point P and to determine the coefficient $t^{(i+1)}$ of the sparse operand $S^{(i+1)}$, respectively.

Recall that the η_T pairing over $\mathbb{F}_{3^m}$ comes in two flavors: the original one involves a cubing over $\mathbb{F}_{3^{6m}}$ after each sparse multiplication. Barreto *et al.* [3] explained how to get rid of that cubing at the price of two cube roots over $\mathbb{F}_{3^m}$ to update the coordinates of point Q. It is essential to consider such an algorithm here in order to minimize the number of arithmetic operations over $\mathbb{F}_{3^m}$ to be performed in the first task (which is the most expensive one).

According to our results, the critical path of the circuit is never located in a cube root operator when pairing-friendly irreducible trinomials or pentanomials [2, 21] are used to define $\mathbb{F}_{3^m}$. If by any chance such polynomials are not available for the considered extension of $\mathbb{F}_3$ and the critical path is in the cube root, it is always possible to pipeline this operation. Therefore, the cost of cube roots is hidden by the first task.

2.2   Final Exponentiation (Line 12)

The final step in the computation of the η_T pairing is the final exponentiation, where the result of Miller's algorithm $R^{((m-1)/2)} = \eta_T(P, Q)$ is raised to the M-th power. This exponentiation is necessary since $\eta_T(P, Q)$ is only defined up to N-th powers in $\mathbb{F}^*_{3^{6m}}$.

In order to compute this final exponentiation, we use the algorithm presented by Beuchat *et al.* in [8]. This method exploits the special form of the exponent M

Algorithm 2. Sparse multiplication over $\mathbb{F}_{3^{6m}}$.

Input: $b \in \{-1, 1\}$; $t^{(i)}$, $y_P^{(i)}$, and $y_Q^{(i)} \in \mathbb{F}_{3^m}$; $R^{(i-1)} \in \mathbb{F}_{3^{6m}}$.

Output: $R^{(i)} = R^{(i-1)} \cdot S^{(i)} \in \mathbb{F}_{3^{6m}}$, where $S^{(i)} = \left(-\left(t^{(i)} \right)^2 + y_P^{(i)} y_Q^{(i)} \sigma - t^{(i)} \rho - \rho^2 \right)$.

1: $p_0^{(i)} \leftarrow r_0^{(i-1)} \cdot t^{(i)}$; $p_1^{(i)} \leftarrow r_1^{(i-1)} \cdot t^{(i)}$; $p_2^{(i)} \leftarrow r_2^{(i-1)} \cdot t^{(i)}$;

2: $p_3^{(i)} \leftarrow r_3^{(i-1)} \cdot t^{(i)}$; $p_4^{(i)} \leftarrow r_4^{(i-1)} \cdot t^{(i)}$; $p_5^{(i)} \leftarrow r_5^{(i-1)} \cdot t^{(i)}$;

3: $p_6^{(i)} \leftarrow t^{(i)} \cdot t^{(i)}$; $p_7^{(i)} \leftarrow -y_P^{(i)} \cdot y_Q^{(i)}$;

4: $s_0^{(i)} \leftarrow -r_0^{(i-1)} - r_1^{(i-1)}$; $s_1^{(i)} \leftarrow -r_2^{(i-1)} - r_3^{(i-1)}$;

5: $s_2^{(i)} \leftarrow -r_4^{(i-1)} - r_5^{(i-1)}$; $s_3^{(i)} \leftarrow p_6^{(i)} + p_7^{(i)}$;

6: $a_0^{(i)} \leftarrow r_2^{(i-1)} + p_4^{(i)}$; $a_2^{(i)} \leftarrow br_4^{(i-1)} + p_0^{(i)} + a_0^{(i)}$; $a_4^{(i)} \leftarrow r_0^{(i-1)} + r_4^{(i-1)} + p_2^{(i)}$;

7: $a_1^{(i)} \leftarrow r_3^{(i-1)} + p_5^{(i)}$; $a_3^{(i)} \leftarrow br_5^{(i-1)} + p_1^{(i)} + a_1^{(i)}$; $a_5^{(i)} \leftarrow r_1^{(i-1)} + r_5^{(i-1)} + p_3^{(i)}$;

8: $p_8^{(i)} \leftarrow r_0^{(i-1)} \cdot p_6^{(i)}$; $p_9^{(i)} \leftarrow r_1^{(i-1)} \cdot p_7^{(i)}$; $p_{10}^{(i)} \leftarrow s_0^{(i)} \cdot s_3^{(i)}$;

9: $p_{11}^{(i)} \leftarrow r_2^{(i-1)} \cdot p_6^{(i)}$; $p_{12}^{(i)} \leftarrow r_3^{(i-1)} \cdot p_7^{(i)}$; $p_{13}^{(i)} \leftarrow s_1^{(i)} \cdot s_3^{(i)}$;

10: $p_{14}^{(i)} \leftarrow r_4^{(i-1)} \cdot p_6^{(i)}$; $p_{15}^{(i)} \leftarrow r_5^{(i-1)} \cdot p_7^{(i)}$; $p_{16}^{(i)} \leftarrow s_2^{(i)} \cdot s_3^{(i)}$;

11: $r_0^{(i)} \leftarrow -ba_0^{(i)} - p_8^{(i)} + p_9^{(i)}$; $r_1^{(i)} \leftarrow -ba_1^{(i)} + p_8^{(i)} + p_9^{(i)} + p_{10}^{(i)}$;

12: $r_2^{(i)} \leftarrow -a_2^{(i)} - p_{11}^{(i)} + p_{12}^{(i)}$; $r_3^{(i)} \leftarrow -a_3^{(i)} + p_{11}^{(i)} + p_{12}^{(i)} + p_{13}^{(i)}$;

13: $r_4^{(i)} \leftarrow -a_4^{(i)} - p_{14}^{(i)} + p_{15}^{(i)}$; $r_5^{(i)} \leftarrow -a_5^{(i)} + p_{14}^{(i)} + p_{15}^{(i)} + p_{16}^{(i)}$;

14: **return** $r_0^{(i)} + r_1^{(i)}\sigma + r_2^{(i)}\rho + r_3^{(i)}\sigma\rho + r_4^{(i)}\rho^2 + r_5^{(i)}\sigma\rho^2$;

(see Table 1) to achieve better performances than with a classical square-and-multiply algorithm. Among other computations, this final exponentiation involves the raising of an element of $\mathbb{F}_{3^{6m}}^*$ to the $3^{(m+1)/2}$-th power, which Beuchat *et al.* [8] perform by computing $(m+1)/2$ successive cubings over $\mathbb{F}_{3^{6m}}$. Each of these cubings requiring 6 cubings and 6 additions over $\mathbb{F}_{3^m}$, the total cost of this step is $3m + 3$ cubings and $3m + 3$ additions.

We present here a new method for computing $U^{3^{(m+1)/2}}$ for $U = u_0 + u_1\sigma + u_2\rho + u_3\sigma\rho + u_4\rho^2 + u_5\sigma\rho^2 \in \mathbb{F}_{3^{6m}}^*$ by exploiting the linearity of the Frobenius map (*i.e.* cubing in characteristic three) to reduce the number of additions. Indeed, noting that $\sigma^{3^i} = (-1)^i\sigma$, $\rho^{3^i} = \rho + ib$ and $(\rho^2)^{3^i} = \rho^2 - ib\rho + i^2$, we obtain the following formulae for U^{3^i}, depending on the value of i:

$$U^{3^i} = \left(u_0 - \epsilon_1 u_2 + \epsilon_2 u_4 \right)^{3^i} + \epsilon_3 \left(u_1 - \epsilon_1 u_3 + \epsilon_2 u_5 \right)^{3^i} \sigma + \left(u_2 + \epsilon_1 u_4 \right)^{3^i} \rho +$$

$$\epsilon_3 \left(u_3 + \epsilon_1 u_5 \right)^{3^i} \sigma\rho + u_4^{3^i} \rho^2 + \epsilon_3 u_5^{3^i} \sigma\rho^2,$$

with $\epsilon_1 = -ib \bmod 3$, $\epsilon_2 = i^2 \bmod 3$, and $\epsilon_3 = (-1)^i$.

Thus, according to the value of $(m+1)/2$ modulo 6, the computation of $U^{3^{(m+1)/2}}$ will still require $3m + 3$ cubings but at most only 6 additions or subtractions over $\mathbb{F}_{3^m}$.

3   Fully Parallel Karatsuba-Ofman Multipliers over $\mathbb{F}_{3^m}$

As mentioned in Section 2.1, our hardware accelerator is based on a pipelined Karatsuba-Ofman multiplier over $\mathbb{F}_{3^m}$. This operator is responsible for the computation of the 17 products involved in the sparse multiplication over $\mathbb{F}_{3^{6m}}$ occuring at each iteration of Miller's algorithm. In the following we give a short description of the multiplier block used in this work.

Let $f(x)$ be an irreducible polynomial of degree m over $\mathbb{F}_3$. Then the ternary extension field $\mathbb{F}_{3^m}$ can be defined as $\mathbb{F}_{3^m} \cong \mathbb{F}_3[x]/\left(f(x)\right)$. Multiplication in $\mathbb{F}_{3^m}$ of two arbitrary elements represented as ternary polynomials of degree at most $m-1$ is defined as the polynomial multiplication of the two elements modulo the irreducible polynomial $f(x)$, i.e. $c(x) = a(x)b(x) \bmod f(x)$. This implies that we can obtain the field product by first computing the polynomial multiplication of $a(x)$ and $b(x)$ of degree at most $2m - 2$ followed by a modular reduction step with $f(x)$. As long as we select $f(x)$ with low Hamming weight (i.e. trinomials, tetranomials, etc.), the modular reduction step can be accomplished at a linear computational complexity $O(m)$ by using a combination of adders and subtracters over $\mathbb{F}_3$. This implies that the cost of this modular reduction step is much lower than the one associated to polynomial multiplication. In this work, due to its subquadratic space complexity, we used a modified version of the classical Karatsuba-Ofman multiplier for computing the polynomial multiplication step as explained next.

3.1   Variations on the Karatsuba-Ofman Algorithm

The Karatsuba-Ofman multiplier is based on the observation that the polynomial product $c = a \cdot b$ (dropping the (x) notation) can be computed as

$$c = a^L b^L + x^n \left[(a^H + a^L)(b^L + b^H) - (a^H b^H + a^L b^L)\right] + x^{2n} a^H b^H,$$

where $n = \lceil \frac{m}{2} \rceil$, $a = a^L + x^n a^H$, and $b = b^L + x^n b^H$.

Notice that since we are working with polynomials, there is no carry propagation. This allows one to split the operands in a slightly different way: For instance Hanrot and Zimmermann suggested to split them into odd and even part [22]. It was adapted to multiplication over $\mathbb{F}_{2^m}$ by Fan et al. [15]. This different way of splitting allows one to save approximatively m additions over $\mathbb{F}_3$ during the reconstruction of the product. This is due to the fact that there is no overlap between the odd and even parts at the reconstruction step, whereas there is some with the higher/lower part splitting method traditionally used.

Another natural way to generalize the Karatsuba-Ofman multiplier is to split the operands not into two, but into three or more parts. That splitting can be done in a classical way, i.e. splitting each operand in ascending parts from the lower to the higher powers of x, or splitting them using a generalized odd/even way, i.e. according to the degree modulo the number of split parts. By applying this strategy recursively, in each iteration each polynomial multiplication is transformed into three or more polynomial multiplications with their degrees progressively reduced, until all the polynomial operands collapse into single coefficients. Nevertheless, practice has shown that is better to truncate the recursion earlier, performing the underlying multiplications using alternative techniques that are more compact and/or faster for low-degree operands (typically the so-called school book method with quadratic complexity has been selected).

3.2 A Pipelined Architecture for the Karatsuba-Ofman Multiplier

The field multiplications involved in the reduced η_T pairing do not present dependencies among themselves and thus, it is possible to compute these products using a pipelined architecture. By following this strategy, once that each stage of the pipeline is loaded, we are able to compute one field multiplication over $\mathbb{F}_{3^m}$ every clock cycle. The pipelined architecture was achieved by inserting registers in-between the computation of the partial product operations associated to the *divide-and-conquer* Karatsuba-Ofman strategy, where the depth of the pipeline can be adjusted according to the complexity of the application at hand. This approach allows us to cut the critical path of the whole multiplier structure.

In order to study a wide range of implementation strategies, we decided to write a VHDL code generator tool. This tool allows us to automatically generate the VHDL description of different Karatsuba-Ofman multiplier versions according to several parameters (field extension degree, irreducible polynomial, splitting method, *etc.*). Our automatic tool was extremely useful for selecting the circuit that showed the best time, the smallest area or a good trade-off between them.

4 A Coprocessor for the η_T Pairing in Characteristic Three

As pointed out by Beuchat *et al.* [9], the computation of $R^{((m-1)/2)}$ and the final exponentiation do not share the same datapath and it seems judicious to pipeline these two tasks using two distinct coprocessors in order to reduce the computation time.

4.1 Computation of Miller's Algorithm

A first coprocessor based on a Karatsuba-Ofman multiplier with seven pipeline stages is responsible for computing Miller's loop (Figure 1). We tried to minimize the amount of hardware required to implement the sparse multiplication over $\mathbb{F}_{3^{6m}}$, while keeping the pipeline busy. Besides the parallel multiplier described in Section 3, our architecture consists of four main blocks:

- *Computation of the coefficients of $S^{(i+1)}$*. The first block embeds four registers to store the coordinates of points P and Q. It is responsible for computing $x_P^{(i+1)}$, $y_P^{(i+1)}$, $x_Q^{(i+1)}$, $y_Q^{(i+1)}$, and $t^{(i+1)}$ at each iteration. It also includes dedicated hardware to perform the initialization step of Algorithm 1 (lines 1 and 2).

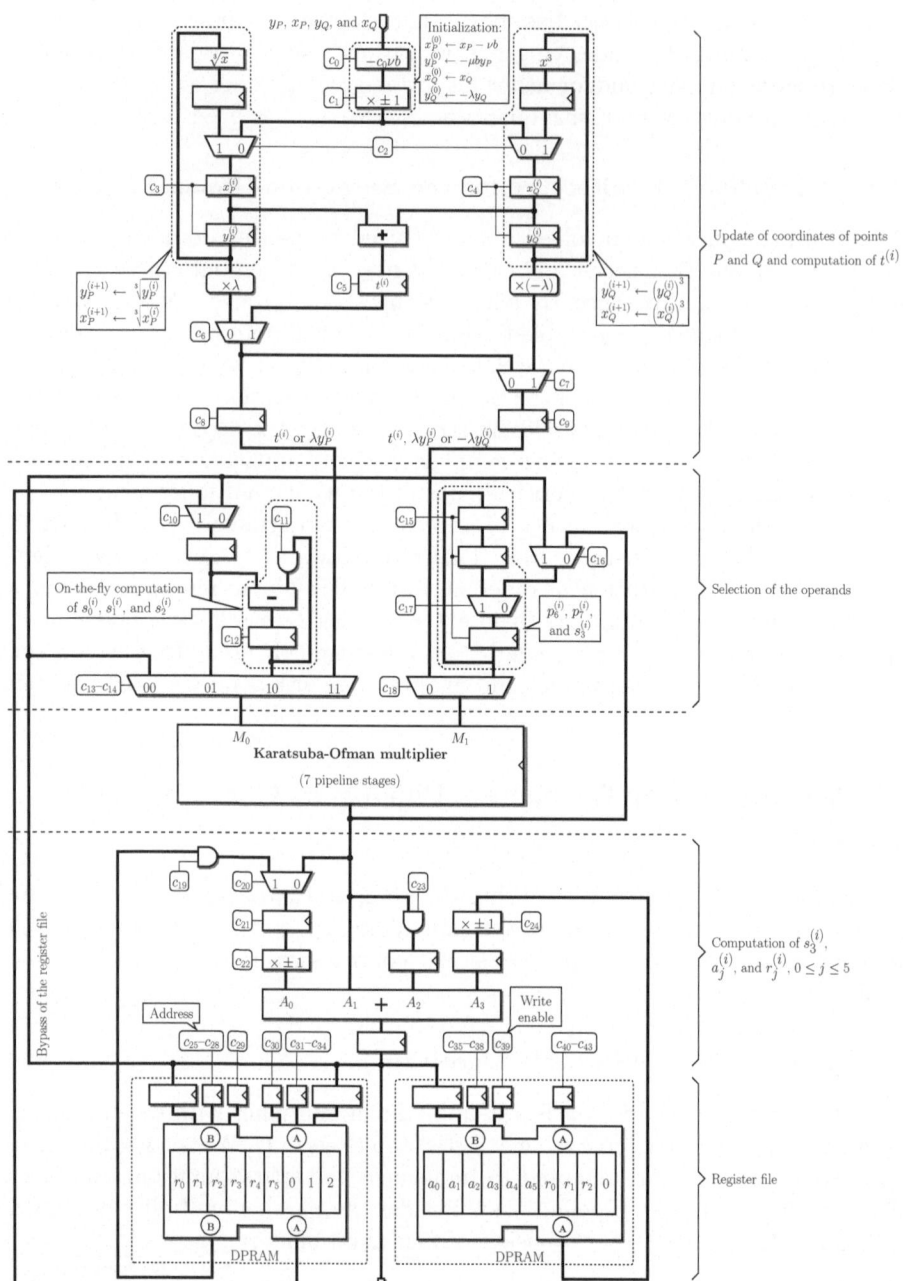

Fig. 1. A coprocessor for the η_T pairing in characteristic three. All control bits c_i belong to $\{0, 1\}$.

- *Selection of the operands of the multiplier.* At each iteration of Miller's algorithm, we have to provide our Karatsuba-Ofman multiplier with $t^{(i)}$, $y_P^{(i)}$, and $y_Q^{(i)}$ in order to compute the coefficients of $S^{(i)}$ (see Algorithm 1, line 6). An accumulator allows us to compute $s_0^{(i)}$, $s_1^{(i)}$, and $s_2^{(i)}$ on-the-fly. We store $p_6^{(i)}$, $p_7^{(i)}$, and $s_3^{(i)}$ in a circular shift register: this approach allows for an easy implementation of lines 8, 9, and 10 of Algorithm 2.
- *Addition over $\mathbb{F}_{3^m}$.* A nice feature of the algorithm we selected for sparse multiplication over $\mathbb{F}_{3^{6m}}$ is that it requires the addition of at most four operands. Thus, it suffices to complement the Karatsuba-Ofman multiplier with a 4-input adder to compute $s_3^{(i)}$, $a_j^{(i)}$, and $r_j^{(i)}$, $0 \leq j \leq 5$. Registers allow us to store several products $p_j^{(i)}$, which is for instance useful when computing $s_3^{(i)} \leftarrow p_6^{(i)} + p_7^{(i)}$.
- *Register file.* The register file is implemented by means of Dual-Ported RAM (DPRAM). In order to avoid memory collisions, we had to split it into two parts and store two copies of $r_0^{(i)}$, $r_1^{(i)}$, and $r_2^{(i)}$.

The initialization step (Algorithm 1, lines 1 to 4) and each iteration of Miller's loop (Algorithm 1, lines 6 to 10) require 17 clock cycles. Therefore, our coprocessor returns $R^{(m-1)/2}$ after $17 \cdot (m+3)/2$ clock cycles.

4.2 Final Exponentiation

Our first attempt at computing the final exponentiation was to use the unified arithmetic operator introduced by Beuchat et al. [8]. Unfortunately, due to the sequential scheduling inherent to this operator, it turned out that the final exponentiation algorithm required more clock cycles than the computation of Miller's algorithm by our coprocessor. We therefore had to consider a slightly more parallel architecture.

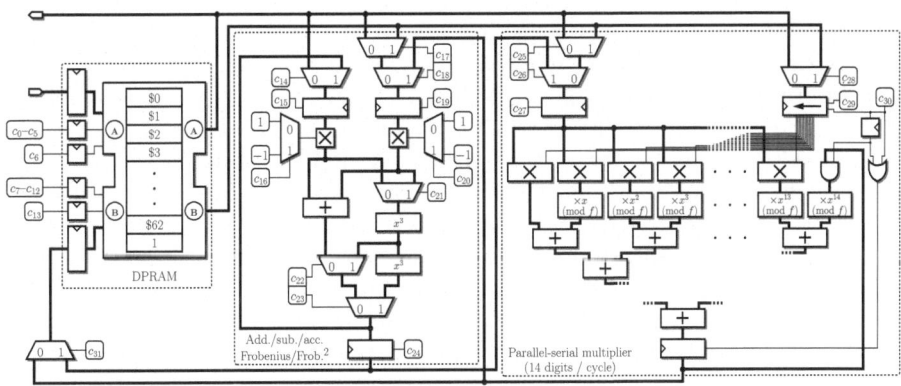

Fig. 2. A coprocessor for the final exponentiation of the η_T pairing in characteristic three

Noticing that the critical operations in the final exponentiation algorithm were multiplications and long sequences of cubings over $\mathbb{F}_{3^m}$, we designed the coprocessor for arithmetic over $\mathbb{F}_{3^m}$ depicted in Figure 2. Besides a register file implemented by means of DPRAM, our coprocessor embeds a parallel-serial multiplier [44] processing 14 coefficients of an operand at each clock cycle, and a novel unified operator for addition, subtraction, accumulation, Frobenius map (*i.e.* cubing), and double Frobenius map (*i.e.* raising to the 9th power). This architecture allowed us to efficiently implement the final exponentiation algorithm described for instance in [8], while taking advantage of the improvement proposed in Section 2.2.

5   Results and Comparisons

Thanks to our automatic VHDL code generator, we designed several versions of the proposed architecture and prototyped our coprocessors on Xilinx FPGAs with average speedgrade. Table 2 provides the reader with a comparison between our work and accelerators for the Tate and the η_T pairing over supersingular (hyper)elliptic curves published in the open literature (our comparison remains fair since the Tate pairing can be computed from the η_T pairing at no extra cost [7]). The third column measures the security of the curve as the key length required by a symmetric-key algorithm of equivalent security. Note that the architecture proposed by Kömürcü & Savas [32] does not implement the final exponentiation, and that Barenghi *et al.* [1] work with a supersingular curve defined over $\mathbb{F}_p$, where p is a 512-bit prime number.

Most of the authors who described hardware accelerators for the Tate pairing over $\mathbb{F}_{3^m}$ considered only low levels of security. Thus, we designed a first architecture for $m = 97$. It simultaneously improves the speed record previously held by Jiang [25], and the Area-Time (AT) product of the coprocessor introduced by Beuchat *et al.* [10].

Then, we studied a more realistic setup and placed-and-routed a second accelerator for $m = 193$, thus achieving a level of security equivalent to 89-bit symmetric encryption. Beuchat *et al.* [7] introduced a unified arithmetic operator in order to reduce the silicon footprint of the circuit to ensure scalability, while trying to minimize the impact on the overall performances. In this work, we focused on the other end of the hardware design spectrum and significantly reduced the computation time reported by Beuchat *et al.* in [7]. A much more unexpected result is that we also improved the AT product. The bottleneck is the usage of the FPGA resources: the unified arithmetic operator allows one to achieve higher levels of security on the same circuit area.

Our architectures are also much faster than software implementations. Mitsunari wrote a very careful multithreaded implementation of the η_T pairing over $\mathbb{F}_{3^{97}}$ and $\mathbb{F}_{3^{193}}$ [37]. He reported a computation time of 92 μs and 553 μs, respectively, on an Intel Core 2 Duo processor (2.66 GHz). Interestingly enough, his software outperforms several hardware architectures proposed by other researchers for low levels of security. When we compare his results with our work,

Table 2. Hardware accelerators for the Tate and η_T pairings

	Curve	Security [bits]	FPGA	Area [slices]	Freq. [MHz]	Calc. time [μs]	AT prod.
Kerins et al. [30]	$E(\mathbb{F}_{3^{97}})$	66	xc2vp125	55616	15	850	47.3
Kömürcü & Savas [32]	$E(\mathbb{F}_{3^{97}})$	66	xc2vp100	14267	77	250.7	3.6
Ronan et al. [39]	$E(\mathbb{F}_{3^{97}})$	66	xc2vp100-6	15401	85	183	2.8
Grabher & Page [19]	$E(\mathbb{F}_{3^{97}})$	66	xc2vp4-6	4481	150	432.3	1.9
Jiang [25]	$E(\mathbb{F}_{3^{97}})$	66	xc4vlx200-11	74105	78	20.9	1.55
Beuchat et al. [7]	$E(\mathbb{F}_{3^{97}})$	66	xc2vp20-6	4455	105	92	0.4
Beuchat et al. [10]	$E(\mathbb{F}_{3^{97}})$	66	xc2vp30-6	10897	147	33	0.36
This work	$E(\mathbb{F}_{3^{97}})$	66	xc2vp30-6	18360	137	6.2	0.11
	$E(\mathbb{F}_{3^{97}})$	66	xc4vlx60-11	18683	179	4.8	0.09
Shu et al. [43]	$E(\mathbb{F}_{2^{239}})$	67	xc2vp100-6	25287	84	41	1.04
Beuchat et al. [7]	$E(\mathbb{F}_{2^{239}})$	67	xc2vp20-6	4557	123	107	0.49
Keller et al. [28]	$E(\mathbb{F}_{2^{251}})$	68	xc2v6000-4	27725	40	2370	65.7
Keller et al. [29]	$E(\mathbb{F}_{2^{251}})$	68	xc2v6000-4	13387	40	2600	34.8
Li et al. [33]	$E(\mathbb{F}_{2^{283}})$	72	xc4vfx140-11	55844	160	590	32.9
Shu et al. [43]	$E(\mathbb{F}_{2^{283}})$	72	xc2vp100-6	37803	72	61	2.3
Ronan et al. [40]	$E(\mathbb{F}_{2^{313}})$	75	xc2vp100-6	41078	50	124	5.1
Ronan et al. [41]	$C(\mathbb{F}_{2^{103}})$	75	xc2vp100-6	30464	41	132	4.02
Barenghi et al. [1]	$E(\mathbb{F}_p)$	87	xc2v8000-5	33857	135	1610	54.5
Beuchat et al. [7]	$E(\mathbb{F}_{2^{459}})$	89	xc2vp20-6	8153	115	327	2.66
Beuchat et al. [7]	$E(\mathbb{F}_{3^{193}})$	89	xc2vp20-6	8266	90	298	2.46
This work	$E(\mathbb{F}_{3^{193}})$	89	xc2vp125-6	46360	130	12.8	0.59
	$E(\mathbb{F}_{3^{193}})$	89	xc4vlx100-11	47433	167	10.0	0.47

we note that we increase the gap between software and hardware when considering larger values of m. The computation of the Tate pairing over $\mathbb{F}_{3^{193}}$ on a Virtex-4 LX FPGA with a medium speedgrade is for instance roughly fifty times faster than software. This speedup justifies the use of large FPGAs which are now available in servers and supercomputers such as the SGI Altix 4700 platform.

Kammler et al. [27] reported the first hardware implementation of the Optimal Ate pairing [45] over a Barreto-Naehrig (BN) curve [5], that is an ordinary curve defined over a prime field $\mathbb{F}_p$ with embedding degree $k = 12$. The proposed design is implemented with a 130 nm standard cell library and computes a pairing in 15.8 ms over a 256-bit BN curve. It is however difficult to make a fair comparison between our respective works: the level of security and the target technology are not the same.

6   Conclusion

We proposed a novel architecture based on a pipelined Karatsuba-Ofman multiplier for the η_T pairing in characteristic three. The main design challenge that we faced was to keep the pipeline continuously busy. Accordingly, we modified the scheduling of the η_T pairing in order to introduce more parallelism in the Miller's algorithm. Note that our careful re-scheduling should allow one to improve the coprocessor described in [10]. We also introduced a faster way to perform the final exponentiation by taking advantage of the linearity of the cubing operation in characteristic three. Both software and hardware implementations can benefit from this technique.

To our knowledge, the place-and-route results on several Xilinx FPGA devices of our designs improved both the computation time and the area-time trade-off of all the hardware pairing coprocessors previously published in the open literature [28,29,1,30,19,32,41,40,39,7,43,10,25]. We are also currently applying the same methodology used in this work to design a coprocessor for the Tate pairing over $\mathbb{F}_{2^m}$, with promising preliminary results.

Since the pipeline depth in the Karatsuba-Ofman multiplier is fixed by our scheduling, one could argue that the clock frequency will decrease dramatically for larger values of m. However, at the price of a slightly more complex final exponentiation, we could increase the number of pipeline stages: it suffices to perform the odd and even iterations of the main loop of Algorithm 1 in parallel (we multiply for instance $R^{(2i-2)}$ by $S^{(2i)}$ and $R^{(2i-1)}$ by $S^{(2i+1)}$ in Algorithm 1), so that the multiplier processes two sparse products at the same time. Then, a multiplication over $\mathbb{F}_{3^{6m}}$, performed by the final exponentiation coprocessor, will allow us to compute the $\eta_T(P,Q)$ pairing. We wish to investigate more deeply such architectures in the near future.

Another open problem of our interest is the design of a pairing accelerator providing the level of security of AES-128. Kammler *et al.* [27] proposed a first solution in the case of an ordinary curve. However, many questions remain open: Is it for instance possible to achieve such a level of security in hardware with supersingular (hyper)elliptic curves at a reasonable cost in terms of circuit area? Since several protocols rely on such curves, it seems important to address this problem.

Acknowledgments

The authors would like to thank Nidia Cortez-Duarte and the anonymous referees for their valuable comments. This work was supported by the Japan-France Integrated Action Program (Ayame Junior/Sakura).

The authors would also like to express their deepest gratitude to all the various purveyors of always fresh and lip-smackingly scrumptious raw fish and seafood delicacies from around つくば. Specials thanks go to 蛇の目寿司, 太丸鮨, and やぐら. どうもありがとうございました！

References

1. Barenghi, A., Bertoni, G., Breveglieri, L., Pelosi, G.: A FPGA coprocessor for the cryptographic Tate pairing over $\mathbb{F}_p$. In: Proceedings of the Fourth International Conference on Information Technology: New Generations (ITNG 2008). IEEE Computer Society Press, Los Alamitos (2008)
2. Barreto, P.S.L.M.: A note on efficient computation of cube roots in characteristic 3. Cryptology ePrint Archive, Report 2004/305 (2004)

3. Barreto, P.S.L.M., Galbraith, S.D., ÓhÉigeartaigh, C., Scott, M.: Efficient pairing computation on supersingular Abelian varieties. Designs, Codes and Cryptography 42, 239–271 (2007)
4. Barreto, P.S.L.M., Kim, H.Y., Lynn, B., Scott, M.: Efficient algorithms for pairing-based cryptosystems. In: Yung, M. (ed.) CRYPTO 2002. LNCS, vol. 2442, pp. 354–368. Springer, Heidelberg (2002)
5. Barreto, P.S.L.M., Naehrig, M.: Pairing-friendly elliptic curves of prime order. In: Preneel, B., Tavares, S. (eds.) SAC 2005. LNCS, vol. 3897, pp. 319–331. Springer, Heidelberg (2006)
6. Bertoni, G., Breveglieri, L., Fragneto, P., Pelosi, G.: Parallel hardware architectures for the cryptographic Tate pairing. In: Proceedings of the Third International Conference on Information Technology: New Generations (ITNG 2006). IEEE Computer Society Press, Los Alamitos (2006)
7. Beuchat, J.-L., Brisebarre, N., Detrey, J., Okamoto, E., Rodríguez-Henríquez, F.: A comparison between hardware accelerators for the modified tate pairing over $\mathbb{F}_{2^m}$ and $\mathbb{F}_{3^m}$. In: Galbraith, S.D., Paterson, K.G. (eds.) Pairing 2008. LNCS, vol. 5209, pp. 297–315. Springer, Heidelberg (2008)
8. Beuchat, J.-L., Brisebarre, N., Detrey, J., Okamoto, E., Shirase, M., Takagi, T.: Algorithms and arithmetic operators for computing the η_T pairing in characteristic three. IEEE Transactions on Computers 57(11), 1454–1468 (2008)
9. Beuchat, J.-L., Brisebarre, N., Shirase, M., Takagi, T., Okamoto, E.: A coprocessor for the final exponentiation of the η_T pairing in characteristic three. In: Carlet, C., Sunar, B. (eds.) WAIFI 2007. LNCS, vol. 4547, pp. 25–39. Springer, Heidelberg (2007)
10. Beuchat, J.-L., Doi, H., Fujita, K., Inomata, A., Ith, P., Kanaoka, A., Katouno, M., Mambo, M., Okamoto, E., Okamoto, T., Shiga, T., Shirase, M., Soga, R., Takagi, T., Vithanage, A., Yamamoto, H.: FPGA and ASIC implementations of the η_T pairing in characteristic three. In: Computers and Electrical Engineering (to appear)
11. Boneh, D., Franklin, M.: Identity-based encryption from the Weil pairing. In: Kilian, J. (ed.) CRYPTO 2001. LNCS, vol. 2139, pp. 213–229. Springer, Heidelberg (2001)
12. Boneh, D., Gentry, C., Waters, B.: Collusion resistant broadcast encryption with short ciphertexts and private keys. In: Shoup, V. (ed.) CRYPTO 2005. LNCS, vol. 3621, pp. 258–275. Springer, Heidelberg (2005)
13. Boneh, D., Lynn, B., Shacham, H.: Short signatures from the Weil pairing. In: Boyd, C. (ed.) ASIACRYPT 2001. LNCS, vol. 2248, pp. 514–532. Springer, Heidelberg (2001)
14. Duursma, I., Lee, H.S.: Tate pairing implementation for hyperelliptic curves $y^2 = x^p - x + d$. In: Laih, C.-S. (ed.) ASIACRYPT 2003. LNCS, vol. 2894, pp. 111–123. Springer, Heidelberg (2003)
15. Fan, H., Sun, J., Gu, M., Lam, K.-Y.: Overlap-free Karatsuba-Ofman polynomial multiplication algorithm. Cryptology ePrint Archive, Report 2007/393 (2007)
16. Frey, G., Rück, H.-G.: A remark concerning m-divisibility and the discrete logarithm in the divisor class group of curves. Mathematics of Computation 62(206), 865–874 (1994)
17. Galbraith, S.D., Harrison, K., Soldera, D.: Implementing the Tate pairing. In: Fieker, C., Kohel, D.R. (eds.) ANTS 2002. LNCS, vol. 2369, pp. 324–337. Springer, Heidelberg (2002)

18. Gorla, E., Puttmann, C., Shokrollahi, J.: Explicit formulas for efficient multiplication in $\mathbb{F}_{3^{6m}}$. In: Adams, C., Miri, A., Wiener, M. (eds.) SAC 2007. LNCS, vol. 4876, pp. 173–183. Springer, Heidelberg (2007)

19. Grabher, P., Page, D.: Hardware acceleration of the Tate pairing in characteristic three. In: Rao, J.R., Sunar, B. (eds.) CHES 2005. LNCS, vol. 3659, pp. 398–411. Springer, Heidelberg (2005)

20. Granger, R., Page, D., Smart, N.P.: High security pairing-based cryptography revisited. In: Hess, F., Pauli, S., Pohst, M. (eds.) ANTS 2006. LNCS, vol. 4076, pp. 480–494. Springer, Heidelberg (2006)

21. Hankerson, D., Menezes, A., Scott, M.: Identity-Based Cryptography. In: Software Implementation of Pairings, ch. 12. Cryptology and Information Security Series, pp. 188–206. IOS Press, Amsterdam (2009)

22. Hanrot, G., Zimmermann, P.: A long note on Mulders' short product. Journal of Symbolic Computation 37(3), 391–401 (2004)

23. Hess, F.: Pairing lattices. In: Galbraith, S.D., Paterson, K.G. (eds.) Pairing 2008. LNCS, vol. 5209, pp. 18–38. Springer, Heidelberg (2008)

24. Hess, F., Smart, N., Vercauteren, F.: The Eta pairing revisited. IEEE Transactions on Information Theory 52(10), 4595–4602 (2006)

25. Jiang, J.: Bilinear pairing (Eta_T Pairing) IP core. Technical report, City University of Hong Kong – Department of Computer Science (May 2007)

26. Joux, A.: A one round protocol for tripartite Diffie-Hellman. In: Bosma, W. (ed.) ANTS 2000. LNCS, vol. 1838, pp. 385–394. Springer, Heidelberg (2000)

27. Kammler, D., Zhang, D., Schwabe, P., Scharwaechter, H., Langenberg, M., Auras, D., Ascheid, G., Leupers, R., Mathar, R., Meyr, H.: Designing an ASIP for cryptographic pairings over Barreto-Naehrig curves. Cryptology ePrint Archive, Report 2009/056 (2009)

28. Keller, M., Kerins, T., Crowe, F., Marnane, W.P.: FPGA implementation of a $GF(2^m)$ Tate pairing architecture. In: Bertels, K., Cardoso, J.M.P., Vassiliadis, S. (eds.) ARC 2006. LNCS, vol. 3985, pp. 358–369. Springer, Heidelberg (2006)

29. Keller, M., Ronan, R., Marnane, W.P., Murphy, C.: Hardware architectures for the Tate pairing over $GF(2^m)$. Computers and Electrical Engineering 33(5–6), 392–406 (2007)

30. Kerins, T., Marnane, W.P., Popovici, E.M., Barreto, P.S.L.M.: Efficient hardware for the Tate pairing calculation in characteristic three. In: Rao, J.R., Sunar, B. (eds.) CHES 2005. LNCS, vol. 3659, pp. 412–426. Springer, Heidelberg (2005)

31. Koblitz, N., Menezes, A.: Pairing-based cryptography at high security levels. In: Smart, N.P. (ed.) Cryptography and Coding 2005. LNCS, vol. 3796, pp. 13–36. Springer, Heidelberg (2005)

32. Kömürcü, G., Savaş, E.: An efficient hardware implementation of the Tate pairing in characteristic three. In: Prasolova-Førland, E., Popescu, M. (eds.) Proceedings of the Third International Conference on Systems – ICONS 2008, pp. 23–28. IEEE Computer Society Press, Los Alamitos (2008)

33. Li, H., Huang, J., Sweany, P., Huang, D.: FPGA implementations of elliptic curve cryptography and Tate pairing over a binary field. Journal of Systems Architecture 54, 1077–1088 (2008)

34. Menezes, A., Okamoto, T., Vanstone, S.A.: Reducing elliptic curves logarithms to logarithms in a finite field. IEEE Transactions on Information Theory 39(5), 1639–1646 (1993)

35. Miller, V.S.: Short programs for functions on curves (1986),
    http://crypto.stanford.edu/miller

36. Miller, V.S.: The Weil pairing, and its efficient calculation. Journal of Cryptology 17(4), 235–261 (2004)
37. Mitsunari, S.: A fast implementation of η_T pairing in characteristic three on Intel Core 2 Duo processor. Cryptology ePrint Archive, Report 2009/032 (2009)
38. Mitsunari, S., Sakai, R., Kasahara, M.: A new traitor tracing. IEICE Trans. Fundamentals E85–A(2), 481–484 (2002)
39. Ronan, R., Murphy, C., Kerins, T., ÓhÉigeartaigh, C., Barreto, P.S.L.M.: A flexible processor for the characteristic 3 η_T pairing. Int. J. High Performance Systems Architecture 1(2), 79–88 (2007)
40. Ronan, R., ÓhÉigeartaigh, C., Murphy, C., Scott, M., Kerins, T.: FPGA acceleration of the Tate pairing in characteristic 2. In: Proceedings of the IEEE International Conference on Field Programmable Technology – FPT 2006, pp. 213–220. IEEE, Los Alamitos (2006)
41. Ronan, R., ÓhÉigeartaigh, C., Murphy, C., Scott, M., Kerins, T.: Hardware acceleration of the Tate pairing on a genus 2 hyperelliptic curve. Journal of Systems Architecture 53, 85–98 (2007)
42. Sakai, R., Ohgishi, K., Kasahara, M.: Cryptosystems based on pairing. In: 2000 Symposium on Cryptography and Information Security (SCIS 2000), Okinawa, Japan, January 2000, pp. 26–28 (2000)
43. Shu, C., Kwon, S., Gaj, K.: FPGA accelerated Tate pairing based cryptosystem over binary fields. In: Proceedings of the IEEE International Conference on Field Programmable Technology – FPT 2006, pp. 173–180. IEEE, Los Alamitos (2006)
44. Song, L., Parhi, K.K.: Low energy digit-serial/parallel finite field multipliers. Journal of VLSI Signal Processing 19(2), 149–166 (1998)
45. Vercauteren, F.: Optimal pairings. Cryptology ePrint Archive, Report 2008/096 (2008)
46. Washington, L.C.: Elliptic Curves – Number Theory and Cryptography, 2nd edn. CRC Press, Boca Raton (2008)

Faster $\mathbb{F}_p$-Arithmetic for Cryptographic Pairings on Barreto-Naehrig Curves*

Junfeng Fan, Frederik Vercauteren**, and Ingrid Verbauwhede

ESAT/SCD-COSIC, Katholieke Universiteit Leuven and IBBT
Kasteelpark Arenberg 10
B-3001 Leuven-Heverlee, Belgium
{jfan,fvercauteren,iverbauwhede}@esat.kuleuven.be

Abstract. This paper describes a new method to speed up $\mathbb{F}_p$-arithmetic for Barreto-Naehrig (BN) curves. We explore the characteristics of the modulus defined by BN curves and choose curve parameters such that $\mathbb{F}_p$ multiplication becomes more efficient. The proposed algorithm uses Montgomery reduction in a polynomial ring combined with a coefficient reduction phase using a pseudo-Mersenne number. With this algorithm, the performance of pairings on BN curves can be significantly improved, resulting in a factor 5.4 speed-up compared with the state-of-the-art hardware implementations. Using this algorithm, we implemented a pairing processor in hardware, which runs at 204 MHz and finishes one ate and R-ate pairing computation over a 256-bit BN curve in 4.22 ms and 2.91 ms, respectively.

Keywords: Pairings, BN curves, Modular reduction.

1   Introduction

A bilinear pairing is a map $\mathbb{G}_1 \times \mathbb{G}_2 \rightarrow \mathbb{G}_T$ where $\mathbb{G}_1$ and $\mathbb{G}_2$ are typically additive groups and $\mathbb{G}_T$ is a multiplicative group and the map is linear in each component. Many pairings used in cryptography such as the Tate pairing [1], ate pairing [11], and R-ate pairing [13] choose $\mathbb{G}_1$ and $\mathbb{G}_2$ to be specific cyclic subgroups of $E(\mathbb{F}_{p^k})$, and $\mathbb{G}_T$ to be a subgroup of $\mathbb{F}_{p^k}^*$.

The selection of parameters has a substantial impact on the security and performance of a pairing. For example, the underlying field, the type of curve, the order of $\mathbb{G}_1$, $\mathbb{G}_2$ and $\mathbb{G}_T$ should be carefully chosen such that it offers sufficient security, but still is efficient to compute. In this paper, we focus on efficient implementation of pairings over BN curves [17]. BN curves are defined over $\mathbb{F}_p$ where $p = 36\bar{t}^4 + 36\bar{t}^3 + 24\bar{t}^2 + 6\bar{t} + 1$ for $\bar{t} \in \mathbb{Z}$ such that p is prime. In this paper, we propose a new modular multiplication algorithm for BN curves. We show that

---

* This work was supported by research grants of Katholieke Universiteit Leuven (OT/06/40) and FWO projects (G.0300.07), by the IAP Programme P6/26 BCRYPT of the Belgian State (Belgian Science Policy), by the EU IST FP6 projects (ECRYPT) and by the IBBT-QoE project of the IBBT.
** Postdoctoral Fellow of the Research Foundation - Flanders (FWO).

C. Clavier and K. Gaj (Eds.): CHES 2009, LNCS 5747, pp. 240–253, 2009.

when choosing $\bar{t} = 2^m + s$, where s is a reasonably *small* number, the modular multiplication in $\mathbb{F}_p$ can be substantially improved. Existing techniques to speed up arithmetic in extension fields (see [7,6] for fast operation in $\mathbb{F}_{p^2}$, $\mathbb{F}_{p^6}$ and $\mathbb{F}_{p^{12}}$) can be used on top of it. The proposed modular reduction algorithm and parameters for BN curves result in a significant improvement on the performance of ate and R-ate pairing.

The remainder of the paper is organized as follows. In Sect. 2 we review cryptographic pairings and their computation. In Sect. 3 we present a new modular multiplication algorithm and compare its complexity with known algorithms. The details of the hardware implementation and results are given in Sect. 4 and Sect. 5, respectively. We conclude the paper in Sect. 6.

2   Previous Works

2.1   Bilinear Pairings

Let $\mathbb{F}_p$ be a finite field and let $E(\mathbb{F}_p)$ be an elliptic curve defined over $\mathbb{F}_p$. Let r be a large prime dividing $\#E(\mathbb{F}_p)$. Let k be the embedding degree of $E(\mathbb{F}_p)$ with respect to r, namely, the smallest positive integer k such that $r|p^k - 1$. We use $E(\mathbb{K})[r]$ to denote the $\mathbb{K}$-rational r-torsion group of the curve for any finite field $\mathbb{K}$. For $P \in E(\mathbb{K})$ and an integer s, let $f_{s,P}$ be a $\mathbb{K}$-rational function with divisor

$$(f_{s,P}) = s(P) - ([s]P) - (s - 1)\mathcal{O},$$

where $\mathcal{O}$ is the point at infinity. This function is also known as Miller function [14,15].

Let $\mathbb{G}_1 = E(\mathbb{F}_p)[r]$, $\mathbb{G}_2 = E(\mathbb{F}_{p^k})/rE(\mathbb{F}_{p^k})$ and $\mathbb{G}_3 = \mu_r \subset \mathbb{F}_{p^k}^*$ (the r-th roots of unity), then the reduced Tate pairing is a well-defined, non-degenerate, bilinear pairing. Let $P \in \mathbb{G}_1$ and $Q \in \mathbb{G}_2$, then the reduced Tate pairing of P, Q is computed as

$$e(P, Q) = (f_{r,P}(Q))^{(p^k-1)/r}.$$

The ate pairing is similar but with different $\mathbb{G}_1$ and $\mathbb{G}_2$. Here we define $\mathbb{G}_1 = E(\mathbb{F}_p)[r]$ and $\mathbb{G}_2 = E(\mathbb{F}_{p^k})[r] \cap \mathrm{Ker}(\pi_p - [p])$, where π_p is the Frobenius endomorphism. Let $P \in \mathbb{G}_1$, $Q \in \mathbb{G}_2$ and let t_r be the trace of Frobenius of the curve, then the ate pairing is also well-defined, non-degenerate bilinear pairing, and can be computed as

$$a(Q, P) = (f_{t_r-1,Q}(P))^{(p^k-1)/r}.$$

The R-ate pairing is a generalization of the ate pairing. For the same choice of $\mathbb{G}_1$ and $\mathbb{G}_2$ as for the ate pairing, the R-ate pairing on BN curves is defined as

$$Ra(Q, P) = (f \cdot (f \cdot l_{aQ,Q}(P))^p \cdot l_{\pi(aQ+Q),aQ}(P))^{(p^k-1)/r},$$

Algorithm 1. Computing the R-ate pairing on BN curves [7]

Input: $P \in E(\mathbb{F}_p)[r]$, $Q \in E(\mathbb{F}_{p^k})[r] \cap Ker(\pi_p - [p])$ and $a = 6\bar{t} + 2$.
Output: $R_a(Q, P)$.

  1: $a = \sum_{i=0}^{L-1} a_i 2^i$.
  2: $T \leftarrow Q$, $f \leftarrow 1$.
  3: **for** $i = L - 2$ downto 0 **do**
  4:    $T \leftarrow 2T$.
  5:    $f \leftarrow f^2 \cdot l_{T,T}(P)$.
  6:    **if** $a_i = 1$ **then**
  7:       $T \leftarrow T + Q$.
  8:       $f \leftarrow f \cdot l_{T,Q}(P)$.
  9:    **end if**
10: **end for**
11: $f \leftarrow (f \cdot (f \cdot l_{aQ,Q}(P))^p \cdot l_{\pi(aQ+Q),aQ}(P))^{(p^k-1)/r}$.

Return f.

---

where $a = 6\bar{t} + 2$, $f = f_{a,Q}(P)$ and $l_{A,B}$ denotes the line through point A and B.

Due to limited space, we only describe the algorithm to compute the R-ate pairing. The algorithms for Tate and ate pairings are similar, and can be found in [7].

2.2 Choice of Curve Parameters

The most important parameters for cryptographic pairings are the underlying finite field, the order of the curve, the embedding degree, and the order of $\mathbb{G}_1$, $\mathbb{G}_2$ and $\mathbb{G}_T$. These parameters should be chosen such that the best exponential time algorithms to solve the discrete logarithm problem (DLP) in $\mathbb{G}_1$ and $\mathbb{G}_2$ and the sub-exponential time algorithms to solve the DLP in $\mathbb{G}_T$ take longer than a chosen security level. In this paper, we will use the 128-bit symmetric key security level.

Barreto and Naehrig [17] described a method to construct pairing-friendly elliptic curves over a prime field with embedding degree 12. The finite field, trace of Frobenius and order of the curve are defined by the following polynomial families:

$$\begin{aligned} p(t) &= 36t^4 + 36t^3 + 24t^2 + 6t + 1, \\ t_r(t) &= 6t^2 + 1, \\ n(t) &= 36t^4 + 36t^3 + 18t^2 + 6t + 1. \end{aligned}$$

The curve is defined as $E : y^2 = x^3 + v$ for some $v \in \mathbb{F}_p$. The choice of $\bar{t}$ must meet the following requirements: both $p(\bar{t})$ and $n(\bar{t})$ must be prime and $\bar{t}$ must be large enough to guarantee a chosen security level. For the efficiency of pairing computation, $\bar{t}$, $p(\bar{t})$ and $t_r(\bar{t})$ should have small Hamming-weight.

For example, [7] suggested to use $\bar{t} = \text{0x6000000000001F2D}$, which is also used in [10] and [12]. With this parameter, pairings defined over $p(\bar{t})$ achieves 128-bit security.

Table 1. Selection of $\bar{t}$ for BN curves in [7]

$\bar{t}$	HW$(6\bar{t}+2)$	HW(t_r)	HW(p)	$\log(2,p)$
0x600000000001F2D	9	28	87	256

2.3 Multiplication in $\mathbb{F}_p$

We briefly recall the techniques for integer multiplication and reduction. Given a modulus $p < 2^n$ and an integer $c < 2^{2n}$, the following algorithms can be used to compute $c \bmod p$.

Barrett reduction. The Barrett reduction algorithm [2] uses a precomputed value $\mu = \lfloor \frac{2^{2n}}{p} \rfloor$ to help estimate $\frac{c}{p}$, thus integer division is avoided. Dhem [8] proposed an improved Barrett modular multiplication algorithm which has a simplified final correction.

Montgomery reduction. The Montgomery reduction method [16] precomputes $p' = -p^{-1} \bmod r$, where r is normally a power of two. Given c and p, it generates q such that $c + qp$ is a multiple of r. As a result, $(c + qp)/r$ is just a shift operation. Algorithm 2 shows both Barrett and Montgomery multiplication algorithms.

Chung-Hasan reduction. In [4,5], Chung and Hasan proposed an efficient reduction method for low-weight polynomial form moduli $p = f(\bar{t}) = \bar{t}^n + f_{n-1}\bar{t}^{n-1} + .. + f_1\bar{t} + f_0$, where $|f_i| \le 1$. The modular multiplication is shown in Alg. 3.

Algorithm 2. Digit-serial modular multiplication algorithm.

Barrett [8]	**Montgomery [16]**
Input: $a = (a_{n-1}, .., a_0)_d,$	**Input:** $a = (a_{n-1}, .., a_0)_d,$
$b = (b_{n-1}, .., b_0)_d,$	$b = (b_{n-1}, .., b_0)_d,$
$p = (p_{n-1}, .., p_0)_d, 0 \le a, b < p,$	$p = (p_{n-1}, .., p_0)_d, 0 \le a, b < p,$
$2^{(n-1)w} \le p < 2^{nw}, d = 2^w.$	$r = d^n,$
Precompute $\mu = \lfloor d^{n+3}/p \rfloor.$	Precompute $p' = -p^{-1} \bmod d, d = 2^w.$
Output: $c = ab \bmod p.$	**Output:** $c = abr^{-1} \bmod p.$
1: $c \leftarrow 0.$	1: $c \leftarrow 0.$
2: **for** $i = n-1$ downto 0 **do**	2: **for** $i = 0$ to $n-1$ **do**
3: $\quad c \leftarrow c \cdot d + a \cdot b_i.$	3: $\quad c \leftarrow c + ab_i.$
4: $\quad \hat{q} \leftarrow \lfloor (\lfloor c/d^{n-2} \rfloor \cdot \mu)/2^{w+5} \rfloor.$	4: $\quad u \leftarrow c \bmod d, q \leftarrow (up') \bmod d.$
5: $\quad c \leftarrow c - \hat{q} \cdot p.$	5: $\quad c \leftarrow (c + qp)/d.$
6: **end for**	6: **end for**
7: **if** $c \ge p$ **then**	7: **if** $c \ge p$ **then**
8: $\quad c \leftarrow c - p.$	8: $\quad c \leftarrow c - p.$
9: **end if**	9: **end if**
Return $c.$	**Return** $c.$

Algorithm 3. Chung-Hasan multiplication algorithm [4].

Input: positive integers $a = \sum_{i=0}^{n-1} a_i t^i$, $b = \sum_{i=0}^{n-1} b_i t^i$, modulus $p = f(t) = t^n + f_{n-1} t^{n-1} + .. + f_1 t + f_0$.
Output: $c(t) = a(t)b(t) \bmod p$.
  1: **Phase I: Polynomial Multiplication**
  2: $c(t) \leftarrow a(t)b(t)$.
  3: **Phase II: Polynomial Reduction**
  4: **for** $i = 2n - 2$ down to n **do**
  5:     $c(t) \leftarrow c(t) - c_i f(t) t^{i-n}$.
  6: **end for**
     Phase III: Coefficient Reduction
  7: $c_n \leftarrow \lfloor c_{n-1}/\bar{t} \rceil$, $c_{n-1} \leftarrow c_{n-1} - c_n \bar{t}$.
  8: $c(t) \leftarrow c(t) - c_n f(t) t$.
  9: **for** $i = 0$ to $n - 1$ **do**
 10:     $q_i \leftarrow \lfloor c_i/\bar{t} \rceil$, $r_i \leftarrow c_i - q_i \bar{t}$.
 11:     $c_{i+1} \leftarrow c_{i+1} + q_i$, $c_i \leftarrow r_i$.
 12: **end for**
 13: $c(t) \leftarrow c(t) - q_n f(t) t$.
Return $c(t)$.

The polynomial reduction phase is rather efficient since $f(t)$ is monic, making the polynomial long division (step 3) simple. Barrett reduction is used to perform divisions required in Phase III. The overall performance is more efficient than traditional Barrett or Montgomery reduction algorithm [4]. In [5], this algorithm is further extended to monic polynomials with $|f_i| \leq s$ where $s \in (0, \bar{t})$ is a small number. Note that the polynomial reduction phase is efficient only when $f(t)$ is monic.

3   Fast Modular Reduction Algorithm for BN Curves

Instead of using a general modular reduction algorithm such as Montgomery or Barrett algorithm, we explore the special characteristics of the prime p. Note that the polynomial $p(t) = 36t^4 + 36t^3 + 24t^2 + 6t + 1$ defined by BN is not monic, but has the following characteristics:

  1. $p(t)$ has small coefficients.
  2. $p^{(-1)}(t) = 1 \bmod t$.

The second condition implies via Hensel's lemma that $p^{(-1)}(t) \bmod t^n$ has integer coefficients. This suggests that multiplication and reduction with Montgomery's algorithm in the polynomial ring could be efficient. We first present a modular multiplication algorithm for polynomial form primes that satisfy $p^{(-1)}(t) = 1 \bmod t$ and then apply this method to BN curves.

3.1   Hybrid Modular Multiplication

Algorithm 4 describes a modular multiplication algorithm for polynomial form moduli. The algorithm is composed of three phases, i.e. polynomial multiplication (step 3), polynomial reduction (step 4-6), and coefficient reduction phase (step 9). Note that we present the algorithm in a digit-serial manner. The polynomial reduction uses the Montgomery reduction, while the coefficient reduction uses division. We call this algorithm Hybrid Modular Multiplication (HMM).

Note that algorithm 4 works for any irreducible polynomial $p(t)$ satisfying the condition $p^{(-1)}(t) = 1 \bmod t$ or equivalently, $p(t) = 1 \bmod t$. It can also be easily modified to support $p(t)$ satisfying $p^{(-1)}(t) = -1 \bmod t$.

Algorithm 4 requires division by $\bar{t}$ in both step 4 and step 9. Like Chung-Hasan's algorithm, division can be performed with the Barrett reduction algorithm [4]. However, the complexity of division can be reduced if $\bar{t}$ is a pseudo-Mersenne number. Algorithm 5 transfers division by $\bar{t}$ to multiplication by s for $\bar{t} = 2^m + s$ where s is small.

3.2   Modular Multiplication for BN Curves

In order to apply Alg. 4 and Alg. 5 to BN curves, we select $\bar{t} = 2^m + s$ where s is small. Note that any choice of $\bar{t}$ which makes p and n primes of the required size will suffice. As such we can choose $\bar{t} = 2^m + s$ where s is small; an example is shown in Table 2.

Algorithm 4. Hybrid Modular Multiplication Algorithm

Input: $a(t) = \sum_{i=0}^{n-1} a_i t^i$, $b(t) = \sum_{i=0}^{n-1} b_i t^i$, and modulus $p(t) = \sum_{i=1}^{n-1} p_i t^i + 1$.
Output: $r(t) = a(t)b(t)t^{-n} \bmod p(t)$.

1: $c(t)(= \sum_{i=0}^{n-1} c_i t^i) \leftarrow 0$.
2: **for** $i = 0$ to $n - 1$ **do**
3:     $c(t) \leftarrow c(t) + a(t)b_i$.
4:     $\mu \leftarrow c_0$ div $\bar{t}$, $\gamma \leftarrow c_0 \bmod \bar{t}$.
5:     $g(t) \leftarrow (p_{n-1}t^{n-1} + .. + p_1 t + 1)(-\gamma)$.
6:     $c(t) \leftarrow (c(t) + g(t))/t + \mu$.
7: **end for**
8: **for** $i = 0$ to $n - 2$ **do**
9:     $c_{i+1} \leftarrow c_{i+1} + (c_i$ div $\bar{t})$, $c_i \leftarrow c_i \bmod \bar{t}$.
10: **end for**
Return $r(t) \leftarrow c(t)$.

Algorithm 5. Division by $\bar{t} = 2^m + s$

Input: a, $\bar{t} = 2^m + s$ with $0 < s < 2^{\lfloor k/2 \rfloor}$.
Output: μ and γ with $a = \mu t + \gamma$, $|\gamma| < \bar{t}$.

1: $\mu \leftarrow 0$, $\gamma \leftarrow a$.
2: **while** $|\gamma| \geq \bar{t}$ **do**
3:     $\rho \leftarrow \gamma$ div 2^m, $\gamma \leftarrow \gamma \bmod 2^m$.
4:     $\mu \leftarrow \mu + \rho$, $\gamma \leftarrow \gamma - s\rho$.
5: **end while**
Return μ, γ.

Table 2. Selection of $\bar{t} = 2^m + s$ for BN curves

$\bar{t}$	HW$(6\bar{t}+2)$	HW(t_r)	HW(p)	log$(2,p)$
$2^{63} + 2^9 + 2^8 + 2^6 + 2^4 + 2^3 + 1$	6	20	68	257

Algorithm 6. Hybrid Modular Multiplication Algorithm for BN curves

---

Input: $a(t) = \sum_{i=0}^{4} a_i t^i$, $b(t) = \sum_{i=0}^{4} b_i t^i$. $p(t) = 36t^4 + 36t^3 + 24t^2 + 6t + 1$, $p^{-1}(t) = 1 \bmod t$, $\bar{t} = 2^m + s$.

Output: $r(t) = a(t)b(t)t^{-5} \bmod p(t)$.

1: $c(t)(= \sum_{i=0}^{4} c_i t^i) \leftarrow 0$.
2: **for** $j = 0$ to 4 **do**
3:     $c(t) \leftarrow c(t) + a(t)b_j$.
4:     $\mu \leftarrow c_0$ div 2^m, $\gamma \leftarrow (c_0 \bmod 2^m) - s\mu$.
5:     $g(t) \leftarrow (36t^4 + 36t^3 + 24t^2 + 6t + 1)(-\gamma)$.
6:     $c(t) \leftarrow (c(t) + g(t))/t + \mu$.
7: **end for**
8: **for** $i = 0$ to 3 **do**
9:     $\mu \leftarrow c_i$ div 2^m, $\gamma \leftarrow (c_i \bmod 2^m) - s\mu$.
10:     $c_{i+1} \leftarrow c_{i+1} + \mu$, $c_i \leftarrow \gamma$.
11: **end for**
12: Repeat step 8-11.
Return $r(t) \leftarrow c(t)$.

With $\bar{t} = 2^m + s$ as shown in Table 2, Algorithm 6 describes a modular multiplication algorithm for BN curves which we call HMMB.

The following lemma provides bounds on the input value such that Algorithm 6 gives a bounded output. The proof is in the appendix.

Lemma 1. *Given* $\bar{t} = 2^m + s$ *and* $\xi = (36s + 1) < 2^{m/2-7}$ *(i.e.* $m \geq 26$*), if the input* $a(t)$ *and* $b(t)$ *satisfy*

$$0 \leq |a_i|, |b_i| < 2^{m/2}, \quad i = 4,$$
$$0 \leq |a_i|, |b_i| < 2^{m+1}, \quad 0 \leq i \leq 3,$$

then $r(t)$ *calculated by Alg. 6 satisfies*

$$0 \leq |r_i| < 2^{m/2}, \quad i = 4,$$
$$0 \leq |r_i| < 2^{m+1}, \quad 0 \leq i \leq 3.$$

This algorithm is suitable for high performance implementation on multi-core systems. One can see that the first loop of HMMB algorithm can be easily parallelized. This is an intrinsic advantage of this algorithm, i.e. no carry propagation occurs during polynomial multiplication. The coefficient reduction phase can also be parallelized. We modify the last two loops of the HMMB algorithm and give a parallel version.

loop1 $\mu_i \leftarrow c_i$ div $2^m, \gamma_i \leftarrow (c_i \bmod 2^m) - s\mu_i$ $0 \leq i \leq 3$.
$\quad\quad c_i \leftarrow \gamma_i + \mu_{i-1}$ $1 \leq i \leq 3$.
$\quad\quad c_0 \leftarrow \gamma_0, c_4 \leftarrow \mu_3$.
loop2 $\mu_i \leftarrow c_i$ div $2^m, \gamma_i \leftarrow (c_i \bmod 2^m) - s\mu_i$ $0 \leq i \leq 3$.
$\quad\quad c_i \leftarrow \gamma_i + \mu_{i-1}$ $1 \leq i \leq 3$.
$\quad\quad c_0 \leftarrow \gamma_0, c_4 \leftarrow c_4 + \mu_3$.

From the proof of Lemma 1 we know $c_i < 5 \cdot 2^{2m+3}$ for $0 \leq i \leq 4$. Thus, after **loop1**, we have

$$|c_i| = |\mu_{i-1}| + |\gamma_i| < 5 \cdot 2^{m+3} + 5 \cdot 2^{m+3}s + 2^m < s2^{m+6}.$$

Furthermore, after **loop2**, we have

$$|c_i| = |\mu_{i-1}| + |\gamma_i| < s2^6 + 2^6 s^2 + 2^m < 2^{m+1}.$$

3.3 Complexity of Algorithm 6

We compare the complexity of Alg. 6 with Montgomery's and Barrett's algorithm for 256-bit BN curves. We assume a digit-serial method is used with digit-size 64-bit. Note that Alg. 6 requires four 64-bit words together with one 32-bit word to represent a 256-bit integer.

For Montgomery multiplication, nine 64x64 multiplications are required in each iteration, resulting in 36 subword multiplications in total. Barrett multiplication has the same complexity as Montgomery algorithm.

For HMMB, in the first loop four 64x64 and one 32x64 multiplications are required in step 3, one $\lceil \log_2(s) \rceil \times \lceil \log_2(\mu) \rceil$ multiplication is required in step 4. The last iteration takes four 32x64 and one 32x32 multiplications. In total, the first loop takes one 32x32, eight 32x64, sixteen 64x64, and five $\lceil \log_2(s) \rceil \times \lceil \log_2(\mu) \rceil$ multiplications, where $\mu < 2^{k+6}$ as shown in the proof of Lemma 1. Note that $p(t)\gamma$ can be performed with addition and shift operation, e.g. $36\gamma = 2^5\gamma + 2^2\gamma$.

The coefficient reduction phase requires eight $\lceil \log_2(s) \rceil \times \lceil \log_2(\mu) \rceil$ multiplications. From the proof of Lemma 1 we know that $c_i < 5 \cdot 2^{2k+3}$, thus $\mu < 2^{k+6}$ in the first **for** loop (step 8-10). In the second **for** loop (step 12), as shown in the end of section 3.2, we have $\mu < s2^6$.

Table 3 compares the number of multiplications required by the Barrett, Montgomery and the HMMB algorithm. Compared to Barrett and Montgomery reduction, HMMB has much lower complexity. One can see that $s\mu$ can be efficiently computed if s is small (see Table 2). Especially, if s is of low Hamming-weight, $s\mu$ can be performed with shift and addition operations.

Table 3. Complexity comparison of different modular multiplication algorithms

Algorithm	32x32	32x64	64x64	$\lceil \log_2(s) \rceil \times \lceil \log_2(\mu) \rceil$
Barrett			36	
Montgomery			36	
HMMB	1	8	16	13

4 Hardware Design of Pairings

As a verification of the efficiency of the HMMB algorithm, we made a hardware implementation of the ate and R-ate pairing using this algorithm. We chose $\bar{t} = 2^{63}+s$, where $s=2^9+2^8+2^6+2^4+2^3+1$. With this setting, the implementation achieves 128-bit security.

4.1 Multiplier

Figure 1 shows the realization of the HMMB algorithm. A row of multipliers is used to carry out step 3, namely, $a_i b_j$ for $0 \le i \le 4$. We used a 64x16 bit multiplier, thus four cycles are required for each iteration. One can adjust the size of multiplier for different design purposes, i.e. high clock frequency, small area and so on. The partial product is then reduced by the "Mod_t" component. The "Mod_t" component, which is composed of a multiplier and a subtracter, generates μ and γ from c_i, namely, $\mu \leftarrow c_i$ div 2^m and $\gamma \leftarrow (c_i \bmod 2^m) - s\mu$. Note that the "Mod_t" component below rc_0 is slightly different, where $\gamma \leftarrow (s\mu - (rc_0 \bmod 2^m))$.

The dataflow of this implementation is slightly different from that in Alg. 6. Instead of performing coefficient reduction in the end, we reduce the coefficients before polynomial reduction. This reduces the length of c_i and rc_i in Fig. 1, and one instead of two coefficient reduction loop is required in the end. The "Mod_t" components are reused to perform the final reduction loop. After that, $r(t)$ is ready in the accumulators.

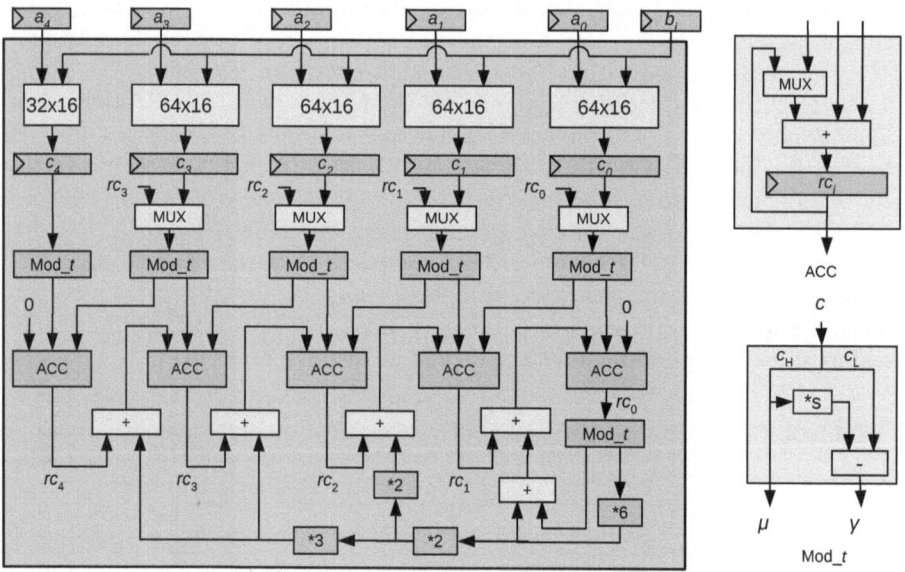

Fig. 1. $\mathbb{F}_p$ multiplier using algorithm HMMB

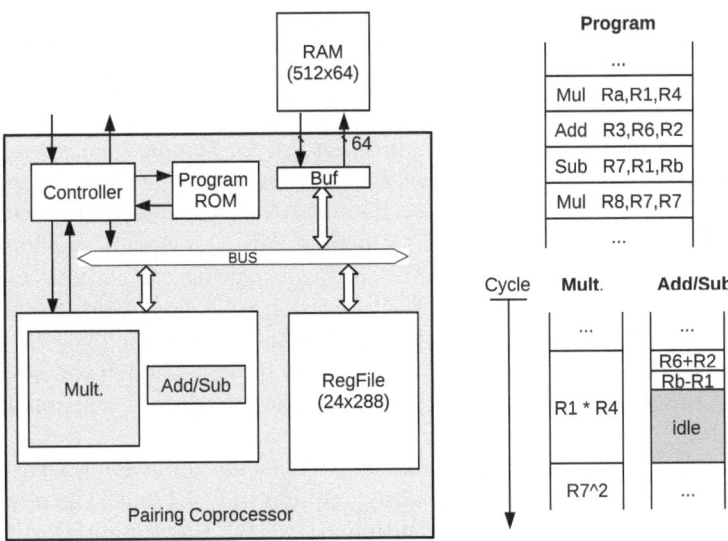

Fig. 2. Block diagram of the system architecture

4.2   Pairing Processor Architecture

Using the multiplier described above, we built a pairing processor. Figure 2 shows the block diagram of the processor. It consists of a micro-controller, a program ROM, an $\mathbb{F}_p$ multiplier and adder/subtracter, a register file and an IO buffer. The data is stored in a 64-bit single port RAM. The program ROM contains subroutines that are used in Miller's loop, such as point addition, point doubling, line evaluation, multiplication in $\mathbb{F}_{p^{12}}$, and so on. The micro-controller realizes Miller's loop by calling the corresponding sub-routines.

The ALU is able to execute multiplication and addition/subtraction instructions in parallel. A simple example is shown in Fig. 2. When performing the mul operation, the micro-controller fetches the next instruction and checks if it is an add or sub instruction. If it is, then it is executed in parallel if there is no data dependency on the ongoing mul instruction. Table 4 gives the number of clock cycles that are required for each subroutine and pairing.

Table 4. Number of clock cycles required by different subroutines

	2T	T+Q	$l_{T,T}(P)$	$l_{T,Q}(P)$	f^2	$f \cdot l$	$f^{(p^k-1)/r}$	ate	R-ate
#Cycles	574	984	422	260	1541	1239	281558	861724	592976

5   Implementation Results

The whole system is synthesized using 130 nm standard cell library. It can run at a maximum frequency of 204 MHz. The pairing processor alone uses around

183 kGates, including 70 kGates used by Register File and 25 kGates used by controller and Program ROM. It finishes one ate and R-ate pairing computation in 4.22 ms and 2.91 ms, respectively. Table 5 compares the result with the state-of-the-art implementations.

Kammler *et al.* [12] reported the first, and so far the only, hardware implementation of cryptographic pairings achieving a 128-bit security. They chose $\bar{t}$=0x6000000000001F2D to generate a 256-bit BN curve. The Montgomery algorithm is used for $\mathbb{F}_p$ multiplication. Compared with this design, our implementation is about 5 times faster in terms of R-ate pairing calculation. The main speedup comes from fast modular multiplication in $\mathbb{F}_p$ and larger register file. For an $\mathbb{F}_p$ multiplication, the multiplier shown in Fig. 1 takes 23 cycles excluding memory access, while 68 cycles are required in [12]. Though the area of our design is around 1.9 times larger, the area-latency product is still smaller than that in [12].

The results of software implementations [10,9] are quite impressive. On an Intel 64-bit core2 processor, R-ate pairing requires only 4.17 ms. The advantages of Intel core2 is that it has a fast multiplier (two full 64-bit multiplication in 8 cycles) and relatively high clock frequency. Though it takes 16 times more clock cycles (10^7 cycles for R-ate [10]) than our hardware implementation, the overall speed is only 1.4 times lower.

There are also some hardware implementations [18,3] for η_T pairing over binary or cubic curves. Note that the security achieved using the reported parameters is much lower than 128-bit, which makes a fair comparison difficult.

Table 5. Performance comparison of software and hardware implementations of pairing

Design	Pairing	Security [bit]	Platform	Area	Frequency [MHz]	Performance [ms]
this design	ate	128	130 nm ASIC	183 kGates	204	4.22
	R-ate					2.91
[12]	Tate	128	130 nm ASIC	97 kGates	338	34.4
	ate					22.8
	R-ate					15.8
[10]	ate	128	64-bit core2	-	2400	6.25
	R-ate					4.17
[9]	ate	128	64-bit core2	-	2400	6.01
[18]	η_T over $\mathbb{F}_{2^{239}}$	67	XC2VP100-6	25278 slices	84	0.034
	η_T over $\mathbb{F}_{2^{283}}$	72		37803 slices	72	0.049
[3]	η_T over $\mathbb{F}_{3^{97}}$	66	XC4VLX60-11	18683 slices	N/A	0.0048
	η_T over $\mathbb{F}_{3^{193}}$	89	XC4VLX100-11	47433 slices	N/A	0.010

6   Conclusions

In this paper, we studied a new fast implementation of cryptographic pairings using BN curves. We introduce a new modular multiplication algorithm and a method to select multiplication-friendly parameters. We show that with careful selection of parameters the proposed algorithm has much lower computational complexity than traditional Barrett or Montgomery methods.

As a verification, we also implemented ate and R-ate pairing in hardware using this algorithm. Our results outperform previous hardware implementations by a factor of roughly 5. Note that smaller digit size can be used when targeting a compact hardware implementation. For future work, it is also definitely interesting to see the performance of this algorithm implemented in software. Finally, we remark that the described algorithms also generalize to other pairing friendly finite fields and even more generally, to other types of finite fields.

Acknowledgments

The authors would like to thank the anonymous referees for detailed review. We adequately appreciate their observations and helpful suggestions.

References

1. Barreto, P.S.L.M., Kim, H.Y., Lynn, B., Scott, M.: Efficient Algorithms for Pairing-Based Cryptosystems. In: Yung, M. (ed.) CRYPTO 2002. LNCS, vol. 2442, pp. 354–369. Springer, Heidelberg (2002)
2. Barrett, P.: Implementing the Rivest Shamir and Adleman Public Key Encryption Algorithm on a Standard Digital Signal Processor. In: Odlyzko, A.M. (ed.) CRYPTO 1986. LNCS, vol. 263, pp. 311–323. Springer, Heidelberg (1987)
3. Beuchat, J., Detrey, J., Estibals, N., Okamoto, E., Rodríguez-Henríquez, F.: Hardware Accelerator for the Tate Pairing in Characteristic Three Based on Karatsuba-Ofman Multipliers. Cryptology ePrint Archive, Report 2009/122 (2009), http://eprint.iacr.org/
4. Chung, J., Hasan, M.A.: Low-Weight Polynomial Form Integers for Efficient Modular Multiplication. IEEE Trans. Comput. 56(1), 44–57 (2007)
5. Chung, J., Hasan, M.A.: Montgomery Reduction Algorithm for Modular Multiplication Using Low-Weight Polynomial Form Integers. In: ARITH 2007: Proceedings of the 18th IEEE Symposium on Computer Arithmetic, Washington, DC, USA, 2007, pp. 230–239. IEEE Computer Society Press, Los Alamitos (2007)
6. Dahab, R., Devegili, A., Ó'hÉigeartaigh, C., Scott, M.: Multiplication and Squaring on Pairing-Friendly Fields. Cryptology ePrint Archive, Report 2006/ /471, http://eprint.iacr.org
7. Devegili, A.J., Scott, M., Dahab, R.: Implementing Cryptographic Pairings over Barreto-Naehrig Curves. In: Takagi, T., Okamoto, T., Okamoto, E., Okamoto, T. (eds.) Pairing 2007. LNCS, vol. 4575, pp. 197–207. Springer, Heidelberg (2007)
8. Dhem, J.-F.: Design of an efficient public-key cryptographic library for RISC-based smart cards. PhD thesis, Universite catholique de Louvain, Louvain-la-Neuve, Belgium (1998)
9. Grabher, P., Großschädl, J., Page, D.: On Software Parallel Implementation of Cryptographic Pairings. In: Avanzi, R., Keliher, L., Sica, F. (eds.) SAC 2008. LNCS, vol. 5381, pp. 35–50. Springer, Heidelberg (2008)

10. Hankerson, D., Menezes, A., Scott, M.: Software implementation of Pairings. In: Joye, M., Neven, G. (eds.) Identity-Based Cryptography (2008)
11. Hess, F., Smart, N.P., Vercauteren, F.: The Eta Pairing Revisited. IEEE Transactions on Information Theory 52(10), 4595–4602 (2006)
12. Kammler, D., Zhang, D., Schwabe, P., Scharwaechter, H., Langenberg, M., Auras, D., Ascheid, G., Leupers, R., Mathar, R., Meyr, H.: Designing an ASIP for Cryptographic Pairings over Barreto-Naehrig Curves. Cryptology ePrint Archive, Report 2009/056 (2009), http://eprint.iacr.org/
13. Lee, E., Lee, H.-S., Park, C.-M.: Efficient and Generalized Pairing Computation on Abelian Varieties. Cryptology ePrint Archive, Report 2009/040, http://eprint.iacr.org/
14. Miller, V.S.: Short Programs for Functions on Curves (unpublished manuscript) (1986), http://crypto.stanford.edu/miller/miller.pdf
15. Miller, V.S.: The Weil Pairing, and Its Efficient Calculation. Journal of Cryptology 17(4), 235–261 (2004)
16. Montgomery, P.: Modular Multiplication without Trial Division. Mathematics of Computation 44(170), 519–521 (1985)
17. Barreto, P.S.L.M., Naehrig, M.: Pairing-friendly elliptic curves of prime order. In: Preneel, B., Tavares, S. (eds.) SAC 2005. LNCS, vol. 3897, pp. 319–331. Springer, Heidelberg (2006)
18. Shu, C., Kwon, S., Gaj, K.: FPGA Accelerated Tate Pairing Based Cryptosystems over Binary Fields. In: Proceedings of IEEE International Conference on Field Programmable Technology (FPT), pp. 173–180 (2006)

APPENDIX

A: Proof of lemma 1

Proof. The proof proceeds in two parts. The first part proves a bound on the coefficients of $c(t)$ after Step 7 and the second part analyzes the two coefficient reduction loops (Step 8-12).

Denote $c_{i,j}$ the coefficients of $c(t)$ at the beginning of the j-th iteration, so $c(t)$ in Step 7 has coefficients $c_{i,5}$ (i.e. $j \leq 5$). Let $\Delta = 2^{2m+3}$ then we first show by induction on j that

$$|c_{i,j}| \leq j\Delta. \tag{1}$$

Clearly Equation (1) holds for $j = 0$, since $c_{i,0} = 0$. Now assume that (1) holds for j, then we will show the inequality holds for $j + 1$. In Step 3, $c_{i,j}$ increases by maximum 2^{2m+2}. In Step 4, we thus obtain

$$|\mu| \leq 2^{m+2} + \frac{j\Delta}{2^m} \quad \text{and} \quad |\gamma| \leq 2^m + s|\mu|.$$

In Step 5, we have $|g_i| \leq 36|\gamma|$, so in Step 6 we finally obtain

$$|c_{i,j+1}| \leq 2^{2m+2} + j\Delta + 36|\gamma| + |\mu| = (j + 1)\Delta - 2^{2m+2} + 36|\gamma| + |\mu|,$$

so it suffices to prove that $36|\gamma| + |\mu| \leq 2^{2m+2}$. Rewriting this leads to

$$36 \cdot 2^m + (36s + 1)|\mu| = 36 \cdot 2^m + \xi|\mu| \le 2^{2m+2},$$

which concludes the proof of (1).

For $c_{3,5}$ we need to obtain a better bound since the bound on the final r_4 is also smaller. Note that coefficient $c_{3,5}$ is computed as $c_{3,5} = a_4 b_4 + 36\gamma$ where γ can be bounded by $2^m + 2^{m+6}s$. This finally leads to the bound

$$c_{3,5} < 2^m + 36(2^m + 2^{m+5}s) < 37 \cdot 2^m + 2^{m+m/2-1}.$$

For the first coefficient reduction step, it is easy to see that for $i = 0, 1, 2$ we have $|\mu| \le 5 \cdot 2^{m+3} + 5 \cdot 2^3$, so after the first reduction we obtain for $i = 0, \ldots, 3$

$$|c_i| \le 2^m + s|\mu| < 2^m + (5 \cdot 2^{m+3} + 5 \cdot 2^3)s < 2^{m+6}s.$$

For c_3 however, we obtain $|\mu| < 37 + 2^{m/2-1} + 2^6 s$ which becomes c_4.

For the second coefficient reduction step, it is again easy to see that $i = 0, 1, 2$ we have $|\mu| \le 2^6 s$ and thus $|c_i| \le 2^m + 2^6 s^2 < 2^{m+1}$. For c_4 we obtain, $c_4 = 37 + 2^{m/2-1} + 2^7 s < 2^{m/2}$, since $m > 26$. $\qquad\square$

Designing an ASIP for Cryptographic Pairings over Barreto-Naehrig Curves[*]

David Kammler[1], Diandian Zhang[1], Peter Schwabe[2], Hanno Scharwaechter[1], Markus Langenberg[3], Dominik Auras[1], Gerd Ascheid[1], and Rudolf Mathar[3]

[1] Institute for Integrated Signal Processing Systems (ISS),
RWTH Aachen University, Aachen, Germany
kammler@iss.rwth-aachen.de
[2] Department of Mathematics and Computer Science
Eindhoven University of Technology, Eindhoven, Netherlands
peter@cryptojedi.org
[3] Institute for Theoretical Information Technology (TI),
RWTH Aachen University, Aachen, Germany
mathar@ti.rwth-aachen.de

Abstract. This paper presents a design-space exploration of an application-specific instruction-set processor (ASIP) for the computation of various cryptographic pairings over Barreto-Naehrig curves (BN curves). Cryptographic pairings are based on elliptic curves over finite fields—in the case of BN curves a field $\mathbb{F}_p$ of large prime order p. Efficient arithmetic in these fields is crucial for fast computation of pairings. Moreover, computation of cryptographic pairings is much more complex than elliptic-curve cryptography (ECC) in general. Therefore, we facilitate programming of the proposed ASIP by providing a C compiler.

In order to speed up $\mathbb{F}_p$ arithmetic, a RISC core is extended with additional scalable functional units. Because the resulting speedup can be limited by the memory throughput, utilization of multiple data-memory banks is proposed.

The presented design needs 15.8 ms for the computation of the Optimal-Ate pairing over a 256-bit BN curve at 338 MHz implemented with a 130 nm standard cell library. The processor core consumes 97 kGates making it suitable for the use in embedded systems.

Keywords: Application-specific instruction-set processor (ASIP), design-space exploration, pairing-based cryptography, Barreto-Naehrig curves, elliptic-curve cryptography (ECC), $\mathbb{F}_p$ arithmetic.

---

[*] This work has been supported by the UMIC Research Centre, RWTH Aachen University. The third author was supported by the European Commission through the ICT Programme under Contract ICT–2007–216499 CACE and through the ICT Programme under Contract ICT-2007-216646 ECRYPT II. Permanent ID of this document: 7e38974d56cc76a7f572f328ee4a3761. Date: 2009/06/15.

C. Clavier and K. Gaj (Eds.): CHES 2009, LNCS 5747, pp. 254–271, 2009.

1 Introduction

Pairings were first introduced to cryptography as a means to break cryptographic protocols based on the elliptic-curve discrete-logarithm problem (ECDLP) [1], [2]. Joux showed in 2000 that they can also be used constructively for tripartite key agreement [3]; Subsequently, different cryptographic protocols have been presented involving cryptographic pairings, including identity-based encryption [4] and short digital signatures [5]. A discussion of various applications that would be impossible or very hard to realize without pairings is given in [6].

Cryptographic pairings are based on elliptic curves. To meet both, security requirements and computational feasibility, only elliptic curves with special properties can be considered as basis for cryptographic pairings. State-of-the-art curves for high-security applications are 256-bit Barreto-Naehrig curves (BN curves), introduced in [7]. They achieve 128-bit security according to [8] or 124-bit security according to [9]. Fast arithmetic on these curves demands for fast finite field arithmetic in a field $\mathbb{F}_p$ of prime order p, where p is determined by the curve construction.

Several high-performance software implementations of pairings over BN curves exist for general-purpose desktop and server CPUs [10,11,12]. However, the so far only implementation targeting an embedded system was published by Devegili et al. in [10] (updated in [13]) for a Philips HiPerSmart™ smart card; a complete pairing computation requires 5.17 s at 20.57 MHz, certainly too much time for interactive processes.

This result shows that in order to make state-of-the-art pairing applications available to the embedded domain we need dedicated hardware to accelerate pairing computations. However, the variety and complexity of pairing applications demand for a flexible and programmable solution, that cannot be satisfied by a static hardware implementation. Application-specific instruction-set processors (ASIPs) are a promising candidate to find a good trade-off between these contradicting demands of speed, flexibility and ease of programmability.

This paper shows a design-space exploration of an ASIP for pairing computations over BN curves. We describe how to trade off execution time against area making the ASIP suitable for use in the embedded domain. Dedicated scalable functional units are introduced that speed up general $\mathbb{F}_p$ arithmetic. Moreover, their critical path delay can be modified in order to be integrated with any existing RISC-like architecture without compromising its clock frequency. We show that the speedup from the special functional units is limited by a memory system with a single memory port. Hence, we introduce a memory system utilizing multiple memory banks. The number of banks can be altered without modification to the pipeline or the target architecture tools including the C compiler. This enables fast design-space exploration. The proposed ASIP thus offers a flexible and scalable implementation for pairing applications.

We are—up to our knowledge—the first to implement and time a complete implementation of high-security cryptographic pairings on dedicated specialized hardware.

We would like to thank Jia Huang for supporting the implementation. We furthermore thank Daniel J. Bernstein, Tanja Lange, Ernst Martin Witte, Filippo Borlenghi, and the anonymous reviewers for suggesting many improvements to our explanations.

Related work. Several architectures for the computation of cryptographic pairings have been proposed in the literature [14, 15, 16, 17, 18, 19, 20, 21, 22, 23, 24, 25, 26]. All these implementations use supersingular curves over fields of characteristic 2 or 3, achieving only very low security levels, sometimes even below 80 bit.

Barenghi et al. recently proposed a hardware architecture for cryptographic pairings using curves defined over fields of large prime characteristic [27]. They use a supersingular curve (with embedding degree 2) defined over a 512-bit field and thus achieve only 72-bit security, according to [9].

Another architecture targeting speedup of pairings and supporting fields of large prime characteristic has been proposed in [28]. The instruction set of a SPARC V8 processor is extended for acceleration of arithmetic in $\mathbb{F}_{2^n}$, $\mathbb{F}_{3^m}$ and $\mathbb{F}_p$. However, the focus is put on minor modifications of the datapath resulting in a performance gain for multiplications in $\mathbb{F}_p$ which is two-fold only.

The architectures closest to the one proposed in this paper are accelerating arithmetic in general $\mathbb{F}_p$ for elliptic-curve cryptography (ECC) [29,30]. However, these designs have not been reported to be used for pairing computations.

Some other architectures for ECC over prime fields limit their support to a prime p which allows for particularly fast modular reduction (see i.e. [31]). These approaches are not adequate for pairing-based cryptography where additional properties of the elliptic curves are required. Thus, a detailed comparison with these architectures is omitted here.

Organization of the paper. Section 2 of the paper gives an overview of cryptographic pairings and Barreto-Naehrig curves. Section 3 describes our approach of an ASIP suitable for pairing computation. In Section 4 we discuss the results. The paper is concluded and future work is outlined in Section 5.

2   Background on Cryptographic Pairings

We only give a short overview of the notion of cryptographic pairings, a comprehensive introduction is given in [32, chapter IX].

For three groups G_1, G_2 (written additively) and G_3 (written multiplicatively) of prime order r a cryptographic pairing is a map $e : G_1 \times G_2 \to G_3$,

- Bilinearity:
  $e(kP, Q) = e(P, kQ) = e(P, Q)^k$ for $k \in \mathbb{Z}$.
- Non-degeneracy:
  For all nonzero $P \in G_1$ there exists $Q \in G_2$ such that $e(P, Q) \neq 1$ and for all nonzero $Q \in G_2$ there exists $P \in G_1$ such that $e(P, Q) \neq 1$.
- Computability:
  There exists an efficient algorithm to compute $e(P, Q)$ given P and Q.

We consider the following construction of cryptographic pairings: Let E be an elliptic curve defined over a finite field $\mathbb{F}_p$ of prime order. Let r be a prime dividing the group order $\#E(\mathbb{F}_p) = n$ and let k be the smallest integer, such that $r \mid p^k - 1$. We call k the embedding degree of E with respect to r. Let t denote the trace of Frobenius fulfilling the equation $n = p + 1 - t$.

Let $P_0 \in E(\mathbb{F}_p)$ and $Q_0 \in E(\mathbb{F}_{p^k})$ be points of order r such that $Q_0 \notin \langle P_0 \rangle$, let $\mathcal{O} \in E(\mathbb{F}_p)$ denote the point at infinity. Define $G_1 = \langle P_0 \rangle$ and $G_2 = \langle Q_0 \rangle$. Let $G_3 = \mu_r$ be the group of r-th roots of unity in $\mathbb{F}_{p^k}^*$.

For $i \in \mathbb{Z}$ and $P \in E$ a Miller function [33] is an element $f_{i,P}$ of the function field of E, such that the principal divisor of $f_{i,P}$ is $\operatorname{div}(f_{i,P}) = i(P) - ([i]P) - (i-1)\mathcal{O}$.

Using such Miller functions, we can define the map

$$e_s : G_1 \times G_2 \to \mu_r; (P, Q) \mapsto f_{s,P}(Q)^{(p^k-1)/r}.$$

For certain choices of s the map e_s is non-degenerate and bilinear. For $s = r$ we obtain the reduced-Tate pairing τ and for $s = T = t - 1$ we obtain the reduced-Ate pairing α by switching the arguments [34]. Building on work presented in [35], Vercauteren introduced the Optimal-Ate pairing in [36] which for BN curves can be computed using $s \approx \sqrt{t}$ and a few additional computations (see also [37]).

Using twists of elliptic curves we can further define the generalized reduced-η pairing [34], [38]. In [12] a method to compute the Tate and η pairing keeping intermediate results in compressed form is introduced. We refer to the resulting algorithms as Compressed-Tate and Compressed-η pairing, respectively.

2.1   Choice of an Elliptic Curve

For cryptographic protocols to be secure on the one hand and the pairing computation to be computationally feasible on the other hand, the elliptic curve E must have certain properties: Security of cryptographic protocols based on pairings relies on the hardness of the discrete logarithm problem in G_1, G_2 and G_3. For the 128-bit security level, the National Institute of Standards and Technology (NIST) recommends a prime group order of 256 bit for $E(\mathbb{F}_p)$ and of 3072 bit for the finite field $\mathbb{F}_{p^k}$ [8].

Barreto-Naehrig curves, introduced in [7], are elliptic curves over fields of prime order p with embedding degree $k = 12$. The group order $n = r$ of $E(\mathbb{F}_p)$ is prime by construction, the values p and n can be given as polynomial expressions in an integer u as follows:

$$p = p(u) = 36u^4 + 36u^3 + 24u^2 + 6u + 1 \text{ and}$$
$$n = n(u) = 36u^4 + 36u^3 + 18u^2 + 6u + 1.$$

For our implementation we follow [10] and set $u = $ 0x6000000000001F2D, yielding two primes $p(u)$ and $n(u)$ of $l = 256$ bit. The field size of $\mathbb{F}_{p^k}$ then has $256 \cdot k = 3072$ bit. Note, that according to [9], a finite fields of size 3072 bit offers only 124-bit security. In this paper we follow the more conservative estimations of [9] and claim only 124-bit security for pairings over 256-bit BN curves.

2.2   Computation of Pairings

The computation of cryptographic pairings consists of two main steps: the computation of $f_{s,P}(Q)$ for Tate and η pairings or of $f_{s,Q}(P)$ when considering the Ate pairing and the final exponentiation with $(p^k - 1)/r$.

The first part is usually done iteratively using variants of Miller's algorithm [33]. Several optimizations of this algorithm have been presented in [39]. The resulting algorithm is often referred to as BKLS algorithm. For BN curves even more optimizations can be applied by exploiting the fact that such curves have sextic twists. A detailed description of efficient computation of pairings over BN curves, including the computation of Miller functions and the final exponentiation is given in [10]. Our implementation follows this description in large parts.

Finite field computations constitute the bulk of the pairing computation – in software implementations typically more than 90% of the time is spent on modular multiplication, inversion and addition, the number of these operations for the implemented pairing algorithms is the following:

Number of	Opt. Ate	Ate	η	Tate	Comp. η	Comp. Tate
multiplications	17,913	25,870	32,155	39,764	75,568	94,693
additions		84,956	121,168	142,772	174,974	155,234	193,496
inversions	3	2	2	2	0	0

Throughout the pairing computation we keep points on elliptic curves in Jacobian coordinates and can thus almost entirely avoid field inversions; our targets for hardware acceleration are thus multiplication and addition in $\mathbb{F}_p$, inversion is implemented as exponentiation with $p - 2$.

3   An ASIP for Cryptographic Pairings

To implement various pairing algorithms (Optimal Ate, Ate, η, Tate, Compressed η and Compressed Tate), a programmable and therefore flexible architecture is targeted in this paper. Standard architectures like embedded RISC cores are flexible, but they are lacking sufficient computational performance for specific applications. Therefore, we apply the ASIP concept to cryptographic-pairing applications in order to reduce the computation time while maintaining programmability. Development and implementation of our ASIP have been carried out using the Processor Designer from CoWare [40].

Keeping control over the data flow on the higher layers of the pairing computation, like $\mathbb{F}_{p^{12}}$ or $E(\mathbb{F}_{p^2})$ arithmetic, is a rather complex task. This calls for a convenient programming model. However, on the lower level realizing the $\mathbb{F}_p$ arithmetic, computational performance is of highest priority. Therefore, we decided to extend a basic 5-stage 32-bit RISC core with special $\mathbb{F}_p$ instructions for modular multiplication, addition and subtraction. Inversions are not considered for special instructions as they are used very seldom ($\leq 3\times$) in any of the targeted applications. The available C compiler enables convenient application development on higher levels, while the computational intensive tasks are mapped to the specialized instructions accessible via intrinsics[1].

Among the targeted $\mathbb{F}_p$ operations, the most challenging one to implement is fast modular multiplication, especially for a large word width (e.g. 256 bit). In general, multiplication in $\mathbb{F}_p$ can be done by first multiplying the two 256-bit factors and then reducing the 512-bit product. This might indeed be the fastest approach, if p could be chosen of a special form as for example specified in [41] or [42]. However, due to the construction of BN curves (see [7]) we cannot use such primes. Therefore, our approach uses Montgomery arithmetic [43].

3.1 Data Processing: A Scalable Montgomery-Multiplier Unit

In 1985 Montgomery introduced an algorithm for modular multiplication of two integers A and B modulo an integer M [43]. The idea of the algorithm is to represent A as $\hat{A} = AR \mod M$ and B as $\hat{B} = BR \mod M$ for a fixed integer $R > M$ with $\gcd(R, M) = 1$. This representation is called Montgomery representation. To multiply two numbers in Montgomery representation we have to compute $\widehat{AB} = \hat{A}\hat{B}R^{-1} \mod M$. For certain choices of R this computation can be carried out much more efficiently than usual modular multiplication: Let us assume that M is odd and let l be the bit length of M. Choosing $R = 2^l$ clearly fulfills the requirements on R and allows for modular multiplication that replaces division operations by shifts, allowing for an efficient hardware implementation.

In the context of $\mathbb{F}_p$-multiplication the modulus M corresponds to p. All $\mathbb{F}_p$ operations can be performed in Montgomery representation. Therefore, all values can be kept in Montgomery representation throughout the whole pairing computation.

Nibouche et al. introduced a modified version of the Montgomery multiplication algorithm in [44]. It splits the algorithm into two multiplication operations, that can be carried out simultaneously, and allows for using carry save (CS) multipliers. This results in a fast architecture that can be pipelined and segmented easily. Therefore, it is chosen as basis for our development. A 4×4-bit example is shown in Fig. 1.

The actual multiplication is carried out in the left half of the architecture, while the reduction is performed in the right part simultaneously. The left part is a conventional multiplier built of gated full adders (gFAs), whereas the right part consists of a multiplier with special cells for the least-significant bits (LSBs). The LSB cells are built around a half adder (HA). Their overall delay is comparable to that of a gFA. A more detailed description can be found in [44].

Due to area constraints we decided to implement only subsets of the regular structures of the multiplier and perform the computation in multiple cycles. The CS-based design provides the opportunity to not only make horizontal but also vertical cuts while the critical path of the multiplier unit depends on its height

---

[1] An adoption of our code to general purpose processors using the GMP library instead of intrinsics is available from http://cryptojedi.org/crypto/

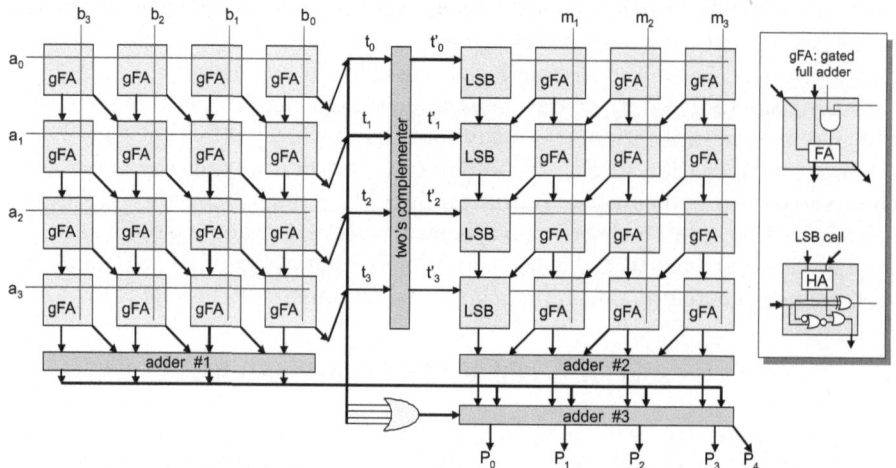

Fig. 1. Montgomery-multiplier based on Nibouche et al. [44]

(H) only. This makes the design *adaptable* to existing cores in terms of *timing* maintaining the performance of their general instruction set. Once the height of the multiplier unit is chosen (in our case $H = 8$), the width (W) can be selected to *adapt* the design to the desired *computational performance* and to trade off *area vs. execution time* of the multiplication.

Multiplication and reduction are carried out simultaneously starting from the most-significant bit (MSB) of their second operand (B and M) first. However, the reduction cannot be started until the incoming data for the LSB cells are available from the two's complementer. Therefore, reduction starts after the first H lines of multiplication have been executed and remains delayed for $\left\lceil \frac{l}{W} \right\rceil$ cycles (required for the computation of H lines). Eventually, the CS results need to be transformed back to two's complement number representation (by *addition #1* and *addition #2*) before they are combined to the result by *addition #3*. This is necessary since the result lies in the range of 0 to $2M - 1$, and requires a final comparison against M, which is difficult to handle in CS representation. The comparison including a necessary subtraction of M is performed in another functional unit introduced later. Equation (1) gives the number of required cycles c_{MM} to perform a Montgomery multiplication with the proposed multi-cycle architecture for the general case.

$$c_{MM} = \left(\left\lceil \frac{l}{H} \right\rceil + 1 \right) \cdot \left\lceil \frac{l}{W} \right\rceil + 2 \tag{1}$$

For evaluation, we implemented this multi-cycle Montgomery-multiplier (MMM) in three different sizes ($W \times H$): 32×8 bit, 64×8 bit and 128×8 bit, resulting in an execution time of 266, 134 and 68 cycles respectively. However, the area savings for smaller (and slower) architectures do not scale as well as the execution time. This results from the increased complexity of the required multiplexing for smaller MMM units. In order to keep the amount of multiplexers small, we

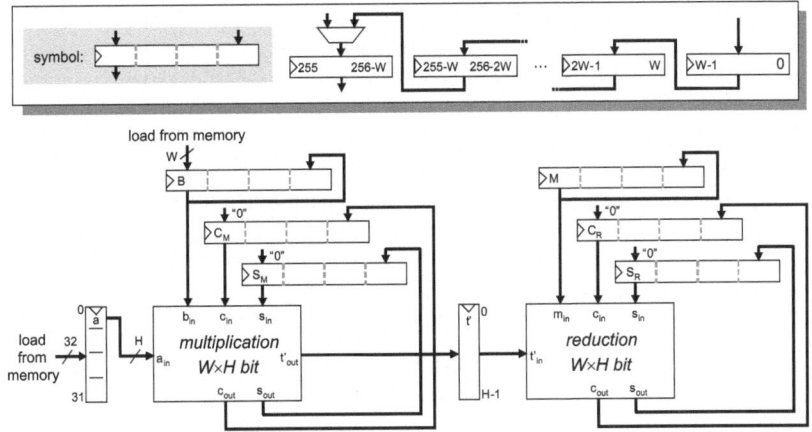

Fig. 2. Structure of the multi-cycle Montgomery-multiplier (MMM)

designed special 256-bit shift registers, that enable a circular shift by W bits for the operands B, M and the corresponding intermediate CS values. This solution is suitable, since the input values are accessed in consecutive order by blocks of W bits. Still, area savings when scaling a 128×8-bit architecture down to 32×8-bit are about 50%.

Fig. 2 shows the overall resulting structure of the MMM unit. The two's complementer is included in the *multiplication unit*, while the *reduction unit* contains additional LSB cells that produce input for the gFA cells on the fly (as depicted in Fig. 1). The input shift registers are initialized step by step during the first $\lceil \frac{l}{W} \rceil$ cycles. After the whole process, the result is stored in the registers for temporary CS values (C_M, S_M, C_R, S_R). The adders for the final summations are not depicted.

An advantage of stepwise executing the multiplication is that the total multiplication width l can be configured at runtime in steps of W. The overall dependence of the execution time on l is quadratic. Modular multiplication is thus significantly faster for smaller multiplication width. This may be interesting for ECC applications requiring lower security.

Similar to the MMM unit we developed a *multi-cycle adder unit* for modular additions and subtractions, which reads two operands block-wise and simultaneously. For evaluation, a 32-bit and a 64-bit version of this unit have been implemented. Details are omitted here since the implementation is straightforward.

Both, MMM and *adder unit* require a final subtraction of M whenever the result exceeds this prime number. A special *writeback unit* takes care of this subtraction right before writing back the data, operating block-wise in multiple cycles as well. This unit has been implemented with a width of 32, 64 and 128 bit.

During the execution of multi-cycle operations for modular addition, subtraction and multiplication the pipeline is stalled. Three special instructions are implemented triggering these operations. Instruction arguments are registers containing the starting address of each of the three 256-bit operands. Since the modulus M is not changed during an application run, a special register is utilized and implicitly accessed by the instructions. This register is initialized with p at the beginning of an application via another dedicated instruction.

3.2   Data Access: An Enhanced Memory Architecture

Due to the large width of the operands, the existing 16x32-bit general purpose register file could only store two operands simultaneously. This results in frequent memory accesses consuming additional cycles and thus decreasing the overall performance of the architecture significantly. Enlarging the register file would be very costly in terms of area consumption. Hence, the instructions triggering the multi-cycle operations for modular addition, subtraction and multiplication are implemented as memory-to-memory instructions. This way, the memory accesses can be almost completely hidden in the actual computation.

The resulting throughput demands on the memory system are quite high. Especially the modular addition/subtraction requires a throughput higher than one 32-bit word per cycle. The following two evident mechanisms to increase memory throughput for ASIP designs are not well suited here: First, using memories with multiple ports is costly. The number of ports is limited to two for SSRAMs and the required area is roughly doubled. Second, designing a dedicated system with several (often specialized) memories targets highest performance, but is a complex task. The data memory space gets segmented irregularly, making it difficult to access and manage for a compiler.

Due to the drawbacks of these two approaches we apply a different technique, which we would like to introduce as *transparent interleaved memory segmentation (TIMS)*. Its basic principle is to extend the number of ports to the memory system in order to increase the throughput by using memory banks. These banks are selected on the basis of address bits and can be accessed in parallel. In case of our ASIP, the LSBs of the address are used for the memory bank selection. This results in an addressing scheme, where the memory is selected by calculating the address modulo the number of memories m_d, which has to be a power of two.

In principle, the distribution of accesses to a banked memory system can be handled in software or hardware. However, memory-access conflicts can occur when simultaneous accesses refer to the same memory. Solving these conflicts in software requires an extension of the C compiler in order to avoid multiple simultaneous accesses to the same memory bank. If these conflicts can be ruled out at compile time, this approach results in very efficient code. However, if the conflicts do occur at runtime, additional code to resolve the conflict needs to be included in the target software at the cost of increased execution time. Especially when pointers are used and function calls have a substantial degree of nesting (which is the case for the targeted pairing applications), detecting the conflicts at compile time is often impossible. It requires a significant extension of the C compiler functionality and comes at the cost of increased code size and execution time.

However, due to fairly simple mechanisms and regularity, the distribution of accesses to the memories and resolution of access conflicts can be handled efficiently at runtime by a dedicated hardware block, the *memory-access unit*

(MAU) that distributes the memory accesses from the pipeline to the correct memory. Memory accesses are requested concurrently by the pipeline on demand resulting in multiple independent read or write connections (unidirectional) between pipeline and MAU. The MAU takes care of granting accesses. Therefore, a simple handshaking protocol is used between pipeline and MAU, which is able to confirm a request within the same cycle in order not to cause any delay cycles when trying to access the fast SSRAMs.

One advantage of this mechanism is, that from the perspective of the core, the memory space remains unchanged, regardless of the number of memories. Existing load and store instructions are sufficient to access the whole data memory space. Even when special instructions perform concurrent memory accesses, a modification in the memory system (e.g. changing number of memories) does not result in a change of the core or the C compiler. This enables orthogonal implementation and modification of the base architecture and the memory system.

A priority-based hardware resolution of access conflicts is implemented in the MAU in two ways. Static priorities can be used if certain accesses always have higher priority than others. For instance write accesses from later pipeline stages should always have higher priority than read accesses from prior stages. When the priority is changing at runtime, dynamic priority management is required. Then, dedicated additional communication lines between core and MAU indicate a change of priority. In our design this is required by the *adder unit*.

Fig. 3 depicts the four different connection schemes between MAU and pipeline. The number and type of connections between MAU and pipeline are determined by the number and type of independent memory accesses initiated by the pipeline, while the number of actual memory connections depends on the number of attached data memories m_d ($m_d = 2$ in this example). For sake of clarity, the actual interconnections within the MAU have been omitted in Fig. 3. The *access-control* block combines the *enable* and *priority* signals with the $\log_2(m_d)$ LSBs of the *address* signals from the read and write connections in order to produce the *grants*. At the same time the *enable* signals for the SSRAMs are set accordingly by this unit. It also controls the crossbars that are switching the correct *address*, *read data* and *write data* signals to the memory ports. Please note, that the *read data* crossbar is switched with one cycle delay compared to the *address* crossbar in order to be in sync with the single cycle read latency of the SSRAMs. Effects of TIMS on physical parameters like timing and area consumption are discussed in detail in the result section.

Overall, the TIMS approach enables to extend any existing architecture with memory banks without altering the existing pipeline or the basic instruction set. The specialized memory-to-memory instructions can take full advantage of parallel accesses to these banks reducing execution time. It is not necessary to extend the C compiler of the base architecture with more than intrinsics for the special instructions. The compiled executable can be used on any architecture variant independently from the number of memory banks. As the MAU hides the actual memory system from the pipeline, the number of memory banks can be changed without modification to the pipeline enabling a fast design space exploration.

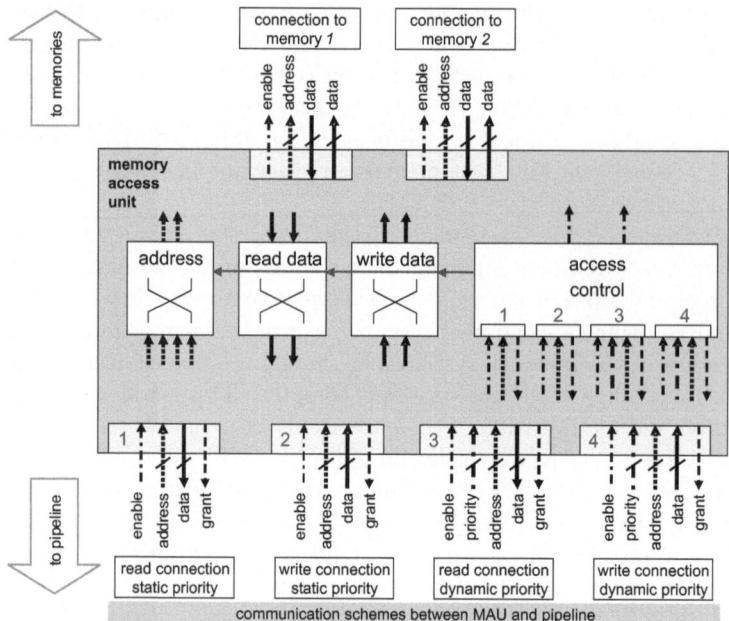

Fig. 3. Interconnect of memory-access unit (MAU)

Still, memory access collisions decrease the performance of the system and cannot be avoided completely due to the automatic address management of the C compiler. However, in our case this effect is kept minimal due to the good distribution of the 256-bit words. For additions and multiplications this causes a maximum additional delay of one cycle only. This results in a maximum performance degradation caused by memory-access conflicts of less than 2% for any of the implemented pairing applications.

4   Results

Overall, we have implemented nine variants of our ASIP with different design parameters regarding number of data memories and width of the computational units for modular multiplication, modular addition and multi-cycle writeback (Table 1). The number of data memories is closely coupled with the width of the *adder* and the *writeback unit*. Other combinations would operate functionally correctly, but would waste either performance or area. The implementation of a 16-bit adder for the single memory case would not significantly reduce area due to additional multiplexing and is therefore neglected. All synthesis results have been obtained with Synopsys Design Compiler [45] using a 130 nm CMOS standard cell library with a supply voltage of 1.2 V and are given before place and route. The memories are synchronous single-port SRAMs with a latency of one cycle. The total data-memory size is 2048 words for each of the design

Table 1. Implemented design variants of the ASIP for pairings

Variant	128m4	64m4	32m4	128m2	64m2	32m2	128m1	64m1	32m1
mod mul size *(bit)*	128×8	64×8	32×8	128×8	64×8	32×8	128×8	64×8	32×8
mod add width *(bit)*	64	64	64	32	32	32	32	32	32
writeback width *(bit)*	128	128	128	64	64	64	32	32	32
# data memories	4	4	4	2	2	2	1	1	1
total area[a] *(kGates)*	195	186	182	164	153	148	145	134	130
core area[b] *(kGates)*	96	87	83	97	86	81	93	83	79
timing *(ns)*	3.69	3.65	3.52	2.96	2.97	3.02	2.95	3.03	3.09
Optimal Ate *(ms)*	17.5	21.8	29.9	15.8	19.4	27.3	19.2	23.4	32.0
Ate *(ms)*	25.3	31.4	42.6	22.8	27.9	38.9	27.6	33.5	45.6
η *(ms)*	32.3	39.5	52.8	28.8	35.0	48.1	34.6	41.6	56.2
Tate *(ms)*	38.5	47.0	62.7	34.4	41.6	57.1	41.1	49.5	65.3
Compressed η *(ms)*	38.6	55.0	86.2	34.5	48.2	77.1	41.6	56.5	85.8
Compressed Tate *(ms)*	48.2	68.9	107.8	43.2	60.3	96.5	52.0	70.7	107.3

[a] Including area for data memories.
[b] Without area for memories, but including area for MAU.

variants. The program memory is not included in the area reports, since it is not changing through the different designs and could be implemented differently (as ROM, RAM, synthesized logic etc.) depending on the final target system. The plain RISC (32-bit, 5-stage pipeline, 32-bit integer multiplier) without memories and extensions consumes 26 kGates and achieves a timing of 2.89 ns.

Fig. 4 shows the area distribution of the different ASIP variants. While the basic core only shows moderate area increase from 17 to 21 kGates for all

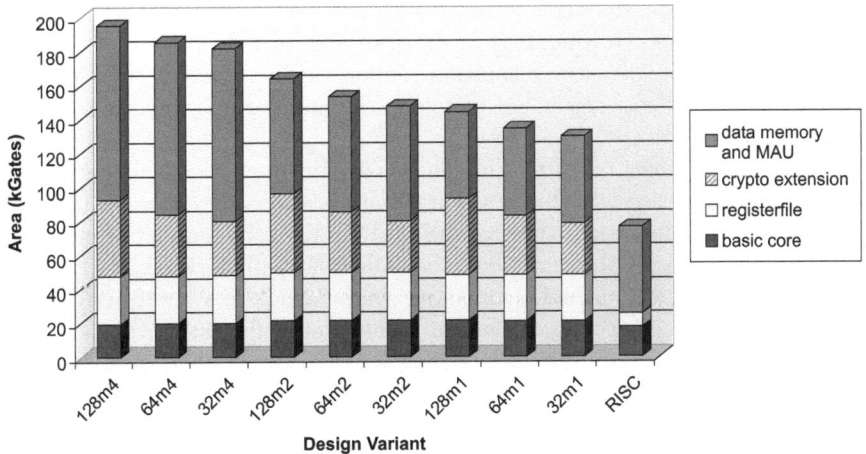

Fig. 4. ASIP area consumption and distribution

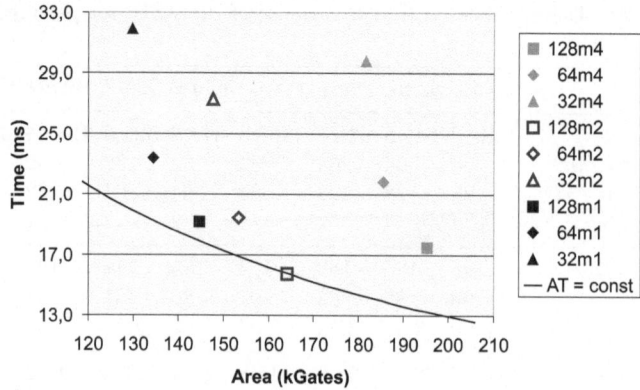

Fig. 5. Area-time trade-off for different ASIP variants (Optimal Ate pairing)

variants (resulting from decoder extensions and additional pipeline registers), the area for the register file increases from 9 to 28 kGates compared to the plain RISC. The reason are specialized 256-bit registers storing the prime number and intermediate results of the modular operations. These registers are independent from the width of any of the additional functional units. The area of the cryptographic extensions is dominated by the MMM unit.

Observe that splitting the memory into two of half the size results in a data-memory area increase of 31%. Utilizing a dual port memory instead would increase area by over 83%. The area overhead due to the MAU lies between only 0.5 and 1.2 kGates, when two memories are attached. Even for four attached memories it is below 3.5 kGates.

However, limitations of TIMS utilizing the proposed MAU become visible when looking at the timing of the different variants of the ASIP. While attaching one or two data memories barely affects the critical path with respect to the original RISC architecture (within design tool accuracy, see Table 1), an increased delay is observed when four memories are attached. This delay is caused by the complexity of priority resolution for four attached memories combined with four independent memory accesses with dynamic priority, which are necessary to implement the 64-bit adder.

The execution times of all six implemented pairing applications on all nine ASIP variants are shown in Table 1. For all applications performance improves significantly with increasing width of the MMM. Also, the number of cycles decreases when increasing the number of connected data memories. Unfortunately, the longer critical path of the four-memory system leads to a lower performance than for the designs with two memories. The overall fastest design is variant *128m2*, executing the Optimal-Ate pairing in 15.8 ms. With the smallest and slowest variant completing the task in 32.0 ms, the user is offered a quite broad design space enabling trade-offs.

In order to evaluate the efficiency of the different design variants, Fig. 5 shows the area-time trade-off for the Optimal-Ate pairing. It can be seen clearly that the best AT product is obtained by the *128m2* design. This shows the importance of investigating the memory architecture of ASIPs during design-space exploration. In our case the best results are obtained with TIMS and *two* data memories in spite of the considerable area increase due to the memory splitting.

4.1   Performance Comparison

To our best knowledge there exists no literature reporting performance figures resulting from actual implementations of cryptographic pairings on dedicated hardware achieving a 124-bit security level. Hardware implementations for lower security levels can obviously be much faster than the proposed design.

Table 2 gives an overview of performance and area consumption for various pairing implementations on dedicated hardware; the given security levels are according to [9]. Whenever more than one design variant is given in a publication, the fastest one with the highest security level is listed in the table. The previous results listed in Table 2 are hardly comparable to the design proposed in this paper. Not only do they achieve lower security levels, they also mainly focus on FPGAs rather than standard cells and mostly use curves over binary or ternary fields. In the following we therefore give a comparison with standard cell designs which accelerate $\mathbb{F}_p$ arithmetic for elliptic-curve cryptography and finally discuss our design in the context of smart cards.

Comparison with standard cell designs for ECC. Other publications describing dedicated-hardware implementations for ECC over fields of large prime

Table 2. Performance and area comparison for pairings

Design	Technology	Area	Freq. (MHz)	Pairing	Field Characteristic	Security (bit)	Time (ms)
this work	130 nm std. cell	97 kGates	338	Opt. Ate	256-bit prime	124	15.8
[14]	Xilinx xc2vp20	8 kSlices	90	Tate	3	97	0.298
[14]	Xilinx xc2vp20	8 kSlices	115	Tate	2	97	0.327
[21]	Xilinx xc2vp100	44 kSlices	33	Tate	2	80	0.146
[18], [46]	Xilinx xc2vp100	38 kSlices	72	Tate	2	76	0.049
[20]	Xilinx xc2v6000	25 kSlices	47	Tate	2	76	2.81
[19]	Xilinx xc2v6000	15 kSlices	40	Tate	2	76	3.0
[27]	Xilinx xc2v8000	34 kSlices	135	Tate	512-bit prime	72	1.61
[23]	Xilinx xc4vlx200	74 kSlices	199	η_T	3	68	0.008
[16]	Altera ep2c35	19 kLEs	147	η_i	3	68	0.027
[17]	180 nm std. cell	194 kGates	200	η_T	3	68	0.047
[15]	Xilinx xc4vlx15	2 kSlices	203	η_T	3	68	0.137
[25]	Xilinx xc2vp100	15 kSlices	85	η_T	3	68	0.183
[26]	Xilinx xc2vp200	14 kSlices	77	Tate	3	68	0.251
[22]	Xilinx xc2vp4	4 kSlices	150	Tate	3	68	0.432
[24]	Xilinx xc2vp125	56 kSlices	15	Tate	3	68	0.85

Table 3. Performance and area comparison for scalar multiplication

Design	Technology	Area	Freq. (MHz)	Scalar Mult. Alg.	$\log_2(|\mathbb{F}_p|)$	Time (ms)
this work	130 nm std. cell	97 kGates	338	NAF recoding	256	0.998
[30]	130 nm std. cell	122 kGates	556	NAF recoding	256	1.01
[29]	130 nm std. cell	107 kGates	138	NAF recoding	256	2.68

characteristic give performance figures in terms of time needed for a scalar multiplication with a scalar k of a certain size, i.e. the computation of $[k]P$ for some $P \in E(\mathbb{F}_p)$. An overview is presented in Table 3.

In order to compare the results of this work with these architectures we implemented scalar multiplication on the 256-bit Barreto-Naehrig curve that we also used for pairing computation. Our design does not accelerate field inversion through hardware, so we use Jacobian projective coordinates to represent the points on the curve, trading inversions for several multiplications.

A scalar multiplication with a 256-bit scalar takes 0.998 ms for the *128m2* variant of the proposed design. This number includes transformation of the scalar into NAF and a transformation from Jacobian into affine coordinates at the end. Note that ASIP variant *128m2* is not only slightly faster than the designs in [29] and [30], but also consumes less area.

Application to smart cards. In [10] (updated in [13]), Devigili et al. report 5.17 s for the computation of the Ate pairing over a 256-bit Barreto-Naehrig curve on a Philips HiPerSmart™ smart card operating at 20.57 MHz. This smart card contains a SmartMIPS-based 32-bit architecture and is manufactured in 180 nm technology. For interactive processes this execution time is not sufficient even when the smart card operates at its maximum frequency of 38MHz. Our design achieves—synthesized in a 180 nm CMOS standard cell library with a supply voltage of 1.8 V—over 230 MHz. Even running our smallest design variant *32m1* at the clock speed of 20.57 MHz (leaving a substantial margin for place and route and implementation of protection mechanisms against side channel attacks), the Ate pairing takes 0.71 s and the Optimal Ate pairing is executed in 0.50 s, which is already sufficient for interactive processes. Depending on the design variant used, speedups of over 20× could be achieved. This gives an impression of the achievable performance increase for the computation of cryptographic pairings in the embedded domain when more specialized hardware is used.

5   Conclusion and Outlook

In this paper we presented a design-space exploration of an ASIP for computation of cryptographic pairings over BN curves. The design is based on extensions of an existing RISC core, which are completely transparent and independent from the original pipeline. Therefore, they could be applied to any RISC-like architecture, which can stall the pipeline during multi-cycle operations. The extensions are adaptable in terms of timing and allow for a trade-off between execution time and area. A flexible and transparent memory-architecture extension making use of multiple memories (TIMS) enables fast design space exploration and the usage of existing compilers, since the address space remains unsegmented. We are—up to our knowledge—the first to implement and time a complete implementation of high-security cryptographic pairings on dedicated specialized hardware.

Future objectives include countermeasures against side-channel attacks, which are not implemented in the current design, either in hard- or in software.

References

1. Menezes, A.J., Okamoto, T., Vanstone, S.A.: Reducing elliptic curve logarithms to logarithms in a finite field. IEEE Trans. Information Theory 39(5), 1639–1646 (1993)
2. Frey, G., Rück, H.G.: A remark concerning m-divisibility and the discrete logarithm in the divisor class group of curves. Math. of Computation 62(206), 865–874 (1994)
3. Joux, A.: A one round protocol for tripartite Diffie-Hellman. In: Bosma, W. (ed.) ANTS 2000. LNCS, vol. 1838, pp. 385–394. Springer, Heidelberg (2000)
4. Boneh, D., Franklin, M.: Identity based encryption from the Weil pairing. In: Kilian, J. (ed.) CRYPTO 2001. LNCS, vol. 2139, pp. 213–229. Springer, Heidelberg (2001)
5. Boneh, D., Lynn, B., Shacham, H.: Short signatures from the Weil pairing. J. Cryptology 17(4), 297–319 (2004)
6. Boneh, D.: A brief look at pairings based cryptography. In: Proceedings of the 48th Annual IEEE Symposium on Foundations of Computer Science – FOCS 2007, pp. 19–26 (2007)
7. Barreto, P.S.L.M., Naehrig, M.: Pairing-friendly elliptic curves of prime order. In: Preneel, B., Tavares, S. (eds.) SAC 2005. LNCS, vol. 3897, pp. 319–331. Springer, Heidelberg (2006)
8. Barker, E., Barker, W., Burr, W., Polk, W., Smid, M.: Recommendation for key management – part 1: General (revised). National Institute of Standards and Technology, NIST Special Publication 800-57 (2007)
   http://csrc.nist.gov/publications/nistpubs/800-57/
   sp800-57-Part1-revised2_Mar08-2007.pdf
9. Näslund, M.: Ecrypt yearly report on algorithms and keysizes (2007-2008) (2008),
   http://www.ecrypt.eu.org/ecrypt1/documents/D.SPA.28-1.1.pdf
10. Devegili, A.J., Scott, M., Dahab, R.: Implementing cryptographic pairings over Barreto-Naehrig curves. In: Takagi, T., Okamoto, T., Okamoto, E., Okamoto, T. (eds.) Pairing 2007. LNCS, vol. 4575, pp. 197–207. Springer, Heidelberg (2007)
11. Grabher, P., Großschädl, J., Page, D.: On software parallel implementation of cryptographic pairings. Cryptology ePrint Archive, Report 2008/205 (2008),
    http://eprint.iacr.org/2008/205
12. Naehrig, M., Barreto, P.S.L.M., Schwabe, P.: On compressible pairings and their computation. In: Vaudenay, S. (ed.) AFRICACRYPT 2008. LNCS, vol. 5023, pp. 371–388. Springer, Heidelberg (2008)
13. Devegili, A.J., Scott, M., Dahab, R.: Implementing cryptographic pairings over Barreto-Naehrig curves. Cryptology ePrint Archive, Report 2007/309 (2007),
    http://eprint.iacr.org/2007/390

14. Beuchat, J.-L., Brisebarre, N., Detrey, J., Okamoto, E., Rodríguez-Henríquez, F.: A comparison between hardware accelerators for the modified Tate pairing over $\mathbb{F}_{2^m}$ and $\mathbb{F}_{3^m}$. In: Galbraith, S.D., Paterson, K.G. (eds.) Pairing 2008. LNCS, vol. 5209, pp. 297–315. Springer, Heidelberg (2008)

15. Beuchat, J.-L., Brisebarre, N., Detrey, J., Okamoto, E., Shirase, M., Takagi, T.: Algorithms and arithmetic operators for computing the η_t pairing in characteristic three. IEEE Trans. Comput. 57(11), 1454–1468 (2008)

16. Beuchat, J.-L., Shirase, M., Takagi, T., Okamoto, E.: An algorithm for the η_t pairing calculation in characteristic three and its hardware implementation. In: Proc. 18th IEEE Symp. Computer Arithmetic – ARITH 2007, pp. 97–104 (2007)

17. Beuchat, J.-L., Doi, H., Fujita, K., Inomata, A., Kanaoka, A., Katouno, M., Mambo, M., Okamoto, E., Okamoto, T., Shiga, T., Shirase, M., Soga, R., Takagi, T., Vithanage, A., Yamamoto, H.: FPGA and ASIC implementations of the η_t pairing in characteristic three. Cryptology ePrint Archive, Report 2008/280 (2008), http://eprint.iacr.org/2008/280

18. Shu, C., Kwon, S., Gaj, K.: FPGA accelerated Tate pairing based cryptosystems over binary fields. In: Proc. IEEE Int'l Conf. Field Programmable Technology – FPT 2006, pp. 173–180 (2006)

19. Keller, M., Ronan, R., Marnane, W., Murphy, C.: Hardware architectures for the Tate pairing over $GF(2^m)$. Computers & Electrical Eng. 33(5-6), 392–406 (2007)

20. Keller, M., Kerins, T., Crowe, F., Marnane, W.: FPGA implementation of a $GF(2^m)$ Tate pairing architecture. In: Bertels, K., Cardoso, J.M.P., Vassiliadis, S. (eds.) ARC 2006. LNCS, vol. 3985, pp. 358–369. Springer, Heidelberg (2006)

21. Ronan, R., Ó hÉigeartaigh, C., Murphy, C., Scott, M., Kerins, T.: FPGA acceleration of the Tate pairing in characteristic 2. In: Proc. IEEE Int'l Conf. Field Programmable Technology, pp. 213–220 (2006)

22. Grabher, P., Page, D.: Hardware acceleration of the Tate pairing in characteristic three. In: Rao, J.R., Sunar, B. (eds.) CHES 2005. LNCS, vol. 3659, pp. 398–411. Springer, Heidelberg (2005)

23. Jiang, J.: Bilinear pairing (Eta_T pairing) IP core. Technical report (2007), http://www.cs.cityu.edu.hk/~ecc/doc/etat_datasheet_v2.pdf

24. Kerins, T., Marnane, W.P., Popovici, E.M., Barreto, P.S.L.M.: Efficient hardware for the Tate pairing calculation in characteristic three. In: Rao, J.R., Sunar, B. (eds.) CHES 2005. LNCS, vol. 3659, pp. 412–426. Springer, Heidelberg (2005)

25. Ronan, R., Murphy, C., Kerins, T., Ó hÉigeartaigh, C., Barreto, P.S.L.M.: A flexible processor for the characteristic 3 η_t pairing. Int'l J. High Performance Systems Architecture 1(2), 79–88 (2007)

26. Kömürcü, G., Savas, E.: An efficient hardware implementation of the Tate pairing in characteristic three. In: Proc. Third Int'l Conf. Systems – ICONS 2008, pp. 23–28 (2008)

27. Barenghi, A., Bertoni, G., Breveglieri, L., Pelosi, G.: A FPGA coprocessor for the cryptographic Tate pairing over $\mathbb{F}_p$. In: Proc. Fifth Int'l Conf. Information Technology: New Generations – ITNG 2008, pp. 112–119 (2008)

28. Vejda, T., Page, D., Großschädl, J.: Instruction set extensions for pairing-based cryptography. In: Takagi, T., Okamoto, T., Okamoto, E., Okamoto, T. (eds.) Pairing 2007. LNCS, vol. 4575, pp. 208–224. Springer, Heidelberg (2007)

29. Satoh, A., Takano, K.: A scalable dual-field elliptic curve cryptographic processor. IEEE Trans. Computers 52(4), 449–460 (2003)

30. Chen, G., Bai, G., Chen, H.: A high-performance elliptic curve cryptographic processor for general curves over $GF(p)$ based on a systolic arithmetic unit. IEEE Trans. Circuits and Systems II: Express Briefs 54(5), 412–416 (2007)

31. Güneysu, T., Paar, C.: Ultra high performance ECC over NIST primes on commercial FPGAs. In: Oswald, E., Rohatgi, P. (eds.) CHES 2008. LNCS, vol. 5154, pp. 62–78. Springer, Heidelberg (2008)
32. Galbraith, S.: Pairings. In: Blake, I.F., Seroussi, G., Smart, N.P. (eds.) Advances in Elliptic Curve Cryptography. London Mathematical Society Lecture Note Series, Cambridge University Press, Cambridge (2005)
33. Miller, V.S.: The Weil pairing, and its efficient calculation. J. Cryptology 17, 235–261 (2004)
34. Hess, F., Smart, N.P., Vercauteren, F.: The Eta pairing revisited. IEEE Trans. Information Theory 52(10), 4595–4602 (2006)
35. Lee, E., Lee, H.S., Park, C.M.: Efficient and generalized pairing computation on Abelian varieties. Cryptology ePrint Archive, Report 2008/040 (2008), http://eprint.iacr.org/2008/040
36. Vercauteren, F.: Optimal pairings. Cryptology ePrint Archive, Report 2008/096 (2008), http://eprint.iacr.org/2008/096
37. Hess, F.: Pairing lattices. In: Galbraith, S.D., Paterson, K.G. (eds.) Pairing 2008. LNCS, vol. 5209, pp. 18–38. Springer, Heidelberg (2008)
38. Barreto, P.S.L.M., Galbraith, S.D., Ó hÉigeartaigh, C., Scott, M.: Efficient pairing computation on supersingular Abelian varieties. Designs, Codes and Cryptography 42(3), 239–271 (2007)
39. Barreto, P.S.L.M., Kim, H.Y., Lynn, B., Scott, M.: Efficient algorithms for pairing-based cryptosystems. In: Yung, M. (ed.) CRYPTO 2002. LNCS, vol. 2442, pp. 354–368. Springer, Heidelberg (2002)
40. CoWare: Processor Designer (2009), http://www.coware.com/products/processordesigner.php
41. National Institute of Standards and Technology, NIST: FIPS 186-2: Digital Signature Standard (DSS) (2000), http://csrc.nist.gov/publications/fips/fips186-2/fips186-2-change1.pdf
42. Bernstein, D.J.: Curve25519: new Diffie-Hellman speed records. In: Yung, M., Dodis, Y., Kiayias, A., Malkin, T.G. (eds.) PKC 2006. LNCS, vol. 3958, pp. 207–228. Springer, Heidelberg (2006)
43. Montgomery, P.: Modular multiplication without trial division. Mathematics of Computation 44(170), 519–521 (1985)
44. Nibouche, O., Bouridane, A., Nibouche, M.: Architectures for Montgomery's multiplication. IEE Proc. – Computers and Digital Techniques 150(6), 361–368 (2003)
45. Synopsys: Design Compiler (2009), http://www.synopsys.com/products/logic/design_compiler.html
46. Shu, C., Kwon, S., Gaj, K.: FPGA accelerated Tate pairing based cryptosystems over binary fields. Cryptology ePrint Archive, Report 2006/179 (2006), http://eprint.iacr.org/2006/179

KATAN and KTANTAN — A Family of Small and Efficient Hardware-Oriented Block Ciphers

Christophe De Cannière[1], Orr Dunkelman[1,2,*], and Miroslav Knežević[1,**]

[1] Katholieke Universiteit Leuven
Department of Electrical Engineering ESAT/SCD-COSIC
and
Interdisciplinary Center for Broad Band Technologies
Kasteelpark Arenberg 10, B-3001 Leuven-Heverlee, Belgium
{christophe.decanniere,miroslav.knezevic}@esat.kuleuven.be
[2] École Normale Supérieure
Département d'Informatique,
CNRS, INRIA
45 rue d'Ulm, 75230 Paris, France
orr.dunkelman@di.ens.fr

Abstract. In this paper we propose a new family of very efficient hardware oriented block ciphers. The family contains six block ciphers divided into two flavors. All block ciphers share the 80-bit key size and security level. The first flavor, KATAN, is composed of three block ciphers, with 32, 48, or 64-bit block size. The second flavor, KTANTAN, contains the other three ciphers with the same block sizes, and is more compact in hardware, as the key is burnt into the device (and cannot be changed).

The smallest cipher of the entire family, KTANTAN32, can be implemented in 462 GE while achieving encryption speed of 12.5 KBit/sec (at 100 KHz). KTANTAN48, which is the version we recommend for RFID tags uses 588 GE, whereas KATAN64, the largest and most flexible candidate of the family, uses 1054 GE and has a throughput of 25.1 Kbit/sec (at 100 KHz).

1 Introduction

Low-end devices, such as RFID tags, are deployed in increasing numbers each and every day. Such devices are used in many applications and environments, leading to an ever increasing need to provide security (and privacy). In order to satisfy these needs, several suitable building blocks, such as secure block ciphers, have to be developed.

The problem of providing secure primitives in these devices is the extremely constrained environment. The primitive has to have a small footprint (where any additional gate might lead to the solution not being used), reduced power

---

* This author was partially supported by the France Telecom Chaire.
** This author is supported in part by IAP Programme P6/26 BCRYPT of the Belgian State (Belgian Science Policy).

C. Clavier and K. Gaj (Eds.): CHES 2009, LNCS 5747, pp. 272–288, 2009.

consumption (as these devices either rely on a battery or on an external electromagnetic field to supply them the required energy), and with sufficient speed (to allow the use of the primitive in real protocols).

The raising importance as well as the lack of secure and suitable candidates, has initiated a research aiming to satisfy these requirements. The first candidate block cipher for these devices was the DESL algorithm [19]. DESL is based on the general structure of DES, while using a specially selected S-box. DESL has key size of 56 bits and a footprint of 1848 GE. The second candidate for the mission is the PRESENT block cipher [4]. PRESENT has an SP-Network structure, and it can be implemented using the equivalent of 1570 GE. A more dedicated implementation of PRESENT in 0.35μm CMOS technology reaches 1000 GE [20].[1] The same design in 0.25μm and 0.18μm CMOS technology consumes 1169 and 1075 GE, respectively.

Some stream ciphers, such as grain [11] and trivium [6] may also be considered fit for these constrained environments, with 1293 and 749 GE[2] implementations, respectively. However, some protocols cannot be realized using stream ciphers, thus, leaving the issue of finding a more compact and secure block cipher open.

In this paper we propose a new family of block ciphers composed of two sets. The first set of ciphers is the KATAN ciphers, KATAN32, KATAN48 and KATAN64. All three ciphers accept 80-bit keys, and have a different block size (n-bit for KATANn). These three block ciphers are highly compact and achieve the minimal size (while offering adequate security). The second set, composed of KTANTAN32, KTANTAN48, and KTANTAN64, realize even smaller block ciphers in exchange for agility. KTANTANn is more compact than KATANn, but at the same time, is suitable only for cases where the device is initialized with one key that can never be altered, i.e., for the KTANTAN families, the key of the device is burnt into the device. Thus, the only algorithmic difference between KATANn and KTANTANn is the key schedule (which may be considered slightly more secure in the KATANn case).

While in this paper we put emphasis on the smallest possible variants, it can be easily seen that increasing the speed of the implementation is feasible with only a small hardware overhead. Therefore, we provide more implementation results in Appendix B. We implemented all six ciphers of the family using an $fsc0l\_d\_sc\_tc$ 0.13μm family standard cell library tailored for UMC's 0.13μm Low Leakage process. We compare our results with previous constructions in Table 1. We note here that some of the implementations achieve an amazingly low gate count due to the number of GE per bit of memory used. This is the issue inherent not only to the encryption algorithm, but also a matter of the technology that is used. Thus, we give a detailed explanation addressing the possible bit representation.

---

[1] Comparing to 0.13μm CMOS technology we note here that the physical size of the chip (in μm^2) is about 8 times bigger than the design with the same number of gate equivalents in 0.13μm CMOS technology.

[2] This work is a full-custom design implemented with C^2MOS dynamic logic [16]. The die size is equivalent to 749 standard CMOS logic NAND gates. The clock frequency required for this solution is far from being suitable for constrained environments.

Table 1. Comparison of Ciphers Designed for Low-End Environments (optimized for size)

Cipher	Block (bits)	Key (bits)	Size (GE)	Gates per Memory Bit	Throughput* (Kb/s)	Logic Process
AES-128 [8]	128	128	3400	7.97	12.4	0.35 μm
AES-128 [10]	128	128	3100	5.8	0.08	0.13 μm
HIGHT [12]	64	128	3048	N/A	188.25	0.25 μm
mCrypton [15]	64	64	2420	5	492.3	0.13 μm
DES [19]	64	56	2309[†]	12.19	44.4	0.18 μm
DESL [19]	64	56	1848[†]	12.19	44.4	0.18 μm
PRESENT-80 [4]	64	80	1570	6	200	0.18 μm
PRESENT-80 [20]	64	80	1000	N/A	11.4	0.35 μm
Grain [9]	1	80	1294	7.25	100	0.13 μm
Trivium [16]	1	80	749	2[◊]	100[‡]	0.35 μm
KATAN32	32	80	802	6.25	12.5	0.13 μm
KATAN48	48	80	927	6.25	18.8	0.13 μm
KATAN64	64	80	1054	6.25	25.1	0.13 μm
KTANTAN32	32	80	462	6.25	12.5	0.13 μm
KTANTAN48	48	80	588	6.25	18.8	0.13 μm
KTANTAN64	64	80	688	6.25	25.1	0.13 μm

[*] — A throughput is estimated for frequency of 100 KHz.
[†] — Fully serialized implementation (the rest are only synthesized).
[‡] — This throughput is projected, as the chip requires higher frequencies.
[◊] — This is a full-custom design using C^2MOS dynamic logic.

We organize this paper as follows: In Section 2 we describe the design criteria used in the construction of the KATAN family. Section 3 presents the building blocks used in our construction as well as the implementation issues related to them. In Sections 4 and 5 we present the KATAN and the KTANTAN families, respectively. The security analysis results are given in Section 6. Several compartive tradeoffs concerning the implemtnation speed and size of the KATAN and KTANTAN families are reported in Appendix B. Finally, we summarize our results in Section 7.

2   Motivation and Design Goals

Our main design goal was to develop a secure 80-bit block cipher with as minimal number of gates as possible. Such ciphers are needed in many constrained environments, e.g., RFID tags and sensor networks.

While analyzing the previous solutions to the problem, we have noticed that the more compact the cipher is, a larger ratio of the area is dedicated for storing the intermediate values and key bits. For example, in grain [11], almost all of the 1294 gates which are required, are used for maintaining the internal state. This phenomena also exist in DESL [19] and PRESENT [4], but to a lesser degree. This follows two-fold reasoning: First, stream ciphers need an internal state of at least twice the security level while block ciphers are exempt from this requirement. Second, while in stream ciphers it is possible to use relatively compact highly nonlinearity combining function, in block ciphers the use of S-box puts a burden on the hardware requirements.

Another interesting issue that we have encountered during the analysis of previous results is the fact that various implementations not only differ in the basic gate technology, but also in the number of gate equivalents required for storing a bit. In the standard library we have used in this work, a simple flip-flop implementation can take between 5 and 12 GE. This, of course, depends on the type of the flip-flop that is used (scan or standard D flip-flop, with or without set/reset signals, input and output capacitance, etc). Typical flip-flops that are used to replace a combination of a multiplexer and a flip-flop are, so called, scan flip-flops of which the most compact version, in our library, has a size equivalent to 6.25 GE. These flip-flops basically act as a combination of a simple D flip-flop and a MUX2to1. Use of this type of flip-flops is beneficial both for area and power consumption.

Here, we can notice that in PRESENT [4], the 80-bit key is stored in an area of about 480 GE, i.e., about 6 GE for one bit of memory, while in DESL, the 64-bit state is stored in 780 GE (about 12 GE for a single bit). As we have already discussed, this is related to many different factors such as the type of flip-flops, technology, library, etc. Finally, we note that in some cases (which do not necessarily fit an RFID tag due to practical reasons) it is possible to reduce the area required for storing one memory bit to only 8 transistors (i.e., about 2 GE) [16]. This approach achieves a much better comparison between different implementations, as usually changing the memory technology we can relatively easily counter the effects of implementers knowledge (or lack of), and discuss the true size of the proposed algorithm.

An additional issue which we observed is that in many low-end applications, the key is loaded once to the device and is never changed. In such instances, it should be possible to provide an encryption solution which can handle a key which is not stored in memory, preferably in a more efficient manner.

A final issue related to reducing the area requirements of the cipher is the block size. By decreasing the block size, it is possible to further reduce the memory complexity of the cipher. On the other hand, reducing the plaintext size to less than 32 bits has strong implications on the security of the systems using this cipher. For example, due to the birthday bound, a cipher with block size smaller than 32 bits is distinguishable from a family of random permutations after 2^{16} blocks.

The life span of a simple RFID tag indeed fits this restriction, but some RFID tags and several devices in sensor networks may need to encrypt larger amounts of data (especially if the used protocols require the encryption of several values in each execution). Thus, we decided to offer 3 block sizes to implementers — 32 bits, 48 bits, and 64 bits.

Our specific design goals are as follows:

- For an n-bit block size, no differential characteristic with probability greater than 2^{-n} exists for 128 rounds (about half the number of rounds of the cipher).
- For an n-bit block size, no linear approximation with bias greater than $2^{-n/2}$ exists for 128 rounds.
- No related-key key-recovery or slide attack with time complexity smaller than 2^{80} exists on the entire cipher.
- High enough algebraic degree for the equation describing half the cipher to thwart any algebraic attack.

We note that the first two conditions ensure that no differential-linear attack (or a boomerang attack) exist for the entire cipher as well. We also had to rank the possible design targets as follows:

- Minimize the size of the implementation.
- Keeping the critical path as short as possible.
- Increase the throughput of the implementation (as long as the increase in the foot print is small).
- Decrease the power consumption of the implementation.

3   General Construction and Building Blocks

Following the design of KeeLoq [17], we decided to adopt a cipher whose structure resembles a stream cipher. To this extent we have chosen a structure which resembles trivium [6], or more precisely, its two register variant bivium as the base for the block cipher. While the internal state of trivium was 288 bits to overcome the fact that each round, one bit of internal state is revealed, in the block cipher this extra security measure is unnecessary. Hence, we select the block size and the internal state of the cipher to be equal.

The structure of the KATAN and the KTANTAN ciphers is very simple — the plaintext is loaded into two registers (whose lengths depend on the block size). Each round, several bits are taken from the registers and enter two nonlinear Boolean functions. The output of the Boolean functions is loaded to the least significant bits of the registers (after they were shifted). Of course, this is done in an invertible manner. To ensure sufficient mixing, 254 rounds of the cipher are executed.

We have devised several mechanisms used to ensure the security of the cipher, while maintaining a small foot print. The first one is the use of an LFSR instead of a counter for counting the rounds and to stop the encryption after 254 rounds. As there are 254 rounds, an 8-bit LFSR with as sparse polynomial feedback can be used. The LFSR is initialized with some state, and the cipher has to stop running the moment the LFSR arrives to some predetermined state.

We have implemented the 8-bit LFSR counter, and the result fits a gate equivalent of 60 gates, while using an 8-bit counter (the standard alternative) took 80 gate equivalents. Moreover, the expected speed of the LFSR (i.e. the critical path) is shorter than the one for the 8-bit counter.

Another advantage for using LFSR is the fact that when considering one of the bits taken from it, we expect a sequence which keeps on alternating between 0's and 1's in a more irregular manner than in a counter (of course the change is linear). We use this feature to enhance the security of our block ciphers as we describe later.

One of the problems that may arise in such a simple construction is related to self-similarity attacks such as the slide attacks. For example, in KeeLoq [17] the key is used again and again. This made KeeLoq susceptible to several slide attacks (see for example [5,13]). A simple solution to the problem is to have the key loaded into an LFSR with a primitive feedback polynomial (thus, altering the subkeys used in the cipher). This solution helps the KATAN family to achieve security against the slide attack.

While the above building block is suitable when the key is loaded into memory, in the KTANTAN family, it is less favorable (as the key is hardcoded in the device). Thus, the only means to prevent a slide attack is by generating a simple, non-repetitive sequence of bits from the key. To do so, we use the "round counter" LFSR, which produces easily computed bits, that at the same time follow a non-repetitive sequence.

The third building block which we use prevents the self-similarity attacks and increases the diffusion of the cipher. The cipher actually has two (very similar but distinct) round functions. The choice of the round function is made according to the most significant bit of the round-counting LFSR. This irregular update also increases the diffusion of the cipher, as the nonlinear update affects both the differential and the linear properties of the cipher.

Finally, both KATAN and KTANTAN were constructed such that an implementation of the 64-bit variants can support the 32-bit and the 48-bit variants at the cost of small extra controlling hardware. Moreover, given the fact that the only difference between a KATANn cipher and KTANTANn is the way the key is stored and the subkeys are derived, it is possible to design a very compact circuit that support all six ciphers.

4   The KATAN Set of Block Ciphers

The KATAN ciphers compose of three variants: KATAN32, KATAN48 and KATAN64. All the ciphers in the KATAN family share the key schedule which accepts an 80-bit key and 254 rounds as well as the use of the same nonlinear functions.

We start by describing KATAN32, and describe the differences for KATAN48 and KATAN64 later. KATAN32, the smallest of this family has a plaintext and ciphertext size of 32 bits. The plaintext is loaded into two registers L_1, and L_2 (of respective lengths of 13 and 19 bits) where the least significant bit of the plaintext is loaded to bit 0 of L_2, while the most significant bit of the plaintext is loaded to bit 12 of L_1. Each round, L_1 and L_2 are shifted to the left (bit i is shifted to position $i + 1$), where the new computed bits are loaded in the least significant bits of L_1 and L_2. After 254 rounds of the cipher, the contents of the registers are then exported as the ciphertext (where bit 0 of L_2 is the least significant of the ciphertext).

KATAN32 uses two nonlinear function $f_a(\cdot)$ and $f_b(\cdot)$ in each round. The nonlinear function f_a and f_b are defined as follows:

$$f_a(L_1) = L_1[x_1] \oplus L_1[x_2] \oplus (L_1[x_3] \cdot L_1[x_4]) \oplus (L_1[x_5] \cdot IR) \oplus k_a$$

$$f_b(L_2) = L_2[y_1] \oplus L_2[y_2] \oplus (L_2[y_3] \cdot L_2[y_4]) \oplus (L_2[y_5] \cdot L_2[y_6]) \oplus k_b$$

where IR is irregular update rule (i.e., $L_1[x_5]$ is XORed in the rounds where the irregular update is used), and k_a and k_b are the two subkey bits. For round i, k_a is defined to be k_{2i}, whereas k_b is k_{2i+1}. The selection of the bits $\{x_i\}$ and $\{y_j\}$ are defined for each variant independently, and listed in Table 2.

After the computation of the nonlinear functions, the registers L_1 and L_2 are shifted, where the MSB falls off (into the corresponding nonlinear function), and the LSB is loaded with the output of the second nonlinear function, i.e., after the round the LSB of L_1 is the output of f_b, and the LSB of L_2 is the output of f_a.

The key schedule of the KATAN32 cipher (and the other two variants KATAN48 and KATAN64) loads the 80-bit key into an LFSR (the least significant bit of the key is loaded to position 0 of the LFSR). Each round, positions 0 and 1 of the LFSR are generated as the round's subkey k_{2i} and k_{2i+1}, and the LFSR is clocked twice. The feedback polynomial that was chosen is a primitive polynomial with minimal hamming weight of 5 (there are no primitive polynomials of degree 80 with only 3 monomials):

$$x^{80} + x^{61} + x^{50} + x^{13} + 1.$$

We note that these locations compose a full difference set, and thus, are less likely to lead to a guess and determine attacks faster than exhaustive key search.

In other words, let the key be K, then the subkey of round i is $k_a || k_b = k_{2 \cdot i} || k_{2 \cdot i+1}$ where

$$k_i = \begin{cases} K_i & \text{for } i = 0 \ldots 79 \\ k_{i-80} \oplus k_{i-61} \oplus k_{i-50} \oplus k_{i-13} & \text{Otherwise} \end{cases}$$

The differences between the various KATAN ciphers are:

- The plaintext/ciphertext size,
- The lengths of L_1 and L_2,
- The position of the bits which enter the nonlinear functions,
- The number of times the nonlinear functions are used in each round.

While the first difference is obvious, we define in Table 2 the lengths of the registers and the positions of the bits which enter the nonlinear functions used in the ciphers. The selection of the bits $\{x_i\}$ and $\{y_j\}$ are defined for each variant independently, and are listed in Table 2.

For KATAN48, in one round of the cipher the functions f_a and f_b are applied twice. The first pair of f_a and f_b is applied, and then after the update of the

Table 2. Parameters defined for the KATAN family of ciphers

Cipher	$|L_1|$	$|L_2|$	x_1	x_2	x_3	x_4	x_5
KATAN32/KTANTAN32	13	19	12	7	8	5	3
KATAN48/KTANTAN48	19	29	18	12	15	7	6
KATAN64/KTANTAN64	25	39	24	15	20	11	9

Cipher	y_1	y_2	y_3	y_4	y_5	y_6
KATAN32/KTANTAN32	18	7	12	10	8	3
KATAN48/KTANTAN48	28	19	21	13	15	6
KATAN64/KTANTAN64	38	25	33	21	14	9

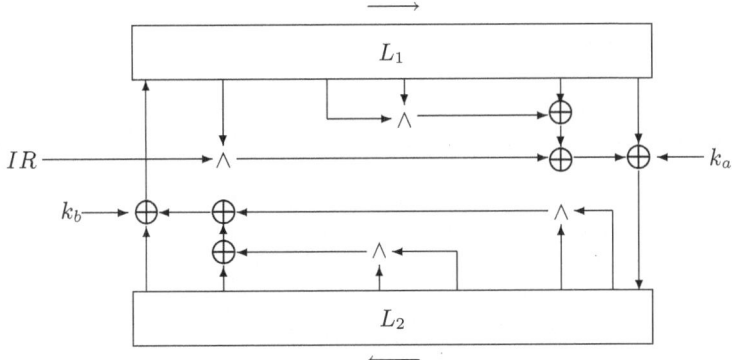

Fig. 1. The Outline of a round of the KATAN/KTANTAN ciphers

registers, they are applied again, using the same subkeys. Of course, an efficient implementation can implement these two steps in parallel. In KATAN64, each round applies f_a and f_b three times (again, with the same key bits).

We outline the structure of KATAN32 (which is similar to the round function of any of the KATAN variants or the KTANTAN variants) in Figure 1.

Finally, specification-wise, we define the counter which counts the number of rounds. The round-counter LFSR is initialized to the all 1's state, and clocked once using the feedback polynomial $x^8 + x^7 + x^5 + x^3 + 1$. Then, the encryption process starts, and ends after 254 additional clocks when the LFSR returns to the all 1's state. As mentioned earlier, we use the most significant bit of the LFSR to control the irregular update (i.e., as the IR signal). For sake of completeness, in Table 3 in the Appendix we give the sequence of irregular rounds.

We note that due to the way the irregular update rule is chosen, there are no sequences of more than 7 rounds that share the pattern of the regular/irregular updates, this ensures that any self-similarity attack cannot utilize more than 7 rounds of the same function (even if the attacker chooses keys that suggest the same subkeys). Thus, it is easy to see that such attacks are expected to fail when applied to the KATAN family.

We implemented KATAN32 using Synopsys Design Compiler version Y-2006.06 and the *fsc0l_d_sc_tc* 0.13μm CMOS library. Our implementation requires 802 GE, of which 742 are used for the sequential logic, and 60 GE are used for the combinational logic. The power consumption at 100 KHz, and throughput of 12.5 Kbps is only 381 nW. This is a gate level power estimation obtained using Synopsys Design Compiler[3].

For KATAN48 the implementation size is 927 GE (of which 842 are for the sequential logic) and the total power consumption is estimated to 439 nW. For the 64-bit variant, KATAN64, the total area is 1054 GE (of which 935 are for the sequential logic) and the power consumption 555 nW.

Here we would like to note that the further area reduction for KATAN48 and KATAN64 is possible by utilizing a clock gating technique. As explained above, the only difference between KATAN32 on one hand and KATAN48 and KATAN64 on the other, is the number of nonlinear functions f_a and f_b applied with the same subkeys per single round. Therefore, we can clock the key register and the counter such that they are updated once in every two (three) cycles for KATAN48 (KATAN64). However, this approach reduces the throughput two (three) times respectively, and is useful only when the compact implementation is an ultimate goal. An area of 916 GE with the throughput of 9.4 Kb/s (at 100 KHz) is obtained for KATAN48 and 1027 GE with the throughput of 8.4 Kb/s (at 100 KHz) for KATAN64.

At the cost of little hardware overhead, a throughput of the KATAN family of block ciphers can be doubled or even tripled. To increase the speed of the cipher, we double (triple) the logic for the nonlinear functions f_a and f_b as well as the logic for the feedback coefficients of the counter and the key register. The implementation results are given in Appendix B.

5   The KTANTAN Family

The KTANTAN family is very similar to the KATAN family up to the key schedule (i.e., the only difference between KATANn and KTANTANn is the key schedule part). While in the KATAN family, the 80-bit key is loaded into a register which is then repeatedly clocked, in the KTANTAN family of ciphers, the key is burnt (i.e., fixed) and the only possible "flexibility" is the choice of subkey bits. Thus, the design problem in the KTANTAN ciphers is choosing a sequence of subkeys in a secure, yet an efficient manner.

In order to minimize the hardware size, while maintaining the throughput, we treat the key as 5 words of 16 bits each. From each 16-bit word we pick the same bit (using a MUX16to1) according to the four most significant bits of the round controlling LFSR. Then, out of the five bits we choose one using the four least significant bits of the round-counting LFSR.

Formally, let $K = w_4||w_3||w_2||w_1||w_0$, where the least significant bit of w_0 is the least significant bit of K, and the most significant bit of w_4 is the most significant bit of K. We denote by T the round-counting LFSR (where T_7 is the most

---

[3] Although the gate level power estimation gives a rough estimate, it is useful for comparison with related work reported in the literature.

significant bit), then, let $a_i = \text{MUX16to1}(w_i, T_7T_6T_5T_4)$, where $\text{MUX16to1}(x, y)$ gives the yth bit of x. Then, the key bits which are used are

$$k_a = \overline{T_3} \cdot \overline{T_2} \cdot (a_0) \oplus (T_3 \vee T_2) \cdot \text{MUX4to1}(a_4a_3a_2a_1, T_1T_0),$$

$$k_b = \overline{T_3} \cdot T_2 \cdot (a_4) \oplus (T_3 \vee \overline{T_2}) \cdot \text{MUX4to1}(a_3a_2a_1a_0, \overline{T_1T_0})$$

(where $\text{MUX4to1}(x, y)$ is a MUX with 4 input bits and 1 output bit).

When considering k_a or k_b, of the 80-bit key, only one bit is used only twice, 15 are used four times, and the remaining 64 bits are used 3 times (but in total each key bit is used at least 5 times). Moreover, even if an attacker tries to pick two keys which realize the same subkey sequence for either k_a or k_b, the maximal length of such a sequence for either k_a of k_b is 35 rounds (i.e., necessarily after 35 rounds the sequences differ). We also note that due to the irregular update, during these 35 rounds, the round function is going to be different in any case.

The last issue concerning the KTANTAN key schedule is finding the most efficient way to implement it. One possible solution is to have the entire selection logic in one round. This approach requires 5 parallel MUX16to1 and our hardware implementations show that the total area consumed by the MUXes is about 180 GE. A second approach is to use one MUX16to1 and re-use it over 5 clock cycles. At a first glance, this approach may lead to a smaller circuit (while the implementation is slower). However, due to the cost of the extra control logic, this approach is not only slower, but leads to a larger circuits.

We implemented KTANTAN32 using the same *fsc0l_d_sc_tc* 0.13μm CMOS library. Our implementation requires 462 GE, of which 244 are used for the sequential logic, and 218 GE are used for the combinational logic. The simulated power consumption at 100 KHz, and throughput of 12.5 Kbps is only 146 nW. For the synthesis and the power estimation we have again used the same version of Synopsys Design Compiler.

For KTANTAN48 the implementation size of 588 GE (of which 344 are used for the sequential logic) is obtained together with the estimated power consumption of 234 nW. For the 64-bit variant, KTANTAN64, the total area is 688 GE (of which 444 are for the sequential logic) and the power consumption 292 nW. By using the clock gating as explained above, the area of 571 GE (684 GE) and the throughput of 9.4 Kb/s (8.4 Kb/s) for KATAN48 (KATAN64) is achieved.

Similar to KATAN family, we can also double (triple) a throughput for all the versions of KTANTAN family. To do that, we double (triple) the number of MUX16to1, MUX4to1, round functions f_a and f_b, and all the logic used for the feedback coefficients of the counter. Additionally, a few more gates are necessary to perform the key schedule efficiently.

6   Security Analysis

Our design philosophy was based on offering a very high level of security. To do so, we designed the ciphers with a very large security margins. For example, as a design target we have set an upper bound for the differential probability of any 128-round differential characteristic at 2^{-n} for an n-bit block size.

6.1 Differential and Linear Cryptanalysis

We have analyzed all ciphers under the assumption that the intermediate encryption values are independent. While this assumption does not necessarily hold, it simplifies the analysis and is not expected to change the results too much. Moreover, in our analysis we always take a "worst case" approach, i.e., we consider the best scenario for the attacker, which is most of the times do not happen. Hence, along with the large security margins, even if the assumption does not hold locally, it is expected that our bounds are far from being tight.

To simplify the task of identifying high probability differentials, we used computer-aided search. Our results show that depending on the used rounds, the best 42-round differential characteristic for KATAN32 has probability of 2^{-11} (it may even be lower for different set of rounds). Hence, any 126-round differential characteristic must have probability no more than $(2^{-11})^3 = 2^{-33}$. Similar results hold for linear cryptanalysis (the best 42-round linear approximation has a bias of 2^{-6}, i.e., a bias of 2^{-16} for 126-round approximation).

For KATAN48, the best 43-round differential characteristic has probability of at most 2^{-18}. Hence, any 129-round differential characteristic has probability of at most $(2^{-18})^3 = 2^{-54}$. As the probability of an active round is at least 2^{-4} this actually proves that our design criteria for 128-round differential characteristics is satisfied. The corresponding linear bias is 2^{-10} (for 43 rounds) or 2^{-28} (for 129 rounds).

Finally, repeating the analysis for KATAN64, our computer-aided search found that the best 37-round differential characteristic has probability 2^{-20}. Hence, any 111-round differential characteristic has probability of at most 2^{-60}, along with the fact that the best 18-round differential characteristic has probability of at most 2^{-5}, then the best 129-round differential characteristic has probability of no more than 2^{-65}. The linear bounds are 2^{-11} for 37 rounds and 2^{-31} for 111 rounds.

Hence, we conclude that the KATAN family is secure against differential and linear attacks. As there is no difference between the KATAN and the KTANTAN families with respect to their differential and linear behaviors, then the above is also true for the KTANTAN family.

6.2 Combined Attacks

As shown in the previous section, the probability of any differential characteristic of 128 rounds can be bounded by 2^{-n} for KATANn. Moreover, even for 64 rounds, there are no "good" characteristics. Hence, when trying to combine these together, it is unexpected to obtain good combined attacks.

For example, consider a differential-linear approximation. As noted before, the differential characteristic of 42-round KATAN32 has probability at most 2^{-11}. The bias of a 42-round KATAN32 is at most 2^{-6}. Hence, the best differential-linear property for 120 rounds is expected to have bias of at most $2 \cdot 2^{-11} \cdot (2^{-6})^2 = 2^{-22}$ (we assume a worst case assumption that allows the attacker to gain some free rounds in which the differential is truncated). Of course, an attacker may try to construct the differential-linear approximation using a different division of rounds. However, as both the probability and bias drop at least exponentially with the number of rounds, a different division is not expected to lead to better differential-linear approximations.

The same goes for the (amplified) boomerang attack. The attack (just like the differential-linear attack) treats the cipher as composed of two sub-ciphers. The probability of constructing a boomerang quartet is $\hat{p}^2\hat{q}^2$, where $\hat{p} = \sqrt{\sum_\beta \Pr{}^2[\alpha \to \beta]}$ where α is the input difference for the quartet, and β is the output difference of the characteristic in the first sub-cipher. Again, as $\hat{p}^2 \le \max_\beta \Pr[\alpha \to \beta]$ which is bounded at 2^{-22} for 84-round KATAN32. The same goes with respect to $\hat{q}$, and thus, the probability of a boomerang quartet in 128-round KATAN32 is at most 2^{-44}.

The same rationale can be applied to KATAN48 and KATAN64, obtaining similar bounds. Specifically, the bounds for differential-linear bias is 2^{-37} (for 140 rounds) and 2^{-50} (for 160 rounds), respectively. The bounds for constructing a boomerang quartet for 128 rounds are 2^{-54} and 2^{-65}, respectively.

Another combined attack which may be considered is the impossible differential attack. This attack is based on finding a differential which has probability zero of as many rounds as possible. The most common way to construct such a differential is in a miss-in-the-middle manner, which is based on finding two (truncated) differentials with probability 1 which cannot co-exist. Due to the quick diffusion, changing even one bit would necessarily affects all bits after at most 42 rounds (37 for KATAN48 and 38 for KATAN64), and thus, there is no impossible differential of more than 168 rounds (after 42 rounds, change of any bit may affect all bits, and thus, after 84 rounds, each differential may have any output difference).

Hence, we conclude that the KATAN family (as well as the KTANTAN family) of block ciphers is secure against combined attacks.

6.3   Slide and Related-Key Attacks

As mentioned before, the way KATAN and KTANTAN were designed to foil self-similarity attacks by using two types of rounds which are interleaved in a non-repeating manner. First, consider the slide attack, which is based on finding two messages such that they share most of the encryption process (which are some rounds apart). Given the fact that there is a difference between the deployed round functions, this is possible only for a very small number of rounds. Even if we allow these relations to be probabilistic in nature (i.e., assume that the bit of the intermediate value is set to 0 thus preventing the change in the function to change the similarity between the states). For example, when considering KATAN32, there is no slide property with probability 2^{-32} starting from the first round of the cipher. The first round from which such a property can be constructed is round 19. If an attacker achieves the same intermediate encryption value after round 19 and round 118, he may find a "slid" pair which maintains the equality with probability 2^{-31} until the end of the cipher (i.e., the output of the second encryption process will be the same as the intermediate encryption value of the first encryption at round 155). This proves that there are no good slid properties in the cipher family (we note that this probability is based on the assumption that the subkeys are the same, which is not the case, unless the key is the all zeros key). When it comes to KATAN48 or KATAN64, this probability is even lower (as there are more bits which need to be equal to zero), i.e., 2^{-62} and 2^{-93}, respectively, rendering slide attacks futile against the KATAN and the KTANTAN families (these values are actually an upper bound as they assume that all the subkeys are the same).

Now consider a related-key attack. In the related-key settings, the attacker searches for two intermediate encryption values as well as keys which develop in the same manner for as many rounds as possible. As noted before, there are no "good" relations over different rounds, which means that the two intermediate encryption values have to be in the same round. However, by changing even one singly bit of the key causes a difference after at most 80 rounds of similar encryption process. Hence, no related-key plaintext pairs (or intermediate encryption values) exist for more than 80 rounds (similarity in 80 rounds would force the key and the intermediate encryption value to be the same). As this is independent of the actual key schedule algorithm, it is easy to see that both the KATAN and the KTANTAN families are secure against this attack.

The only attack in this category which remains is related-key differential attack. This is the only attack where there is a difference between the two families of ciphers according to their key schedule algorithm. We first consider the case of the KATAN family. The key schedule of the KATAN family expands linearly the 80-bit key into 508 subkey bits (each is used once in KATAN32, twice in KATAN48, and thrice in KATAN64). We note that the probability of the differential is reduced any time a difference enters one of the nonlinear functions (i.e., the AND operation). Thus, it is evident that good related-key differentials have as little active bits as possible. Moreover, we can relate the number of active bits throughout the encryption process to the issue of active bits in the key schedule. Each active bit of the subkey (i.e., a subkey bit with a difference) either causes a difference in the internal state (which in turn incurs probability and activation of more bits), or is being canceled by previous differences. We note that each active subkey bit which is not canceled, necessarily induces probability "penalty" of 2^{-2} in KATAN32, 2^{-4} in KATAN48, and 2^{-6} in KATAN64. Moreover, due to the way the cipher works, each active bit can "cancel" at most four other active subkey bits.[4] Hence, if the weight of the expanded subkey difference is more than 80, then it is assured that the probability of any related-key differential of KATAN32 is at most 2^{-32} (this follows the fact that each active bit in the intermediate encryption value may cancel up to four subkey bit differences injected, and we shall assume a worst case assumption that the positions align "correctly"). For KATAN48, due to the increased penalty, it is sufficient that the minimal weight of the expanded subkey difference is more than 60, and for

---

[4] We note that five or six can be canceled, but in this case, the probability penalty of an active bit is increased by more than the "gain" offered by using this active bit more times.

KATAN64 the minimal weight needs to be at least 54. We were analyzing the minimal weight using the MAGMA software package, and the current bounds are between 72 and 84. Hence, we conclude that the KATAN family of block ciphers is expected to be resistant to related-key differential attacks.

For the KTANTAN family, due to the fixed key, the concept of related-key attacks is of theoretical interest. Still, we can follow a more detailed analysis using the same ideas as we used for regular differential searches. While the search space is huge, our current results show that there is no related-key differential characteristic for more than 150 rounds of KTANTAN32 with probability greater than 2^{-32}. Similar results are expected to hold for KTANTAN48 and KTANTAN64.

6.4 Cube Attacks and Algebraic Attacks

Given the low algebraic degree of the combining function, it may look as if KATAN and KTANTAN are susceptible to algebraic attacks or cube attack [7]. However, when considering the degree of the expressions involving the plaintext, one can see that after 32 rounds (for KATAN32) the degree of each internal state bit is at least 2, which means that after 160 rounds, the degree of each internal state bit can reach 32. For KATAN48, the degree is at least 2 after 24 rounds, (or about 48 after 144 rounds), and for KATAN64 it is 2 after 22 rounds and can reach 64 after 132 rounds). Hence, as the degree can reach to the maximal possible value (and there are some more rounds to spare), it is expected that the KATAN and KTANTAN families are secure against algebraic attacks.

Another possible attack is the cube attack, which was successful against reduced-round variants of Trivium (with less initialization rounds than in the Trivium). We note that in trivium the internal state is clocked four full cycles (i.e., each bit traverse all the registers exactly four times). In KATAN32, most bits traverse the registers eight times (where a few does so only seven times), going through more nonlinear combiners (each Trivium round uses only one AND operation per updated bit), and thus is expected to be more secure against this attack than Trivium. The same is also true for KATAN48 (where about half of the bits traverse the registers 10 times, and the other bits do so 11 times) and KATAN64 (where most of the bits traverse the registers 12 times, and a few do that only 11 times).

7 Summary

In this paper we have presented two families of hardware efficient encryption algorithms. The family of cipher is suitable for low-end devices, and even offer algorithmic security level of 80 bits in cases where the key is burnt into the device (of course, the implementation has to be protected as well). As the proposal have a simple structure and use very basic components, it appears that common techniques to protect the implementation should be easily adopted.

References

1. Biham, E.: New Types of Cryptanalytic Attacks Using Related Keys. Journal of Cryptology 7(4), 229–246 (1994)
2. Biham, E., Shamir, A.: Differential Cryptanalysis of the Data Encryption Standard. Springer, Heidelberg (1993)
3. Biryukov, A., Wagner, D.: Slide Attacks. In: Knudsen, L.R. (ed.) FSE 1999. LNCS, vol. 1636, pp. 245–259. Springer, Heidelberg (1999)
4. Bogdanov, A.A., Knudsen, L.R., Leander, G., Paar, C., Poschmann, A., Robshaw, M.J.B., Seurin, Y., Vikkelsoe, C.: PRESENT: An Ultra-Lightweight Block Cipher. In: Paillier, P., Verbauwhede, I. (eds.) CHES 2007. LNCS, vol. 4727, pp. 450–466. Springer, Heidelberg (2007)
5. Courtois, N.T., Bard, G.V., Wagner, D.: Algebraic and Slide Attacks on KeeLoq. In: Nyberg, K. (ed.) FSE 2008. LNCS, vol. 5086, pp. 97–115. Springer, Heidelberg (2008)
6. De Canniére, C., Preneel, B.: Trivium Specifications, eSTREAM submission, http://www.ecrypt.eu.org/stream/triviump3.html
7. Dinur, I., Shamir, A.: Cube Attacks on Tweakable Black Box Polynomials. In: Joux, A. (ed.) EUROCRYPT 2009. LNCS, vol. 5479, pp. 278–299. Springer, Heidelberg (2009); IACR ePrint report 2008/385
8. Feldhofer, M., Wolfkerstorfer, J., Rijmen, V.: AES implementation on a grain of sand. In: IEE Proceedings of Information Security, vol. 152(1), pp. 13–20. IEE (2005)
9. Good, T., Benaissa, M.: Hardware results for selected stream cipher candidates. In: Preproceedings of SASC 2007, pp. 191–204 (2007)
10. Hämäläinen, P., Alho, T., Hännikäinen, M., Hämäläinen, T.D.: Design and Implementation of Low-Area and Low-Power AES Encryption Hardware Core. In: Ninth Euromicro Conference on Digital System Design: Architectures. IEEE Computer Society, Los Alamitos (2006)
11. Hell, M., Johansson, T., Meier, W.: Grain — A Stream Cipher for Constrained Environments, eSTREAM submission, http://www.ecrypt.eu.org/stream/p3ciphers/grain/Grain_p3.pdf
12. Hong, D., Sung, J., Hong, S.H., Lim, J.-I., Lee, S.-J., Koo, B.-S., Lee, C.-H., Chang, D., Lee, J., Jeong, K., Kim, H., Kim, J.-S., Chee, S.: HIGHT: A New Block Cipher Suitable for Low-Resource Device. In: Goubin, L., Matsui, M. (eds.) CHES 2006. LNCS, vol. 4249, pp. 46–59. Springer, Heidelberg (2006)
13. Indesteege, S., Keller, N., Dunkelman, O., Biham, E., Preneel, B.: A Practical Attack on KeeLoq. In: Smart, N.P. (ed.) EUROCRYPT 2008. LNCS, vol. 4965, pp. 1–18. Springer, Heidelberg (2008)
14. Langford, S.K., Hellman, M.E.: Differential-Linear Cryptanalysis. In: Desmedt, Y.G. (ed.) CRYPTO 1994. LNCS, vol. 839, pp. 17–25. Springer, Heidelberg (1994)
15. Lim, C.H., Korkishko, T.: mCrypton – A Lightweight Block Cipher for Security of Low-Cost RFID Tags and Sensors. In: Song, J.-S., Kwon, T., Yung, M. (eds.) WISA 2005. LNCS, vol. 3786, pp. 243–258. Springer, Heidelberg (2006)
16. Mentens, N., Genoe, J., Preneel, B., Verbauwhede, I.: A low-cost implementation of Trivium. In: Preproceedings of SASC 2008, pp. 197–204 (2008)
17. Microchip Technology Inc. KeeLoq® Authentication Products, http://www.microchip.com/keeloq/
18. Matsui, M.: Linear Cryptanalysis Method for DES Cipher. In: Helleseth, T. (ed.) EUROCRYPT 1993. LNCS, vol. 765, pp. 386–397. Springer, Heidelberg (1994)

19. Poschmann, A., Leander, G., Schramm, K., Paar, C.: New Light-Weight DES Variants Suited for RFID Applications. In: Biryukov, A. (ed.) FSE 2007. LNCS, vol. 4593, pp. 196–210. Springer, Heidelberg (2007)
20. Rolfes, C., Poschmann, A., Leander, G., Paar, C.: Ultra-lightweight implementations for smart devices – security for 1000 gate equivalents. In: Grimaud, G., Standaert, F.-X. (eds.) CARDIS 2008. LNCS, vol. 5189, pp. 89–103. Springer, Heidelberg (2008)

A The Irregular Update Sequence

Table 3. The sequence of the irregular updates. 1 means that the irregular update rule is used in this round, while 0 means that this is not the case.

Rounds	0–9	10–19	20–29	30–39	40–49	50–59
Irregular	1111111000	1101010101	1110110011	0010100100	0100011000	1111000010
Rounds	60–69	70–79	80–89	90–99	100–109	110–119
Irregular	0001010000	0111110011	1111010100	0101010011	0000110011	1011111011
Rounds	120–129	130–139	140–149	150–159	160–169	170–179
Irregular	1010010101	1010011100	1101100010	1110110111	1001011011	0101110010
Rounds	180–189	190–199	200–209	210–219	220–229	230–239
Irregular	0100110100	0111000100	1111010000	1110101100	0001011001	0000001101
Rounds	240–249	250–253				
Irregular	1100000001	0010				

B   Implementation Trade-Offs

Table 4. Area-Throughput Trade-Offs

Cipher	Block (bits)	Key (bits)	Size (GE)	Gates per Memory Bit	Throughput* (Kb/s)	Logic Process
KATAN32	32	80	802	6.25	12.5	0.13 μm
KATAN32	32	80	846	6.25	25	0.13 μm
KATAN32	32	80	898	6.25	37.5	0.13 μm
KATAN48[†]	48	80	916	6.25	9.4	0.13 μm
KATAN48	48	80	927	6.25	18.8	0.13 μm
KATAN48	48	80	1002	6.25	37.6	0.13 μm
KATAN48	48	80	1080	6.25	56.4	0.13 μm
KATAN64[†]	64	80	1027	6.25	8.4	0.13 μm
KATAN64	64	80	1054	6.25	25.1	0.13 μm
KATAN64	64	80	1189	6.25	50.2	0.13 μm
KATAN64	64	80	1269	6.25	75.3	0.13 μm
KTANTAN32	32	80	462	6.25	12.5	0.13 μm
KTANTAN32	32	80	673	6.25	25	0.13 μm
KTANTAN32	32	80	890	6.25	37.5	0.13 μm
KTANTAN48[†]	48	80	571	6.25	9.4	0.13 μm
KTANTAN48	48	80	588	6.25	18.8	0.13 μm
KTANTAN48	48	80	827	6.25	37.6	0.13 μm
KTANTAN48	48	80	1070	6.25	56.4	0.13 μm
KTANTAN64[†]	64	80	684	6.25	8.4	0.13 μm
KTANTAN64	64	80	688	6.25	25.1	0.13 μm
KTANTAN64	64	80	927	6.25	50.2	0.13 μm
KTANTAN64	64	80	1168	6.25	75.3	0.13 μm

* — A throughput is estimated for frequency of 100 KHz.
† — Using clock gating.

Programmable and Parallel ECC Coprocessor Architecture: Tradeoffs between Area, Speed and Security

Xu Guo[1], Junfeng Fan[2], Patrick Schaumont[1], and Ingrid Verbauwhede[2]

[1] Bradley Department of Electrical and Computer Engineering
Virginia Tech, Blacksburg, VA 24061, USA
[2] ESAT/SCD-COSIC, Katholieke Universiteit Leuven and IBBT
Kasteelpark Arenberg 10, B-3001 Leuven-Heverlee, Belgium
{xuguo,schaum}@vt.edu,{Junfeng.Fan,Ingrid.Verbauwhede}@esat.kuleuven.be

Abstract. Elliptic Curve Cryptography implementations are known to be vulnerable to various side-channel attacks and fault injection attacks, and many countermeasures have been proposed. However, selecting and integrating a set of countermeasures targeting multiple attacks into an ECC design is far from trivial. Security, performance and cost need to be considered together. In this paper, we describe a generic ECC coprocessor architecture, which is scalable and programmable. We demonstrate the coprocessor architecture with a set of countermeasures to address a collection of side-channel attacks and fault attacks. The programmable design of the coprocessor enables tradeoffs between area, speed, and security.

1 Introduction

Elliptic-curve cryptography (ECC) is the algorithm-of-choice for public-key cryptography in embedded systems. Performance, security and cost are the three important dimensions of ECC implementations. ECC accelerators should support multiple security and performance levels, allowing the system to adjust its security-performance to application-specific needs [1]. To achieve these goals, much research has been conducted targeting different aspects, and most research topics fall into two categories: efficient implementations and security analysis.

The computational intensive kernel of ECC is well suited for hardware acceleration, and many Hardware/Software (HW/SW) codesigns have been proposed to evaluate the tradeoffs between cost and performance. The challenge is how to perform optimizations at multiple abstraction levels (e.g. how to devise more efficient scalar multiplication algorithms or how to minimize the communication overhead for the HW/SW interface) and how to map the ECC system architecture to various platforms (e.g. resource-constrained 8-bit platforms or more powerful 32-bit microprocessor with bus systems).

For security analysis, ECC implementations are known to be vulnerable to various side-channel attacks (SCA), including power analysis (PA) attacks, electromagnetic attacks (EMA) and fault attacks (FA). Since Kocher *et al.* [2] showed

C. Clavier and K. Gaj (Eds.): CHES 2009, LNCS 5747, pp. 289–303, 2009.

the first successful PA attacks, there have been dozens of proposals for new SCA attacks and countermeasures. These attacks and countermeasures all tend to concentrate on a single abstraction level at a time [7]. For example, the Smart Card software is developed on fixed hardware platforms, so the results in that area are software-based solutions. At the same time, many special circuit styles [32,33] have been developed to address PA at the hardware level. Such circuit-level solutions are treated independently from the software-level solutions.

From the above descriptions, we have found two gaps in current ECC research. First, security has been generally treated as a separate dimension in designs and few researchers have proposed countermeasures targeting at system integration. For example, some of the fault attack countermeasures or fault detection methods are just like software patches applied to the original algorithms (e.g. perform Point Verification (PV) [23] or Coherence Check [3,34] during computation). The fault model is hypothesized without considering how to introduce faults in an actual platform. Further, the impact of circuit-level PA countermeasures on area, performance and power consumption in large designs remains unclear. Second, most published papers proposed their own attacks with corresponding countermeasures and very few researchers discussed countermeasures targeting multiple attacks. Since a cryptosystem will fail at its weakest link [4] it is not surprising to see how a given countermeasure can actually introduce a new weakness, and thus enable another attack [5]. Although there has been some effort to connect the PA and FA countermeasures [6], solutions at system integration level are unexplored.

Therefore the question now becomes how to fill both of the gaps in one system design. Specifically, we try to move these two research topics to the next step by building a flexible ECC coprocessor architecture with the ability to consider efficiency and security simultaneously and provide a unified countermeasure which can be easily integrated into system designs.

The contributions of this research are three-fold. First, we propose a generic programmable and parallel ECC coprocessor architecture. The architecture is scalable and can be adapted to different bus interfaces. Since it is programmable, both of the efficient ECC scalar multiplication algorithms and algorithmic level countermeasures can be uploaded to the coprocessor without hardware modifications. Second, after review of the security risks for ECC implementations, we suggest a set of countermeasure to protect the the coprocessor against different types of passive attacks and fault injection attacks. Finally, we implement a programmable and parallel ECC coprocessor on an FPGA to show the feasibility of the method. The implementation is scalable over area, cost, and security. The resulting platform allows us to quantify and compare the performance overhead of various algorithmic-level countermeasures.

The remainder of this paper is as follows. Section 2 gives a brief description of ECC implementation and related attacks with corresponding countermeasures. Our proposed generic programmable and parallel coprocessor architecture will be . discussed in section 3. The unified countermeasure is analyzed in section 4. The

FPGA implementation of our proposed architecture with unified countermeasure is described in section 5. Section 6 concludes the paper.

2 ECC Background

2.1 Implementation of ECC over $GF(2^m)$

A non-supersingular elliptic curve E over $GF(2^m)$ is defined as the set of solutions $(x, y) \in GF(2^m) \times GF(2^m)$ of the equation:

$$E: \ y^2 + xy = x^3 + ax^2 + b \ , \tag{1}$$

where $a, b \in GF(2^m)$, $b \neq 0$, together with the point at infinity.

A basic building block of ECC is the Elliptic Curve Scalar Multiplication (ECSM), an operation of the form $K \cdot P$ where K is an integer and P is a point on an elliptic curve. A scalar multiplication can be realized through a sequence of point additions and doublings (see Fig.1). This operation dominates the execution time of cryptographic schemes based on ECC.

2.2 Types of Attacks

Several kinds of attacks on cryptographic devices have been published. They can be categorized into two types: passive attacks and active attacks [15]. Passive attacks are based on the observation of side-channel information such as the power consumption of the chip or its electromagnetic emanations. Examples of passive attacks include Simple Power Analysis (SPA), Differential Power Analysis (DPA), Simple Electromagnetic Analysis (SEMA) and Differential Electromagnetic Analysis (DEMA). On the other hand, active attacks, including fault injection attacks, deliberately introduce abnormal behavior in the chip in order to recover internal secret data.

As mentioned earlier, a cryptosystem will fail at its weakest link and one countermeasure against one SCA attack may benefit another attack. Therefore, in this paper we want to consider active as well as passive attacks, and define a unified countermeasure that resists a collection of published attacks. Before

ECC Scalar Multiplication (double-and-add-always)	ECC Scalar Multiplication (Montgomery Ladder)

Input: P, $K=\{k_{n-1},..,k_0\}$
 Output: $Q=K\bullet P$
 1: $Q[0]\leftarrow P$
 2: **for** $i=n-2$ to 0 **do**
 3:     $Q[0] \leftarrow 2Q[0]$
         $Q[1] \leftarrow Q[0] + P$
         $Q[0] \leftarrow Q[k_i]$
 Return $Q[0]$
	Input: P, $K=\{k_{n-1},..,k_0\}$
	 Output: $Q = K\bullet P$
	 1: $Q[0]\leftarrow P$, $Q[1]\leftarrow 2P$;
	 2: **for** $i=n-2$ to 0 **do**
	 3: $Q[1-k_i] \leftarrow Q[0] + Q[1]$
	         $Q[k_i] \ \ \ \leftarrow 2Q[k_i]$
	
	 Return $Q[0]$

Fig. 1. Elliptic Curve Scalar Multiplication (ECSM) algorithms

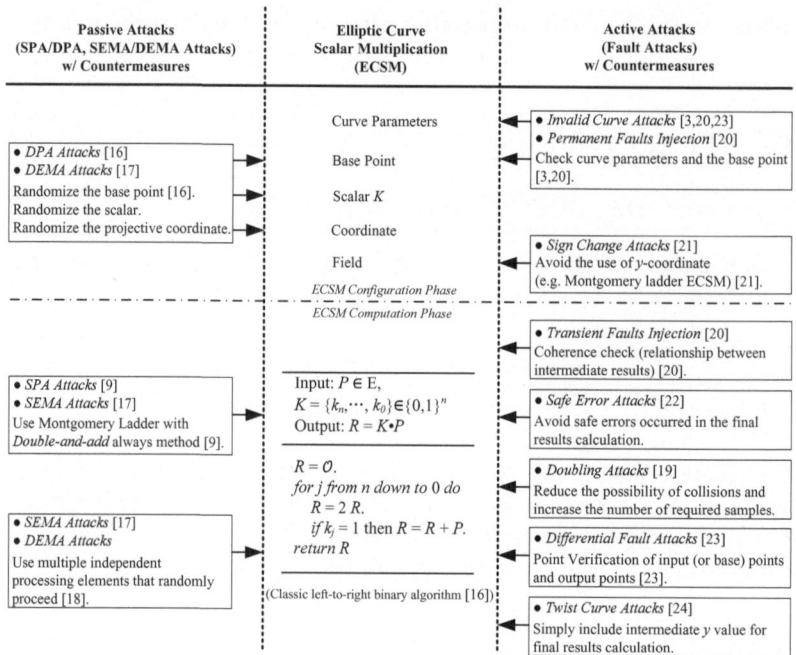

Fig. 2. Summary of attacks and corresponding countermeasures on ECC

proposing our unified countermeasure we first review the known security risks of ECC, as well as the corresponding countermeasures. A brief discussion of each attack and corresponding countermeasure will be provided in section 4.2.

As shown in Fig. 2, we divide all the countermeasures into two categories: protection at the ECSM configuration phase and computation phase. Most fault injection attacks are specific to the ECC algorithm and most of them are combined with passive attacks. Besides, some of the active attacks are very powerful. Even if countermeasures of standard passive attack are used, attackers can still easily retrieve the secret scalar K with only a few power traces (e.g. two power traces for the doubling attacks [19]).

3   Proposed Programmable and Parallel Coprocessor Architecture

Integration of various countermeasures into an existing ECC coprocessors can be very challenging. First, for many proposed ECC coprocessors with single datapath, the added countermeasures will sometimes largely sacrifice its efficiency. Second, the research of side-channel attacks keeps on evolving. Thus, how to devise a flexible ECC coprocessor which can support security updates is also very important. Therefore, a novel generic ECC coprocessor architecture design is proposed to solve the above problems. The architecture of this coprocessor is shown in Fig. 3 and all the design considerations will be discussed below.

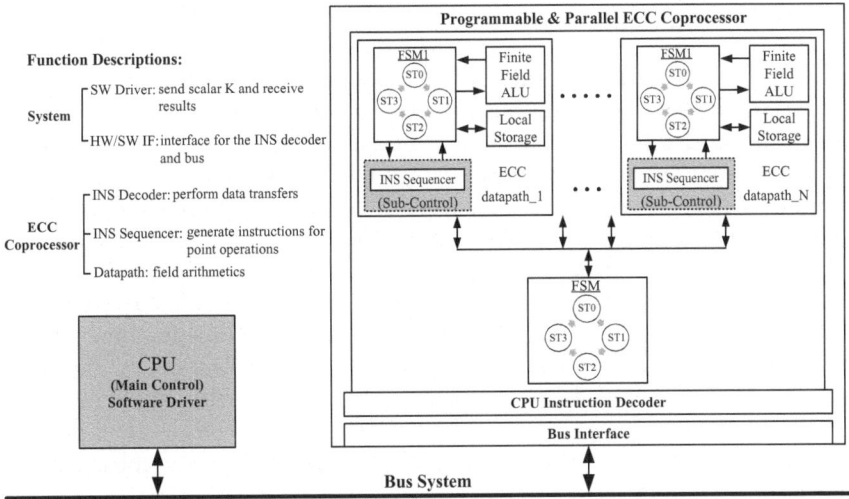

Fig. 3. The structure of generic programmable and parallel ECC coprocessor

The hardware/software partitioning method adopted in this design is trying to offload the field arithmetic operations from the CPU and execute them in a dedicated ECC coprocessor [13,14]. For traditional ECC coprocessor designs, all other higher level point operations, such as point addition/doubling, are implemented in software running on CPU. However, this partitioning may result in a HW/SW communication bottleneck since the lower-level field multiplication function will always be called by upper-level point operations, including a large amount of instruction and data transfers.

Targeting the above communication bottleneck problem we tried to optimize the HW/SW boundary in two steps: reducing data transfers as well as accelerating instruction transfers. As a result, the CPU is only in charge of sending initialization parameters and receiving final results, and the instruction sequencer will issue all the required instructions for a full ECSM. A further optimization has been made to make the ECC coprocessor programmable, which is out of two concerns. First, in general the field operations can already be very fast (a digit-serial multiplier with digit size of 82 can finish one multiplication in $GF(2^{163})$ in 2 clock cycles [11]) and big performance gain of the whole ECC system can only be obtained if new point operation algorithms are proposed. In this case, by fixing the lowest level field operations in hardware, updating an ECC system is just replacing the software assembly codes in the instruction sequencer with the new point operation algorithms without the need to rewrite the HDLs and synthesize the whole design. Second, this method can also enable the integration of the latest countermeasures against side-channel attacks for security updates.

3.1   Programmability

For our proposed structure, the CPU (main control) is not only able to send data/instructions through the bus, like the controller in most of the HW/SW codesigns, but also to program the instruction sequencer as a sub-controller in the coprocessor. The coprocessor consists of a CPU instruction decoder and single/multiple ECC datapaths, and each ECC datapath is composed of an instruction sequencer, a dedicated instruction decoder, ALU and local memory. Each ECC datapath can be programmed to carry out field operations independently.

However, the design of an instruction sequencer in the ECC datapath can be tricky. Since we have defined it to support the programmable feature, the direct use of hardware FSMs does not work. Another option is using a microcoded controller. However, the design of a dedicated controller with FSMs to dispatch instructions from microcoded controller itself can still be complex and inflexible. Finally, we come to a solution by customizing an existing low-end microcontroller to meet our requirements.

This programmable architecture gives us the freedom to efficiently utilize various countermeasures against different side-channel attacks. For example, we can program the sub-controller component so that it performs Montgomery ladder in order to thwart SPA attacks. We can easily add base point randomization to it in order to thwart DPA attacks. Finally, if the implementation requires resistance to fault attacks, we can update the program in the sub-controller to add coherence check [3] and so on. In short, the flexibility of programmable sub-controller makes the coprocessor able to update with the latest countermeasures.

3.2   Scalability

As shown in Fig.3, our proposed generic coprocessor architecture is scalable for parallel implementations because the ECC datapath can execute the scalar multiplication almost independent of the bus selections and CPU. Once the CPU sends the scalar K to each ECC datapath to initialize the computation, the datapath will work independently. The scalability here means the maximum number of independent ECC datapaths attached to the CPU instruction decoder is purely dependent on the bus latency. As shown in Fig.4, the CPU can control one ECC coprocessor with N datapaths, and N point multiplications can be performed at the same time.

According to the iteration structure shown in Fig. 4, we can derive an equation to express the relation between the maximum number of parallel ECC datapaths and bus latency. The basic idea is to overlap the communication time with the computation time. Here, we assume the bus latency is T_{delay} cycles per transfer, and scalar K and results (X, Y) each needs the same M times bus transfers (including both instruction and data transfers), and the ECSM on one ECC datapath requires T_{comp} cycles to complete, so the effective maximum number, N_{max}, of parallel ECC datapath can be expressed as

$$N_{max} = (T_{comp}/MT_{delay}) + 1. \tag{2}$$

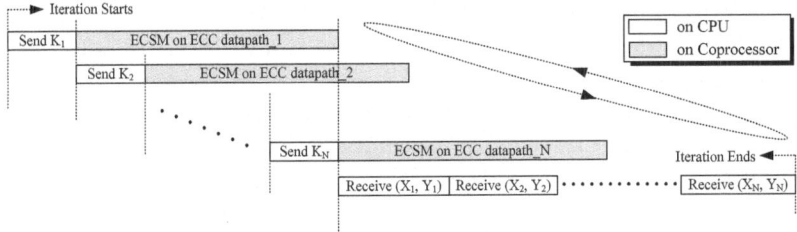

Fig. 4. Exploration of the parallelism within the proposed generic coprocessor architecture

Due to this massive parallel architecture, we can get the fastest implementation with $T_{avg\_min} = 3MT_{delay}$ cycles. This means for the fastest ECC coprocessor configuration with maximum number of parallel ECC datapath, the minimum computation time in average is only related to the bus latency. The system configuration for meeting this upper bound is to make the ECSM computation time be exactly overlapped with the communication time. For actual operations shown in Fig. 4, let the CPU keep sending scalar K and initiating ECSM until the first ECSM computation ends and then start to receive results. Also, we can have tradeoff designs between area and speed with different number of parallel ECC datapath to fit for different embedded applications, and then we can get the computation time in average, T_{avg} as

$$T_{avg} = \frac{(2N+1)MT_{delay} + T_{comp}}{N}. \tag{3}$$

4   Selection of Countermeasures

Indeed, one can not simply integrate all the countermeasures targeting different attacks, as shown in Fig. 2, to thwart multiple attacks due to complexity, cost and the fact that a countermeasure against one attack may benefit another one [5]. For example, the double-and-add-always method to prevent SPA can be used for Safe Error attacks [22]. Unified countermeasures to tackle both the passive and active attacks are attractive. Kim *et al.* proposed a unified countermeasure for RSA-CRT [25]. Baek *et al.* extended Shamir's trick, which was proposed for RSA-CRT, to secure ECC from DPA and FA [6]. However, Joye showed in [26] non-negligible portion of faults was undetected with the unified countermeasure and settings in [6].

In this paper, we suggest a set of existing countermeasures to thwart both passive and active attacks on ECC. Especially, we focus on ECC over binary extension field. We try to take into account as many attacks/countermeasures as possible. Three countermeasures are selected.

1. *Montgomery Ladder Scalar Multiplication* [9]. The Montgomery Powering Ladder, shown in Fig. 1, performs both point addition and poing doubling for each key bit. In theory, it is able to prevent SPA and TA.

2. **Random Splitting of Scalar** [27]. This method randomly splits $K = K_1 + K_2$, and performs $Q_1 = K_1 \cdot P$ and $Q_2 = K_2 \cdot P$. Eventually, $K \cdot P$ is calculated as $Q_1 + Q_2$.

3. **Point Verification** [3,23]. PV checks if a point lies on an curve or not. Let E: $y^2 + xy = x^3 + ax^2 + b$ be an elliptic curve defined over $GF(2^m)$, one can check the validity of a point $P(x_p, y_p)$ by simply verifying the equation $y_p^2 + x_p y_p = x_p^3 + ax_p^2 + b$.

4.1   Security Analysis of the Proposed Unified Countermeasure

SPA/SEMA attacks [9,17] use a single measurement to guess the secret bits. The use of Montgomery Scalar Multiplication computes point addition and doubling each time without depending on the value of each bit of scalar K. Therefore, it is secure against SPA attacks. For SEMA, though in [28] the authors indicated that this can also resist SEMA, we think it is too early to conclude that since in [17] the capability of multi-channel EMA attacks has not been comprehensively investigated.

DPA/DEMA attacks [16,17] use statistical models to analyze multiple measurements. The random scalar splitting was proposed in [27] and the idea behind it is very similar to Coron's [16] first countermeasure against DPA. As for DEMA, the random splitting of K can be considered as a signal information reduction countermeasure [17] against statistical attacks, including DEMA.

Doubling attacks [19] explore the operand reuse in scalar multiplication for two ECSMs with different base point. Borrowing the authors' analysis on Coron's first countermeasure in [19], the random splitting of 163 bit scalar K can simply extend the search space to 2^{81} (birthday paradox applies here), which is enough to resist the doubling attack.

Safe Error attacks [22] introduce safe errors when redundant point operations are used (e.g. double-and-add-always ECSM). No safe errors can be introduced based on our proposed scheme. In order to pass the PV, the outputs from the coordinate conversion must be correct, which means both intermediate results P_1 and P_2 must be used in order to calculate the y value at the last step of coordinate conversion. So, no safe errors can be introduced in either P_1 or P_2.

Invalid Curve attacks [20,23] need to compute the ECSM on a cryptographically weaker group. When using López-Dahab coorindates and Montgomery Scalar Multiplication, all the curve parameters will be used for calculating the final results in affine coordinates. So, if any of the curve parameters is changed in order to force the computation on a weaker elliptic curve, it cannot pass the final point verification.

Differential Fault attacks [23] try to make the point leave the cryptographically strong curve. The use of PV at the final step is the standard countermeasure to resist this kind of attacks.

Permanent Faults Injection attacks [20] target the non-volatile memory for storing the system parameters. All the curve parameters and base points in our design can be hardwired to prevent malicious modifications. If needed, the curve parameters can also be made flexible as well, but then their integrity will need to be verified. This can be done, for example, using a hash operation.

Transient Faults Injection attacks [20] try to modify the system parameters when they are transferred into working memory. By using the random scalar splitting scheme ($K = K_1 + K_2$) the final results have to be obtained through two steps of point operations. For the first step, we use López-Dahab projective coordinates and Montgomery Scalar Multiplication to get two results for $K_1 \cdot P$ and $K_2 \cdot P$. If transient faults are inserted during this step, before coordinate conversion the invariant relation of intermediate point $Q[0]$ and $Q[1]$ (see Fig. 1) will be destroyed. As a result, $K \cdot P = (K_1 \cdot P + K_2 \cdot P)$ cannot pass the final PV since the errors will propagate in the affine point addition.

Twist Curve attacks [24] apply to the case when performing Montgomery Scalar Multiplication, the y-coordinate is not used. However, for our case, to obtain the final results both x and y are needed for final affine point addition.

Sign Change attacks [21] change the sign of a point when the scalar is encoded in Non-adjacent form (NAF). The point sign change implies only a change of sign of its y-coordinate. The use of Montgomery Scalar Multiplication and López-Dahab coordinates can resist this attack.

4.2 Map the Unified Countermeasure to the Coprocessor

In Fig. 5, the dataflow of the ECSM using our proposed unified countermeasures is illustrated. After randomly splitting the scalar K into K_1 and K_2 on the CPU,

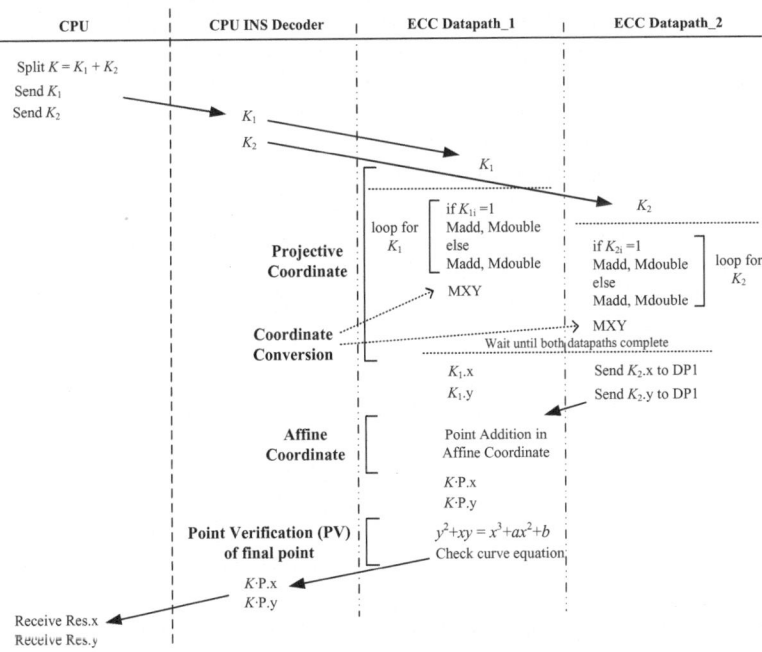

Fig. 5. Dataflow of ECSM with splitting K and PV countermeasures

K_1 and K_2 will be sent to the ECC coprocessor. Then, the calculation of K_1 · P and K_2 · P are processed concurrently, and after that the two intermediate resulting points in Affine Coordinate from both datapaths will be added, generating K · P. The point verification will be performed before the output is generated.

4.3   Operation Modes

In order to support the countermeasures suggested above, it is obvious that two point multiplications have to be computed with an extra final affine point addition. Both introduces performance overhead. Therefore, we can make the ECC coprocessor work under different operation modes. Below, four modes have been defined with different security requirements:

1. *Normal Operation Mode.* Two ECC datapaths can calculate different K and return two separate point multiplication results. Note that this normal mode can also implement the basic duplication [31] or concurrent processing comparison [3] to detect faults by sending the same K to the two datapaths.

2. *Safe Mode with the splitting K countermeasure.* Two ECC datapaths compute the split K values in projective coordinate and then perform one affine point addition in the end. PV is not used here.

3. *Safe Mode with the PV countermeasure.* Based on the normal operation mode, we can add the PV to both datapaths before outputting the final results.

4. *Safe Mode with both splitting K and PV countermeasures.* This is the operation mode with the highest security level described in Fig. 5.

Combining the aforementioned programmable feature with the above defined four operation modes, we can customize a *protocol* of how to efficiently select modes to reduce the performance overhead.

5   FPGA Implementation

In order to prove the feasibility of the proposed generic ECC coprocessor architecture and give a concrete example with quantitative experimental results, we have also built the generic coprocessor on an FPGA based SoC platform.

When comparing the actual implementation in Fig. 6 with the generic architecture in Fig. 3, the CPU becomes the 32-bit MicroBlaze, PLB bus is used, the instruction sequencer is replaced with programmable Dual-PicoBlaze microcontrollers, the finite field ALU is implemented with addition, multiplication and square function units, and the 163-bit register array is used as local storage.

There are many design options for ECC implementations, and different approaches differ in the selection of coordinate system, field and type of curves [8]. In our design we will use Montgomery Scalar Multiplication and López-Dahab projective coordinates [9]. For hardware implementations of the lowest level field arithmetic, the field multiplication is implemented both as bit-serial [10] and digit-serial multipliers [11] with different digit sizes; the field addition is simply logic XORs; the field square is implemented by dedicated hardware with

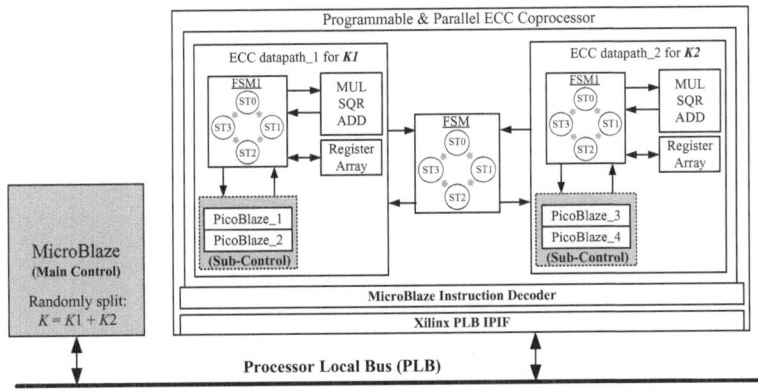

Fig. 6. Block diagram of the proposed ECC SoC architecture on FPGA

square and reduction circuits [12]; the field inversion consists of a sequence of field multiplications and squares based on Fermat's Theorem [12].

For the implementation of the instruction sequencer, 8-bit PicoBlaze microcontrollers are used. The PicoBlaze has a predictable performance. It takes always two clock cycles per instruction. It costs 53 slices and 1 block RAM on the Virtex-5 XC5VLX50 FPGA. The use of PicoBlaze as a new control hierarchy was first proposed in [29] and based on that we in [30] proposed a Dual-PicoBlaze based design to achieve a high instruction rate of 1 instruction per cycle by interleaving instructions from two PicoBlazes. However, by applying the Dual-Picoblaze architecture we can only get performance enhancement. Therefore, we decided to customize the Picoblaze by replacing the instruction ROM with a dual-port RAM to incorporate the programmable feature.

Based on the discussion on Section 3.2, the timing profile of parallel implementations of multiple ECC datapaths based on the Processor Local Bus (PLB) on a Xilinx FPGA platform can be obtained. The PLB interface is a memory-mapped interface for peripheral components with typical bus transfer latency, T_{delay}, of 9 clock cycles. Instruction and data transfers for one operand require $M = 20$ times bus transfers. As we use digit size 82 (denoted as $D82$), one scalar multiplication on ECC defined on $GF(2^{163})$ takes 24,689 clock cycles, i.e. $T_{comp} = 24,689$. If we just consider the ideal case for data/instruction transfers without any software overhead (e.g. function calls), the minimum average time, T_{avg_min}, is about of 540 clock cycles. However, 139 ECC datapaths in parallel are required to achieve this minimum delay, which makes it impractical. Still, we can find reasonable tradeoffs through Equation 3.

Following the discussion in Section 4.2, we also implement a ECC coprocessor with two datapath to integrate our proposed unified countermeasure. It is shown in Fig. 6. Moreover, we designed the coprocessor to support different operation modes as defined in section 4.3.

Since the PLB bus is 32-bit wide and our targeting implementation is based on $GF(2^{163})$, we need 6 times bus transfers to send one scalar K to the coprocessor, which means we still have 29 bits left when transfer the last word. Therefore, we

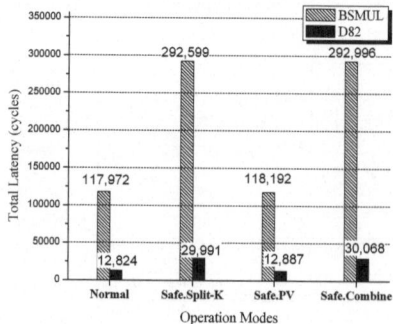

Fig. 7. Comparison of timing profiling between different operation modes

encode these bits as function bits and they can be interpreted by the MicroBlaze Instruction Decoder to make the coprocessor work under different modes.

Another advantage of this proposed architecture is that we can quantify the security cost when different countermeasures are used. Without changing the hardware, we can simply make changes in the software driver running on MicroBlaze to turn the ECC coprocessor into different operation modes. This is similar with the workload characterization in software-only implementations [35]. Hence, the security cost can be expressed in terms of pure performance overhead.

In Fig. 7, we compare the timing profiling of the design working in four operation modes. Here, we select two extreme coprocessor configurations, the smallest (with bit-serial multiplier, BSMUL) and the fastest ($D82$), for detailed comparison. The whole system works at 100MHz, and the maximum frequency of both coprocessors with BSMUL and $D82$ is 165.8MHz. The logic critical path mainly goes through the programmable PicoBlaze. The hardware cost for the bit-serial multiplier based design is 2,179 slices and 6,585 slices for the $D82$-based design when implemented on a Vertex-5 XC5VLX50 FPGA.

Note that the cycle counts for normal mode and safe mode with only PV are the average speed for two ECSMs in parallel. From [31], the authors conclude that if fair performance evaluations are performed, many fault attack countermeasures are not better than the naive solutions, namely duplication or repetition. So, this means it is also important to find a universal platform to quantify the security cost. From Fig. 7, it is easy to see that based on our proposed generic ECC coprocessor architecture we can quantify the overhead of these countermeasures in terms of a single metric – cycle counts.

To compare with other similar ECC codesigns [13,14,29,30,36,37], our proposed ECC coprocessor architecture considers optimizations for performance, flexibility and security at the same time. For performance, the designs described in [29,30], same as the base design of the ECC coprocessor architecture proposed in this paper with single ECC datapath, have already shown good tradeoffs between area and speed. For flexibility, the programmable coprocessor is addressed for its advantages of both performance and security enhancement, and the massively parallel architecture can be explored to meet various application requirements. For security, unlike [14,36,37] which only consider passive attacks (e.g. power analysis attacks and timing attacks), our design can defend most existing passive and active attacks. Besides, it can be easily updated with new algorithmic-level countermeasures to resist new attacks without hardware changes.

6   Conclusions

In this paper we have presented a generic ECC coprocessor architecture, which can fill the gap between efficient implementation and security integration at the architecture level. For security, a unified countermeasure is proposed by combining the random scalar splitting [27] and Point Verification [3,23]. For performance, the introduction of distributed storage and new control hierarchy into the ECC coprocessor datapath can greatly reduce the communication overhead faced by a traditional centralized control scheme. Scalable parallelism can also be explored to achieve tradeoff designs between area and speed. The feasibility and efficiency of our proposed generic ECC coprocessor architecture and unified countermeasure are verified and shown from an actual implementation on FPGA. Experimental results show that the proposed programmable and parallel ECC coprocessor architecture can be suitable for a wide range of embedded applications with different user defined security requriments.

Acknowledgments. This project was supported in part by the US National Science Foundation through grant 0644070 and 0541472, by Virginia Tech Pratt Fund, by IAP Programme P6/26 BCRYPT of the Belgian State (Belgian Science Policy), by FWO projects G.0300.07, by the European Comission through the IST Programme under Contract IST-2002-507932 ECRYPT NoE, and by the K.U. Leuven-BOF.

References

1. Alrimeih, H., Rakhmatov, D.: Security-Performance Trade-offs in Embedded Systems Using Flexible ECC Hardware. IEEE Design & Test of Computers 24(6), 556–569 (2007)
2. Kocher, C., Jaffe, J., Jun, B.: Differential power analysis. In: Wiener, M. (ed.) CRYPTO 1999. LNCS, vol. 1666, pp. 388–397. Springer, Heidelberg (1999)
3. Dominguez-Oviedo, A.: On Fault-based Attacks and Countermeasures for Elliptic Curve Cryptosystems. PhD Thesis, University of Waterloo (2008)
4. Verbauwhede, I., Schaumont, P.: Design Methods for Security and Trust. In: Proceedings of the conference on Design, automation and test in Europe –DATE 2007, pp. 1–6 (2007)
5. Yen, S.-M., Kim, S., Lim, S., Moon, S.-J.: A Countermeasure against One Physical Cryptanalysis May Benefit Another Attack. In: Kim, K.-c. (ed.) ICISC 2001. LNCS, vol. 2288, pp. 414–427. Springer, Heidelberg (2002)
6. Baek, Y.-J., Vasyltsov, I.: How to prevent DPA and fault attack in a unified way for ECC scalar multiplication – ring extension method. In: Dawson, E., Wong, D.S. (eds.) ISPEC 2007. LNCS, vol. 4464, pp. 225–237. Springer, Heidelberg (2007)

7. Schaumont, P., Hwang, D., Yang, S., Verbauwhede, I.: Multilevel Design Validation in a Secure Embedded System. IEEE Transactions on Computers 55(11), 1380–1390 (2006)
8. Hankerson, D., Menezes, A., Vanstone, S.: Guide to Elliptic Curve Cryptography. Springer, Heidelberg (2004)
9. López, J., Dahab, R.: Fast multiplication on elliptic curves over GF(2^m). In: Koç, Ç.K., Paar, C. (eds.) CHES 1999. LNCS, vol. 1717, pp. 316–327. Springer, Heidelberg (1999)
10. Großschädl, J.: A low-power bit-serial multiplier for finite fields GF(2^m). In: ISCAS 2001, vol. IV, pp. 37–40. IEEE, Los Alamitos (2001)
11. Kumar, S., Wollinger, T., Paar, C.: Optimum Digit Serial GF(2^m) Multipliers for Curve-Based Cryptography. IEEE Transactions on Computers 55(10), 1306–1311 (2006)
12. Rodríguez-Henríquez, F., Saqib, N.A., Díaz-Pérez, A., Koç, Ç.K.: Cryptographic Algorithms on Reconfigurable Hardware. Springer, Heidelberg (2006)
13. Koschuch, M., Lechner, J., Weitzer, A., Großschädl, J., Szekely, A., Tillich, S., Wolkerstorfer, J.: Hardware/Software co-design of elliptic curve cryptography on an 8051 microcontroller. In: Goubin, L., Matsui, M. (eds.) CHES 2006. LNCS, vol. 4249, pp. 430–444. Springer, Heidelberg (2006)
14. Sakiyama, K., Batina, L., Preneel, B., Verbauwhede, I.: Superscalar Coprocessor for High-Speed Curve-Based Cryptography. In: Goubin, L., Matsui, M. (eds.) CHES 2006. LNCS, vol. 4249, pp. 415–429. Springer, Heidelberg (2006)
15. Amiel, F., Villegas, K., Feix, B., Marcel, L.: Passive and Active Combined Attacks: Combining Fault Attacks and Side Channel Analysis. In: FDTC 2007, pp. 92–102. IEEE, Los Alamitos (2007)
16. Coron, J.-S.: Resistance against differential power analysis for elliptic curve. In: Koç, Ç.K., Paar, C. (eds.) CHES 1999. LNCS, vol. 1717, pp. 292–302. Springer, Heidelberg (1999)
17. Agrawal, D., Archambeault, B., Rao, J.R., Rohatgi, P.: The EM side-channel(s). In: Kaliski Jr., B.S., Koç, Ç.K., Paar, C. (eds.) CHES 2002. LNCS, vol. 2523, pp. 29–45. Springer, Heidelberg (2003)
18. Ciet, M., Neve, M., Peeters, E., Quisquater, J.: Parallel FPGA implementation of RSA with residue number systems - can side-channel threats be avoided? In: IEEE International Symposium on Micro-NanoMechatronics and Human Science, vol. 2, pp. 806–810. IEEE Computer Society Press, Los Alamitos (2003)
19. Fouque, P.-A., Valette, F.: The Doubling Attack - Why Upwards Is Better than Downwards. In: Walter, C.D., Koç, Ç.K., Paar, C. (eds.) CHES 2003. LNCS, vol. 2779, pp. 269–280. Springer, Heidelberg (2003)
20. Ciet, M., Joye, M.: Elliptic Curve Cryptosystems in the Presence of Permanent and Transient Faults. Design, Codes and Cryptography 36, 33–43 (2005)
21. Blömer, J., Otto, M., Seifert, J.-P.: Sign change fault attacks on elliptic curve cryptosystems. In: Breveglieri, L., Koren, I., Naccache, D., Seifert, J.-P. (eds.) FDTC 2006. LNCS, vol. 4236, pp. 36–52. Springer, Heidelberg (2006)
22. Yen, S.-M., Joye, M.: Checking before output not be enough against fault-based cryptanalysis. IEEE Trans. on Computers 49(9), 967–970 (2000)
23. Biehl, I., Meyer, B., Müller, V.: Differential Fault Attacks on Elliptic Curve Cryptosystems. In: Bellare, M. (ed.) CRYPTO 2000. LNCS, vol. 1880, pp. 131–146. Springer, Heidelberg (2000)
24. Fouque, P.-A., Lercier, R., Real, D., Valette, F.: Fault Attack on Elliptic Curve with Montgomery Ladder Implementation. In: FDTC2008, pp. 92–98. IEEE, Los Alamitos (2008)

25. Kim, C.H., Quisquater, J.-J.: How can we overcome both side channel analysis and fault attacks on RSA-CRT? In: FDTC 2007, pp. 21–29. IEEE, Los Alamitos (2007)
26. Joye, M.: On the Security of a Unified Countermeasure. In: FDTC 2008, pp. 87–91. IEEE, Los Alamitos (2008)
27. Joye, M., Ciet, M. (Virtually) Free Randomization Techniques for Elliptic Curve Cryptography. In: Qing, S., Gollmann, D., Zhou, J. (eds.) ICICS 2003. LNCS, vol. 2836, pp. 348–359. Springer, Heidelberg (2003)
28. De Mulder, E., Ors, S.B., Preneel, B., Verbauwhede, I.: Electromagnetic Analysis Attack on an FPGA Implementation of an Elliptic Curve Cryptosystem. In: EUROCON 2005, vol. 2, pp. 1879–1882. IEEE, Los Alamitos (2005)
29. Guo, X., Schaumont, P.: Optimizing the HW/SW Boundary of an ECC SoC Design Using Control Hierarchy and Distributed Storage. In: DATE 2009, pp. 454–459. EDAA (2009)
30. Guo, X., Schaumont, P.: Optimizing the Control Hierarchy of an ECC Coprocessor Design on an FPGA based SoC Platform. In: Becker, J., Woods, R., Athanas, P., Morgan, F. (eds.) ARC 2009. LNCS, vol. 5453, pp. 169–180. Springer, Heidelberg (2009)
31. Malkin, T.G., Standaert, F.-X., Yung, M.: A Comparative Cost/Security Analysis of Fault Attack Countermeasures. In: Breveglieri, L., Koren, I., Naccache, D., Seifert, J.-P. (eds.) FDTC 2006. LNCS, vol. 4236, pp. 159–172. Springer, Heidelberg (2006)
32. Hwang, D., Tiri, K., Hodjat, A., Lai, B.C., Yang, S., Schaumont, P., Verbauwhede, I.: AES-Based Security Coprocessor IC in 0.18um CMOS with resistance to differential power analysis side-channel attacks. IEEE Journal of Solid-State Circuits 41(4), 781–791 (2006)
33. Chen, Z., Zhou, Y.: Dual-Rail Random Switching Logic: A Countermeasure to Reduce Side-Channel Leakage. In: Goubin, L., Matsui, M. (eds.) CHES 2006. LNCS, vol. 4249, pp. 242–254. Springer, Heidelberg (2006)
34. Giraud, C.: An RSA Implementation Resistant to Fault Attacks and to Simple Power Analysis. IEEE Trans. on Computers 55(9), 1116–1120 (2006)
35. Koschuch, M., Großschädl, J., Payer, U., Hudler, M., Krüger, M.: Workload Characterization of a Lightweight SSL Implementation Resistant to Side-Channel Attacks. In: Franklin, M.K., Hui, L.C.K., Wong, D.S. (eds.) CANS 2008. LNCS, vol. 5339, pp. 349–365. Springer, Heidelberg (2008)
36. Sakiyama, K., Batina, L., Schaumont, P., Verbauwhede, I.: HW/SW Co-design for TA/SPA-resistant Public-Key Cryptosystems. In: ECRYPT Workshop on Cryptographic Advances in Secure Hardware (2005)
37. Batina, L., Mentens, N., Preneel, B., Verbauwhede, I.: Balanced point operations for side-channel protection of elliptic curve cryptography. IEE Proceedings of Information Security 152(1), 57–65 (2005)

Elliptic Curve Scalar Multiplication Combining Yao's Algorithm and Double Bases

Nicolas Méloni and M. Anwar Hasan

Department of Electrical and Computer Engineering
University of Waterloo

Abstract. In this paper we propose to take one step back in the use of double base number systems for elliptic curve point scalar multiplication. Using a modified version of Yao's algorithm, we go back from the popular double base chain representation to a more general double base system. Instead of representing an integer k as $\sum_{i=1}^{n} 2^{b_i} 3^{t_i}$ where (b_i) and (t_i) are two decreasing sequences, we only set a maximum value for both of them. Then, we analyze the efficiency of our new method using different bases and optimal parameters. In particular, we propose for the first time a binary/Zeckendorf representation for integers, providing interesting results. Finally, we provide a comprehensive comparison to state-of-the-art methods, including a large variety of curve shapes and latest point addition formulae speed-ups.

Keywords: Double-base number system, Zeckendorf representation, elliptic curve, point scalar multiplication, Yao's algorithm.

1 Introduction

In order to compute elliptic curve point multiplication, that is to say kP where P is a point on an elliptic curve, defined over a prime field, and k is an integer, a lot of effort has been made to adapt and optimize generic exponentiation methods (such as Non-adjacent form (NAF), window NAF and fractional window NAF). In 1995, Dimitrov and Cooklev [8] have introduced the use the double base number system (DBNS) to improve modular exponentiation speed. The idea is to represent k as a sum of terms of the form $c_i 2^{b_i} 3^{t_i}$ with $c_i = 1$ or -1. The main advantage of this representation is the fewer number of terms it requires. A very interesting case is when the base element x is fixed, so that one can precompute all the $x^{2^{b_i} 3^{t_i}} \mod p$. The DBNS seems to be not that efficient in the case of a randomly chosen element. In order to overcome this problem and adapt the DBNS to elliptic curve point multiplication, Dimitrov, Imbert and Mishra have introduced the concept of double base chains, where the integer k is still represented as a sum of $c_i 2^{b_i} 3^{t_i}$ but with the restriction that (b_i) and (t_i) must be two decreasing sequences [9]. The restriction causes the number of terms to increase, but allows to perform the scalar multiplication using a Horner like scheme. Allowing c_i to belong to a larger set than $\{-1, 1\}$ as well as choosing optimal parameters based on the ratio of the number of doublings to that of triplings also helped to achieve better results.

C. Clavier and K. Gaj (Eds.): CHES 2009, LNCS 5747, pp. 304–316, 2009.

The original double base representation has probably not been utilized as much as it should have been for developing improved exponentiation algorithms. To the end, our contribution is to show that the use of a modified version of Yao's algorithm allows to partly overcome the drawbacks of the DBNS. By imposing a maximum bound on b_i's and t_i's, that is clearly less restrictive than the double base chain condition, we show that our method provides significant improvement even when compared to the most recently optimized double base methods. Moreover, we introduce a binary/Zeckendorf method which, on the classical Weierstrass curve, provides similar results.

2   Background

In this section, we give a brief review of the materials used in the paper.

2.1   Elliptic curves

Definition 1. *An elliptic curve E over a field K denoted by E/K is given by the equation*

$$E : y^2 + a_1 xy + a_3 y = x^3 + a_2 x^2 + a_4 x + a_6$$

where $a_1, a_2, a_3, a_4, a_6 \in K$ are such that, for each point (x, y) on E, the partial derivatives do not vanish simultaneously.

In this paper, we only deal with curves defined over a prime finite field ($K = \mathbb{F}_p$) of characteristic greater than 3. In this case, the equation can be simplified to

$$y^2 = x^3 + ax + b$$

where $a, b \in K$ and $4a^3 + 27b^2 \neq 0$. Points are affine points (x, y) satisfying the curve equation and a point at infinity. The set of points $E(K)$ defined over K forms an abelian group. There exist explicit formulae to compute the sum of two points that involves field inversions. When the field inversion operation is considerably costlier than a field multiplication, one usually uses a projective version of the above equation. In this case, a point is represented by three, or more, coordinates, and many such projective coordinate systems have been proposed to speed up elliptic curve group operations. For a complete overview of those coordinates, one can refer to [7,14].

Another feature of the elliptic curve group law is that it allows fast composite operations as well as different type of additions. To take full advantage of our point scalar multiplication method, and in addition to the classical addition (**ADD**) and doubling (**DBL**) operations, we consider the following operations:

- **tripling (TPL)**: point tripling
- **readdition (reADD)**: addition of a point that has been added before to another point
- **mixed addition (mADD)**: addition of a point in affine coordinate (i.e. $Z = 1$) to another point.

In addition to those coordinate systems and composite operations, many curve shapes have been proposed to improve group operation formulae. In this paper, we will consider a variety of curve shapes including:

- tripling oriented Doche-Icart-Kohel curves (3DIK) [10]
- Edwards curves (Edwards) [13,3] with inverted coordinates [4]
- Hessian curves [6,15,16]
- Extended Jacobi Quartics (ExtJQuartic) [6,12,15]
- Jacobi intersections (JacIntersect) [6,17]
- Jacobian coordinates (Jacobian) with the special case $a_4 = -3$ (Jacobian-3).

Table 1 summarize the cost of those operations on all the considered curves.

Table 1. Elliptic curve operations cost

Curve shape	DBL	TPL	ADD	reADD	mADD
3DIK	2M+7S	6M+6S	11M+6S	10M+6S	7M+4S
Edwards	3M+4S	9M+4S	10M+1S	10M+1S	9M+1S
ExtJQuartic	2M+5S	8M+4S	7M+4S	7M+3S	6M+3S
Hessian	3M+6S	8M+6S	6M+6S	6M+6S	5M+6S
InvEdwards	3M+4S	9M+4S	9M+1S	9M+1S	8M+1S
JacIntersect	2M+5S	6M+10S	11M+1S	11M+1S	10M+1S
Jacobian	1M+8S	5M+10S	11M+5S	10M+4S	7M+4S
Jacobian-3	3M+5S	7M+7S	11M+5S	10M+4S	7M+4S

(1) proposed in this work.

Finally, some more optimizations can be found in [21,19] for the quintupling formulae. In section 4, we use the specific formulae from [20] using the z-coordinate trick to compute Fibonacci number point multiples. One can also refer to [2] for an extensive overview of different formulae, coordinates systems, curve shapes and their latest updates.

2.2   Double Base Number System

Let k be an integer. As mentioned earlier, one can represent k as the sum of terms of the form $c_i 2^{b_i} 3^{t_i}$, where $c_i \in \{-1, 1\}$. Such a representation always exists. In fact, this number system is quite redundant. One of the most interesting properties is that, among all the possible representations for a given integer, some of them are really sparse, that is to say that the number of non-zero terms is quite low.

To compute DBNS representation of an integer, one usually use a greedy algorithm. It consists of the following: find the closest integer of the form $2^{b_i} 3^{t_i}$ to k, subtract it from k and repeat the process with $k' = k - 2^{b_i} 3^{t_i}$ until it is equal to zero.

Performing a point scalar multiplication using this number system is relatively easy. Letting k be equal to $\sum_{i=1}^{n} c_i 2^{b_i} 3^{t_i}$, one just needs to compute $[c_i 2^{b_i} 3^{t_i}]P$ for $i = 1$ to n and then add all the points. If the number of additions is indeed quite low, in practice such a method requires too many doublings and triplings. That is why the general DBNS representation has been considered to be not suitable for point scalar multiplication.

To overcome this problem, Dimitrov, Imbert, and Mishra [9] have introduced the concept of double-base chains. In this system, k is still represented as $\sum_{i=1}^{n} c_i 2^{b_i} 3^{t_i}$, but with the restriction that (b_i) and (t_i) must be two decreasing sequences, allowing a Horner-like evaluation of kP using only b_1 doublings and t_1 triplings. Computing such a representation can be done using Algorithm 1. The main drawback of this method is that it significantly increases the number of point additions.

Algorithm 1. Computing a double-base chain computing k

Input: $k \geq 0$
Output: $k = \sum_{i=1}^{n} s_i 2^{b_i} 3^{t_i}$ with $(b_i, t_i) \searrow$
1: **while** $k \neq 0$ **do**
2:     $s = 1$
3:     Find the best default approximation of k of the form $z = 2^b 3^t$ with $b \leq b_{max}$ and
       $t \leq t_{max}$
4:     Print(s, b, t)
5:     $b_{max} = b; t_{max} = t$
6:     **if** $k < z$ **then** $s = -s$
7:     $k = |k - z|$
8: **end while**

Some improvements have been proposed by applying various modifications including the possibility for c_i to be chosen in a larger set than $\{-1, 1\}$ [11], the use of multiple bases [21], etc. One can finally refer to [1] for a view of the latest optimizations.

3   Modified Yao's Algorithm

3.1   Yao's Algorithm

Published in 1976 [22], Yao's algorithm can be seen as the right-to-left counterpart of the classical Brauer algorithm. Let $k = k_{l-1} 2^{l-1} + \cdots + k_1 2 + k_0$ with $k_i \in \{0, 1, \ldots, 2^w - 1\}$, for some w. The algorithm first computes $2^i P$ for all i lower than $l - 1$ by successive doublings. Then it computes $d(1)P, \ldots, d(2^w - 1)P$, where $d(j)$ is the sum of the 2^i such that $k_i = j$. Said differently, it mainly consists in considering the integer k as

$$1 \times \underbrace{\sum_{k_i=1} 2^i}_{d(1)} + 2 \times \underbrace{\sum_{k_i=2} 2^i}_{d(2)} + \cdots + (2^w - 1) \times \underbrace{\sum_{k_i=2^w-1} 2^i}_{d(2^w-1)}.$$

We can see that $d(1)$ is the sum of all the powers of 2 associated to digit 1, $d(2)$ is the sum of all the powers of 2 associated to digit 2 etc. Finally kP is obtained as $d(1)P + 2d(2)P + \cdots + (2^w - 1)d(2^w - 1)P$. In order to save some group operations, it is usually computed as $d(2^w - 1)P + (d(2^w - 1)P + d(2^w - 2)P) + \cdots + (d(2^w - 1)P + \cdots + d(1)P)$.

Example 1. Let $k = 314159$. We have $\text{NAF}_3(\text{k}) = 100\,0300\,1003\,0000\,5007$, $l = 19$ and $2^w - 1 = 7$. One can compute kP in the following way:

- consider k as $1 \times (2^{18} + 2^{11}) + 3 \times (2^{14} + 2^8) + 5 \times 2^3 + 7 \times 2^0$
- compute $P, 2P, 4P, \ldots 2^{18}P$
- $d(1)P = 2^{18}P + 2^{11}P$, $d(3)P = 2^{14}P + 2^8P$, $d(5)P = 2^3P$, $d(7)P = P$
- $kP = 2(d(7)P) + 2(d(7)P + d(5)P) + 2(d(7)P + d(5)P + d(3)P) + d(7)P + d(5)P + d(3)P + d(1)P = 7d(7)P + 5d(5)P + 3d(3)P + d(1)P$

In this example, we have:

$$d(1) = 100\,0000\,1000\,0000\,0000$$
$$d(3) = 000\,0100\,0001\,0000\,0000$$
$$d(5) = 000\,0000\,0000\,0000\,1000$$
$$d(7) = 000\,0000\,0000\,0000\,0001$$
$$k = 100\,0300\,1003\,0000\,5007$$
$$= 7d(7) + 5d(5) + 3d(3) + d(1)$$

3.2 Modified Yao's Algorithm

Now, we adapt the preceding algorithm in order to take advantage of the DBNS representation. To do so, let us consider k in (one of) its DBNS form: $k = 2^{b_n}3^{t_n} + \cdots + 2^{b_1}3^{t_1}$. As in Yao's original algorithm, we first compute 2^iP for all i lower than $\max(b_j)$. Then, for all j lower than $\max(t_i)$, we define $d(j)P$ as the sum of all the $2^{b_i}P$ such that $t_i = j$. Finally we have $kP = d(0)P + 3d(1)P + \ldots 3^{\max(t_i)}d(\max(t_i))P$.

Example 2. Let $k = 314159$. One of the representations of k in the DBNS is

$$2^{10}3^5 + 2^83^5 + 2^{10}3 + 2^23^2 + 3^2 + 2,$$

$\max(a_i) = 10$ and $\max(b_i) = 5$. One can compute kP in the following way:

- compute $P, 2P, 2^2P, \ldots, 2^{10}P$
- $d(0)P = 2P$, $d(1)P = 2^{10}P$, $d(2)P = 2^2P + P$, $d(5) = 2^{10}P + 2^8P$
- $kP = 3(3(3^3d(5)P+d(2)P)+d(1)P)+d(0)P = 3^5d(5)P+3^2d(2)P+3d(1)P+d(0)P$

We can see that the number of operations is $\max(b_i)$ doublings, $\max(t_i)$ triplings and $n - 1$ additions. With our modified algorithm, we obtain the same complexity as the double-base chain method. However, in our case, the numbers of doublings and triplings are independent, which means that $2^{\max(b_i)}3^{\max(t_i)}$ can be quite larger than k. It can be seen as a waste of operations, as we could expect it to just as large as k. In order to reduce this additional cost, we simply propose to use a maximum bound for both the b_i's and the t_i's so that $2^{max(b_i)}3^{max(t_i)} \sim k$.

4   Extending the Modified Yao's Algorithm

We have seen how Yao's algorithm can be adapted to the double-base number system. In this section, we generalize our approach to different number systems via an extended version of Yao's algorithm.

4.1   Generalization of Yao's Algorithm

We have seen that Yao's algorithm can be efficiently adapted to the double-base number system. We can now derive a general form of Yao's algorithm based on any number system using two sets of integers.

Let $\mathcal{A} = \{a_1, \ldots, a_r\}$ and $\mathcal{B} = \{b_1, \ldots, b_t\}$ be two sets of integers. Let k be an integer that can be written as $\sum_{i=1}^{n} a_{f(i)} b_{g(i)}$ with $f : \{1, \ldots n\} \to \{1, \ldots r\}$ and $g : \{1, \ldots n\} \to \{1, \ldots t\}$. It is possible to use a generalized version of Yao's algorithm to compute kP. To do so, we first compute the $b_i P$'s, for $i = 1 \ldots t$. Then, for $j = 1 \ldots r$, we compute $d(j)P$ as the sum of all the $b_{g(i)}P$ such that $f(i) = j$. In other terms, $d(1)P$ will be the sum of all the $b_{g(i)}P$ associated to a_1, $d(2)P$ will be the sum of all the $b_{g(i)}P$ associated to a_2 etc. Finally, $kP = a_1 d(1)P + a_2 d(2)P + \cdots + a_n d(n)P$.

It is easy to see that with a proper choice of sets, we find again the previous forms of the algorithm. The original version is associated to the sets $\mathcal{A} = \{1, 2, \ldots, 2^n\}$ and $\mathcal{B} = \{1, 3, 5, \ldots, 2^w - 1\}$ and the double-base version to $\mathcal{A} = \{1, 2, \ldots, 2^{b_{max}}\}$ $\mathcal{B} = \{1, 3, \ldots, 3^{t_{max}}\}$. We also remark that both sets can contain negative integers. As the operation $P \to -P$ is almost free on elliptic curves, we always consider signed representation in our experiments.

The aim of the following subsections is to present a different set of integers to improve the efficiency of our method.

4.2   Double-Base System Using Zeckendorf Representation

Let $(F_n)_{n \geq 0}$ be the Fibonacci sequence defined as $F_0 = 0, F_1 = 1, \forall n \geq 0, F_{n+2} = F_{n+1} + F_n$. Any integer can be represented as a finite sum of Fibonacci numbers [23]. Just like in the case of the classical double-base system, we introduce a mixed binary-Zeckendorf number system (BZNS). It simply consists of representing an integer k as $2^{b_n} F_{Z_n} + \cdots + 2^{b_1} F_{Z_1}$. Computing such a representation can be done using the same kind of greedy algorithm as with the classical DBNS.

Remark 1. The choice of such a representation is not arbitrary. It is based on the fact that on elliptic curves in Weierstraßform, the sequence $F_2 P, \ldots F_n P$ can be efficiently computed thanks to the formulae proposed in [20]. In that case, each point addition is performed faster than a doubling.

We now apply our generalized Yao's algorithm to the sets $\{F_2, \ldots, F_{Z_{max}}\}$ and $\{1, 2, \ldots, 2^{b_{max}}\}$. In this case, we first compute $F_i P$ for all i lower than Z_{max}, by consecutive additions. Then, for all j lower than $\max(b_i)$, we define $d(j)P$ as the sum of all the $F_{Z_i} P$ such that $b_i = j$. Finally we have $kP = d(0)P + 2d(1)P + \ldots 2^{\max(b_i)} d(\max(b_i))P$.

Example 3. Let $k = 314159$. One of the representations of k in the BZNS is

$$2^8 F_{16} + 2^8 F_{13} + 2^5 F_{10} + 2F_9 + 2F_5 + F_2$$

$\max(b_i) = 8$ and $\max(Z_i) = 16$. One can compute kP in the following way:

- consider k as $2^8(F_{16} + F_{13}) + 2^5 F_{10} + 2(F_9 + F_5) + F_2$
- compute $P, 2P, 3P, \ldots, F_{16}P$
- $d(0)P = F_2P, d(1)P = F_9P + F_5P, d(5)P = F_{10}P, d(8) = F_{16}P + F_{13}P$
- $kP = 2(2^4(2^3 d(8)P + d(5)P) + d(1)P) + d(0)P = 2^8 d(8)P + 2^5 d(5)P + 2d(1)P + d(0)P.$

5   ECC Implementation and Comparisons

In this section, we provide a comprehensive comparison between our different versions of Yao's algorithm and the most recent double-base chain methods.

5.1   Caching Strategies

Caching intermediate results while computing an elliptic curve group operation is one very important optimization criteria. In this subsection, we show that the use of our generalized algorithm allows some savings that cannot be done with the traditional methods. To better clarify this point, we fully detail our caching strategy for curves in Weierstraßform using jacobian coordinates with parameter $a = 3$ (Jac-3). Similar methods are applicable to all the different curve types.

Addition:
$P = (X_1, Y_1, Z_1), Q = (X_2, Y_2, Z_2)$ and $P + Q = (X_3, Y_3, Z_3)$

$$A = X_1 Z_2^2, \quad B = X_2 Z_1^2, \quad C = Y_1 Z_2^3, \quad D = Y_2 Z_1^3, E = B - A,$$
$$F = 2(D - C), G = (2E)^2, H = E \times G, I = A \times G,$$

and

$$X_3 = F^2 - H - 2I \; Y_3 = F(F - X_3) - 2CH, Z_3 = ((Z_1 + Z_2)^2 - Z_1^2 - Z_2^2)E$$

Doubling:
$2P = (X_3, Y_3, Z_3)$

$$A = X_1 Y_1^2, B = 3(X_1 - Z_1)^2 (X_1 + Z_1)^2$$

and

$$X_3 = B^2 - 8A, Y_3 = -8Y_1^4 + B(4A - X_3), Z_3 = (Y_1 + Z_1)^2 - Y_1^2 - Z_1^2.$$

One can verify that these two operations can be computed using 11M+5S and 3M+5S respectively. It has been shown that some of the intermediate results can be reused under particular circumstances. More precisely, if a point $P = (X_1, Y_1, Z_1)$ is added to any other point, it is possible to store the data Z_1^2 and Z_1^3. During the same scalar multiplication, if the point P is added again to another point, reusing those stored values saves 1M+1S. This is what is usually called a readdition and its cost is 10M+4S instead of 11M+5S. With mixed and, of course, the general addition (one of the added points has its z-coordinate equal to 1), this is the only kind of point additions that can occur in all the traditional scalar multiplication methods.

Our new method allows more variety in caching strategies and point addition situations. From the doubling formulae, we can see that if we store Z_1^2 after the doubling of P and if we have to add P to another point, reusing Z_1^2 saves 1S. Adding a point that has already been doubled will be called $dADD$.

We now apply this to our scalar multiplication algorithm. We first compute the sequence $P \rightarrow 2P \rightarrow \cdots \rightarrow 2^{b_{max}}P$. For each doubled point (i.e. $P \rightarrow 2P \rightarrow \cdots \rightarrow 2^{b_{max}-1}P$), it is possible to store Z^2. Different situations can now occur:

- **addition after doubling (dADD)**: addition of a point that has already been doubled before
- **double addition after doubling (2dADD)**: addition of two points that have already been doubled before
- **addition after doubling + readdition (dreADD)**: addition of a point that has already been doubled before to a point that has been added before
- **double readition (2reADD)**: addition of two points that has been added before
- **addition after doubling + mixed addition dmADD**: addition of a point that has already been doubled before to a point in affine coordinate (i.e. $Z = 1$)
- **mixed readdition (mreADD)**: addition of a point in affine coordinate (i.e. $Z = 1$) to a point that has been added before

Remark 2. It is also possible to cache Z^2 after a tripling. Adding a point that has already been tripled has the same cost as that has been after a doubling. Thus, we will still call this operation $dADD$.

In Table 2 we summarize the costs of the different operations for each considered curve.

Table 2. New elliptic curve operations cost

Curve shape	dADD	2dADD	dreADD	2reADD	dmADD	mreADD
3DIK	11M+6S	11M+6S	10M+6S	9M+6S	7M+4S	6M+4S
Edwards	10M+1S	10M+1S	10M+1S	10M+1S	9M+1S	9M+1S
ExtJQuartic	7M+3S	7M+2S	7M+2S	7M+2S	6M+2S	6M+2S
Hessian	6M+6S	6M+6S	6M+6M	6M+6S	5M+6S	5M+6S
InvEdwards	9M+1S	9M+1S	9M+1S	9M+1S	8M+1S	8M+1S
JacIntersect	11M+1S	11M+1S	11M+1S	11M+1S	10M+1S	10M+1S
Jacobian	11M+4S	10M+4S	10M+3S	9M+3S	7M+3S	6M+3S
Jacobian-3	11M+4S	10M+4S	10M+3S	9M+3S	7M+3S	6M+3S

5.2 Implementations and Results

We have carried out experiments on 160-bit and 256-bit scalars over all the elliptic curves mentioned in section 2.1 and all values of b_{max}, t_{max} and Z_{max} such that $2^{b_{max}}3^{t_{max}}$ and $2^{b_{max}}F_{Z_{max}}$ are 160-bit or 256-bit integers. For each curve and each set of parameters, we have:

- generated 10000 pseudo random integers in $\{0, \ldots, 2^m - 1\}$ $m = 160, 256$,
- converted each integer into the DBNS/BZNS systems using the corresponding parameters,
- counted all the operations involved in the point scalar multiplication process.

Table 3. Optimal parameters and operation count for 160-bit scalars

Curve shape	Method	b_{max}	t_{max}	Z_{max}	# multiplications
3DIK	4-NAF	-	-	-	1645.8
	DB chain	80	51	-	1502.4
	Yao-DBNS	44	74	-	1477.3
Edwards	4-NAF	-	-	-	1321.6
	DB chain	156	3	-	1322.9
	Yao-DBNS	140	13	-	1283.3
ExtJQuartic	4-NAF	-	-	-	1308.5
	DB chain	156	3	-	1311.0
	(2,3,5)NAF	131	12	-	1226.0
	Yao-DBNS	140	13	-	1210.9
Hessian	4-NAF	-	-	-	1601.9
	DB chain	100	38	-	1565.0
	Yao-DBNS	113	30	-	1501.8
InvEdwards	4-NAF	-	-	-	1287.8
	DB chain	156	3	-	1290.3
	(2,3,5)NAF	142	9	-	1273.8
	Yao-DBNS	140	13	-	1258.6
JacIntersect	4-NAF	-	-	-	1389.4
	DB chain	150	7	-	1438.8
	Yao-DBNS	143	11	-	1301.2
Jacobian	4-NAF	-	-	-	1573.8
	DB chain	100	38	-	1558.4
	Yao-DBNS	131	19	-	1534.9
	Yao-BZNS	142	-	28	1534.8
Jacobian-3	4-NAF	-	-	-	1511.9
	DB chain	100	38	-	1504.3
	(2,3,5)NAF	131	12	-	1426.8
	Yao-DBNS	131	19	-	1475.3
	Yao-BZNS	142	-	28	1476.9

In Tables 3 and 4, we report the best results obtained for each case, with the best choice of parameters. Results are given in number of base field multiplications. To do so and in order to ease the comparison with previous works, we assume that $S = 0.8M$. However, different ratios could give slightly different results.

As the efficiency of any method is directly dependent on that of the curve operations, in appendix, we give in Tables 5, 6, 7 and 8 the curve operation count of our methods, in order to ease comparisons with future works that might use improved formulae. However, one has to be aware that those operation counts are only valid for the parameters they correspond to. A significant improvement of any of those curve operations may significantly change the optimal parameters for a given method.

As shown in Tables 3 and 4, our new method is very efficient compared to previously reported optimized double-base chains approaches [1] or optimized w-NAF methods [5], whatever the curve is. We obtain particularly good results on extended

Table 4. Optimal parameters and operation count for 256-bit scalars

Curve shape	Method	b_{max}	t_{max}	Z_{max}	# multiplications
3DIK	4-NAF	-	-	-	2603.3
	DB chain	130	80	-	2393.2
	Yao-DBNS	63	122	-	2319.2
Edwards	4-NAF	-	-	-	2088.5
	DB chain	252	3	-	2089.7
	Yao-DBNS	220	23	-	2029.8
ExtJQuartic	4-NAF	-	-	-	2068.9
	DB chain	253	2	-	2071.2
	Yao-DBNS	215	26	-	1911.4
Hessian	4-NAF	-	-	-	2542.4
	DB chain	150	67	-	2470.6
	Yao-DBNS	185	45	-	2374.0
InvEdwards	4-NAF	-	-	-	2038.7
	DB chain	252	3	-	2041.2
	Yao-DBNS	220	23	-	1993.3
JacIntersect	4-NAF	-	-	-	2185.4
	DB chain	246	7	-	2266.1
	Yao-DBNS	236	13	-	2050.5
Jacobian	4-NAF	-	-	-	2492.1
	DB chain	160	61	-	2466.2
	Yao-DBNS	185	45	-	2416.2
	Yao-BZNS	227	-	44	2419.8
Jacobian-3	4-NAF	-	-	-	2391.8
	DB chain	160	61	-	2379.0
	Yao-DBNS	185	45	-	2316.2
	Yao-BZNS	22	-	44	2329.2

Jacobi Quartics, with which we improve the best results found in the literature, even taking into account the recent multi-base chains (or (2,3,5)NAF) [18]. However, one should note that part of those improvements are due to the fact that [1] and [5] uses older formulae for Extended JQuartic and Hessian curves. As an example, doubling is performed using 3M+4S instead of the actual 2M+5S. This saves 0.2M per doublings, that is to say around 32M (160×0.2M) for 160-bit scalars.

We can also see the interest of our new binary/Zeckendorf number system, for each curve where Euclidean additions are fast, it gives similar results as the classical double-base number system. It could be really interesting to generalize this number system to other curves and find more specific optimizations.

Finally, growing in size makes our algorithm even more advantageous, for every curves. Considering the (2,3,5)NAF method, no data are given for 256-bit scalars in the original paper. Due to lack of time, we have not been able to implement by ourselves this algorithm but we expect a similar behavior.

6   Conclusions

In this paper we have proposed an efficient generalized version of Yao's algorithm, less restrictive than the double-base chain method, to perform the point scalar multiplication on elliptic curves defined over prime fields. The main advantage of this representation is that it takes advantage of the natural sparseness of the double-base number system without any additional and unnecessary computations. In the end, our method performs faster than all the previous double-base chains methods, over all types of curves. On the extended Jacobi Quartics, it also provides the best result found literature, faster than the (2,3,5)NAF, recently claimed as the fastest scalar multiplication algorithm. Finally we have proposed a new number system, mixing binary and Zeckendorf representation. On curves providing fast Euclidean addition, the BZNS provides very good results.

Acknowledgment. This work was supported in part by NSERC grants awarded to Dr. Hasan. We also are very grateful to the anonymous referees of CHES 2009 for their invaluable comments.

References

1. Bernstein, D.J., Birkner, P., Lange, T., Peters, C.: Optimizing Double-Base Elliptic-Curve Single-Scalar Multiplication. In: Srinathan, K., Rangan, C.P., Yung, M. (eds.) INDOCRYPT 2007. LNCS, vol. 4859, pp. 167–182. Springer, Heidelberg (2007)
2. Bernstein, D.J., Lange, T.: Explicit-formulas database,
   http://hyperelliptic.org/EFD
3. Bernstein, D.J., Lange, T.: Faster addition and doubling on elliptic curves. In: Kurosawa, K. (ed.) ASIACRYPT 2007. LNCS, vol. 4833, pp. 29–50. Springer, Heidelberg (2007)
4. Bernstein, D.J., Lange, T.: Inverted Edwards coordinates. In: Boztaş, S., Lu, H.-F(F.) (eds.) AAECC 2007. LNCS, vol. 4851, pp. 20–27. Springer, Heidelberg (2007)
5. Bernstein, D.J., Lange, T.: Analysis and optimization of elliptic-curve single-scalar multiplication. In: Finite fields and applications: proceedings of Fq8, pp. 1–19 (2008)

6. Chudnovsky, D.V., Chudnovsky, G.V.: Sequences of numbers generated by addition in formal groups and new primality and factorization tests. Adv. Appl. Math. 7(4), 385–434 (1986)
7. Cohen, H., Frey, G. (eds.): Handbook of Elliptic and Hyperelliptic Cryptography. Chapman and Hall, Boca Raton (2006)
8. Dimitrov, V., Cooklev, T.: Two algorithms for modular exponentiation using nonstandard arithmetics. IEICE Transactions on Fundamentals of Electronics, Communications and Computer Sciences 78(1), 82–87 (1995)
9. Dimitrov, V., Imbert, L., Mishra, P.K.: Efficient and secure elliptic curve point multiplication using double-base chains. In: Roy, B. (ed.) ASIACRYPT 2005. LNCS, vol. 3788, pp. 59–78. Springer, Heidelberg (2005)
10. Doche, C., Icart, T., Kohel, D.R.: Efficient scalar multiplication by isogeny decompositions. In: Yung, M., Dodis, Y., Kiayias, A., Malkin, T.G. (eds.) PKC 2006. LNCS, vol. 3958, pp. 191–206. Springer, Heidelberg (2006)
11. Doche, C., Imbert, L.: Extended Double-Base Number System with Applications to Elliptic Curve Cryptography. In: Barua, R., Lange, T. (eds.) INDOCRYPT 2006. LNCS, vol. 4329, pp. 335–348. Springer, Heidelberg (2006)
12. Duquesne, S.: Improving the arithmetic of elliptic curves in the Jacobi model. Inf. Process. Lett. 104(3), 101–105 (2007)
13. Edwards, H.M.: A normal norm for elliptic curves. Bulletin of the American Mathematical Society 44, 393–422 (2007)
14. Hankerson, D., Menezes, A., Vanstone, S.: Guide to Elliptic Curve Cryptography. Springer, Heidelberg (2004)
15. Hisil, H., Carter, G., Dawson, E.: New formulae for efficient elliptic curve arithmetic. In: Srinathan, K., Rangan, C.P., Yung, M. (eds.) INDOCRYPT 2007. LNCS, vol. 4859, pp. 138–151. Springer, Heidelberg (2007)
16. Hisil, H., Koon-Ho Wong, K., Carter, G., Dawson, E.: An intersection form for jacobi-quartic curves. Personal communication (2008)
17. Liardet, P., Smart, N.P.: Preventing SPA/DPA in ECC systems using the jacobi form. In: Koç, Ç.K., Naccache, D., Paar, C. (eds.) CHES 2001. LNCS, vol. 2162, pp. 391–401. Springer, Heidelberg (2001)
18. Longa, P., Gebotys, C.: Setting speed records with the (fractional) multibase non-adjacent form method for efficient elliptic curve scalar multiplication. Technical report, Department of Electrical and Computer Engineering University of Waterloo, Canada (2009)
19. Longa, P., Miri, A.: New composite operations and precomputation scheme for elliptic curve cryptosystems over prime fields. In: Cramer, R. (ed.) PKC 2008. LNCS, vol. 4939, pp. 229–247. Springer, Heidelberg (2008)
20. Meloni, N.: New point addition formulae for ECC applications. In: Carlet, C., Sunar, B. (eds.) WAIFI 2007. LNCS, vol. 4547, pp. 189–201. Springer, Heidelberg (2007)
21. Mishra, P.K., Dimitrov, V.S.: Efficient quintuple formulas for elliptic curves and efficient scalar multiplication using multibase number representation. In: Garay, J.A., Lenstra, A.K., Mambo, M., Peralta, R. (eds.) ISC 2007. LNCS, vol. 4779, pp. 390–406. Springer, Heidelberg (2007)
22. Yao, A.C.: On the evaluation of powers. SIAM Journal on Computing 5(1), 100–103 (1976)
23. Zeckendorf, E.: Représentations des nombres naturels par une somme de nombre de Fibonacci ou de nombres de Lucas. Bulletin de la Soci. Royale des Sciences de Liège, pp. 179–182 (1972)

A Detailed Operation Counts

Table 5. Detailed operation count for the Yao-DBNS scalar multiplication using 160-bit scalar

Curve shape	DBL	TPL	ADD	reADD	dADD	2dADD	2reADD	dreADD	mADD	dmADD	mreADD
3DIK	43.50	73.43	1.20	0.64	16.10	3.49	0.01	0.29	0.66	0.45	0.01
Edwards	139.12	12.84	1.68	1.55	18.48	0.97	0	0.01	1.59	0.22	0.01
ExtJQuartic	139.12	12.84	1.68	1.55	18.48	0.97	0	0.01	1.59	0.22	0.01
Hessian	112.22	29.73	1.26	1.07	17.40	1.63	0.01	0.17	1.07	0.28	0.03
InvEdwards	139.12	12.84	1.68	1.55	18.48	0.97	0	0.01	1.59	0.22	0.01
JacIntersect	142.19	10.94	2.40	1.64	17.71	0.81	0	0.13	2.22	0.29	0.03
Jacobian	130.10	18.71	1.43	1.09	18.36	1.11	0	0.14	1.31	0.25	0.03
Jacobian-3	130.10	18.71	1.43	1.09	18.36	1.11	0	0.14	1.31	0.25	0.03

Table 6. Detailed operation count for the Yao-BZNS scalar multiplication using 160-bit scalars

Curve shape	DBL	ZADD	ADD	reADD	mADD	2reADD	mreADD
Jacobian	141.27	25.53	19.55	0.48	1.72	0	0
Jacobian-3	141.27	25.53	19.55	0.48	1.72	0	0

Table 7. Detailed operation count for the Yao-DBNS scalar multiplication using 256-bit scalar

Curve shape	DBL	TPL	ADD	reADD	dADD	2dADD	2reADD	dreADD	mADD	dmADD	mreADD
3DIK	62.48	121.72	1.33	1.07	23.78	6.17	0.01	0.49	0.66	0.52	0.03
Edwards	219.17	22.89	1.87	2.37	28.46	1.61	0	0.23	1.48	0.35	0.06
ExtJQuartic	214.29	25.86	1.93	1.93	27.88	1.92	0.01	0.29	1.73	0.32	0.01
Hessian	184.33	44.73	1.38	1.47	26.58	2.57	0.01	0.21	1.18	0.34	0.03
InvEdwards	219.17	22.89	1.87	2.37	28.46	1.61	0	0.23	1.48	0.35	0.06
JacIntersect	235.26	12.94	2.37	3.04	29.31	1.39	0.02	0.24	2.25	0.33	0.05
Jacobian	184.33	44.78	1.38	1.47	26.58	2.57	0.01	0.21	1.18	0.34	0.03
Jacobian-3	184.33	44.78	1.38	1.47	26.58	2.57	0.01	0.21	1.18	0.34	0.03

Table 8. Detailed operation count for the Yao-BZNS scalar multiplication using 256-bit scalars

Curve shape	DBL	ZADD	ADD	reADD	mADD	2reADD	mreADD
Jacobian	226.3	41.4	30.01	0.74	1.20	0	0.02
Jacobian-3	226.3	41.4	30.01	0.74	1.20	0	0.02

The Frequency Injection Attack on Ring-Oscillator-Based True Random Number Generators

A. Theodore Markettos and Simon W. Moore

Computer Laboratory, University of Cambridge, UK
theo.markettos@cl.cam.ac.uk

Abstract. We have devised a frequency injection attack which is able to destroy the source of entropy in ring-oscillator-based true random number generators (TRNGs). A TRNG will lock to frequencies injected into the power supply, eliminating the source of random jitter on which it relies. We are able to reduce the keyspace of a secure microcontroller based on a TRNG from 2^{64} to 3300, and successfully attack a 2004 EMV ('Chip and PIN') payment card. We outline a realistic covert attack on the EMV payment system that requires only 13 attempts at guessing a random number that should require 2^{32}. The theory, three implementations of the attack, and methods of optimisation are described.

1 Introduction

Random numbers are a vital part of many cryptographic protocols. Without randomness, transactions are deterministic and may be cloned or modified. In this paper we outline an attack on the random number generators used in secure hardware. By injecting frequencies into the power supply of a device we can severely reduce the range of random numbers used in cryptography. Fig. 1 illustrates the patterns from our attack on a secure microcontroller.

Consider an example in the EMV banking protocol (initiated by Europay, MasterCard and Visa, marketed as 'Chip and PIN' in the UK) [1]. For cash withdrawal an automatic telling machine (ATM) picks an unpredictable number from four billion possibilities. Imagine if an insider can make a small covert modification to an ATM to reduce this to a small number, R.

He could then install a modified EMV terminal in a crooked shop. A customer enters and pays for goods on their card. While the modified terminal is doing the customer's EMV transaction with their secret PIN, it simulates an ATM by performing and recording $\sqrt{R}$ ATM transactions. The customer leaves, unaware that extra transactions have been recorded.

The crooked merchant then takes a fake card to the modified ATM. The ATM will challenge the card with one of R random numbers. If the shop recorded a transaction with that number, he can withdraw cash. If not, the fake card terminates the transaction (as might happen with dirty card contacts) and starts again. By the Birthday Paradox we only need roughly $\sqrt{R}$ attempts at the ATM

C. Clavier and K. Gaj (Eds.): CHES 2009, LNCS 5747, pp. 317–331, 2009.

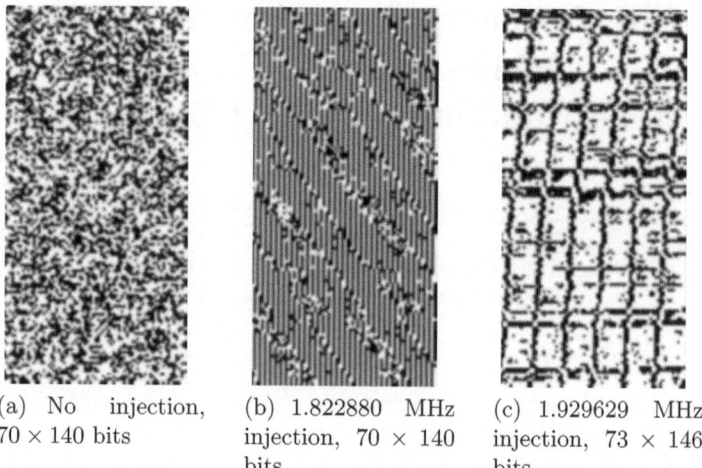

(a) No injection, 70 × 140 bits

(b) 1.822880 MHz injection, 70 × 140 bits

(c) 1.929629 MHz injection, 73 × 146 bits

Fig. 1. TRNG bitstream from secure microcontroller with frequency injection, raster scanning left-to-right then top-to-bottom. Bit-widths chosen to illustrate sequences found. Recording into SRAM of 28KB of sequential random bytes at maximum rate, later replayed through the serial port as a hexadecimal string.

before, at least, a 50% chance of success. The customer has no defence: both their card and PIN were used in the transaction, just not at the time they expected.

In this attack we have reduced the ability of a microcontroller known to be used in ATMs to produce 4 billion (2^{32}) random numbers, to just 225 ($< 2^8$) (illustrated in Fig. 1). For more than a 50% chance of a successful attack we only need to record 13 transactions in the shop and try 13 transactions at the ATM. Our attack is based on injecting signals into the power supply of a smartcard or secure microcontroller. It is made possible by adding a small number of extra components costing tens of dollars.

2   Random Number Generation

A true random number generator (TRNG) must satisfy two properties: *uniform statistics* and contain a source of *entropy*. Non-uniform statistics might enable the attacker to guess common values or sequences. Entropy comprises a source of uncertainty in a normally predictable digital system. Failure of these properties in even subtle ways leads to weaknesses in cryptographic systems ([2], [3]).

A common implementation of a TRNG is provided by comparing free-running oscillators. These are designed to be sensitive to thermal, shot or other types of random noise, and present it as timing variations. Such timing variations can be measured by a digital system, and the entropy collected. An oscillator that is easy to fabricate on a CMOS digital integrated circuit is the ring oscillator (see Fig. 2), which is used in many TRNG designs.

Practical TRNG sources are typically *whitened* by post-processing before cryptographic use, to ensure uniform statistics. Typically whitening functions

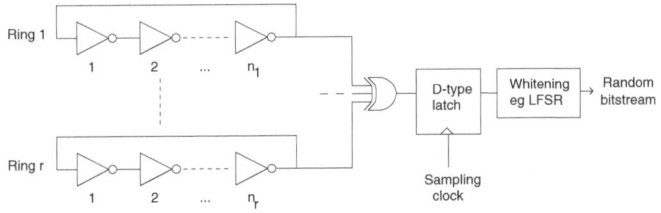

Fig. 2. Outline of the basic ring oscillator TRNG

include calculating the remainder of a polynomial division using a linear-feedback shift register (LFSR), or hash functions. If the entropy source is removed, TRNG outputs revert to a repeating sequence from the whitening function.

In this paper we examine the operation of the ring oscillator, and explain how the principle of *injection locking* may be used by an attacker to take control of this entropy source.

3   Theory

3.1   Ring Oscillator TRNG Operation

Hajimiri et al.[4] give the frequency of a single-ended[1] CMOS ring oscillator formed from N inverters with equal-length NMOS and PMOS transistors to be:

$$f_0 \equiv \frac{\omega_0}{2\pi} \approx \frac{\mu_{\text{eff}} W_{\text{eff}} C_{\text{ox}}(\frac{V_{\text{DD}}}{2} - V_{\text{T}})}{8\eta N L q_{\text{max}}} \tag{1}$$

This relates the fundamental frequency f_0 to the gate-oxide capacitance per unit area C_{ox}, transistor length L, power supply voltage V_{DD}, gate threshold voltage V_T and proportionality constant $\eta \approx 1$. q_{max} is the amount of charge a node receives during one switching period. We consider both NMOS and PMOS transistors together, giving effective permeability μ and transistor width W:

$$W_{\text{eff}} = W_{\text{n}} + W_{\text{p}} \tag{2}$$

$$\mu_{\text{eff}} = \frac{\mu_{\text{n}} W_{\text{n}} + \mu_{\text{p}} W_{\text{p}}}{W_{\text{n}} + W_{\text{p}}} \tag{3}$$

These are all physical constants determined in the construction of the oscillator. A ring oscillator with no other effects would be completely predictable.

Oscillators do not have a perfectly stable output. In the time domain, random noise means they sometimes transition before or after the expected switching time. In the frequency domain this implies small random fluctuations in the phase of the wave, slightly spreading its spectrum. This same effect is referred to as *jitter* in the time domain and as *phase noise* in the frequency domain. These are both cumulative over time (seen in Fig. 3).

---

[1] In a *single-ended* ring the connection between each node is a single unbalanced signal, as opposed to a *differential* ring, in which each connection is a balanced pair.

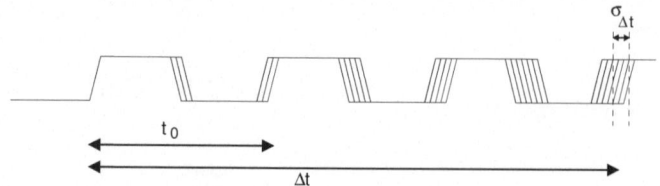

Fig. 3. Jitter in the time domain causes increasing uncertainty in the timing of transitions

In a single-ended ring oscillator, a time Δt after the starting, Hajimiri derives that the jitter due to thermal noise will have a standard deviation:

$$\sigma_{\Delta t} \approx \sqrt{\frac{8}{3\eta}} \sqrt{\frac{kT}{P} \frac{V_{\text{DD}}}{V_{\text{char}}}} \tag{4}$$

where P is the power consumption and kT the Boltzmann constant multiplied by temperature. V_{char} is the characteristic voltage across a MOSFET – in the long-channel mode it is $\frac{2}{3}((V_{\text{DD}}/2) - V_T)$.

This is equivalently written as a phase noise spectrum:

$$L\{\omega\} \approx \frac{8}{3\eta} \frac{kT}{P} \frac{V_{\text{DD}}}{V_{\text{char}}} \frac{\omega_0^2}{\omega^2} \tag{5}$$

where ω_0 is the natural angular frequency of the oscillator and variable ω is some deviation from it (ie $\omega = 0$ at ω_0).

In a TRNG based on ring oscillators jitter is converted into entropy by measuring the timing of transitions: jitter causes the exact timing to be unpredictable. There are two main ways to construct such a TRNG: relatively prime ring lengths ([5] and several patents) and identical ring lengths [6]. Both employ a topology based on that of Fig. 2. The combined signals from the rings are sampled at some uncorrelated frequency, producing a stream of bits, which is then whitened before cryptographic use.

In the identical rings context, we have two or more rings running at the same frequency. Entropy is wasted when jitter from one transition overlaps jitter from another since only one transition is measured. Sunar et al.[6] extends this to tens or hundreds of rings to increase the probability that at time t there will be a ring R that does not transition. Cumulative jitter is measured as the phase drift between each ring.

With relatively prime rings, the outputs slide past each other, minimising the likelihood of two rings transitioning together. Transition timing is based on a prime factor and the integral of past jitter. Sunar points out that fabrication of relatively prime rings to produce more concentrated entropy is expensive. In our experimental work we concentrate on relatively prime rings, since, we suggest, these are more difficult to lock to an input frequency (or frequencies). For identical rings it should be much simpler.

3.2   Frequency Injection Attacks

Bak [7] describes how a dynamical system will, at certain frequencies, resonate, and, at others, be chaotic. A resonator, such as a pendulum, with natural frequency m, will lock when driven by any frequency n forming a rational m/n. Adler [8] describes the conditions for lock as applied to a vacuum tube LC electronic oscillator. This effect is known as *injection locking*.

Our attack constitutes injecting a signal of frequency $f_i \equiv \omega_i/2\pi$ and magnitude V_i into the ring oscillators, causing them to lock to the injected frequency. Locking is a steady state: at lock the relative phase ϕ between the two oscillators is constant, so $d\phi/dt = 0$. Once lock has been achieved, the ring's natural frequency is irrelevant; jitter in the injecting signal will be received equally by all the rings, impairing any TRNG that compares jitter between oscillators.

Mesgarzadeh and Alvandpour [9] analyse this for a three-ring CMOS oscillator deliberately made asymmetric by the forcing input being an additional gate overdriving one signal. They prove Adler's work also applies to their ring oscillator. Rearranging their condition for lock in Adler's form, we have:

$$2Q\left|\left(\frac{\omega_i}{\omega_0} - 1\right)\right| < \frac{V_i}{V_0} \tag{6}$$

where V_0 is the amplitude of the oscillator at its natural frequency and Q is its quality factor, a measure of the damping of the oscillator. From our experiments, Fig. 4 shows the difference between rings sliding past each other and in lock.

To achieve injection locking, we must ensure our interference can reach the ring oscillators in a secure circuit. In this paper we achieve it by coupling the injection frequency on to the power supply of the device.

The difficulty in proceeding with an analytic solution is determining Q. Adler originally derived the formulation in the context of an LC tank that has a natural sinusoidal operating frequency. Such an oscillator converts energy between two

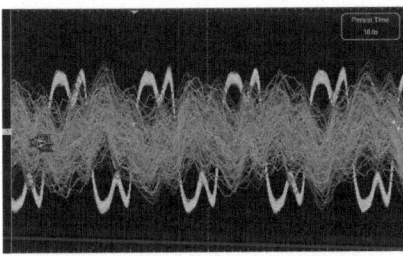 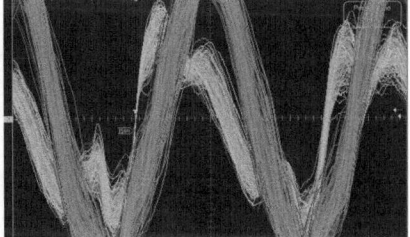

(a) No injected signal, rings slide past each other   (b) Strong injection, traces lock phase

Fig. 4. Injection locking effect shown on oscilloscope persistent display (discrete inverter experiment from Sec. 4). View dimensions $8\,V \times 200\,ns$; $5\,V$ power supply. Triggering on 3-element ring, with 5-element-ring trace in front. Note in particular how resonances are set up in the ring oscillators that increase the amplitude above the power rails from $5\,Vp\text{-}p$ to $10\,Vp\text{-}p$.

forms, voltage and current in this case. It synchronises to an external signal of suitable frequency to maximize the energy extracted from this driving source. For instance, a pendulum will adjust phase so an external periodic displacement will accelerate its own velocity.

A ring oscillator lacks a clear system-wide change between two alternating states, being just a circle where a rising and a falling edge chase each other, without any natural point defining where a new cycle starts. An idealised 3-element ring consists of three identical inverters, connected via three identical transmission lines. All three inverters and transmission lines oscillate in exactly the same way, but 120° out of phase. A waveform applied via the power supply or an EM field is a *global stimulus* that affects all three inverters equally. It will, therefore, encourage the ring oscilator to synchronise simultaneously with three versions of the stimulus, all 120° apart in phase. Their synchronising effect is thus largely cancelled out.

A global stimulus can only be effective if the three parts are not exactly identical. In a real-world ring oscillator layout asymmetries, device variations, and loading due to the output tap all break this 120° symmetry and will allow one of the 120° alternatives to win over the other two. How quickly the ring will lock on to a global stimulus will largely depend on the size of this asymmetry.

Unlike pendula or LC tanks, ring oscillators are also non-linear. In short rings, such as $N = 3$, each gate is in a constant state of transition, so operates linearly, and the output more clearly resembles a sinusoid. But in longer rings, where $N \gg 10$, each gate spends a small fraction of the time in transition, so the ring output is more like a square wave. Adler's model fits this case less well.

3.3   Effect of Injection on Jitter

Mesgarzadeh and Alvandpour indicate their injection can operate on the ring signals as a first-order low-pass filter with a single pole located at:

$$p = 2\pi\omega_i \ln \frac{1}{1+S} \tag{7}$$

where S is the injection ratio V_i/V_0 or, in power terms, $\sqrt{P_i/P_0}$. In other words, the function in the domain of the Laplace transform is:

$$H(j\omega) = \frac{1}{1 + (2\pi\omega_i \ln \frac{1}{1+S})j\omega} \tag{8}$$

where $j = \sqrt{-1}$. It is analogous to a series R-C filter with $RC = p$.

If we can locate pole p close enough to the origin we can filter out high frequency and large cycle-to-cycle jitter. Increased injection power S reduces this filtering effect.

A successful attack is the intersection of two regions. From (6), if the injection power is too low the system will not lock. From (8), if the power is too high jitter will not be filtered out. For the TRNG system a weaker condition is required: if the jitter of the rings mirrors that of the driving oscillator, the attack is a success. The TRNG measures differences in jitter between rings, so will not measure jitter common to all rings.

We analyse the attack on relatively prime rings but demonstrate our attack in 'black box' experiments with no knowledge of device construction. Therefore we assume that we are reducing jitter by the effect of equalising jitter between rings, rather than a reduction of jitter in the whole system.

Yoo et al.[10] describes locking effects due to poor layout but generally not in an adversarial manner. They investigate changing the DC supply voltage, but not its AC components. Sunar [6] considers active glitch attacks and concludes these are only able to attack a finite number of bits due to their limited duration. The frequency injection attack is much more powerful, since it can attack all bits simultaneously for as long as desired.

4   Discrete Logic Measurements

We set out to measure phase differences in two relatively prime rings. Given their primality, the ring outputs should drift past each other, based on a combination of cumulative jitter and the underlying ring frequency differences. For non-locked rings, we expect phase lag to be uniformly distributed. When locked, phase lag will be concentrated on one value.

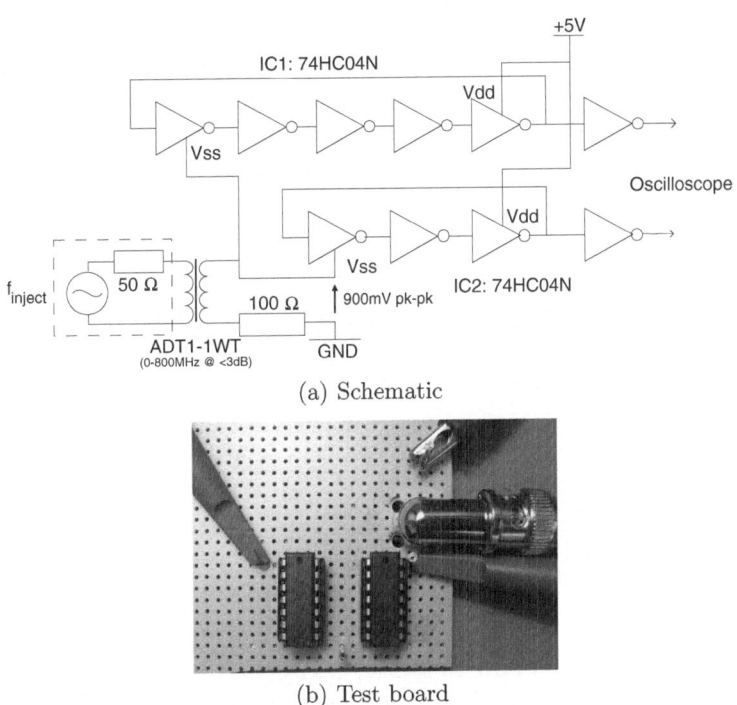

(a) Schematic

(b) Test board

Fig. 5. Measurement experiment using 74HC04 inverter ICs

Injection locking is very difficult to simulate in a transient analysis tool such as SPICE [11]. It requires very small timesteps for each oscillation cycle, and a high Q oscillator may require thousands of cycles to lock. When close to the natural frequency the beat frequency may be a few tens of Hertz. To measure this accurately in a simulation with picosecond-scale steps requires an infeasibly long simulation time. In addition, the asymmetries of a real system will not be modelled in a simulated ideal design.

Due to I/O buffering, it is difficult to measure such behaviour of fast analogue signals inside an FPGA or ASIC. We first measured the effect in discrete logic. With limited complexity possible, we investigated the simplest ring oscillators: the outputs from three- and five- element rings, with and without frequency injection in the power supply. We used the 74HC04 inverter chip to construct the two mutually-prime rings seen in Fig. 5(a). Phase lag was measured by triggering an oscilloscope on the rising edge of the three-element ring, and measuring the time up to the rising edge of the five-element ring. Such short rings are used in real TRNGs – though they may have a greater region of linear operation than longer rings.

We set up a Tektronix AFG3252 function generator to inject a sine wave at 900 mV pk-pk into the 5 V power rails and by sweeping frequency we observed locking at 24 MHz. A Tektronix TDS7254B oscilloscope measured the phase lag between the two rings when injecting and the resulting histograms are plotted in Fig. 6. A very clear clustering around 10 ns can be seen, indicating a lock. This effect is visible in the time domain traces seen in Fig. 4, which show a marked reduction in the variability of the 5-element output. The slight clustering seen in Fig. 6(a) is, we believe, due to slightly non-uniform oscilloscope triggering.

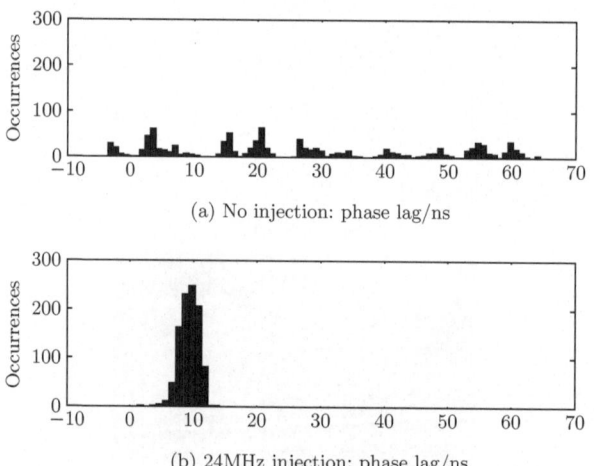

(a) No injection: phase lag/ns

(b) 24MHz injection: phase lag/ns

Fig. 6. Phase delay between 74HC04 3- and 5- element rings. (a) with no injection, (b) with 24 MHz at 900 mV pk-pk injected into power supply. (25000 samples).

5   Secure Microcontroller

We tested an 8051-compatible secure microcontroller which has been used in ATMs and other security products. It includes features such as an anti-probing coating and tamper detection and at release claimed to be the most secure product on the market. Our example had a date code of 1995 but the device is still recommended for new payment applications by the manufacturer.

It provides a hardware TRNG based on frequency differences between two ring oscillators and timing from the user's crystal (11.059 MHz here), and produces 8 bits every 160 μs. 64 bits from the TRNG may be used as the internal encryption key. No further operation details are documented.

We programmed the device to deliver the random bitstream as hexadecimal digits through the serial port and displayed it in realtime as a two dimensional black and white image. We adjusted the function generator to inject a sinusoid at 500 mV peak-peak into the chip's power supply as shown in Fig. 7.

By sweeping the frequency we spotted changes in the patterns produced by the TRNG. The most interesting f_{inject} was at about 1.8 MHz. Obviously periodic sequences were visible: see Fig. 1(a)–1(c). In particular the sequence length of the TRNG was controlled by the injected frequency. With another choice of f_{inject} we could also prevent the TRNG returning any values. At no time during any of these tests did the microcontroller otherwise misbehave or detect a fault condition. The device is designed to operate at 5 V with a minimum operating voltage of 4.25 V so it is running within specification.

Fig. 1(b) indicates a distinct 15-bit long texture, on top of a longer 70-bit sequence. Uncertainty is only present at the overlap between these textures. In a 420-bit sample, we estimate 18 bits of noise. In a 15-bit sequence that means 0.65 bits may be flipped. The attacker knows the base key is the sequence 010101010..., but not where the 15-bit sequence starts (15 possibilities) or the noise. In most cases noise is constrained to 3 bits at the start of the 15 bit sequence. In the full 64-bit sequence, the bit flips are $0.65 \times 4 = 2.6$. 3 bits flipped over the whole 64-bit sequence in one of 12 positions gives $\binom{12}{3} = 220$ combinations. Thus we estimate that the total keyspace is smaller than $220 \times 15 = 3300$. In a key length of 32 bits there are 1.3 bits of noise; the equivalent calculation with 2 bits gives a keyspace of less than $\binom{6}{2} \times 15 = 225 \approx 2^8$.

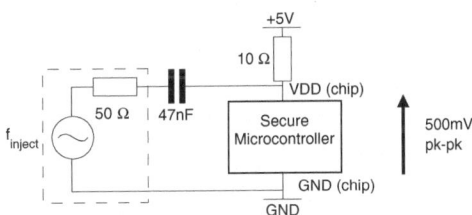

Fig. 7. Frequency injection to secure microcontroller

6   EMV Smartcard Attack

We applied this methodology to an EMV payment card issued in 2004 by a British High Street bank. We picked the first available; with no knowledge of the internals we treated it as a 'black box'. This is a non-invasive attack where no modifications are required to the card under attack.

First we needed to deduce the operating frequency of the card's TRNG. We assumed that such a card would have power analysis protection, so we performed an electromagnetic assessment. An electric field antenna was constructed on a test card. Copper foil was attached beneath the chip as shown in Figs. 8 and 9, with foil traces between the backside foil patch and the ground pad of the chip. The card was inserted into a Chipdrive Micro 100 card reader, and standard ISO7816-4 GET CHALLENGE commands used to read the RNG.

We measured three different spectra: (a) not powered or attached to reader (background interference); (b) powered, attached to reader but held in reset and not clocked; and (c) when reading random numbers.

Since a ring oscillator is likely to remain powered even when the card is not initialised, we looked for frequencies that exist when the card is inserted into the reader and unclocked, but not present when the card is removed. We found four such frequencies in the range 0–500 MHz: we chose f_{inject} to be 24.04 MHz, the only frequency below 100 MHz. As this is a black-box system, we do not know if this is optimal; it is merely the first frequency we tried.

We modified the reader to inject a signal as shown in Fig. 10, and observed the random number statistics. Sweeping the injected frequency and graphically viewing random bits, we saw no obvious pattern changes. However statistical analysis of the random data revealed injecting f_{inject} at 1 V pk-pk across the card powered at 5 V caused the random function to skew. At all times during measurement the card continued to respond correctly to ISO7816-4 commands and would perform EMV transactions while under attack.

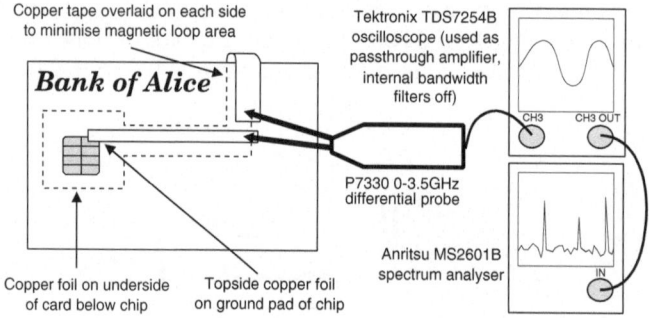

Fig. 8. Electric field characterisation of EMV smartcard

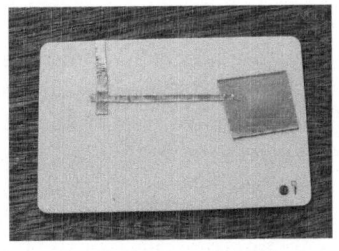

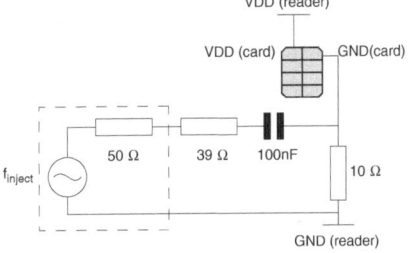

Fig. 9. Electric field antenna on underside of example smartcard

Fig. 10. Smartcard frequency injection circuit

Table 1. Statistical test results from injection into EMV card

NIST	Pass			Fail
No injection	187			1
Injection	28			160
Dieharder	**Pass**	**Poor**	**Weak**	**Fail**
No injection	86	6	6	9
Injection	28	16	5	58

The statistics were analysed using all the tests in the NIST [12] and Dieharder [13][2] test suites using 1.56×10^9 bits. An outline of the results are shown in Table 1, with tabulated NIST results in the Appendix. By failing most of the tests we can see that the sequence has become non-random. The FFT test reveals periodicities of around 2000 and 15000 bits. The Rank test, where a 32×32 matrix of random bits should have a rank > 28 (true for our control data), fails with many ranks as low as 19 implying rows or columns are not independent.

7 Recommendations and Further Work

7.1 Optimisation of the Attack

In the Introduction we outlined an attack on the EMV payment system, which works whether the smartcard uses Static or Dynamic Data Authentication protocols.

An ATM is a computer in a safe, comprising either a PC or custom circuit boards. At least one ATM uses the secure microcontroller we tested as its cryptoprocessor. ATM physical security focuses on preventing access to the money held inside; this attack needs no access to the cash compartment. Adding injection hardware involves adding a tap to one wire on the PCB – this could be done by an insider or simply by picking the mechanical locks. June 2009 reports [14] uncovered malware in ATMs installed by insiders, while in another case [15] an attacker bought up 'white-label' ATMs (normally found in shops and bars), fitted internal recording devices and resold them.

---

[2] Dieharder version 2.28.1, a superset of the DIEHARD suite.

The required number of transactions is small and unlikely to raise alerts at the bank, which is afraid of false alarms. Customers complain about false positives, so there is commercial pressure to be lenient. If the cash withdrawal is performed before the card is used again by the customer, the bank has no way of knowing the transaction was recorded earlier. ATMs are typically only serviced when they go wrong. Even if our proposed frequency injector could be spotted by a technician, it may be many months before they are on site.

While we developed our attack with laboratory equipment, the cost of characterising each smartcard, card reader or ATM processor can be made very low. An electric field antenna may be fitted inside a commercial reader, so that nothing is fixed to the card surface. A commercial tunable radio receiver may be attached to the antenna to scan for frequencies of interest, while the frequency synthesiser in a similar receiver may be modified as a cheap way to generate injection frequencies. Given a quantity of identical cards (cheaply acquired on the black market, having little value once expired or cancelled by the bank) the search is easy to parallelise.

Attacking the TRNG on a card can be optimised by listening to other commands it performs. Each card provides an Answer To Reset – a response from the card software which can also be used to fingerprint its manufacturer [16]. We found cards with the same ATR emitted the same frequencies, most likely if they were built on the same operating system/hardware combination. After characterisation, the attacker can decide which frequencies to inject to a live card based on the ATR. This logic can be built into a terminal or ATM tap; interception of the card serial data line will reveal the ATR.

Due to electromagnetic interference (EMI) regulations, devices are designed to operate in the presence of interference. Neither of the commercial devices tested failed to operate at any point during these attacks. It is difficult to see how TRNGs could actively detect such attacks without compromising their EMI immunity.

7.2   Defences

We have demonstrated this attack allows the keyspace to be reduced to a size easily brute-forced. As soon as the attacker knows some plaintext, the key may be easily found. The simplest defence is to prevent a brute-force attack. Therefore the best system allows few permitted guesses, which raises the risks for the attacker. Preventing the attacker gaining large quantities of random material would also prevent device characterisation.

To prevent interference a device can filter injected frequencies. Voltage regulation or merely extra power supply smoothing may prevent locking, or shielding may be required for electromagnetic attacks. Devices could refuse to operate at their known-vulnerable frequencies. While this may render them less EMI-immune, it may be acceptable in high security applications. TRNG designs where the feedback loop is combined with logical or register elements [17] may be sufficient to break the locking effect.

Designers can work towards preventing locking by reducing asymmetries in the rings. Carefully balanced transistors may be used, as may equal tapping points on each node. Also, the differential ring oscillator is less affected by supply and substrate noise [18]. It may be feasible to use this instead of the single-ended ring commonly used. Careful design is required to ensure reducing susceptibility does not destroy the source of entropy – Hajimiri [4] indicates the differential oscillator increases the phase noise, which may be beneficial.

7.3   Further Work

We have outlined the principles of this attack but there are many ways in which it could be refined.

Further analysis of the effect of power supply injection is necessary. In the literature, injection locking has mostly been analysed through direct coupling to the signal undergoing oscillation, while, here, we use a different mode of coupling, by co-ordinated biasing of the gates it passes through. It would be instructive to determine the minimum power required for this attack and, in particular, how much it can be reduced by on-chip filtering. There are some well-known defences against passive power analysis; it would be interesting to evaluate these for protecting against frequency injection attacks.

In addition, it may also be feasible to perform this attack via high-powered electromagnetic radiation, which is more easily focused and more difficult to mitigate than a power attack. This could be done using magnetic loops to induce currents into the device at the appropriate frequencies, or using the device itself to demodulate the injected frequency (such as a 3 GHz carrier amplitude-modulated by 1.8 MHz); the carrier will more readily propagate, but be filtered away by parasitic capacitance on the chip leaving the 1.8 MHz harmonic.

Systems with identical ring lengths may be particularly vulnerable due to their shared resonant frequencies. There is further scope for directing this attack if the ring geometries are known. Fig. 1 shows some texture of our TRNG; it may be interesting to use this approach to reverse engineer a TRNG's design from the bitstream.

8   Conclusion

In this paper we have outlined an attack on ring-oscillator based random number generators. We have described the effect, and measured its consequences on a security microcontroller used in the EMV system, and in an EMV card. We believe this is an important effect, which all designers of random number generators should test.

Acknowledgements. Markus Kuhn suggested the experiments with the secure microcontroller and provided many components of the experimental setup plus valuable feedback on the paper. Steven Murdoch provided the Python EMV protocol library used to drive communication with the smartcard.

References

1. EMVCo, LLC: EMV 4.2 specification (June 2008) http://www.emvco.com/
2. Bellare, M., Goldwasser, S., Micciancio, D.: "Pseudo-random" number generation within cryptographic algorithms: The DSS case. In: Kaliski Jr., B.S. (ed.) CRYPTO 1997. LNCS, vol. 1294, pp. 277–291. Springer, Heidelberg (1997)
3. Bello, L.: DSA-1571-1 openssl – predictable random number generator. Debian Security Advisory (2008), http://www.debian.org/security/2008/dsa-1571
4. Hajimiri, A., Limotyrakis, S., Lee, T.H.: Jitter and phase noise in ring oscillators. IEEE J. Solid-State Circuits 34(6), 790–804 (1999)
5. Eastlake, D., Schiller, J., Crocker, S.: Best Common Practice 106: Randomness requirements for security. Technical report, IETF (2005)
6. Sunar, B., Martin, W.J., Stinson, D.R.: A provably secure true random number generator with built-in tolerance to active attacks. IEEE Trans. Computers 56(1), 109–119 (2007)
7. Bak, P.: The Devil's staircase. Physics Today 39(12), 38–45 (1986)
8. Adler, R.: A study of locking phenomena in oscillators. In: Proc. IRE and Waves and Electrons, vol. 34, pp. 351–357 (1946)
9. Mesgarzadeh, B., Alvandpour, A.: A study of injection locking in ring oscillators. In: Proc. IEEE International Symposium on Circuits and Systems, vol. 6, pp. 5465–5468 (2005)
10. Yoo, S.K., Karakoyunlu, D., Birand, B., Sunar, B.: Improving the robustness of ring oscillator TRNGs, http://ece.wpi.edu/~sunar/preprints/rings.pdf
11. Lai, X., Roychowdhury, J.: Analytical equations for predicting injection locking in LC and ring oscillators. In: IEEE 2005 Custom Integrated Circuits Conference, pp. 461–464 (2005)
12. Rukhin, A., et al.: A statistical test suite for random and pseudorandom number generators for cryptographic applications. Technical Report SP800-22, National Institute of Standards and Technology, USA (2008)
13. Brown, R.G., Eddelbuettel, D.: Dieharder: A random number test suite, http://www.phy.duke.edu/~rgb/General/dieharder.php (accessed 2009-03-03)
14. Mills, E.: Hacked ATMs let criminals steal cash, PINs. ZDNet UK (June 2009), http://news.zdnet.co.uk/security/0,1000000189,39660339,00.htm
15. Bogdanich, W.: Stealing the code: Con men and cash machines; criminals focus on A.T.M.'s, weak link in banking system. The New York Times (August 2003), http://query.nytimes.com/gst/fullpage.html?res=9803E6DD103EF930A3575BC0A9659C8B63
16. Rousseau, L.: pcsc_tools package: ATR table, http://ludovic.rousseau.free.fr/softwares/pcsc-tools/smartcard_list.txt (accessed 2009-03-03)
17. Sunar, B.: True random number generators for cryptography. In: Koç, Ç.K. (ed.) Cryptographic Engineering, pp. 55–74. Springer, Heidelberg (2009)
18. Herzel, F., Razavi, B.: A study of oscillator jitter due to supply and substrate noise. IEEE Trans. Circuits and Systems II 46(1), 36–42 (1999)

Appendix

Tabulated NIST Test Results from EMV Smartcard

Table 2. NIST results from EMV smartcard

	No injection			Apply f_{inject}		
	χ^2 P-value	Passes	Overall	χ^2 P-value	Passes	Overall
Frequency	0.3215	97.44%	PASS	0.0000	21.54%	FAIL
Block Frequency	0.6262	98.97%	PASS	0.0000	0.51%	FAIL
Cumulative Sums	0.2904	97.95%	PASS	0.0000	22.05%	FAIL
Cumulative Sums	0.3902	97.95%	PASS	0.0000	21.54%	FAIL
Runs	0.3811	99.49%	PASS	0.0000	40.00%	FAIL
Longest Run	0.3548	98.97%	PASS	0.0000	73.85%	FAIL
Rank	0.5501	100.00%	PASS	0.0000	0.00%	FAIL
FFT	0.0001	100.00%	PASS	0.0000	0.51%	FAIL
Non-Overlapping Template[a]	0.4523	99.00%	PASS	0.0000	90.89%	FAIL
Overlapping Template	0.4470	98.97%	PASS	0.0000	9.23%	FAIL
Universal	0.0211	98.97%	PASS	0.1488	98.46%	PASS
Approximate Entropy	0.1879	98.97%	PASS	0.0000	1.54%	FAIL
Random Excursions[b]	0.3050	99.26%	PASS	0.2836	99.50%	PASS
Random Excursions Variant[c]	0.4922	99.39%	PASS	0.3053	99.56%	PASS
Serial	0.1168	100.00%	PASS	0.0000	0.00%	FAIL
Serial	0.5501	98.97%	PASS	0.0000	0.00%	FAIL
Linear Complexity	0.0358	99.49%	PASS	0.9554	98.46%	PASS

Dataset of 195×10^6 random bytes. NIST performed 195 runs each using fresh 10^6 bytes. Minimum pass rate 96.86% except Random Excursions (96.25% no injection, 93.03% injected)

---

[a] Mean over 148 tests.

[b] Mean over 8 tests.

[c] Mean over 18 tests.

Low-Overhead Implementation of a Soft Decision Helper Data Algorithm for SRAM PUFs

Roel Maes[1], Pim Tuyls[1,2], and Ingrid Verbauwhede[1]

[1] K.U. Leuven ESAT/COSIC and IBBT, Leuven, Belgium
[2] Intrinsic-ID, Eindhoven, The Netherlands
{roel.maes,pim.tuyls,ingrid.verbauwhede}@esat.kuleuven.be

Abstract. Using a Physically Unclonable Function or PUF to extract a secret key from the unique submicron structure of a device, instead of storing it in non-volatile memory, provides interesting advantages like physical unclonability and tamper evidence. However, an additional Helper Data Algorithm (HDA) is required to deal with the *fuzziness* of the PUF's responses. To provide a viable alternative to costly protected non-volatile memory, the PUF+HDA construction should have a very low overhead. In this work, we propose the first HDA design using *soft-decision information* providing an implementation that occupies 44.8% less resources than previous proposals. Moreover, the required size of the used PUF can be reduced upto 58.4% due to the smaller entropy loss.

Keywords: Physically Unclonable Functions, Helper Data Algorithm, FPGA Implementation, Soft-Decision Decoder, Toeplitz Hash.

1 Introduction

The theoretical study of a cryptographic scheme aims to provide a well defined and quantitative understanding of its security. However, when the scheme enters the practical domain, more parameters join in the game. A security application does not only need to be as secure as possible, but also as inexpensive, fast, power-efficient and flexible as possible, which often means that the security is reduced in order to improve these practical characteristics. Moreover, the vast expansion of physical attacks on cryptographic implementations has shown that certain assumptions upon which the theoretical security of a scheme is based do not necessarily hold in practice, *e.g.* the existence of secure key storage. Private keys often need to be stored in publicly accessible devices, *e.g.* smart cards or RFID-tags, allowing adversaries to physically attack the implementation [1]. Tampering attacks [2,3], in which an attacker physically invades the device in order to extract sensitive information, are among the strongest known physical attacks and are in general always able to obtain the private key if no specific countermeasures are taken. Advanced techniques to detect and/or resist tampering in integrated circuits (*e.g.* [4,5]) are indispensable in security-sensitive applications, but unavoidably add to the *overhead* of the security aspect.

C. Clavier and K. Gaj (Eds.): CHES 2009, LNCS 5747, pp. 332–347, 2009.

Among the proposed tampering countermeasures, Physically Unclonable Functions or PUFs [6] take a special place because of their interesting properties and the cost-effective solutions they offer. A PUF implements a functionality that is highly dependent on the exact physical properties of the embedding device, down to a submicron level. PUFs on integrated circuits (ICs) take advantage of the intrinsic physical uniqueness of the device caused by unavoidable random deep-submicron manufacturing variations, which makes their behavior unique and unclonable. Moreover, since tampering attacks are bound to alter the physical integrity of the chip, they will also change the PUF's behavior [7] and PUFs can hence be used as a tamper detection mechanism. Their instance-specific unique behavior and their anti-tampering properties make PUFs on ICs ideal constructions for secure key storage, *i.e.* the PUF responses can be used to generate a unique and physically unclonable device key [8]. In addition, since the unique PUF behavior arises automatically, no non-volatile memory is needed for storing a key. A number of possible PUF implementations on ICs have been proposed, based on delay measurements [9,10] and power up values of memory elements [11,12]. In the latter category, SRAM-based PUFs exhibit convenient qualities: the power up states of SRAM cells are dependent on intrinsically present manufacturing variability which increases with shrinking technology nodes, SRAM cells are small, commonly used and available early in a new manufacturing process.

Since a PUF evaluation implies a physical measurement, the extraction of a key from the responses is not straightforward. Physical measurements are susceptible to noise and the measured random variables often come from a non-uniform source. On the other hand, we expect cryptographic keys to be highly reliable and to have full entropy to be secure. In order to bridge this gap, Helper Data Algorithms (HDAs) have been introduced [13,14], which are able to transform noisy and non-uniform variables into reliable and uniformly distributed bit strings using public *helper data*. This helper data, although it can be made public without disclosing any information about the extracted key, will always leak some entropy on the PUF responses. In short, one always needs to input more entropy into a HDA than the actual extracted key will contain, since part of it is leaked by the helper data. This *entropy loss* is a function of the noise levels of the input, and is an important characteristic of a HDA which should be minimized. In case of an SRAM PUF, the amount of entropy loss in the HDA relates directly to the number of SRAM cells needed in the PUF to extract a key and hence the size of the PUF on silicon. Since it is in our interest to minimize the implementation cost of the key storage, we would like this number to be as small as possible. The HDA itself also causes an overhead cost and its implementation should hence also be resource-optimized.

Contributions. In this work, we propose a new low-overhead design for a HDA that uses available soft-decision information. A practical FPGA implementation of the design is provided with a considerably lower implementation cost than previous HDA implementations [15], concerning both the required PUF size (58.4% smaller) and the resource usage of the HDA (44.8% smaller).

Related Work. SRAM PUFs were introduced in [11] and similar constructions are studied in [16,17,12]. They provide a practical PUF implementation because of the ease of use and general availability of SRAM cells on regular silicon devices. The concept of HDAs has been introduced as shielding functions in [14] and fuzzy extractors in [13]. A first efficient implementation on FPGA of a HDA for key extraction was proposed in [15]. We will refer regularly to this work and compare our results. The use of soft-decision information to improve performance is a long known result in channel coding and its usefulness for HDAs was first demonstrated in [18]. To the best of our knowledge, this work is the first to propose an efficient HDA implementation using soft-decision information.

2   Preliminaries

2.1   Helper Data Algorithms

A noisy and partially random variable, like a PUF response or a biometric, is often referred to as a *fuzzy secret*. Helper Data Algorithms (HDAs) are used to extract cryptographic keys from fuzzy secrets, and have been introduced as *fuzzy extractors* in [13] or *shielding functions* in [14]. We will use the formal definition of a fuzzy extractor from [13]:

Definition 1 (Fuzzy Extractor). *A $(m, n, \delta, \mu, \epsilon)$-fuzzy extractor is a pair of randomized procedures,* generate *(Gen) and* reproduce *(Rep):*

1. *The generation procedure* Gen *on input $X \in \{0,1\}^m$ outputs an extracted string $S \in \{0,1\}^n$ and helper data $W \in \{0,1\}^*$.*
2. *The reproduction procedure* Rep *takes an element $X' \in \{0,1\}^m$ and a bit string $W \in \{0,1\}^*$ as inputs. The correctness property of fuzzy extractors guarantees that if the Hamming distance $dist[X; X'] \leq \delta$ and S, W were generated by $(S, W) \leftarrow$ Gen(X), then* Rep$(X', W) = S$.
3. *The security property guarantees that for any distribution $\mathbb{D}$ on $\{0,1\}^m$ of min-entropy μ, the string S is nearly uniform even for those who observe W: if $(S, W) \leftarrow$ Gen$(X \leftarrow \mathbb{D})$, then it holds that the statistical distance $\Delta[(S, W); (U \leftarrow \mathbb{U}_n, W)] \leq \epsilon$, with $\mathbb{U}_n$ the uniform distribution on $\{0,1\}^n$.*

The used notions of min-entropy and statistical distance are described in Appendix A. The correctness property of fuzzy extractors takes care of possible noise in the fuzzy secret. As long as the distance between the fuzzy secret during generation and reproduction is limited, the same extracted output can be obtained. This is also known as *information reconciliation*. The security property tells us that the extracted output is very close to uniform as long as the fuzzy secret contains a sufficient amount of min-entropy, even when the helper data is observed. This is called *privacy amplification*. The important contribution of HDAs is the ability to extract a *private key* from a fuzzy secret if a *public helper channel* is available. It is however important to guarantee the integrity of the helper data [19]. The information reconciliation and privacy amplification

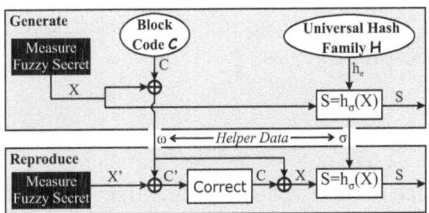 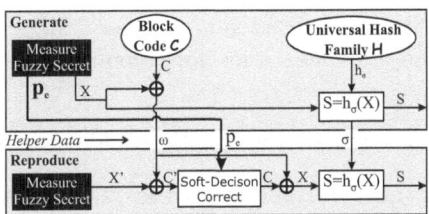

(a) Classical helper data algorithm using the code offset technique and a universal hash function.

(b) Proposed soft-decision helper data algorithm. The additional *soft-decision information*, available as the bit error probabilities p_e, is transfered as helper data.

Fig. 1. Constructions of a helper data algorithm

functionality of a HDA are typically implemented by two separate algorithms. We elaborate on a common construction for both:

Information Reconciliation with the Code Offset Technique [13]. A binary linear block code $\mathcal{C}$ with parameters $[n, k, d]$ contains code words of length n, dimension k and minimal Hamming distance d and is able to correct at least $t = \lfloor (d-1)/2 \rfloor$ bit errors occurring in a single code word. The code offset technique picks a uniformly random code word, denoted as $C \leftarrow \mathcal{C}$, in the generation phase and calculates the offset between the fuzzy secret X and C: $\omega = X \oplus C$. This offset ω is made publicly available as helper data. In the reproduction phase, a new version X' of the fuzzy secret is measured and $C' = X' \oplus \omega$ is calculated. If $\mathbf{dist}[X; X'] \equiv \mathbf{dist}[C; C'] \leq t$ then C' can be corrected to C: $C = \mathsf{Correct}(C')$, which allows the reproduction of $X = C \oplus \omega$. As observed in [13], publishing ω amounts to a min-entropy loss of $n-k$, i.e. $\tilde{\mathbf{H}}_\infty(X|\omega) = \mathbf{H}_\infty(X) - n + k$. We aim to minimize this loss while maintaining an acceptable level of error-correction.

Privacy Amplification with Universal Hash Functions [20]. A universal hash family $\mathcal{H}$ with parameters $[a, b]$ is a set of functions $\{h_i : \{0,1\}^a \rightarrow \{0,1\}^b\}$ such that the collision probability on two distinct inputs is at most 2^{-b} for a randomly picked function from $\mathcal{H}$: $\mathbf{Pr}(h_R(x) = h_R(x')) \leq 2^{-b}, \forall x \neq x' \in \{0,1\}^a$ and $h_R \leftarrow \mathcal{H}$. The *left-over hash lemma* [21] states that universal hash functions can act as a "magnifying glass" for randomness: when taking a random variable with limited min-entropy as input, the output distribution will be close to uniform. Upon generation, a function $h_\sigma \leftarrow \mathcal{H}$ is randomly selected and applied to X to obtain a random output $S = h_\sigma(X)$. The index σ is made available as helper data such that the same hash function can be used in the reproduction procedure to reproduce S from X' after information reconciliation.

Complete helper data algorithm. Using the two techniques mentioned above, a complete HDA can be constructed, as shown in Figure 1(a). The helper data W consists of the code offset ω and the hash function index σ: $W = (\omega, \sigma)$. The two main blocks to be implemented in order to perform this HDA are an error-correcting decoder and a universal hash function. In this work, we carefully select the parameters of these blocks and provide a resource-optimized design and implementation for reconfigurable hardware devices.

2.2   Soft-Decision Error Correction

A classic method to increase the performance of an error-correcting decoder, and hence decrease the code redundancy $(n - k)$, is using *soft-decision information* in the decoding algorithm. This technique could equivalently lower the entropy loss of a HDA. Soft-decision decoding is possible when *reliability measures* for received bits are available, which is called the soft-decision information. Two well known soft-decision decoding algorithms are the Viterbi algorithm for convolutional codes [22] and the belief propagation algorithm for LDPC codes [23]. However, both types are inappropriate for use in the code offset technique since they require very long data streams to work efficiently while the length of a fuzzy secret is often limited. We would like to use a soft-decision decoding algorithm for rather short linear block codes $(n \leq 2^8)$ in order to maintain efficiency. We discuss two such decoders:

Soft-decision Maximum-Likelihood Decoding (SDML) is a straightforward algorithm that selects the code word that was most likely transmitted based on the bit reliabilities. SDML achieves the best error-correcting performance possible, but generally at a decoding complexity exponential in the code dimension k. Repetition codes $(k = 1)$ can still be efficiently SDML decoded. Conversely, if $k = n$, SDML decoding degenerates to making a hard decision on every bit individually based on its reliability, and if $k = n - 1$, the block code is a parity check code and SDML decoding is done efficiently by flipping the least reliable bit to match the parity. This last technique is known as Wagner decoding [24].

Generalized Multiple Concatenated Codes (GMC). An r-th order Reed-Muller code $\mathrm{RM}_{r,m}$, is a linear block code with parameters $n = 2^m$, $k = \sum_{i=0}^{r} \binom{m}{i}$ and $d = 2^{m-r}$. It is well known that $\mathrm{RM}_{r,m}$ can be decomposed in the concatenation of two shorter inner codes, $\mathrm{RM}_{r-1,m-1}$ and $\mathrm{RM}_{r,m-1}$, and a simple length-2 block code as outer code. This decomposition can be applied recursively until one reaches $\mathrm{RM}_{0,m'}$, which is a repetition code, or $\mathrm{RM}_{r'-1,r'}$ (or $\mathrm{RM}_{r',r'}$), which is a parity check (or degenerated) code, all of which can be efficiently soft-decision decoded with SDML. This technique, known as Generalized Multiple Concatenated decoding (GMC) [25], yields a much lower decoding complexity then SDML, but only a slightly decreased error-correcting capability.

2.3   SRAM PUFs and Soft-Decision Helper Data

Extensive experiments in [11] show that the power up value of a randomly selected SRAM cell is random over $\{0, 1\}$, but tends to take the same value at every power up. This is due to the random manufacturing mismatch between the electrical parameters defining the cell's behavior. The power up values of SRAM cells can hence be used as PUF responses, and the function taking an

SRAM cell's address as challenge and returning its power up value as response is called an SRAM PUF. Occasionally, a cell is encountered with no distinct preference toward 0 or 1, introducing noisy bits.

Previous proposals concerning key extraction from an SRAM PUF [11,15] assume that the bit error probability of a response is constant, *i.e.* every response bit has the same probability of being measured incorrectly. However, experimental data shows that this is not quite the case, as most cells only very rarely produce a bit error while a minority of cells are faulty more often. In fact, the error probability of a randomly selected cell is itself a random variable drawn according to a certain distribution, and hence not a constant value. A theoretical derivation of this distribution, based on a model for the manufacturing variability in SRAM cells, was given in [18] and is summarized in Appendix B. It is clear that using a block code adapted to the *average* bit error rate, as in [11,15], is overly pessimistic for the majority of the bits, as most of them have an error probability much smaller than the average. The distribution of the error probability in both cases is shown in Figure 2, and from Figure 2(b) it is clear that in this case, around 60% of the bits have an error probability which is smaller than the assumed *fixed* average. A construction that takes into account the specific error probability of the individual bits would achieve a better overall performance, needing less redundancy and hence causing a smaller min-entropy loss. This is precisely what soft-decision decoding achieves. A HDA based on soft-decision decoding in the code-offset technique is shown in Figure 1(b).

In Section 3.1 we present a hardware design for an information reconciliation algorithm that uses the individual error probabilities of the response bits as soft-decision information. The bit error probability is measured during the generation phase and made publicly available as helper data. It is hence important to know the min-entropy leakage caused by revealing the error probabilities. It turns out that revealing P_e does not leak any min-entropy on the response X, *i.e.* $\widetilde{\mathbf{H}}_\infty(X|P_e) = \mathbf{H}_\infty(X)$. A proof for this statement is given in [18]. Measuring the bit error probability amounts to performing multiple measurements of every

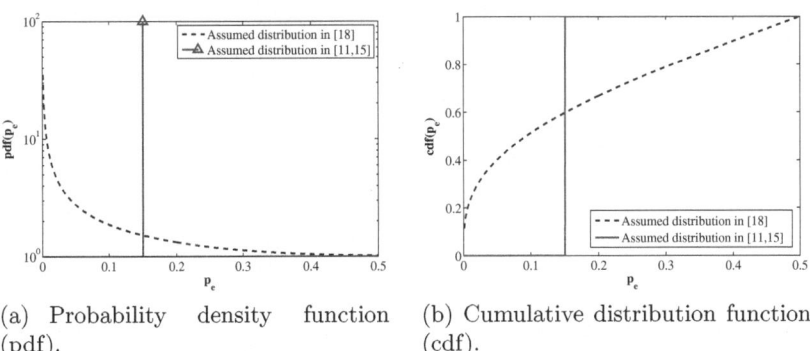

(a) Probability   density   function (pdf).

(b) Cumulative distribution function (cdf).

Fig. 2. Distributions of the bit error probability as assumed respectively in [18] and [11,15]. The expected value for P_e is set equal for both cases: $\mathbb{E}[P_e] = 15\%$.

response bit and estimating the most-likely value, which could be inefficient. Our simulations show that an estimate based on a limited amount of measurements, in the order of 10 to 100, already greatly improves the decoder performance. Moreover, this measurement should be performed only once for every PUF.

3   Designing a Soft-Decision HDA for FPGA

This section provides the main contribution of this work, *i.e.* a low-overhead design of a soft-decision helper data algorithm for a reconfigurable hardware device. A first efficient HDA implementation for FPGAs was given in [15]. We will build upon this work and try to improve their results. Sections 3.1 and 3.2 describe the respective design choices for the information reconciliation and the privacy amplification algorithm that we choose to implement.

3.1   Soft-Decision Information Reconciliation Design

The code offset technique as described in Section 2.1 is an efficient technique for turning an error-correcting decoder into an information reconciliation algorithm. As motivated in Section 2.3, we will use a soft-decision decoder to reduce the min-entropy loss of the information reconciliation.

Representing the Soft-Decision Information. As stated in Section 2.3 and shown in Fig. 1(b), the error probability p_{e_i} of an individual SRAM cell i will be used as soft-decision helper data. In general, p_{e_i} takes real values in $]0, \frac{1}{2}[$ and we need to determine a way of representing p_{e_i} in binary format, such that it can be efficiently used in a soft-decision decoder. We denote a codeword C from a block code $\mathcal{C}$ of length n as $C = (C_0, \ldots, C_{n-1})$, an n-bit SRAM PUF response as $X = (X_0, \ldots, X_{n-1})$ and the corresponding vector with error probabilities as $p_e = (p_{e_0}, \ldots, p_{e_{n-1}})$. When receiving an n-bit possibly noisy code word $C' = X' \oplus \omega$, with ω the code offset helper data as defined in 2.1, a soft-decision decoder tries to find the corrected code word C^* which maximizes the (log-)likelihood:

$$C^* = \underset{C \in \mathcal{C}}{\operatorname{argmax}} \prod_{i=0}^{n-1} (1 - p_{e_i})^{(C_i' \oplus C_i)} \cdot p_{e_i}^{(1 \oplus C_i' \oplus C_i)},$$

$$= \underset{C \in \mathcal{C}}{\operatorname{argmax}} \sum_{i=0}^{n-1} (-1)^{C_i} \cdot (-1)^{C_i'} \cdot \left(\log_\beta (1 - p_{e_i}) - \log_\beta (p_{e_i}) \right),$$

with $\beta > 1$ a design parameter. For convenience, we work with the log-likelihood and choose the soft-decision helper data s_i of an SRAM PUF response bit i to be:

$$s_i \overset{\text{def}}{=} \lfloor \log_\beta (1 - p_{e_i}) - \log_\beta (p_{e_i}) \rfloor, \qquad (1)$$

which is a deterministic function of the error probability and an integer approximation of the magnitude of the log-likelihood of bit i. For a noisy PUF

Algorithm 1. SDML-DECODE-Repetition$_n(L)$ with *soft output*

$L^* := \sum_{i=0}^{n-1} L_i$
return $(L^*, \ldots, L^*)_n$

Algorithm 2. GMC-DECODE-RM$_{r,m}(L)$ with *soft output*

---

define $F(x,y) := \text{sign} (x \cdot y) \cdot \min \{|x|, |y|\}$
define $G(s,x,y) := \lfloor \frac{1}{2} (\text{sign} (s) \cdot x + y) \rfloor$
if $r = 0$ **then**
    $L^* = \text{SDML-DECODE-Repetition}_{2^m}(L)$
else if $r = m$ **then**
    $L^* = L$
else
    $L_j^{(1)} = F(L_{2j-1}, L_{2j}), \forall j = 0 \ldots 2^{m-1} - 1$
    $L^{(1)*} = \text{GMC-DECODE-RM}_{r-1,m-1}(L^{(1)})$
    $L_j^{(2)} = G(L_j^{(1)*}, L_{2j-1}, L_{2j}), \forall j = 0 \ldots 2^{m-1} - 1$
    $L^{(2)*} = \text{GMC-DECODE-RM}_{r,m-1}(L^{(2)})$
    $L^* = \left(F(L_0^{(1)*}, L_0^{(2)*}), L_0^{(2)*}, \ldots, F(L_{2^{m-1}-1}^{(1)*}, L_{2^{m-1}-1}^{(2)*}), L_{2^{m-1}-1}^{(2)*} \right)$
end if
return L^*

---

response X', the soft-decision information that enters the decoder is calculated as: $L_i = (-1)^{X_i' \oplus \omega_i} \cdot s_i$. The decoder tries to find the code word $C^* = \text{argmax}_{C \in \mathcal{C}} \sum_{i=0}^{n-1} (-1)^{C_i} \cdot L_i$. In the remainder of the text, s_i and L_i will be represented by 8-bit signed (2's-complement) integers $\in [-128, 127]$. The log-base β is a design parameter that is chosen large enough to avoid overflows in the decoder algorithm, but as small as possible to keep the approximation error small.

Choosing a Soft-Decision Decoder Algorithm. Among the linear block codes, Reed-Muller codes have a relatively high error-correcting performance similar to BCH codes, and are easier to decode. As explained in Section 2.2, there exists also a relatively efficient algorithm for soft-decision decoding of Reed-Muller codes based on GMC. Bösch et al. [15] demonstrate that using *code concatenation*, where the decoded words from an *inner code* form a code word from an *outer code*, can substantially reduce the min-entropy loss. A balanced concatenation of two different codes, *e.g.* a repetition code and a Reed-Muller code, will achieve a better performance than the case were only a single code is considered. Taking all this into account, we decide to implement a soft-decision decoder as a concatenation of a SDML repetition decoder for the inner code and a GMC Reed-Muller decoder for the outer code.

SDML repetition decoding of soft-decision information L amounts to calculating $L^* = \sum_{i=0}^{n-1} L_i$. The most-likely transmitted code word was all zeros if $L^* > 0$ and all ones if $L^* < 0$. Moreover, the magnitude of L^* gives a reliability for this decision which allows to perform a second soft-decision decoding for the outer code. Algorithm 1 outlines the simple operation for the SDML decoding of a repetition code. As an outer code, we use a $RM_{r,m}$ code and decode it with an adapted version of the soft-decision GMC decoding algorithm as introduced in [25]. The soft-decision output of the repetition decoder is used as input by the GMC decoder. The operation of the GMC decoder we use is given by Algorithm 2. Note that this a recursive algorithm, calling itself twice if $0 < r < m$.

Decoder Design. We propose a hardware architecture to efficiently execute the soft-decision decoders given by Algorithms 1 and 2. Since our main design goal is providing an as small as possible HDA implementation, we try to minimize the used hardware resources. As a general architecture, we opt for a highly serial execution of the algorithms using a small 8-bit custom datapath. Looking at the algorithms, we identify the following major operations:

- Algorithm 1 performs a summation of n 8-bit integers. We implement this serially using an 8-bit signed accumulator.
- To evaluate the function $F(x, y)$ in Algorithm 2, we propose a 3-cycle execution. In the first two cycles, $x + y$ and $x - y$ are computed and their signs c_+ and c_- are stored. In the third cycle, the output is computed as $F(x, y) = c_+ \cdot (x \cdot (c_+ \neq c_-) + y \cdot (c_+ = c_-))$. This last operation amounts to choosing between x and y and possibly changing the sign based on the values of c_+ and c_- and can be done with an adder/subtracter with one of the inputs set to zero.
- For $G(s, x, y)$ in Algorithm 2, we propose a 2-cycle execution. In the first cycle, the sign of s is loaded and in the second cycle, $G(s, x, y)$ can be calculated as an addition or subtraction of x and y based on sign (s), followed by a bit shift.

To be able to execute these operations, we propose the arithmetic unit (AU) depicted in gray in Figure 3. The signed adder/subtracter can change the sign of any of its inputs, or put them to zero. The sign bits of the two previous AU

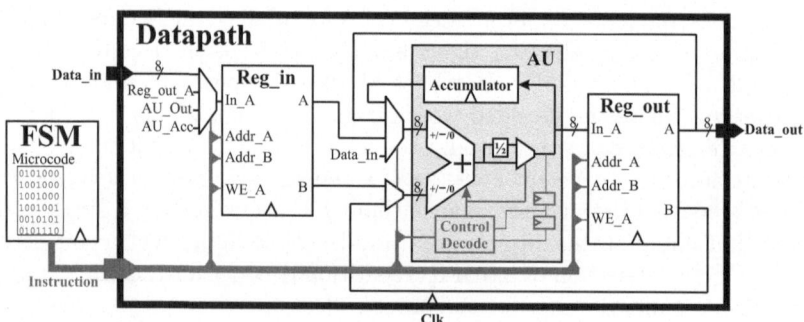

Fig. 3. Details of the soft-decision decoder architecture. The datapath consists of an Arithmetic Unit (AU) and an input and output register file. The controller contains the microcode to execute the decoder algorithm.

outputs are used as control signals. The AU is combined with an input and output dual port register file into a custom 8-bit datapath as shown in Figure 3. Dual port register files can be efficiently implemented on an FPGA using SRAM-based Lookup Tables (LUTs). The depth of the register files depends on the choice of the decoder parameters. The algorithm execution is controlled by an FSM applying the consecutive algorithm steps that are stored as microcode. An operational example of a soft-decision decoder using this design is presented in Section 4, providing detailed implementation parameters and performance results.

3.2 Privacy Amplification Design

In Section 2.1, it was mentioned that privacy amplification amounts to applying a universal hash function. Krawczyk [26] proposed an LFSR-based Toeplitz universal hash algorithm which performs the multiplication of a random Toeplitz matrix with the hash argument. As Krawczyk already showed, this algorithm can be efficiently implemented in hardware, since the columns of the pseudorandom Toeplitz matrix can be generated by an LFSR, and the resulting product of each column with the hash input is accumulated to calculate the full matrix product. This construction is shown in Figure 4(a) and was implemented on an FPGA in [15]. However, the need for an LFSR and an accumulator register of the same size as the key (e.g. 128 bit) and an input register of the argument size (e.g. 64 bit) yields a relatively expensive implementation on an FPGA, when regular flip-flops are used to implement them. This is because the number of available flip-flops on typical FPGAs is rather low. In [15], this results in the hash algorithm occupying the major part of the used resources for the HDA. More resource-efficient methods for implementing shift registers on FP-GAs exist [27], however, they cannot be used directly in Krawczyk's algorithm,

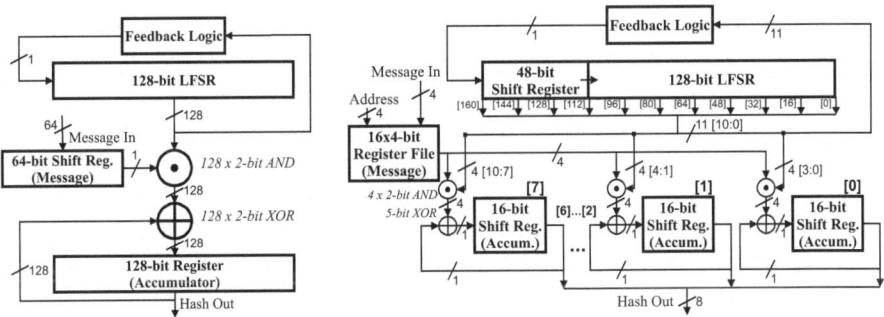

(a) The fully parallel datapath as proposed in [26] and implemented in [15].

(b) Our serialized datapath optimized for implementation with 16-bit shift registers.

Fig. 4. Datapath for (64-bit in/128-bit out) Toeplitz hash implementation

since parallel access to all the bits in the LFSR is required. The implementation from [27] only allows parallel access to every 16th bit of the LFSR state. We reworked the algorithm such that it can be executed in a serial way, using the resource-efficient shift register implementations. This required some modifications to the datapath, but the functional behavior of the algorithm is preserved. The basic idea behind the serialization is that, in stead of accumulating an entire 128-bit product in every cycle, only a partial product (using the accessible bits) is calculated and accumulated in 16-bit rotation shift registers. The resulting datapath is shown in Figure 4(b). This drastically decreases the resource usage of the FPGA implementation as shown by the implementation results in Section 4.

4   Implementation Parameters and Results

In this section, details of a full HDA implementation are provided and compared to the results from [15]. In order to make a fair comparison, the same values for the *average* bit error probability (15%), the amount of min-entropy in the SRAM PUF responses (78%) and for the decoder failure rate ($\leq 10^{-6}$) are chosen.

Determining the decoder parameters. We simulated our decoder proposal in software with SRAM PUF responses sampled according to the proposed distribution from [18]. The bit error probabilities are estimated from 64 measurements. We compared the number of SRAM PUF response bits that were necessary to obtain $\lceil 128/0.78 \rceil = 171$ non-redundant bits after decoding with a failure rate $\leq 10^{-6}$ for different parameters (n, r, m) of the decoder. The best decoder we tested is the one with code parameters $(n = 3, r = 2, m = 6)$ and the design parameter $\beta = 1.8$, and uses $\lceil 171/22 \rceil \times 3 \times 64 = 1536$ SRAM PUF response bits.

FPGA implementation. We described our design in VHDL and synthesized and implemented it on a Xilinx Spartan-3E500 FPGA using Xilinx ISE Design Suite 10.1. The implementation results concern the routed netlist. The functional correctness and the cycle count is tested by applying test benches with ModelSim.

Soft-Decision Decoder: The decoder takes 192×8-bit log-likelihoods as input and every three consecutive values are accumulated (repetition decoded) to obtain 64×8-bit inputs for the $RM_{2,6}$-decoder which outputs a 64-bit error-corrected code word. To execute Algorithm 2, the input and output register file size are respectively set to 64×8-bit and 32×8-bit. The instructions to carry out Algorithm 2 are stored as embedded microcode. The FPGA implementation occupies 164 slices and 2×16-kbit Block RAMs. The critical path is 19.9ns and one complete decoding cycle (input + decode + output) finishes in 1248 cycles.

LFSR-based Toeplitz hash: The universal hash function accepts 64-bit message blocks and hashes them in a 128-bit value. Our implementation occupies 59 slices. The critical path is 9.2ns and one complete hash cycle (input seed + input message + hash + output) finishes in 432 cycles.

Table 1. Implementation and performance results on a Xilinx Spartan-3E500 FPGA compared to the results from [15]. The given results concern HDA implementations which take SRAM PUF response bits with a 15% average error probability and 78% min-entropy as an input and produce a full-entropy 128-bit key with failure rate $\leq 10^{-6}$.

		(1)	(2)	(3)
	(1) The soft-decision HDA implementation as proposed in this section.			
	(2) The HDA implementation from [15] with the lowest SRAM usage.			
	(3) The HDA implementation from [15] with the lowest HDA resource overhead.			
Decoder (1 round)	Slices	164	580	110
	Block RAMs	2	?*	?*
	Cycles	1248	1716	855
	SRAM Usage	192 bit	264 bit	176 bit
Toeplitz Hash (1 round)	Slices	59	327	319
	Cycles	432	96	64
Complete HDA	Slices (Spartan-3E500)	237 (5.1%)	$\geq$ **907** ($\geq$ 19.5%)	$\geq$ **429** ($\geq$ 9.2%)
	Block RAMs (Spartan-3E500)	2 (10%)	?* (?*)	?* (?*)
	Critical Path	19.9 ns	6.6 ns	5.7 ns
128-bit Key Extraction	Rounds	8	14	35
	Cycles	10298	$\geq$ 24024	$\geq$ 29925
	Performance	205 μs @ 50.2 MHz	$\geq$ 159 μs @ 151.5 MHz	$\geq$ 171 μs @ 175.4 MHz
	SRAM Usage	**1536 bit**	**3696 bit**	**6160 bit**
	Helper Data Size	13952 bit	3824 bit	6288 bit

* The results from [15] for (2) and (3) do *not* include the resources for the controller, hence the number of Block RAMs needed for algorithm control cannot be compared.

Complete HDA: The complete HDA executes the decoder $\lceil 171/22 \rceil = 8$ times and hashes the 8 corrected 64-bit words into a 128-bit key. The implementation of the full HDA + control occupies 237 slices and 2×16-kbit Block RAMs for the microcode. The critical path is 19.9ns and the complete key generation (initialize + 8× decode and hash + output) finishes in 10298 cycles.

Discussion with respect to previous results. The two main parameters we want to optimize are the SRAM usage of the SRAM PUF and the resource overhead of the HDA implementation. Table 1 compares our implementation results (1) to two different implementations from [15]: (2) the implementation with the lowest SRAM usage, implementing a concatenation of a Golay[24,13] code and a Repetition[11,1] code[1] and (3) the implementation with the lowest HDA resource overhead, implementing a concatenation of a $RM_{1,4}$ code and a Repetition[11,1] code. It is clear from Table 1 that our soft-decision based implementation outperforms the previous implementations on both characteristics. The construction

---

[1] We remark that a decoder with an even lower SRAM usage than (2), but still higher than our implementation, is proposed in [15], based on BCH codes. However, no implementation is provided and no fair comparison can be made.

proposed in this section uses 58.4% less SRAM bits and over 44.8% less slices than the respective optimized implementations from [15]. These improvements come at the cost of an increased helper data size ($\times 3.6$) and the need to perform multiple measurement during generation to obtain the soft-decision information. On the other hand, more helper data is not necessarily a problem in many applications, since the helper data can be (externally) stored and transfered in plain without revealing information about the key, only its integrity should be guaranteed. The actual PUF+HDA implementation, *e.g.* residing on an embedded device, remains small. Measuring the error probability of the SRAM PUF cells can be done together with the regular functional testing of the IC right after manufacturing. Performing 10 to 100 measurements can be done relatively fast. Even when very few (< 10) measurements are available, the reconfigurable decoder allows to use stronger codes and remains more efficient than hard decision decoding. We also note that an average error of 15%, as assumed here and in [15] is very safe. Experiments on SRAM PUFs show error probabilities as low as 5%, requiring less initial measurements for the soft-decision decoder to be effective.

5   Conclusion

The bit error probability of an SRAM PUF is not a constant value, but a random variable for every individual response bit. This observation suggests the use of soft-decision information to lower the min-entropy loss of the helper data algorithm, resulting in a more efficient use of the SRAM PUF. We propose a design of a soft-decision helper data algorithm and implement it on an FPGA. A soft-decision Reed-Muller decoder is implemented using a small custom 8-bit datapath which can be easily reconfigured to work with different code parameters depending on the noise levels. The privacy amplification is performed by a serialized LFSR-based Toeplitz hash implementation that makes optimal use of the available FPGA resources. Both constructions constitute to a HDA which has a considerably lower implementation overhead than previous proposals and can even be of independent interest in other domains. The drawbacks of having to store more helper data and having to perform multiple initial measurements are no issue in many applications and should be considered as trade-offs. In any case, this work presents a new direction in the exploration of the design space of efficient helper data algorithms.

Acknowledgments

This work was supported by the IAP Program P6/26 BCRYPT of the Belgian State and by K.U.Leuven-BOF funding (OT/06/04). The first author's research is funded by IWT-Vlaanderen under grant number 71369.

References

1. Verbauwhede, I., Schaumont, P.: Design methods for security and trust. In: Proc. of Design Automation and Test in Europe (DATE 2008), NICE,FR, p. 6 (2007)
2. Anderson, R.J., Kuhn, M.G.: Low Cost Attacks on Tamper Resistant Devices. In: Christianson, B., Lomas, M. (eds.) Security Protocols 1997. LNCS, vol. 1361, pp. 125–136. Springer, Heidelberg (1998)
3. Skorobogatov, S.P.: Semi-invasive attacks - A new approach to hardware security analysis. University of cambridge, computer laboratory: Technical report (April 2005)
4. Yang, J., Gao, L., Zhang, Y.: Improving Memory Encryption Performance in Secure Processors. IEEE Trans. Comput. 54(5), 630–640 (2005)
5. Posch, R.: Protecting Devices by Active Coating. Journal of Universal Computer Science 4(7), 652–668 (1998)
6. Ravikanth, P.S.: Physical one-way functions. PhD thesis, Chair-Benton, Stephen, A. (2001)
7. Tuyls, P., Schrijen, G.-J., Škorić, B., van Geloven, J., Verhaegh, N., Wolters, R.: Read-Proof Hardware from Protective Coatings. In: Goubin, L., Matsui, M. (eds.) CHES 2006. LNCS, vol. 4249, pp. 369–383. Springer, Heidelberg (2006)
8. Tuyls, P., Batina, L.: RFID-Tags for Anti-Counterfeiting. In: Pointcheval, D. (ed.) CT-RSA 2006. LNCS, vol. 3860, pp. 115–131. Springer, Heidelberg (2006)
9. Gassend, B., Clarke, D., van Dijk, M., Devadas, S.: Silicon physical random functions. In: CCS 2002: Proceedings of the 9th ACM conference on Computer and communications security, pp. 148–160. ACM, New York (2002)
10. Lee, J.W., Lim, D., Gassend, B., Suh, G.E., van Dijk, M., Devadas, S.: A technique to build a secret key in integrated circuits for identification and authentication applications. In: VLSI Circuits, 2004. Technical Papers, pp. 176–179 (June 2004)
11. Guajardo, J., Kumar, S.S., Schrijen, G.-J., Tuyls, P.: FPGA Intrinsic PUFs and Their Use for IP Protection. In: Paillier, P., Verbauwhede, I. (eds.) CHES 2007. LNCS, vol. 4727, pp. 63–80. Springer, Heidelberg (2007)
12. Kumar, S.S., Guajardo, J., Maes, R., Schrijen, G.J., Tuyls, P.: Extended abstract: The butterfly PUF protecting IP on every FPGA. In: IEEE International Workshop on Hardware-Oriented Security and Trust (HOST-2008), pp. 67–70 (2008)
13. Dodis, Y., Ostrovsky, R., Reyzin, L., Smith, A.: Fuzzy extractors: How to generate strong keys from biometrics and other noisy data. SIAM Journal on Computing 38(1), 97–139 (2008)
14. Linnartz, J.P.M.G., Tuyls, P.: New shielding functions to enhance privacy and prevent misuse of biometric templates. In: Kittler, J., Nixon, M.S. (eds.) AVBPA 2003. LNCS, vol. 2688, pp. 393–402. Springer, Heidelberg (2003)
15. Bösch, C., Guajardo, J., Sadeghi, A.-R., Shokrollahi, J., Tuyls, P.: Efficient Helper Data Key Extractor on FPGAs. In: Oswald, E., Rohatgi, P. (eds.) CHES 2008. LNCS, vol. 5154, pp. 181–197. Springer, Heidelberg (2008)
16. Su, Y., Holleman, J., Otis, B.: A Digital 1.6 pJ/bit Chip Identification Circuit Using Process Variations. IEEE Journal of Solid-State Circuits 43(1), 69–77 (2008)
17. Holcomb, D.E., Burleson, W.P., Fu, K.: Initial SRAM state as a fingerprint and source of true random numbers for RFID tags. In: Proceedings of the Conference on RFID Security (2007)
18. Maes, R., Tuyls, P., Verbauwhede, I.: A Soft Decision Helper Data Algorithm for SRAM PUFs. In: IEEE International Symposium on Information Theory (2009)

19. Boyen, X.: Reusable Cryptographic Fuzzy Extractors. In: ACM CCS 2004, pp. 82–91. ACM Press, New York (2004)
20. Carter, J.L., Wegman, M.N.: Universal classes of hash functions. In: STOC 1977: Proceedings of the 9th ACM symposium on Theory of computing, pp. 106–112. ACM Press, New York (1977)
21. Bennett, C.H., Brassard, G., Robert, J.-M.: Privacy Amplification by Public Discussion. SIAM J. Comput. 17(2), 210–229 (1988)
22. Viterbi, A.: Error bounds for convolutional codes and an asymptotically optimum decoding algorithm. IEEE Trans. on Information Theory 13(2), 260–269 (1967)
23. Gallager, R.G.: Low Density Parity-Check Codes. IRE Trans. Inform. Theory 8, 21–28 (1962)
24. Silverman, R.A., Balser, M.: Coding for Constant-Data-Rate Systems-Part I. A New Error-Correcting Code. Proceedings of the IRE 42(9), 1428–1435 (1954)
25. Schnabl, G., Bossert, M.: Soft-decision decoding of Reed-Muller codes as generalized multiple concatenated codes. IEEE Trans. on Information Theory 41(1), 304–308 (1995)
26. Krawczyk, H.: LFSR-based Hashing and Authentication. In: Desmedt, Y.G. (ed.) CRYPTO 1994. LNCS, vol. 839, pp. 129–139. Springer, Heidelberg (1994)
27. George, M., Alfke, P.: Linear Feedback Shift Registers in Virtex Devices (April 2007)

A   Measures of Randomness

We briefly describe some concepts from information theory which are used to quantify the notion of the *amount of randomness* present in a measured variable, *i.e.* statistical distance and min-entropy. Let X and Y be two (discrete) possibly correlated random variables taking values from a set $\mathcal{S}$. We define:

- The statistical distance between (the distributions of) X and Y as:
  $\Delta[X;Y] \overset{\text{def}}{=} \frac{1}{2} \sum_{s \in \mathcal{S}} |\mathbf{Pr}(X = s) - \mathbf{Pr}(Y = s)|$.
- The min-entropy of (the distribution of) X as:
  $\mathbf{H}_\infty(X) \overset{\text{def}}{=} -\log_2 \max\{\mathbf{Pr}(X = s) : s \in \mathcal{S}\}$.
- The average conditional min-entropy [13] of (the distribution of) X given Y as: $\widetilde{\mathbf{H}}_\infty(X|Y) \overset{\text{def}}{=} -\log_2 \mathbb{E}_y\left[2^{-\mathbf{H}_\infty(X|Y=y)}\right]$.

B   SRAM PUF Response Model and Distribution

As is clear from the construction of an SRAM PUF as described in Section 2.3, and also in [11], the generation of an SRAM PUF response bit is determined by the stochastic mismatch of the electrical parameters in an SRAM cell. A simple model for this mismatch is proposed in [18] and summarized here. Let M and N be two normally distributed random variables with respective probability density functions $\varphi_{\mu_M, \sigma_M}$ and φ_{0, σ_N}. $\varphi_{\mu, \sigma}$ is the probability density function of a normal distribution with mean μ and standard deviation σ. A value $m_i \leftarrow M$ is i.i.d. sampled every time a new SRAM cell i is manufactured and represents the random device mismatch in the cell caused by manufacturing variation. A value $n_i^{(j)} \leftarrow N$ is i.i.d. sampled at the j-th power up of cell i and represents the amplitude of the stochastic noise voltage acting on cell i at the time of the power up. The power up state of SRAM cell i after the j-th power up is denoted as $x_i^{(j)} \in \{0, 1\}$, and it is assumed that $x_i^{(j)}$ is fully determined by m_i and $n_i^{(j)}$:

$$
x_i^{(j)} = \begin{cases} 0 & , \text{if } m_i + n_i^{(j)} > T, \\ 1 & , \text{if } m_i + n_i^{(j)} \leq T, \end{cases} \tag{2}
$$

with T a *threshold parameter* for a specific SRAM technology.

The power up behavior of an SRAM cell i is described by the probability p_{x_i} that this cell powers up as '1', and the related probability p_{e_i} that this cell produces a bit error. Both parameters are themselves random variables. They are sampled for a particular SRAM cell at manufacturing time according to their respective distributions:

$$
\mathbf{pdf}_{P_r}(x) = \frac{\lambda_1 \cdot \varphi\left(\lambda_2 - \lambda_1 \cdot \Phi^{-1}(x)\right)}{\varphi\left(\Phi^{-1}(x)\right)}, \text{ and}
$$
$$
\mathbf{pdf}_{P_e}(x) = \mathbf{pdf}_{P_r}(x) + \mathbf{pdf}_{P_r}(1 - x),
$$

with $\lambda_1 = \sigma_N/\sigma_M$ and $\lambda_2 = (T - \mu_M)/\sigma_M$ and $\varphi = \varphi_{0,1}$. The derivation of these distributions and an experimental validation thereof are given in [18].

CDs Have Fingerprints Too*

Ghaith Hammouri[1], Aykutlu Dana[2], and Berk Sunar[1]

[1] CRIS Lab, Worcester Polytechnic Institute
100 Institute Road, Worcester, MA 01609-2280
{hammouri,sunar}@wpi.edu
[2] UNAM, Institute of Materials Science and Nanotechnology
Bilkent University, Ankara, Turkey
aykutlu@unam.bilkent.edu.tr

Abstract. We introduce a new technique for extracting unique fingerprints from identical CDs. The proposed technique takes advantage of manufacturing variability found in the length of the CD lands and pits. Although the variability measured is on the order of 20 nm, the technique does not require the use of microscopes or any advanced equipment. Instead, we show that the electrical signal produced by the photodetector inside the CD reader is sufficient to measure the desired variability. We investigate the new technique by analyzing data collected from 100 identical CDs and show how to extract a unique fingerprint for each CD. Furthermore, we introduce a technique for utilizing fuzzy extractors over the Lee metric without much change to the standard code offset construction. Finally, we identify specific parameters and a code construction to realize the proposed fuzzy extractor and convert the derived fingerprints into 128-bit cryptographic keys.

Keywords: Optical discs, fingerprinting, device identification, fuzzy extractor.

1 Introduction

According to the Business Software Alliance about 35% of the global software market, worth \$141 Billion, is counterfeit. Most of the counterfeit software is distributed in the form of a compact disc (CD) or a digital video disc (DVD) which is easily copied and sold in street corners all around the world but mostly in developing countries. Given the severity of the problem at hand, a comprehensive solution taking into account the manufacturing process, economical implications, ease of enforcement, and the owner's rights, needs to be developed. While this is an enormous undertaking requiring new schemes at all levels of implementation, in this work, we focus only on a small part of the problem, i.e. secure fingerprinting techniques for optical media.

To address this problem the SecuRom technology was introduced by Sony DADC. The technology links the identifiers produced to executable files which may only be accessed when the CD is placed in the reader. The main advantage of

---

* This material is based upon work supported by the National Science Foundation under Grant No. CNS-0831416.

C. Clavier and K. Gaj (Eds.): CHES 2009, LNCS 5747, pp. 348–362, 2009.

this technology is that it can be used with existing CD readers and writers. While the specifics of the scheme are not disclosed, in practice, the technology seems to be too fragile, i.e. slightly overused CDs become unidentifiable. Another problem is at the protocol level. The digital rights management (DRM) is enforced too harshly, therefore significantly curtailing the rights of the CD owner.

In this paper we take advantage of CD manufacturing variability in order to generate unique CD fingerprints. The approach of using manufacturing variability to fingerprint a device or to build cryptographic primitives has been applied in several contextes. A popular example is a new hardware primitives called *Physical Unclonable Functions* (PUFs). These primitives were proposed for tamper-detection at the physical level by exploiting *deep-submicron and nano-scale* physical phenomena to build low-cost tamper-evident key storage devices [7,8,6,12]. PUFs are based on the subtleties of the operating conditions as well as random variations that are imprinted into an integrated circuit during the manufacturing process. This phenomenon, i.e., manufacturing variability, creates minute differences in circuit parameters, e.g., capacitances, line delays, threshold voltages etc., in chips which otherwise were manufactured to be logically identical. Therefore, it becomes possible to use manufacturing variability to uniquely fingerprint circuits. More recently, another circuit fingerprinting technique was introduced. The technique exploits manufacturing variability in integrated chips to detect Trojan circuits inserted during the manufacturing process [5].

Another secure fingerprinting technology named RF-DNA was developed by Microsoft Research [1]. The RF-DNA technology provides unique and unclonable physical fingerprints based on the subtleties of the interaction of devices when subjected to an electromagnetic wave. The fingerprints are used to produce a cryptographic certificate of authenticity (COA) which when associated with a high value good may be used to verify the authenticity of the good and to distinguish it from counterfeit goods. Another application of manufacturing variability is fingerprinting paper objects. In [4] the authors propose Laser Surface Authentication which uses a high resolution laser microscope to capture the image texture from which the fingerprint is developed. In a more recent proposal, a cheap commodity scanner was used to identify paper documents [3]. While most of the results cited above were developed in the last decade, the idea of using physical fingerprints to obtain security primitives is not new at all. According to [1], access cards based on physical unclonable properties of media have been proposed decades ago by Bauder in a Sandia National Labs technical report [2].

Our Contribution: We introduce a method which exploits CD manufacturing variability to generate unique fingerprints from logically identical CDs. The biggest advantage of our approach is that it uses the electrical signal generated by the photodiode of a CD reader. Thus no expensive scanning or imaging equipment of the CD surface is needed. This means that regular CD readers can implement the proposed method with minimal change to their design. We investigate the new approach with a study of over 100 identical CDs. Furthermore, we introduce a new technique, called the threshold scheme, for utilizing fuzzy extractors over the Lee metric without much change to the standard code offset construction [10]. The threshold scheme allows us to use error correcting codes working under the Hamming metric for samples which are close under the Lee metric. The threshold scheme is not restricted to CDs, and therefore can serve in any noisy fingerprinting application where the Lee metric is relevant. With the aid of the proposed fuzzy extractor we give specific parameters and a code construction to convert the derived fingerprints into 128-bit cryptographic keys.

The remainder of the paper is organized as follows. In Section 2, we discuss the physical aspects of CD storage, the sources of manufacturing variability and the statistical model capturing the CD variability. Section 3 presents experimental data to verify our statistical model. In Section 4 we discuss the fingerprint extraction technique and determine the parameters necessary for key generation. We discuss the robustness of the fingerprint in Section 5 and finally conclude in Section 6.

2   Pits and Lands

On a typical CD data is stored as a series of lands and pits formed on the surface of the CD. The pits are bumps separated by the lands to form a spiral track on the surface of the CD. The spiral track starts from the center of the CD and spirals outward. It has a width of about 0.5 μm and a 1.6 μm separation. The length of the land or pit determines the stored data. The encoding length can assume only one of nine lengths with minimum value in the range 833 to 972 nm up to a maximum of 3054 to 3563 nm with increments ranging from 278 to 324 nm. Note that the range is dependent on the speed used while writing the CD. To read the data on the CD the reader shines a laser on the surface of the CD and collects the reflected beam. When the laser hits the pits it will reflect in a diffused fashion thus appearing relatively dark compared to the lands. Upon the collection of the reflected beam, the reader can deduce the location and length of the lands and pits which results in reading the data on the CD.

CDs are written in two ways, pressing and burning. In pressed CDs a master template is formed with lands and pits corresponding to the data. The master template is then pressed into blank CDs in order to form a large number of copies. In burned CDs, the writing laser heats the dye layer on the CD-R to a

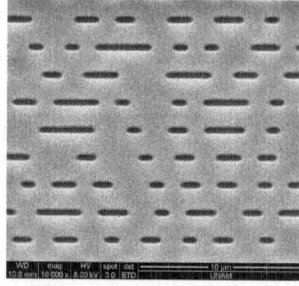

Fig. 1. Lands and pits image using an optical microscope

Fig. 2. Lands and pits image using a scanning electron microscope

point where it turns dark, thus reflecting the reading laser in a manner consistent with physical lands. Note that burned CDs will not have physical lands and pits but will act as if they had these features. Figures 1 and 2 show the lands and pits of a pressed CD. We captured Figure 1 using an optical microscope and Figure 2 using a scanning electron microscope.

2.1   Source of Variation

Similar to any physical process, during the writing process CDs will undergo manufacturing variation which will directly affect the length of the lands and pits. For burned CDs this variability will be a direct result of the CD velocity while writing takes place. This velocity is assumed to be at a fixed rate between 1.2 and 1.4 m/s where the velocity variation during writing should be within $\pm 0.01 m/s$ [11]. Pressed CDs are manufactured by molding thermoplastics from a micro or nanostructured master prepared by lithographic methods. The molding process itself is optimized for replication fidelity and speed with typical replication variations on the order of tens of nanometers [17]. The molding process involves contacting the thermoplastic with the master slightly above the glass transition temperature of the material, with a preset pressure for a brief amount of time, cooling the master and the thermoplastic to below the glass transition temperature and demoulding. Local variations of polymer material's mechanical and thermal properties, local variations of the temperature and pressure all potentially lead to variations in the imprinted structures. The thermal stresses induced during cooling and demoulding also potentially lead to variations. In this paper we aim at using the small variation in the length of lands and pits in order to form a unique fingerprint for each CD. In the next section we characterize the length features of lands and pits.

2.2   Single Location Characterization

Together lands and pits form the full spiral track. Therefore, it makes sense to fingerprint only lands or pits. The length of both lands and pits will follow similar distributions which is why we will simply use the term *location* to refer to either of them. We label the lengths of n consecutive locations by starting from a reference point on the track, as $L_1, L_2, \ldots, L_n$. In the ideal setting $L_i = c_i \cdot L$ for a small constant integer $c_i \in [3, 4, \ldots, 11]$ and $L \approx 300$ nm. However, due to the subtle variations we discussed in the previous section we expect $L_i = c_i \cdot L + \ell_i$. The variable ℓ_i is expected to be quite small compared to L_i, and therefore difficult to measure precisely. Still our measurements should be centered around the ideal length. Hence, quite naturally across all identical CDs we model L_i as a random variable drawn from a Gaussian distribution $\mathcal{H}_i = N(M_i, \Sigma)$ where $M_i = c_i \cdot L$ and Σ denotes the mean and the standard deviation respectively[1].

---

[1] $N(\mu, \sigma)$ is a normal distribution with mean μ and standard deviation σ.

Here we are assuming that regardless of the location, the standard deviation Σ will be the same. This is a quite a realistic assumption since Σ essentially captures the manufacturing variability which should affect all locations similarly. The more precise the manufacturing process is, the less of a standard deviation we would expect $\mathcal{H}_i$ to have. A perfect manufacturing process would yield $\Sigma = 0$ and would therefore give all CDs the same exact length of a specific location across all identical CDs. On the other hand, for better identification of CDs we would like $\mathcal{H}_i$ to have a relatively large Σ.

In a typical CD reader, the reading laser is reflected from the CD surface back into a photodiode which generates an electrical signal that depends on the intensity of the reflected laser. Therefore, the electrical signal is expected to depict the shape of the CD surface. If these electrical signals are used to measure the length of any given location, we expect these measurements to have a certain level of noise following a Gaussian distribution. So for location i on CD_j we denote this distribution by $\mathcal{D}_{ij} = N(\mu_{ij}, \sigma)$. The noise in the length measurements is captured through the standard deviation σ. Since this quantity mainly depends on the readers noise, we assume that its the same for all CDs and all CD locations. Contrary to Σ, to identify different CDs using the length information of CD locations we would like to see a relatively small σ.

3   Experimental Validation

To validate the statistical model outlined in the previous section, we conducted extensive experiments on a number of CDs. We directly probed into the electrical signal coming out of the photodiode constellation inside the CD reader. The intensity of this signal will reflect the CD surface geometry, and therefore can be used to study the length of the CD locations. To sample the waveform we used a 20 GHz oscilloscope. Each CD was read a number of times in order to get an idea of the actual $\mathcal{D}$ distribution. Similarly, we read from the same locations of about 100 identical CDs in order to generate the $\mathcal{H}$ distribution. Each collected trace required about 100 MBytes of storage space. Moreover, synchronizing the different traces to make sure that the data was captured from the same location of the CD was quite a challenge. We had to assign a master trace which represented the locations we were interested in studying and then ran the other traces through multiple correlation stages with the master to finally extract synchronized signals from the same locations on different CDs. Automating the process in order to accurately capture this massive amount of data was a time consuming challenge. However, we note that all this work would be almost trivially eliminated if we had access to the internal synchronization signals of the CD reader chip. The captured signals were then further processed using Matlab to extract the location lengths and obtain the distributions. After processing, we extracted the length of 500 locations (lands) on the CDs. We used commercially pressed CDs for all the experiments reported in this paper.[2]

---

[2] We have verified a similar behavior for burned CDs. Not surprisingly, data coming from burned CDs had a much larger variation and was easier to analyze.

Figure 3 shows the histogram of lengths extracted from 550 reads for a randomly chosen location on one CD. The mean length of the histogram is about $\mu_{ij} = 958$ nm. This histogram captures the $\mathcal{D}$ distribution. The other locations observe similar distributions with different mean lengths which will depend on the encoded information. When considering data coming from different locations and different CDs we obtain $\sigma = 20$ nm (with an average standard deviation of 2 nm on σ). This will be a good estimate for the noise observed during probing of the electrical signals. These results verify the assumption that the noise in the electrical signal can be approximated as Gaussian noise. Note that with Gaussian noise simple averaging can be used to substantially reduce the noise level. As we are interested in studying the behavior of the location lengths across

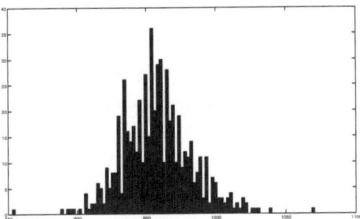
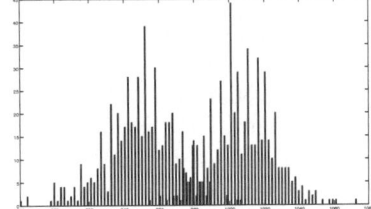

Fig. 3. Histogram of reads coming from the same location on the same CD

Fig. 4. Histograms of reads coming from the same location on two identical CDs

different CDs, we next shift our attention to two CDs before we look at a larger batch of CDs. Figure 4 captures a histogram for the length of the same location on two identical CDs..What is important here is the distance between the two Gaussians. The larger this distance becomes the easier it is to identify CDs. Our basic thesis for fingerprinting CDs is that the length of a single location will vary across multiple identical CDs. As pointed out earlier, this behavior can be modeled with the Gaussian distribution $\mathcal{H}_i$. The histogram in Figure 4 captures this for two CDs. To generalize these results and estimate the $\mathcal{H}_i$ distribution we need a larger sample space. The major problem here is that each data point needs to come from a different CD. Therefore, to obtain a histogram which clearly depicts a Gaussian we would need to test on the order of 500 CDs. This was not possible as each CD required substantial time, computing power and storage space in order to produce final data points. However, we were able to carry out this experiment for about 100 CDs. Each CD was read about 16 times to reduce the noise. Finally, we extracted the lengths of 500 locations for each of the CDs. Figure 5 depicts the histogram over 100 CDs for a randomly chosen location out of the 500 extracted locations. The histogram in Figure 5 has a mean of about 940 nm. Overall locations, Σ had a mean of 21 nm (with an average standard deviation of 1.8 nm on Σ). The histogram in Figure 5 looks similar to a Gaussian distribution generated from 100 data points. However, it would be interesting to get a confirmation that with more data points this histogram would actually yield a Gaussian. To do so, we normalized the lengths

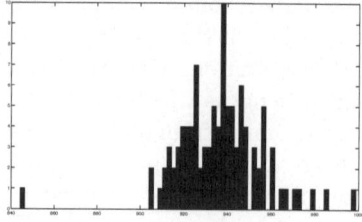
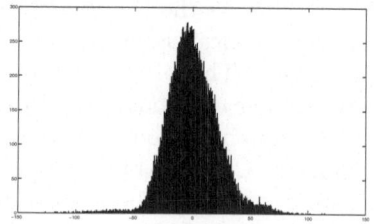

Fig. 5. Histograms of reads coming from the same location on 100 identical CDs

Fig. 6. Histograms of reads coming from 500 locations on 100 identical CDs

of each location by subtracting the average length for that particular location. Since the distribution for each location had roughly the same Σ the normalization process effectively made all these distributions identical with a mean of 0 and a standard deviation of Σ. We then collected all these data points (on the order of 50,000 points) and plotted the corresponding histogram. This is shown in Figure 6. The histogram of Figure 6 strongly supports our thesis of normally distributed location lengths across different CDs. One might observe a slight imbalance on the positive side of the Gaussian. This behavior seems to be a result of the DC offset observed while reading some of the CDs. Fortunately, this will not pose a problem for our fingerprinting technique as we will be normalizing each batch of data to have a mean of zero, thus removing any DC components. We finish this section by showing the histogram in Figure 7. The main purpose of this histogram is to confirm that what we are studying is in fact the length of data locations written on the CD. We elaborated earlier that on a CD data is stored in discrete lengths ranging from about 900 nm to about 3300 nm taking 9 steps in increments of about 300 nm. We build the histogram in Figure 7 using the data collected from 500 locations over the 100 CDs without normalizing each location's length to zero. In Figure 8 we show a similar histogram with data extracted by processing images coming from a scanning electron microscope.

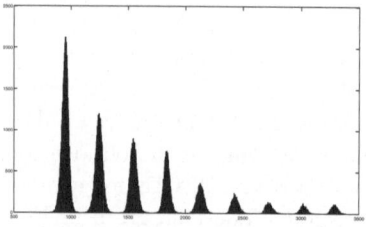
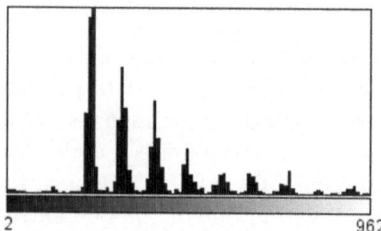

Fig. 7. Histogram of location lengths using the electrical signal

Fig. 8. Histogram of location areas using electron microscope images

4   CD Fingerprinting

There are many challenges in deriving a robust and secure fingerprint. One important issue is the reading noise. Similar to a human fingerprint, we saw in the previous section that the readings used to extract the CD fingerprint are inherently noisy. The extraction of a deterministic and secure fingerprint from noisy data has been previously studied in the literature [15,14,10]. Most relevant to our work is the fuzzy extractor technique proposed by Dodis et al. in [10]. For the remainder of this section we will present a quick review of the fuzzy extractor technique and then discuss how this technique can be modified and applied to the CD setting. Moreover, we will discuss the experimental results and present various bounds needed to achieve high levels of security.

4.1   Fuzzy Extractors

Loosely speaking a fuzzy extractor is a technique to extract an almost uniform random string from a given input such that it is possible to reproduce the same output string from a noisy version of the input. In [10] the authors show how a fuzzy extractor can be built using an error correcting code along with a universal hashing function. Their construction requires that the output of the fingerprint (the biometric data in their language) be represented as an element of $\mathcal{F}^n$ for some field $\mathcal{F}$ and an integer n which represents the size of the fingerprint. Moreover, it is naturally assumed that the noise experienced by the fingerprint is upper bounded by a constant distance from the original fingerprint in order to guarantee identical reproduction of the extracted key. We start by quoting the following theorem introduced in [10], and then give the specific construction which the theorem describes.

Theorem 1. ([10]) *Given any $[n, k, 2t+1]_{\mathcal{F}}$ code $\mathcal{C}$ and any m, ϵ, there exists an average-case $(M, m, \ell, t, \epsilon)$-fuzzy extractor, where $\ell = m + kf - nf - 2\log(\frac{1}{\epsilon}) + 2$. The generation algorithm GEN and the recovery algorithm REP are efficient if $\mathcal{C}$ has efficient encoding and decoding.*

We explain the parameters in the theorem by outlining an actual construction. This construction is proposed in [10] and further explained in [12]. As stated in the theorem, $\mathcal{C}$ is an error correcting code over the field $\mathcal{F}$, where $f = \log(|\mathcal{F}|)$.[3] For the construction we will also need a family of universal hashing functions $\mathbf{H}$.[4] The generation algorithm GEN takes the fingerprint $x \in \mathcal{F}^n$ as input and outputs the triplet (k, w, v). Here, x is drawn from some distribution X over $\mathcal{F}^n$ which has min-entropy m. Note that in our context the parameter m captures the entropy provided by the CD variability. GEN starts by computing $w = x + c$ for a randomly chosen code word $c \in \mathcal{C}$ and then computes the key $k = h_v(x) \in \{0, 1\}^{\ell}$ for some string v chosen uniformly at random such that $h_v \in \mathbf{H}$. The recovery algorithm REP takes in the *helper data* (w, v) along with x', a noisy version of the

---

[3] Note that all logarithms in this paper are with respect to base 2.

[4] For details on universal hashing the reader is referred to [9].

fingerprint x, and returns the key k. REP starts by computing $c' = w - x'$ which is a noisy version of c. If the Hamming distance between x and x' is less than t then so will the Hamming distance between c and c'. Therefore, using the error correcting code $\mathcal{C}$, REP can reproduce c from c'. Next, REP computes $x = w - c$ and consequently compute $k = h_v(x)$ which will conclude the recovery algorithm. All that remains to be defined is the parameter ϵ which captures the security of the fuzzy extractor. Specifically, if the conditional min-entropy[5] $H_\infty(X|I)$ (meaning X conditioned on I)[6] is larger than m then $\mathbf{SD}((k, (w, v), I), (U_\ell, (w, v), I) \le \epsilon)$ where $\mathbf{SD}(A, B) = \frac{1}{2} \sum_v |\Pr(A = v) - \Pr(B = v)|$ is the statistical distance between two probability distributions A and B. Finally, U_ℓ is the uniform distribution over $\{0, 1\}^\ell$ and I is any auxiliary random variable.

With this construction we will have a clear way to build a fuzzy extractor. However, the key size ℓ and the security parameter ϵ will both depend on m and the code used. Moreover, the code will depend on the noise rate in the fingerprint. We finish this section by relating the min-entropy and the error rate of the fingerprint. Recall, that x is required to have a min-entropy of m and at the same time using the above construction x will have n symbols from $\mathcal{F}$. To merge these two requirements we define the average min-entropy in every symbol $\delta = m/n$. We also define ν to be the noise rate in the fingerprint x and $F = |\mathcal{F}|$. With these definitions we can now prove the following simple bound relating the noise rate and the min-entropy rate δ/f.

Proposition 1. *For the fuzzy extractor construction of Theorem 1, and for any meaningful security parameters of $\epsilon < 1$ and $\ell > 2$ we have $H_F(\nu) < \frac{\delta}{f}$. Where H_F is the F-ary entropy function.*

Proof. From Theorem 1 we now that $\ell = m + kf - nf - 2\log(\frac{1}{\epsilon}) + 2$. Let $A = \ell + 2\log(\frac{1}{\epsilon}) - 2 = m + kf - nf$. From the conditions above we now that $A > 0$ and therefore $m + kf - nf > 0$. Let $R = k/n$ which yields $(\delta + Rf - f)n > 0$ and therefore $R > 1 - \delta/f$. Using the sphere packing bound where $R \le 1 - H_F(\nu)$ we immediately get $H_F(\nu) < \frac{\delta}{f}$.

As it is quite difficult to calculate the min-entropy for a physical source we will estimate this quantity over the symbols of x. The bound given above will give us an idea whether the min-entropy in the symbols of x will be sufficient to handle the measured noise rate. Next we shift our attention to the fingerprint extraction technique. Note here that we still did not address how the data extracted from the CDs will be transformed into the fingerprint x.

4.2   Fingerprint Extraction

In Section 3 we described how the empirical data suggests that every CD has unique location lengths. These location lengths as can be seen from Figure 7

---

[5] The definition of min entropy is $H_\infty(A) = -\log(\max_a \Pr[A = a])$.

[6] Typically we use the $|$ operator to mean concatenation. This will be the only part of the paper where it will have a different meaning.

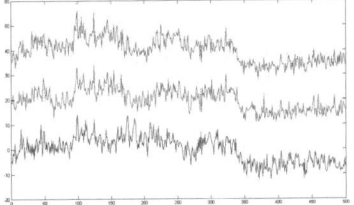 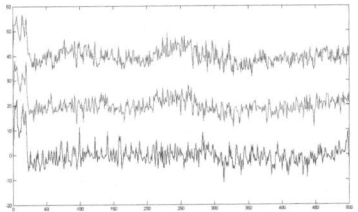

Fig. 9. Length variation over 500 locations from CD1 with the bottom trace taken 3 months after the top two traces

Fig. 10. Length variation over 500 locations from CD2 with the bottom trace taken 3 months after the top two traces

will have different values depending on the encoded information. Moreover, we discussed earlier that the raw data measured from the electrical signal will sometimes have different DC offsets. Therefore, it is important to process the data before the different locations can be combined together in order to produce the final fingerprint x. The first step in processing the data coming from every location on every CD is to remove the signal noise. To achieve this, the length of every location on a CD is averaged over a number of readings. Since we are assuming Gaussian noise, the noise level σ will scale to $\sigma/\sqrt{a}$ where a is the number of readings used for averaging. Next, we normalize the data using the ideal average of each location. As the ideal location lengths are discretized it becomes easy to find the ideal length for every location and subtract it from the measured lengths. This will guarantee that all location lengths have similar distributions as we saw in Figure 6. Finally, to remove the DC component we need a second normalizing step. We subtract the mean of the reading coming from different locations of the same CD. Figures 9 and 10 show the variation in the length of 500 locations for two identical CDs after being averaged and normalized. Each figure contains three traces with an added horizontal shift to set the traces apart. The top two traces in each figure are obtained from readings taken at different times using one CD reader. The bottom trace in each figure was obtained three months after the first two traces using a second CD reader with a different brand and model. The vertical axis represents the variation in nanometers from the ideal length of that location. These figures clearly support the idea of identical CDs having different fingerprints which are reproducible from different readers. We still need to outline a technique to extract a final fingerprint. Even after the previous averaging and normalization steps we will still have errors in the length readings. Although we will be using a fuzzy extractor to correct the errors, the biggest challenge towards achieving an efficient extraction technique will be the nature of these errors. The noise is Gaussian over the real values of the lengths. This means that even when the data is discretized the error will manifest itself more as a shift error from the ideal length rather than a bit flip error. Unfortunately, the Hamming metric does not naturally accommodate for this kind of error. Moreover, if we assume that every location length of the CD will be a symbol in the extracted fingerprint, then the error rate would be very high as it is very difficult to get the same exact

Table 1. Formulation of the threshold scheme for CD fingerprint extraction

---

Threshold Scheme: (GEN,REP) parameterized by $M, m, \ell, t, \epsilon, l, \mathcal{C}, \mathbf{H}, \tau = 2^s$

GEN: $(k, w, v) \leftarrow \text{GEN}(\text{CD}j)$

1. Obtain (a) samples for the length of each of the n locations on $\text{CD}j$.
2. Generate $z = z_n \dots z_1$:
   a. Average the lengths over a samples,
   b. Subtract the ideal mean from the averaged reads,
   c. Normalize the sequence to have a zero mean and set that to z.
3. Find u such that $-2^{u-1} \leq z_i \leq 2^{u-1} - 1$ for all i, and shift z_i to $0 \leq z_i \leq 2^u - 1$.
4. Shift the binary representation of z_i left by l bits, round to an integer and set to $\hat{z}_i$.
5. Form $z_{2,i}$, the lowest $s + 1$ bits of $\hat{z}_i$, and $x_i = z_{1,i}$, the remaining bits of $\hat{z}_i$.
6. Set $x = x_n \dots x_1$ to be the fingerprint template.
7. Choose a random code word $c \in \mathcal{C}$, such that $c = c_n \dots c_1$.
8. Compute $w_i = (x_i | z_{2,i}) + (c_i | \tau)$ and form $w = w_n \dots w_1$.
9. Randomly choose v to compute $k = h_v(x)$ where $h_v \in \mathbf{H}$, and output (k, w, v).

REP: $k \leftarrow \text{REP}(\text{CD}j, w, v)$

1. Generate $z' = z'_n \dots z'_1$ as $\hat{z} = \hat{z}_n \dots \hat{z}_1$ was generated in Steps 1 through 4 of GEN.
2. Set c'_i to be the highest $u + l - s - 1$ bits of $w_i - z'_i$.
3. Use $\mathcal{C}$ to correct $c' = c'_n \dots c'_1$ to $c = c_n \dots c_1$.
4. Compute $x_i = w_i - c_i$.
5. Form $x = x_n \dots x_1$ and return $k = h_v(x)$.

---

length for the CD locations. A more natural distance metric in this situation would be the Lee metric [16]. However, this will require finding long codes that have good decoding performance under the Lee metric. To solve this problem we propose a *threshold* scheme which uses the Hamming distance while allowing a higher noise tolerance level. The threshold scheme also works naturally with the fuzzy extractor construction of Theorem 1. Table 1 shows a formulation of the threshold scheme applied to the CD setting. The threshold τ solves the error correcting problem with respect to the Lee distance. In particular, τ helps control the error rate which arises when treating the real values as symbols over some field. Without a threshold scheme ($\tau = 0$), the error rate will be very high. On the other hand, if τ grows too large then the error rate will be low. However, the Hamming distance between the extracted fingerprint originating from different CDs will decrease thus decreasing distinguishability between CDs. An important aspect about the threshold scheme is that it is very simple to compute and does not require previous knowledge of the distribution average.

4.3   Entropy Estimation and 128-Bit Security

The previous sections dealt with the theoretical aspects of extracting the CD fingerprint. In this section we take more of an experimental approach where we are interested in computing actual parameters. The most important parameters that we need to estimate are the entropy of the source (the CD variability) and the noise level. With these two parameters the rest of the parameters can be determined. The first and hardest task here will be to decide the amount of entropy generated by the source. In [12] and [13] the authors use a universal source coding algorithm in order to estimate the secrecy rate. In particular it was proposed to use the Context-Tree Weighting Method (CTW) [19]. What is quite useful about the CTW algorithm is that in [18] it was shown that for any binary stationary and ergodic source X, the compression rate achieved by CTW is upper bounded by the min-entropy $H_\infty(X)$ as the length of the input sequence approaches infinity. This is a good indication about the entropy produced by the source provided enough bits are fed to the algorithm. To apply this algorithm to our setting we start by using the data coming from the 100 CDs. On each CD we collected data from 500 locations and processed the data with a threshold value of $\tau = 2^2$. The final data came out to be in the range $[0, 2^5 - 1]$ and we did not use any fractional bits so $l = 0$. With these parameters the size of the symbols was $f = 2$. This means that every CD produced 1000 bits. The data was fed into the CTW algorithm which resulted in a compression rate of about 0.83 bits of entropy per extracted bit. Recall here that these samples were not averaged over multiple reads. Therefore the error rate is quite high. When we averaged over 16 samples the combined entropy rate became 0.71. This is expected since the noise will add to the entropy. In order to get a more precise estimate for the min entropy we decided to average over 225 reads. With this many reads we had to restrict our sample to only 14 CDs as the amount of data quickly becomes large. With the new sample the compression rate of the CTW algorithm was about 0.675 which seemed to be a good estimate of our min-entropy. For this sample, the average error rate is $P_e = 0.08$. On the other hand the collision probability P_c, the probability of extracting similar bits between two different CDs, is about 0.46.

Proposition 1 suggests that for a noise rate of 0.08 and $f = 2$ the entropy of the source should be at least 0.40 which translates to $\delta = 0.8 < 1.35$, and therefore we conclude that we have enough entropy in our source. However, with this level of entropy we are placing stringent conditions on R, i.e. the rate of the error correcting code.[7] To relax the restriction on the code rate we took a closer look at our source bits. Ideally the two bits would have the same entropy. However, looking at Figure 9 and 10 and multiple similar figures we clearly see that there is a degree of dependency between the adjacent locations. There is a low probability of a sharp change in the length variability from one location to its neighbor. With this observation we would suspect that the most significant bit will have less entropy as it is less likely to change across adjacent locations. To verify this observation, we applied the CTW algorithm to each of the two extracted bits separately. For the most significant bit, the entropy for the cases of no averaging, averaging over 16 reads, and averaging over 225 reads was $1, 0.9$ and 0.6-bits of entropy, respectively. When we repeated this process for the least significant bit we obtained $1, 1$ and 0.98-bits of entropy, respectively. Clearly, we have more entropy in the least significant bit. It seems reasonable to only use the least significant bit to form the fingerprint and the final key. This would

---

[7] Recall from the prof of Proposition 1 that $R \geq A/nf + (1 - \delta/f)$ for a security level of at least $A = \ell + 2\epsilon - 2$.

effectively increase the entropy of our source while very slightly affecting the error rate and the collision rate. For this least significant bit scheme we obtained $P_e = 0.08$ and $P_c = 0.46$.

We now have $P_e = 0.08$, $\delta = 0.98$ and $f = 1$. With these parameters we can build a fuzzy extractor which can extract secure keys from CD fingerprints. For a 128-bit key we set $\ell = 128$. Similarly, to achieve a fuzzy extractor output which reveals very little information about the fingerprint we set $\epsilon = 64$. Using the equation of Theorem 1 we require that the error correcting code in the fuzzy extractor should satisfy $k \geq 190 + 0.02n$. Note that although $P_e = 0.08$, this is the expected error rate. For a practical scheme we require the fuzzy extractor to correct around a 0.17 error rate. These parameters can now be satisfied using a binary BCH code of $[255, 45, 88]$. More specifically, we define a code word containing 7 code words of this BCH code, which will make $n = 1785$. With this construction the failure probability[8] P_{fail} will be on the order of 10^{-6}. Note here that treating the 7 code words separately to generate separate parts of the key would significantly decrease ϵ but will decrease the failure probability. Therefore, in our failure probability we treat the 7 code words as a single entity. As we noted earlier, our data suffers from higher error rates due to the external connections which we used. With an on-chip process we can expect the error rate to drop significantly.

5   Robustness of the Fingerprint

A CD fingerprint can be used to tie software licenses to individual CDs where the software is stored. Under this use scenario it becomes important to address the robustness of the fingerprint. In all our experiments the data collected came from locations in the same sector of the CD. In a real application readings would typically be collected from different sectors. Thus ensuring that a scratch or any physical damage to a specific location will not render the CD fingerprint useless.

Another important concern regarding the robustness of the fingerprint is that of aging. Although no quantitative estimate of fingerprint durability can be given within the scope of this work, mechanisms related to viscoelastic relaxation in optical disc patterns need to be discussed briefly. Optical discs are printed on polymeric substrates, which have glass transition temperatures typically above 150 °C. The viscosity of such materials are temperature dependent and governed by an Arrhenius type exponential temperature dependence, described by an activation energy defined by the glass transition temperature. In its simplest form, the Arrhenius model assumes that the rate of change is proportional to $e^{\frac{-E_a}{kT}}$ where E_a is the activation energy, k is the Boltzmann constant (an invariant physical parameter) and T is the absolute temperature (temperature in degrees Kelvin). Even at lower temperatures (natural operating and storage temperature range of the optical disc), viscosity of the polymer remains finite. During the molding process, most of the internal stresses are relieved upon cooling, resulting in fluctuations in the nanoscale structure of the bit patterns. The

---

[8] Here, $P_{fail} = 1 - \left(1 - \sum_{i=0}^{t=43} \binom{n}{i} P_e^i (1 - P_e)^{n-i}\right)^7$.

pressed discs have a thin metal coating, which is typically coated on to the polymer disc by evaporation or sputter coating, that results in the increase of the surface temperature by up to 50 °C. This process is also likely to be a source of local thermoelastic stress buildup which relaxes over the lifetime of the CD. In a first order approximation, the disc material can be thought of as a Kelvin-Voigt material, and creep relaxation can be approximated by a single time-constant exponential behavior. In such a case, most of the viscoelastic relaxation will occur at the early stages of disc production, and latter time scales will have less of an effect. It may be speculated that the fingerprints due to length fluctuations of 25 nm upon 300 nm characteristic bit length will persist within at least 10% of the CD lifetime, which is predicted to be 217 years at 25 °C and 40% relative humidity conditions. This gives an estimated 20 year lifetime for the fingerprint [20]. Due to the exponential dependence of the relaxation on time, by recording the signature on a slightly aged optical disc (months old), the persistance of the signature can be increased.

6   Conclusion

In this paper we showed how to generate unique fingerprints for any CD. The proposed technique works for pressed and burned CDs, and in theory can be used for other optical storage devices. We tested the proposed technique using 100 identical CDs and characterized the variability across the studied CDs. We also gave specific parameters and showed how to extract a 128-bit cryptographic keys. This work opens a new door of research in the area of CD IP-protection.

References

1. DeJean, G., Kirovski, D.: RF-DNA: radio-frequency certificates of authenticity. In: Paillier, P., Verbauwhede, I. (eds.) CHES 2007. LNCS, vol. 4727, pp. 346–363. Springer, Heidelberg (2007)
2. Bauder, D.W.: An anti-counterfeiting concept for currency Systems. Research Report PTK-11990, Sandia National Labs, Albuquerque, NM, USA (1983)
3. Clarkson, W., Weyrich, T., Finkelstein, A., Heninger, N., Halderman, J.A., Felten, E.W.: Fingerprinting blank paper using commodity scanners. In: Proceedings of S&P 2009, Oakland, CA, May 2009. IEEE Computer Society, Los Alamitos (to appear, 2009)
4. Cowburn, R.P., Buchanan, J.D.R.: Verification of authenticity. US Patent Application 2007/0028093, July 27 (2006)
5. Agrawal, D., Baktir, S., Karakoyunlu, D., Rohatgi, P., Sunar, B.: Trojan detection using IC fingerprinting. In: Proceedings of S&P 2007, Oakland, California, USA, May 20-23, 2007, pp. 296–310. IEEE Computer Society, Los Alamitos (2007)
6. Lim, D., Lee, J.W., Gassend, B., Suh, G.E., van Dijk, M., Devadas, S.: Extracting secret keys from integrated circuits. IEEE Transactions on VLSI Systems 13(10), 1200–1205 (2005)
7. Ravikanth, P.S.: Physical One-Way Functions. PhD thesis, Department of Media Arts and Science, Massachusetts Institute of Technology, Cambridge, MA, USA (2001)

8. Tuyls, P., Schrijen, G.J., Skoric, B., van Geloven, J., Verhaegh, N., Wolters, R.: Read-proof hardware from protective coatings. In: Goubin, L., Matsui, M. (eds.) CHES 2006. LNCS, vol. 4249, pp. 369–383. Springer, Heidelberg (2006)
9. Carter, L., Wegman, M.: Universal hash functions. Journal of Computer and System Sciences 18(2), 143–154 (1979)
10. Dodis, Y., Ostrovsky, R., Reyzin, L., Smith, A.: Fuzzy extractors: how to generate strong keys from biometrics and other noisy data. SIAM Journal on Computing 38(1), 97–139 (2008)
11. European Computer Manufacturers' Association. Standard ECMA-130: Data interchange on read-only 120mm optical data disks (CD-ROM) (2nd edn.). ECMA, Geneva, Switzerland (1996)
12. Guajardo, J., Kumar, S.S., Schrijen, G.J., Tuyls, P.: FPGA intrinsic PUFs and their use for IP protection. In: Paillier, P., Verbauwhede, I. (eds.) CHES 2007. LNCS, vol. 4727, pp. 63–80. Springer, Heidelberg (2007)
13. Ignatenko, T., Schrijen, G.J., Skoric, B., Tuyls, P., Willems, F.: Estimating the secrecy-rate of physical unclonable functions with the context-tree weighting method. In: Proceedings of ISIT 2006, Seattle, Washington, USA, July 9-14, 2006, pp. 499–503. IEEE, Los Alamitos (2006)
14. Juels, A., Sudan, M.: A fuzzy vault scheme. Designs, Codes and Cryptography 38(2), 237–257 (2006)
15. Juels, A., Wattenberg, M.: A fuzzy commitment scheme. In: Proceedings of CCS 1999, pp. 28–36. ACM Press, New York (1999)
16. Lee, C.: Some properties of nonbinary error-correcting codes. IRE Transactions on Information Theory 4(2), 77–82 (1958)
17. Schift, H., David, C., Gabriel, M., Gobrecht, J., Heyderman, L.J., Kaiser, W., Köppel, S., Scandella, L.: Nanoreplication in polymers using hot embossing and injection molding. Microelectronic Engineering 53(1-4), 171–174 (2000)
18. Willems, F.M.J.: The context-tree weighting method: extensions. IEEE Transactions on Information Theory 44(2), 792–798 (1998)
19. Willems, F.M.J., Shtarkov, Y.M., Tjalkens, T.J.: The context-tree weighting method: basic properties. IEEE Transactions on Information Theory 41(3), 653–664 (1995)
20. Stinson, D., Ameli, F., Zaino, N.: Lifetime of Kodak writable CD and photo CD media. Eastman Kodak Company, Digital & Applied Imaging, NY, USA (1995)

The State-of-the-Art in IC Reverse Engineering

Randy Torrance and Dick James

Chipworks Inc.
3685 Richmond Road, Ottawa, Ontario, Canada K2H 5B7
rtorrance@chipworks.com, djames@chipworks.com

Abstract. This paper gives an overview of the place of reverse engineering (RE) in the semiconductor industry, and the techniques used to obtain information from semiconductor products.

The continuous drive of Moores law to increase the integration level of silicon chips has presented major challenges to the reverse engineer, obsolescing simple teardowns, and demanding the adoption of new and more sophisticated technology to analyse chips. Hardware encryption embedded in chips adds a whole other level of difficulty to IC analysis.

This paper covers product teardowns, and discusses the techniques used for system-level analysis, both hardware and software; circuit extraction, taking the chip down to the transistor level, and working back up through the interconnects to create schematics; and process analysis, looking at how a chip is made, and what it is made of. Examples are also given of each type of RE. The paper concludes with a case study of the analysis of an IC with embedded encryption hardware.

1 Introduction

One of the most basic business requirements is the need to know what the competition is doing. If a company wants to get into a new area of business, the simplest thing to do is buy an existing product and take it apart to see what is inside it. Having done that, we know the parts list involved, and the technological challenges to be faced in manufacturing the new version.

Reverse engineering (RE) can cover objects from as large as aircraft down to the smallest microchip, and the motivations have varied from the paranoia of the Cold War, through commercial piracy, to competitive intelligence, product verification, and courts of patent law. If we look back over the last few decades, reverse engineers around the world have had a significant influence on the dissemination of technology in the electronics industry.

RE is now a recognised part of the competitive intelligence field, and is commonly used to benchmark products and support patent licensing activities. A side area is the need to RE archaic parts that have gone out of service, and need replacing in long-lived equipment such as military systems, nuclear reactors, airliners, and ships.

A fact of life these days is that simple teardowns of products are just not good enough any more. Advances in semiconductor technology, namely the massive

C. Clavier and K. Gaj (Eds.): CHES 2009, LNCS 5747, pp. 363–381, 2009.

integration of billions of individual devices and masses of functions into single components, have forced RE to evolve into a specialised niche of the engineering profession.

2   RE in the Semiconductor Industry

The question most often asked about reverse engineering is "is it legal?" The short answer is – yes! In the case of semiconductors, RE is protected in the US by the Semiconductor Chip Protection Act, which allows it "for the purpose of teaching, analyzing, or evaluating the concepts or techniques embodied in the mask work or circuitry..." There is similar legislation in Japan, the European Union, and other jurisdictions.

In the semiconductor business, RE customers fall into two groups: those who are interested in technical information, and those that are interested in patent-related information. The technical information customers are usually within manufacturing companies, performing product development, or strategic marketing or benchmarking studies. The patent clients are usually patent lawyers or intellectual property (IP) groups within companies. There are also companies that are purely licensing companies, and deal only in IP.

Types of RE

Reverse engineering of semiconductor-based products can broadly take several forms:
- Product teardowns – identify the product, package, internal boards, and components
- System level analysis – analyse operations, functions, timing, signal paths, and interconnections
- Process analysis – examine the structure and materials to see how it is manufactured, and what it is made of
- Circuit extraction – delayer to transistor level, then extract interconnections and components to create schematics and netlists.

3   Product Teardowns

Product teardowns are the simplest type of RE in the electronics arena; the unit is simply disassembled, the boards and sub-assemblies are photographed, and the components are inventoried. Reverse engineers are usually only interested in what components are in the device at this level, but there are also companies that use the data to provide a bill of materials and tentative costing for the manufacture.

Figure 1 shows an Apple 8 GB iPod nano personal media player, partly torn down to expose the internal board and the ICs used [1]. Optical and x-ray analyses (Fig. 2) showed that the 64 Gb flash memories were actually 2 x 32 Gb stacked packages, each containing four 8 Gb dice (total 64 Gb). In this case, we continued with detailed process analyses of the 8 Gb flash chips, since they were leading edge devices from Samsung and Toshiba.

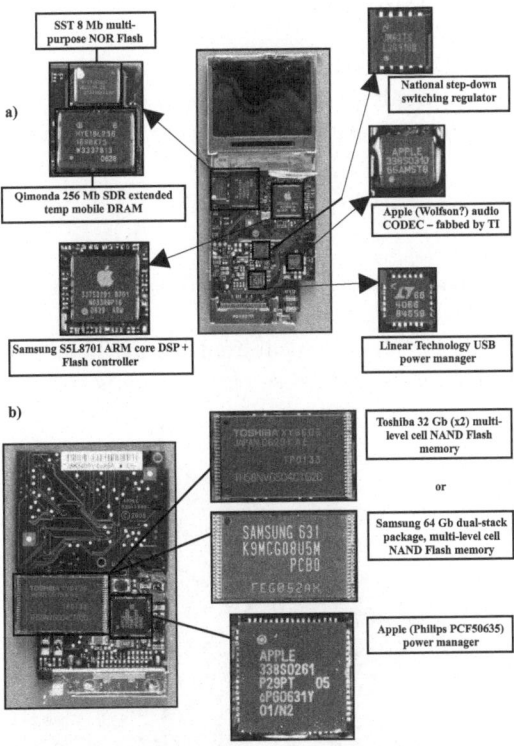

Fig. 1. Partial Teardown of Apple 8 GB iPod Nano: a) Top b) Bottom

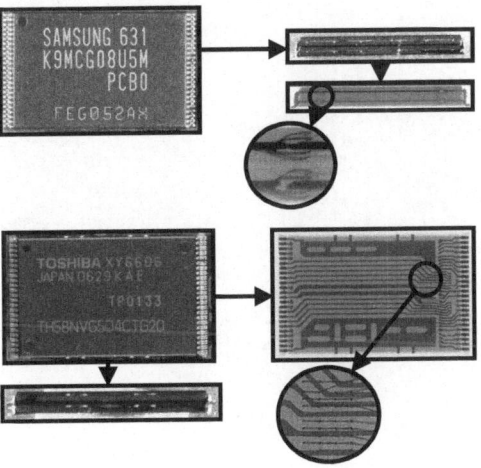

Fig. 2. Optical and X-Ray Images of 64 Gb Flash Devices

4 System Level Analysis

Just as there is a huge variation in electronic systems, there is also a variety of methods for system analysis. Electronic systems can consist of hardware, software, firmware, communications, transducers, etc. System analysis is useful for all of these.

4.1 Hardware

Hardware analysis takes one of two forms: reverse engineering or functional analysis.

Reverse engineering is a hierarchical analysis method. Take the example of a cell phone. The first phase of reverse engineering is to tear down the phone,

Fig. 3. Delayered Nine Layer PCB from Cell Phone

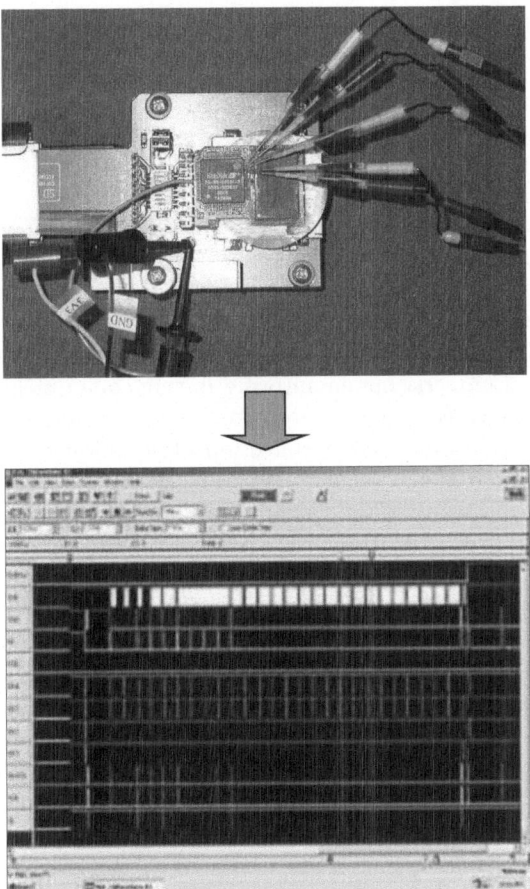

Fig. 4. Probing SanDisk SD Memory Card for Functional Analysis

making notes of all connections between subsystems. Next, the main board is reverse engineered. Photos are taken of the board for future work. All components on the board are catalogued and then selectively removed. If the board is multi-layered, it can be delayered and imaged (Figure 3). The connections between all components are then identified and entered into the board schematic. Alternatively, electrical probing can sometimes be used to find the connections. Either way, a complete schematic of the board can be re-created.

Functional analysis entails system monitoring during functional operation. A system can be instrumented with probes wherever needed (sometimes with great difficulty, but it can usually be done, as shown in Figure 4). Microprobing is used to monitor on-chip signals. Test cases are developed, and stimulus created for operating the system in its functional modes. Signal generators, logic analyzers, and oscilloscopes are used to drive the system and collect the results. The signals and full system are then analyzed. Using the cell phone example once again, the phone can be partially disassembled, but still electrically connected to allow for operation. Probes can be used to monitor key buses, pins of chips, and connectors. The phone can then be operated, and the signals analyzed, to understand the operation.

4.2 Software

As with hardware, software can be analyzed using the same two techniques; reverse engineering and functional analysis.

Software reverse engineering is the process of taking machine code and converting it back into human-readable form. The first task is often extraction of embedded code from an on-chip memory. Many techniques are available, such as EEPROM programmers, bus monitoring during code upload, and schematic extraction. Sometimes the code is protected with software or hardware locks. These can often be disabled via a collection of techniques. A chip's test port can be a good method of accessing its contents. IC microsurgery can be used to modify or bypass hardware locks. Usually these techniques require circuit analysis first, in order to identify the locks and find modifications that will disable them.

Encrypted code requires encryption analysis, followed by decryption. This requires both the keys and an understanding of the encryption algorithm. The keys can often be read from the embedded memory, along with the code, using the techniques described above. The encryption algorithm can sometimes be discovered via documentation or functional analysis. If these methods fail, then circuit extraction can often be used to reverse engineer the algorithm.

Once the code is extracted, disassemblers can be used as long as the processor and instruction set are known. Tools are then available to help take assembly code and structure it into a more C-like format. This structured code can then be analyzed by software experts. Code can be analyzed in either "static" ("dead") mode or "dynamic" ("live") mode. Live analysis is undertaken when it is possible to obtain the full control of the processor: starting and stopping code, inspecting registers, memory, tracing code execution. Live analysis is always preferable to dead code analysis which consists of analyzing just the instructions without the ability to inspect the code while running. Using software simulators enables another mode of software RE which is in between these two.

Software functional analysis is similar to hardware functional analysis. Test cases are designed, stimulus is created, the code can be instrumented, and the software executed. The outputs of this software can take many forms, from creating charts or driving a GUI, to controlling a robot or playing a song. These outputs can be analyzed to better understand the software or system.

5 Process Analysis

Process analysis of chips is straightforward in theory, since micro-analytical tools have been around for some time. Every wafer fab has a range of equipment for process control and failure analysis, and Chipworks uses the lab-scale equivalent.

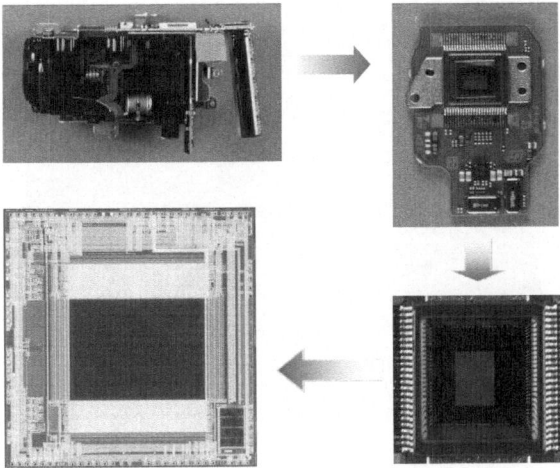

Fig. 5. Disassembly of CMOS Image Sensor from Camera Module

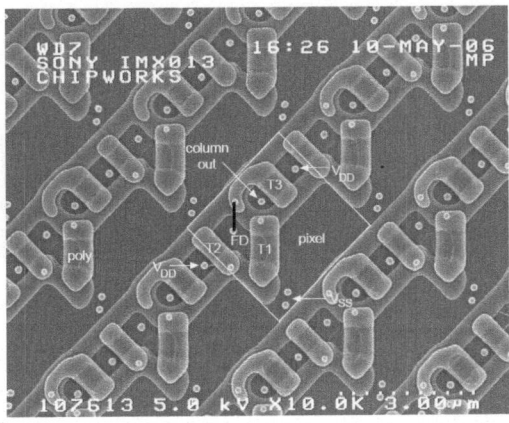

Fig. 6. Plan-View SEM of Pixels at the Polysilicon Level

Using a Sony DCR-DVD505 Handycam as an example, we were interested in the CMOS image sensor in the camera.

We removed the camera module from the unit and took it apart, recording the details as we went, and ended up with the CMOS imager die (Figure 5), which turns out to be a Sony Clearvid IMX013 chip.

Then we get into the actual chip analysis. This part was a fairly leading-edge sensor, with a small pixel size of 2.85 μm x 2.85 μm, so the emphasis was on a detailed examination of the pixel. Figures 6 to 9 show some of the features seen in the pixel area.

When performing process analysis, plan-view imaging gives limited process information, so the primary source of data is cross-sectional analysis,

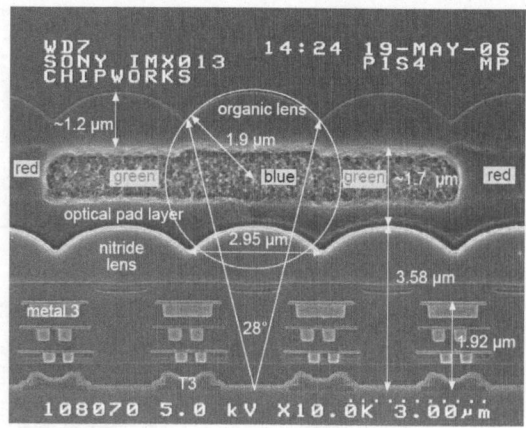

Fig. 7. Cross-Sectional SEM of Pixels

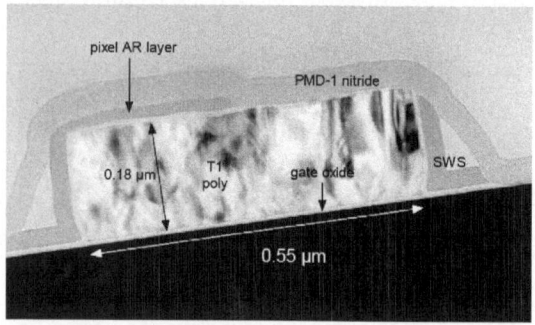

Fig. 8. TEM Cross Section of Pixel Transfer Transistor

usually using SEM, TEM, and scanning capacitance microscopy (SCM). For details of the chemical composition, the most commonly used technique is energy dispersive x-ray analysis, although occasionally we use other methods such as secondary ion mass spectrometry or Auger analysis.

A few words of explanation here with respect to Figures 8 and 9. A TEM looks *through* the sample to give high resolution images of the device structure, and SCM is a way of seeing the positive and negative doping that makes up the actual working transistors, resistors, etc., in the silicon chip.

Looking at Figure 6, we see a plan-view image of part of the pixel array, showing the transfer transistor (T1), and the T2 reset transistor and T3 source follower transistors, comprising the 3 transistor pixel circuit. The short black line in the centre of the image represents a metal 1 strap joining the floating diffusion (FD), between T1 and T2, to the gate of T3.

Figure 7 shows a cross section of the pixel structure, illustrating the organic and nitride lenses, the colour filters, three layers of copper metallization in the array, and the T3 transistors on the substrate. There is also a fourth aluminium metal

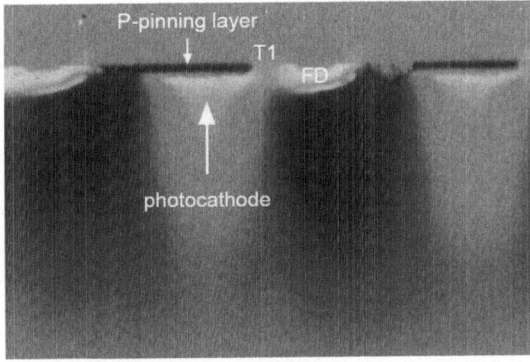

Fig. 9. SCM Cross Section of Pixels

layer, not shown in this section, used for bond pads and as a light shield (the white bars in the die photograph in Figure 4). The 28° angle of acceptance is also shown.

Figure 8 is a TEM image of the transfer transistor gate, and it is clear that the nitride layer used for the sidewall spacer has only been partially etched off the top of the gate; the residual nitride on the photocathode (left) side has been used as an antireflective (AR) layer in the photocathode area.

The doping structure of the pixels is illustrated in the SCM image in Figure 9. Chemical staining has been used for decades to highlight the doped areas in silicon, but even after many years of experiment, it is still more of an art than a science. The development of the SCM allows us to distinguish features such as the P-pinning layer above the photocathode, and the floating diffusion, more clearly. The deeper blue areas are the P-type isolation regions in the N-substrate.

There are two parallel trends in semiconductor processing. There is the well publicized Moores law shrinkage of dimensions, moving to the 45 nm node and below, with the introduction of high-k/metal gate transistors, and there is a drive to more process integration as RF/mixed signal and embedded memory processes are merged into CMOS logic processes.

As can be imagined, examining features deep into the nanometer scale (gate oxides are now 1.2 nm - 1.5 nm thick) stretches analytical capabilities to the limits. They can be imaged with high-resolution electron microscopy, but obtaining details of the chemical composition of the structure is now in the realm of counting atoms [5,6].

Similarly to the other forms of RE, our final documents can take several forms, from reports specifically focused on a feature described in a patent claim, to comprehensive reports detailing the full structural and process analysis of a high-end chip. It all depends on what the customer wants!

6   Circuit Extraction

Circuit extraction of semiconductor chips becomes increasingly more difficult with each new generation. In the "good old days" of 10 to 20 years ago, a circuit

Fig. 10. As RE Used to be Done!

analyst's life was much simpler. A typical IC of those days may have had one layer of metal, and used 1 μm - 2 μm technology. After package removal, all features could usually be seen from the top level metal planar view.

The die could then be put under optical imaging equipment in order to take multiple high-magnification images. The photographs were developed and taped together in an array to recreate an image of the chip. Engineers then used the "crawl-aroundon- the-floor" technique (Figure 10), where they annotated the wires and transistors. This was followed by drawing out the schematic first on paper, then in a schematic editor.

Life has changed since those days. The complexity of devices has followed Moores law, and we are now extracting circuits from 45 nm chips. Moreover, these devices now have up to 12 layers of metal, and use an esoteric combination of materials to create both the conductors and dielectrics [2,3]. They may have hundreds of millions of logic gates, plus huge analog, RF, memory, and other macrocell areas. MEMs, inductors, and other devices are also being integrated onchip.

The circuit extraction flow proceeds as follows:

- Package removal (known in the industry as device "depot")
- Delayering
- Imaging
- Annotation
- Schematic read-back and organization
- Analysis

6.1   Device Depot

Depot may well be the only step of the process that still follows the traditional methods. Typically, packages are etched off in a corrosive acid solution (Figure 11). A variety of acids at various temperatures are used depending on the composition and size of the particular package. These solutions dissolve away the packaging material, but do not damage the die.

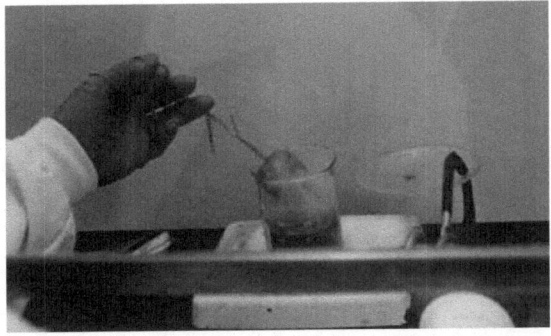

Fig. 11. Into the Acid Bath, My Pretty!

Hermetic and ceramic packages require different techniques that usually involve mechanical or thermal treatment to remove lids, or dice from substrates, or even polish away a ceramic substrate.

6.2 Device Delayering

Modern semiconductor devices range from 1.0 μm single metal bipolar chips, through 0.35 μm BiCMOS diffused MOS (BCDMOS) chips, to 45 nm 12 metal microprocessors, and everything in between. Both aluminum and copper can be used for metal on the same chip. Depending on the process generation, the polysilicon gates and source/drains can use different silicides. A variety of low-k dielectrics are now interspersed with fluorosilicate glass (FSG), phosphosilicate glass (PSG), and SiO_2. Layer thicknesses vary greatly. For instance, on a 7 metal 65 nm Texas Instruments (TI) [4] baseband processor chip we recently analyzed (Figure 12), we found:

- Interconnect layers included Cu, Al, TiN, and TaN
- Metal thicknesses ranged from 0.15 to 1.4 μm
- Dielectrics included silicon nitride, oxynitride, oxide, SiOC, SiONC, and PSG
- Dielectric thicknesses varied from ~ 0.3 μm to 2.6 μm (with individual layers of particular materials as thin as 47 nm), and gate oxide was 2.2 nm thick.

A delayering lab needs to create a single sample of the device at each metal layer, and at the polysilicon transistor gate level. As such, it needs to accurately strip off each layer, one at a time, while keeping the surface planar. This requires detailed recipes for removal of each layer. These recipes include a combination of methods such as plasma (dry) etching, wet etching, and polishing. As the complexity and variation of chips increases, so too does the number of recipes. A modern chipdelayering lab would now have over a hundred such recipes, specific to different processes and materials.

For unknown or unusual chips, it is advisable to start with a cross section (Figure 12). The cross section can be analyzed using scanning electron microscopes (SEM), transmission electron microscopes (TEM), and other techniques

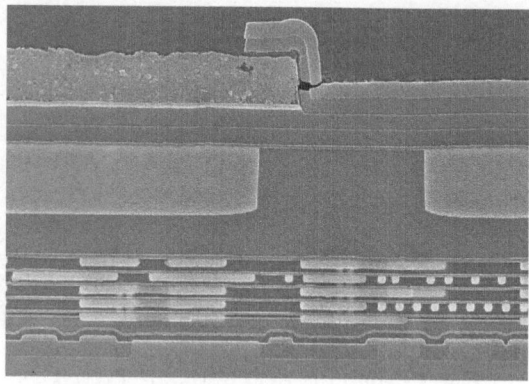

Fig. 12. SEM Cross Section of 65 nm TI Baseband Chip for Nokia

to determine the composition and thickness of all the layers. A delayering technician uses this information to choose the best delayering recipe for a chip. The recipe also varies depending on the type of imaging to be performed. Optical imaging looks best if the transparent dielectric is left on over the layer to be imaged. SEM, due to its operating methodology of electron reflection from a non-planar surface, requires the dielectric to be removed.

6.3   Imaging

Advanced RE labs currently use two types of imaging, optical and SEM. Up to and including the 0.25 μm generation of semiconductor chips, optical imaging was sufficient. However, for 0.18 μm technologies and smaller, optical imaging cannot resolve the smallest features, and SEM must be used (Figure 13).

The size of ICs, and the large magnifications required for the advanced feature sizes, now means that manually shooting images is no longer practical. Imaging systems now must have automated steppers integrated with the microscope. Our twodimensional steppers allow us to set up a shoot in the evening, and come back in the morning to find the entire layer imaged.

Next we use specially developed software to stitch the thousands of images per layer together, with minimal spatial error. Then more software is required to synchronize the multiple layers so that there is no misalignment between layers. Contacts and vias must be lined up with the layers above and below in order for extraction to proceed.

6.4   Annotation

Once all images are stitched and aligned, the actual work of reading back the circuit begins. Full circuit extraction requires taking note of all transistors, capacitors, diodes, and other components, all interconnect layers, and all contacts and vias. This can be done manually or using automation.

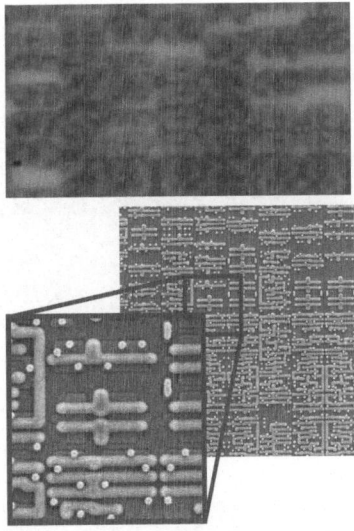

Fig. 13. Optical (top) and SEM images of 130-nm chip

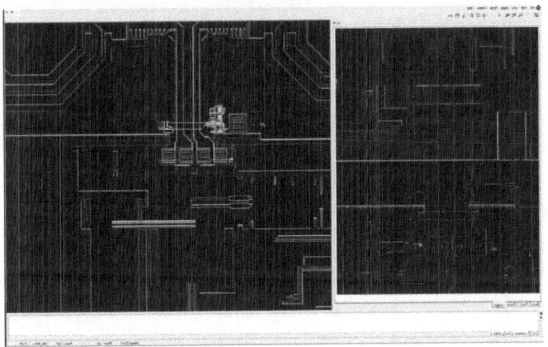

Fig. 14. Power Net of an RF-Switch Inductor and VCO Subsystem

There are multiple tools available to help with this process, including Chipworks' ICWorks Extractor. This tool is used to view all the imaged layers of a chip individually and aligned to each other. In one mode it allows several layers of a chip to be visible in multiple windows simultaneously (Figure 14). Each window shows the same two-dimensional area in each layer. A lock-step cursor allows the engineer to see exactly what lies above or below the feature he is looking at in one layer.

An extraction engineer can then use the tool to annotate and number all wires and devices in his area of interest (Figure 15). 2D and 3D image recognition and processing software can be used (Figure 16), or the engineer may do it manually. Image recognition software can also be used to recognize standard cells in digital logic. This can greatly aid the extraction of large blocks of digital cells.

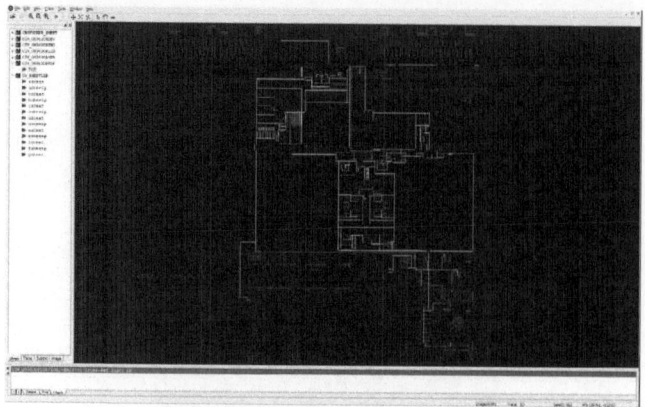

Fig. 15. Annotated VDD net on an RF Transceiver

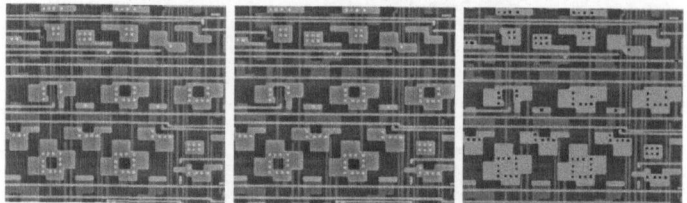

Fig. 16. Automated feature extraction from SEM images

6.5   Verification and Schematic Creation

The annotation process can be error prone. Often the images are not perfect, manual techniques are used, bits of dust fall on the chip during imaging, or the image recognition software introduces an error. Hence, verification is performed at this stage. Design rule checks can find many issues, such as below minimum sized features or spaces, hanging wires, vias without wires, etc.

At this stage the ICWorks tool can automatically extract a netlist from the annotations, and from this netlist create a flat schematic (see Fig. 17). The schematic, netlist, and annotations are all associated with each other, such that one cannot be changed without changing all three.

The netlist and schematic can now be checked for other simple rule violations. Floating gates, shorted outputs, nets with no inputs or outputs, and shorted supplies can be checked.

6.6   Schematic Analysis and Organization

This is one of the steps requiring the most thought, since the schematic organization on a page, or in hierarchy, goes a long way to making a design coherent. Devices placed poorly on a schematic, or a strange hierarchy, can make the design

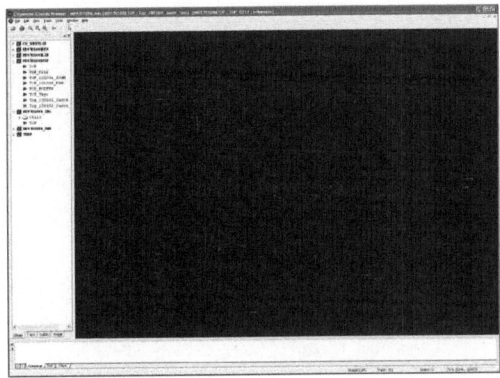

Fig. 17. Flat Schematic Auto-Exported from Annotated Images

very difficult to understand. Hence, this step usually requires very experienced analysts.

The analysis phase can be very iterative, and use many sources of information. Often public information is available for devices. This can take the form of marketing information, datasheets, technical papers, or patents. These can often help with the schematic organization, for instance if block diagrams are available. They can also help in the understanding of architectures and sometimes circuit designs.

Analysis can also be done using typical chip design techniques. A circuit can be hand analyzed using transistor and logic theory. Layout structures are often recognizable, for instance differential pairs, bipolar devices for bandgap references, etc. In fact, The ICWorks tool can find these structures automatically. Hierarchy can also sometimes be seen in the layout. If not, it can be created using a bottom-up schematic organization approach. Functional and timing analysis can be further validated using simulation. Multiple methods are usually used for verification.

The final product of circuit reverse engineering can take many forms. A complete set of hierarchical schematics can be delivered. This set of schematics can be used to also create a hierarchical netlist. Simulated waveforms, block diagrams, timing diagrams, analysis discussion, and circuit equations can be used to round out the report.

Since RE companies analyze so many ICs, they can also create comparative and trend reports. For instance, Chipworks has analyzed many CMOS image sensors over the years. As the technology and circuit designs evolve, they are monitored. The evolution can be shown from both a process point of view and a circuit point of view.

7   A Case Study

Used together, the above techniques can be very powerful. To illustrate that point, lets review a project we just finished; analyzing a digital ASIC with

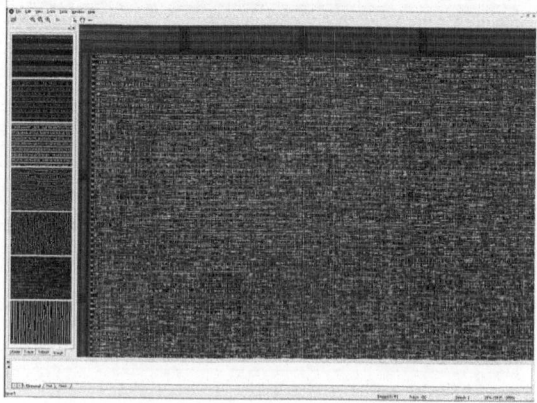

Fig. 18. Annotated digital logic

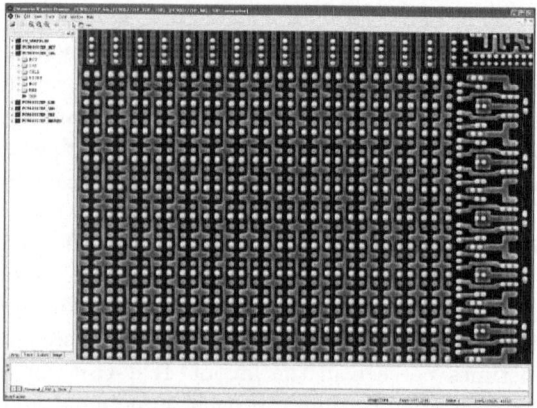

Fig. 19. Metal 1 image of mask programmed ROM

embedded analog and memory blocks, and including embedded encryption hardware. The goal of the project was to fully understand the ASIC, build a model of the ASIC, and get simulations up and running.

The first step was to run system level functional tests while the chip was still in its system. Logic probes were connected, the system was powered up, and vectors were collected which could be used later for simulations.

Next, the chip was depotted, delayered, imaged, stitched, and aligned. We found the chip contained 12,000 gates of digital logic and an embedded EEP-ROM. The entire chip was annotated, and the ICWorks tool created a netlist and flat schematic from this annotation. A portion of the digital logic annotation is shown in Figure 18. Annotation and schematic rule checks were used to verify a quality schematic starting point. In fact, for this project we annotated the entire chip twice, then compared the results to minimize annotation errors.

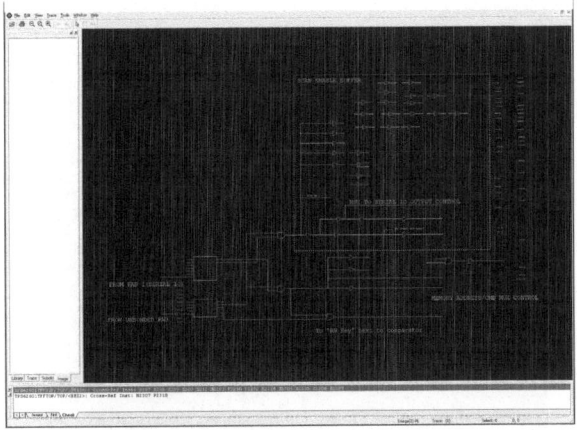

Fig. 20. Logic Scan Control Circuit

The schematics were then partially organized. The memory schematic was completely organized, and the main registers of the digital block were grouped. A few of the major busses were labeled and the I/Os were connected to the major blocks.

In order to run a full chip simulation on the netlist, we would need to extract all the contents of the chip, including both the hardware and memory contents. Different memory types have different challenges in reading them. Embedded SRAMs are the simplest. These memories are volatile, no data is stored in them during power down, so they do not need to be extracted. ROMs can be extracted using traditional RE techniques of physically reading back the mask programming. Figure 19 shows a metal 1 mask programmed ROM. Unfortunately EEPROMs are more difficult than either of these.

We knew up front that this chip included on-chip encryption, and that the keys were stored in the EEPROM. Hence, we anticipated a challenge in being able to access this memory. As expected, the memory was well protected, and much of this memory could not be directly read off-chip. Additionally, the interface to this chip was encrypted, so we had no idea how to generate a memory read command anyhow. The solution to this was to use the test hardware embedded in the chip.

This particular chip had both scan path test circuitry for the digital logic, and memory BIST for the EEPROM. Once we had organized the test and memory circuits, we set to work analyzing them. The scan test control circuit is shown in Figure 20. We found a method where we could almost read out the memory locations using a combination of the digital and memory test circuitry. A single application of microsurgery looked as though it would unlock the bits.

We took a single chip, used jet-etching to remove a portion of the package, then used focused ion beam (FIB) techniques to modify a connection on the chip (Figure 21). Next we used our analysis to create scan path vectors, with the appropriate control signals, and successfully read out the encryption keys and other memory contents via the test port.

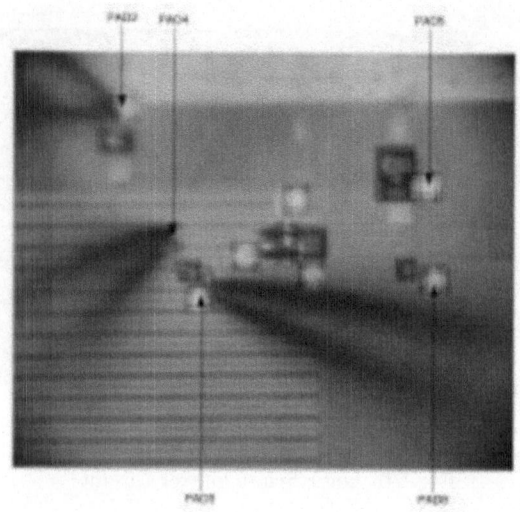

Fig. 21. Microsurgery Altered Chip Being Pico-probed

At this point, we created a memory model to use with our netlist. The vectors collected from the actual system were run on the netlist, and we verified that our chip model gave the same outputs as the actual chip tested. Hence, we confirmed our netlist and memory contents were correct.

The encryption algorithm also needs to be understood to be able to complete the analysis of this chip. This was accomplished via schematic organization and simula- tion. As we organized the chip, we found some interesting structures, such as a 56 bit register. Therefore, we ran our simulations, and monitored the busses in the area of this register. Sure enough, keys were read from our memory model, loaded into this embedded block, and a standard DES algorithm was observed.

Now we understood the encryption, had the keys, and had full chip simulations running. Since we had a full netlist, we were able to run full chip simulations and monitor any internal nodes required. This allowed us to complete the analysis of this chip and understand all the commands it could execute.

8   Summary

In this paper we have reviewed the different types of reverse engineering pertinent to the semiconductor industry. For reverse engineers, life will not get any easier in the electronics business. In semiconductors, the next challenge will be the 32 nm node devices already being ramped up in development fabs. The consumer electronics business keeps bouncing from new toy to yet another new toy, and it is necessary to be aware of all the new products that keep appearing.

As is shown in this paper, the RE business has to keep evolving to keep up with the changes in electronics and design, and it has become a discipline in itself, created by the needs of the global market for competitive intelligence and IP support.

Acknowledgements

We would like to thank Chipworks' laboratory staff and engineers who actually do all the hard work of analyzing these complex devices. Without them, we would have no material for this paper!

References

1. James, D.: A Case Study: Looking Inside Apple's iPOD Nano– a Teardown to the Atomic Scale,
   http://electronics.wesrch.com/Paper/
   paper_details.php?id=EL1SE1KWRX174&paper_type=pdf&type=author
2. Nii, H., et al.: A 45 nm High Performance Bulk Logic Platform Technology (CMOS6) using Ultra High NA (1.07) Immersion Lithography with Hybrid Dual-Damascene Structure and Porous Low-k BEOL. In: IEDM 2006 Technical Digest, pp. 685–688 (2006)
3. Narasimha, S., et al.: High Performance 45 nm SOI Technology with Enhanced Strain, Porous Low-k BEOL, and Immersion Lithography. In: IEDM 2006 Technical Digest, pp. 689–692 (2006)
4. Chatterjee, A., et al.: A 65 nm CMOS Technology for Mobile and Digital Signal Processing Applications. In: IEDM 2004 Technical Digest, pp. 665–668 (2004)
5. 2005 ITRS, Metrology section
6. Vartanian, V., et al.: Metrology Challenges for 45 nm Strained-Si Devices. In: 2005 International Conference on Characterization and Metrology for ULSI Technology (2005)

Trojan Side-Channels: Lightweight Hardware Trojans through Side-Channel Engineering

Lang Lin[1], Markus Kasper[2], Tim Güneysu[2],
Christof Paar[1,2], and Wayne Burleson[1]

[1] Department of Electrical and Computer Engineering
University of Massachusetts, Amherst, USA
[2] Horst Görtz Institute for IT Security
Ruhr University Bochum, Germany
{llin,burleson}@ecs.umass.edu,
{gueneysu,mkasper,cpaar}@crypto.rub.de

Abstract. The general trend in semiconductor industry to separate design from fabrication leads to potential threats from untrusted integrated circuit foundries. In particular, malicious hardware components can be covertly inserted at the foundry to implement hidden backdoors for unauthorized exposure of secret information. This paper proposes a new class of hardware Trojans which intentionally induce physical side-channels to convey secret information. We demonstrate power side-channels engineered to leak information below the effective noise power level of the device. Two concepts of very small implementations of *Trojan side-channels* (TSC) are introduced and evaluated with respect to their feasibility on Xilinx FPGAs. Their lightweight implementations indicate a high resistance to detection by conventional test and inspection methods. Furthermore, the proposed TSCs come with a physical encryption property, so that even a successful detection of the artificially introduced side-channel will not allow unhindered access to the secret information.

Keywords: Trojan Hardware, Side-Channel Analysis, Covert Channel, Trojan Side-Channel, Hardware Trojan Detection.

1 Introduction

Historically the choice to implement cryptographic routines in hardware was mainly driven by high-security applications such as banking or government systems. Nowadays, this has changed since the trend towards system-on-a-chip solutions has facilitated the integration of high-performance cryptography also in commercial off-the-shelf silicon devices. For example, the majority of current PCs and laptops are sold with built-in trusted platform module (TPM) chips. Another pertinent example is the trend towards pervasive embedded computing, bringing hardware cryptography to products such as smartphones, payment cards, RFID-equipped goods or medical devices. Since these personal devices can be physically accessed by their owners, their security often relies on hardware-based security modules. In this context, security modules implemented in silicon are generally considered more trustworthy than software solutions. This is

C. Clavier and K. Gaj (Eds.): CHES 2009, LNCS 5747, pp. 382–395, 2009.

mainly due to the need for expensive tools to modify and probe circuits on the submicron scale, which imposes a significant barrier to attackers and thus implicitly provides a basic level of protection against key extractions and algorithm manipulations.

Due to the recent fabless trends in semiconductor industry, malicious circuit manipulations such as "hardware Trojans" can be furtively implanted into the genuine integrated circuits (IC) to compromise their security. Such attacks on the hardware design of chips lead to serious consequences, as (1) a very large number of devices will be affected, e.g., millions of e-banking authentication tokens or TPM chips, (2) the attack might not be noticed for a very long time and perhaps more importantly, (3) security breaks of this kind are almost impossible to fix because there is no practical hardware equivalent for software updates. Even though there is no solid evidence on malicious manipulations of commercial devices at manufacturing time up to now, "Trojan hardware" is considered a serious threat for security modules of all kinds [1]. Given the complexity of the current semiconductor supply chain including fabless semiconductor companies, there is an urgent need to implement organizational measures [2] to enable trusted IC manufacturing.

Several recent academic works highlighted potential threats by demonstrating concepts of possible hardware Trojans. In [3] a malicious core embedded into a central processing unit (CPU) is proposed. This work rates the detectability of such Trojans as low mainly due to the small hardware overhead (a total of 2300 gates for the Trojan circuits) and timing perturbations. In [5], the register-transfer-level (RTL) netlist of a cryptographic application on reconfigurable hardware is manipulated with additional logic to implement malicious hardware. This Trojan has a complicated triggering pattern that will most likely remain undetected in conventional function tests. Another work highlights the possibilities to use hardware Trojans to covertly leak secret information through wireless channels such as thermal, optical and radio channels [6]. Nevertheless this work still requires trigger circuitry and most of the proposed channels are realized by signals on output pins. This might be a drawback when it comes to the detectability of the malicious circuitry.

Well-designed cryptographic hardware modules are very difficult to be analyzed or modified. However, for the same reasons it is also difficult for chip-designers to detect malicious manipulations that were introduced to their circuitry during the manufacturing process. Modern ICs often contain large blocks of unused circuits, which may be left from previous versions of the design or for temporary testing purposes. Malicious hardware can be hidden in these unused chip-areas. As long as the TSC circuits are tiny, for example less than 100 gates, they cannot be easily distinguished from other hundreds of thousands of gates by basic chip layout inspection.

To detect hardware Trojans, three general approaches [7] have been proposed. The failure analysis community employs sophisticated techniques for visual inspection such as optical and scanning electron microscopy (SEM) or even picosecond imaging circuit analysis [8]. These methods are very dependent on laboratory instrumentation and are often not feasible to be applied to production-run ICs. Other approaches generate test patterns using the standard VLSI fault detection tools to find unexpected device behavior generated by malicious hardware [9,10]. However, this method may not detect Trojans with complicated triggering patterns or carefully hidden channels that are leaking information. The third approach profiles an IC by various analog measurements of, e.g., power traces or internal delays. Then the profile of a trusted Trojan-free IC is used as reference for analyzing suspicious ICs. In [11], the power consumption of ICs is profiled to detect Trojans by means of a subspace projection analysis [12]. Other works based on side-channel and path delay profiles are described in [13,14]. Evaluators following this approach can only detect Trojan hardware circuits if at least 0.1-0.01% of the pristine ICs circuit area is modified [11]. This translates into hundreds of extra gates for most modern devices. To summarize, there are no feasible evaluation techniques to our knowledge that can detect hardware manipulations with very small gate counts.

In general, IC function tests do not include side-channel analysis (SCA). SCA allows to extract information from physical channels inherently existing in electronic devices. The most commonly analyzed side-channels are the power consumption and the electromagnetic radiation [15] of running ICs. In addition, many other physical properties like timing behavior or sound waves [16,17] have been shown to leak exploitable information. During the last decade many different side-channel analyses have been demonstrated, that exploit physical information leakage to compromise the security of cryptographic routines especially in embedded devices.

So far to the best of our knowledge, most research works examine side-channels as undesired phenomena that require special attention to protect devices from sophisticated attacks. In this paper we change this perspective and use intentionally introduced side-channel leakage as a building block for Trojan circuitry. We propose and demonstrate hardware Trojans which are much more subtle than introducing complex logic blocks, but still can completely compromise otherwise secure hardware by leaking exploitable information. In particular, we design Trojans using less than 100 gates to generate artificial power side-channels suitable to covertly leak secret information. We refer to these covert channels as "Trojan side-channels" (TSC).

This paper is organized as follows: Section 2 introduces the basic concepts of TSC. Section 3 presents two different approaches for TSCs that are built upon spread-spectrum theory and artificial leakage functions induced on the key schedule of cryptographic algorithms. In Section 4, we discuss further work and research related to the concept of TSCs, before we finally draw conclusions in Section 5.

2   Introducing the Trojan Side-Channel

Before discussing the concept of Trojan side-channels, we introduce the parties involved in a Trojan side-channel scenario. We refer to the party implanting the

Trojan hardware into the genuine ICs as *attacker*, and the party attempting to detect infected ICs as *evaluator*. An attacker could be for example an untrusted IC foundry, and an evaluator, who attempts to verify correctness and integrity of an IC, a Common Criteria Testing Laboratory (CCTL).

Furthermore we assume that the designers implemented all cryptographic schemes on their device with state-of-the-art countermeasures against SCA attacks, such as the use of side-channel resistant logic styles. Hence, the initial implementation on the chips should be considered as side-channel resistant and not exposing any secret information by side-channel leakage. Note that a device protected at this level is likely to be evaluated according to its side-channel resistance. Thus, a Trojan implanted in such a device needs to be designed to evade detection even during evaluation of the pristine IC's side-channel resistance by sophisticated methods such as higher-order power analysis or template attacks.

Based on these requirements we define the following design goals for a circuit-level implementation of a TSC:

- Detectability:
  - Size: The required amount of logic gates has to be minimized to evade detection of the TSC by evaluators.
  - Side-Channel Leakage: The TSC must not be detected when performing power analyses targeting the pristine ICs functionality. As a minimum requirement, the relative power consumption of the TSC circuit with respect to the overall power should be negligible so that it cannot be obtained from the power traces just by visual inspection.
  - Trigger: The TSC must not effect functionality of the device in any way to avoid detection by extensive functionality testing. This also prohibits triggering and communication using external pins.
- Usability:
  - The TSC must not be exploited by anyone else than the attacker, who knows all details of the modification. We call this "encryption property".

The principle of a TSC is visualized in Figure 1. The compromised device is modeled by an IC having an embedded crypto core. Without TSC, the secret key K cannot be recovered by means of SCA. During the IC manufacturing process, the attacker covertly inserts a TSC circuit that encodes K into physical leakage. We model this encoding by an encryption function[1] $e(K)$, which is designed to reserve usage and detection of the side-channel only to the implementing attacker. Once the IC is deployed in the field, an evaluator must not be able to detect the TSC. Furthermore, the encryption property of the Trojan is designed to avoid usage of the Trojan even by an evaluator that is aware of the existence of the TSC.

TSCs that incorporate such an encryption property require special attention during IC security evaluation. A good encryption property requires an evaluator to overcome an infeasible computational or experimental effort, e.g., 2^{80} calculations or 2^{40} measurements, to access the secret information that is exposed.

---

[1] Note that the notion of encryption in this context does not necessarily imply the strong security properties as commonly assumed for cryptographic schemes.

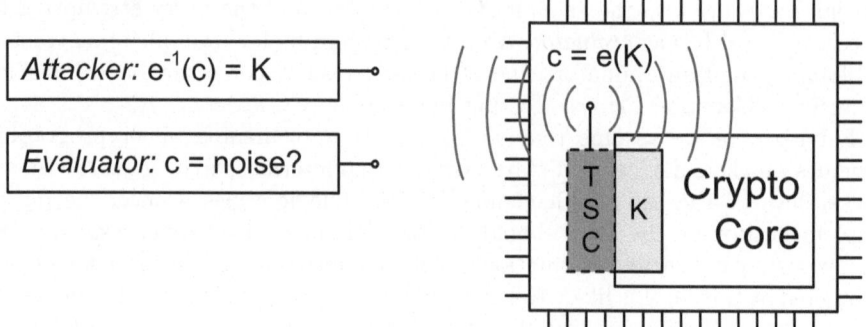

Fig. 1. Principle of Trojan side-channels

On the other side the attacker (who designed the TSC) needs to have some advantageous knowledge, allowing him to make use of the TSC within feasible efforts.

The class of TSCs introduced in this work use an internal state (artificially introduced as part of the TSC or an inherently existing internal state) to encode the secret information to be leaked by means of a logic combination function. The output of this logic is then connected to a building block acting as an artificial leakage source. For the side-channel based on power consumption, such a leakage circuit can be realized, for example, using big capacitances, toggling logic or pseudo-NMOS gates. Note that the amount of generated leakage is part of the TSC design space and can be engineered to take any desired signal-to-noise ratio (SNR). This choice of a SNR affects both, attacker and evaluator, as it determines the amount of measurement samples required to detect and use the TSC. In addition, it might also affect the size of the TSC.

3   Implementations of TSCs

In the following sections, we demonstrate two very different ways to implement TSCs. These examples shall provide a first impression of the flexibility of the introduced concept of TSCs and highlight the wide range of design options available to attackers. Although TSCs aim to be implemented in ASICs, we preliminary demonstrate their capability by practical experiments on FPGA implementations. Note that these FPGA implementations should be regarded as proof-of-concept implementations only, because FPGAs are (due to their course-grain logic elements) not able to fully mimic the design options and structural characteristics that apply for ASICs.

3.1   TSCs Based on Spread-Spectrum Theory

Our first TSC design is a side-channel adapting the concepts from spread-spectrum communications (also known as code-division multiple access

(CDMA)) to distribute the leakage of single bits over many clock cycles. The basics of the CDMA encoding are very similar to conventional stream-ciphers. A bit sequence (the code) is used to modulate information bits using XOR. Contrary to the stream-cipher concept, CDMA uses many code bits to transfer single bits of information, i.e., the code bits are changing much faster than the information bits. This strategy spreads the information contained in a single bit along a longer bit (or code) sequence which allows transmission and recovery of information in subliminal channels even below noise level. This property makes CDMA the method of choice to implement hidden military communication channels. The demodulation used to decode CDMA channels helps to understand how CDMA can establish channels in this sub-noise domain. The process of decoding using a correlation demodulator is very close to what the community of cryptographers knows as correlation power analysis. The demodulator uses subsequent power measurements and correlates them to the synchronized code sequence. If the correct code has been used, this leads to a positive correlation coefficient for encoded zeros and a negative correlation coefficient for encoded ones. The more power traces the demodulator analyzes, the more "process gain" (which is the ratio of code sequence length to the bit information length in spread-spectrum encoding) is available to overcome a low demodulation SNR. Note that the CDMA channel can only be demodulated using the correct code sequence and demodulation with different code sequences will not lead to any significant correlation. Therefore it is possible to transfer bits on multiple CDMA channels simultaneously, as each CDMA channel is indistinguishable from noise for all other channels.

Our TSC employs this method by using a pseudo-random number generator (PRNG) to create a CDMA code sequence. This sequence is then used to XOR-modulate the secret information bits. The modulated sequence is forwarded to a leakage circuit (LC) to set up a covert CDMA channel in the power side-channel. In this model, the advantage of the attacker is the knowledge about the exact setup of the code generator (more precisely, the initialization vector and feedback coefficients of the implemented PRNG) which are required to predict the code sequence. Knowing all details of the PRNG used gives the attacker the essential advantage over evaluators who cannot distinguish the covert channel from noise. For decoding this channel, the attacker performs a correlation demodulation on measurement points of subsequent clock cycles as described above. When evaluating side-channel leakage of the compromised device, the leakage due to the TSC will not be detected during attacks on the pristine IC core. Note that depending on the leakage generating circuit, it might be necessary to consider a mapping of the used code with respect to a suitable power model prior to correlation-based demodulation. Note further that to amplify the quality of results, the attacker can repeat measurements several times and average the corresponding traces to reduce the overall impact of noise.

Experimental Results. To demonstrate the applicability of the proposed TSC, we implement an entire AES cryptosystem and a TSC using the linear feedback shift register (LFSR) shown in Figure 2 on a Xilinx Spartan-3E FPGA

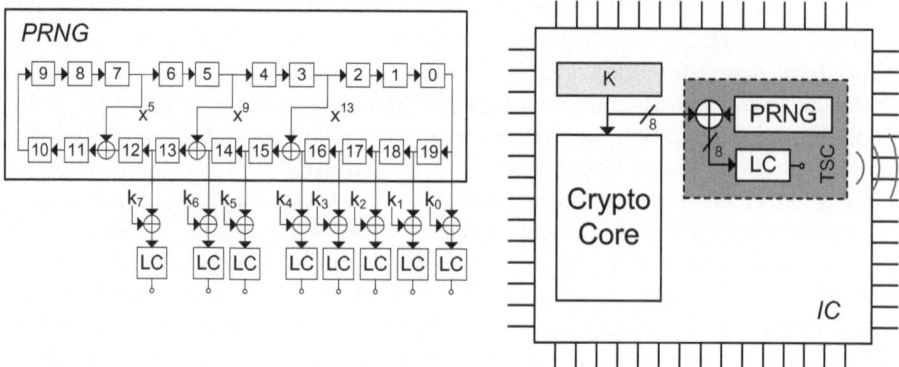

Fig. 2. Diagram of a spread-spectrum TSC circuit based on a pseudo-random number generator (PRNG) and a separate leakage circuit (LC)

running at 50 MHz. We use a 1 Ω serial current-sensing resistor to probe the power consumption of the FPGA core, similar to the experimental setups described in [22,23]. The transient power traces are measured by an Agilent Infiniium 54832D oscilloscope. In this experiment, we used a single CDMA channel to transmit a single key bit. The leakage generating circuit (LC) was realized by connecting eight identical flip-flop elements to the single output of the XOR gate to mimic a large capacitance. To demodulate the CDMA channel, the attacker has to take into account that the flip-flops do not leak a good Hamming weight signal. Therefore, the power consumption behavior of the leakage circuit has to be modeled to map the CDMA code to the leakage circuit. In this case the flip-flops will cause short circuit currents when toggling, and additional currents from discharging their load capacitances on one-to-zero transitions. The latter can be used to distinguish the transmitted data during demodulation: For encoded bits which are one, the circuit will have a higher leakage on all zero-to-one transitions of the code, while for encoded zeros the circuit will generate more leakage currents on one-to-zero transitions of the code. The attacker uses the resulting two different code sequences for demodulation. The higher correlation coefficient will identify the actually transferred bit.

In our setup, the PRNG is based on a simple LFSR using the primitive polynomial $x^{20} + x^{13} + x^9 + x^5 + 1$. This PRNG thus generates a code sequence with a maximum order of 2^{20}-1. We successfully extracted the secret key bit with code lengths of at least 1030 bits. In Figure 3, the correlation coefficient for the code sequence identifying the transmitted bit is with 0.04 considerably larger than the correlation coefficients resulting from wrong choices of code sequences and false guesses of the transmitted bit. The FPGA implementation of this TSC circuit requires 42 flip-flops (FF) and 6 look-up tables (LUT) occupying 23 slices (49 gates would be required for the corresponding ASIC circuit). Note that these resource consumption is nearly negligible with respect to our iterative AES implementation that requires 531 slices utilizing 442 FFs and 825 LUTs (this does even not take additional resources for control or I/O into account).

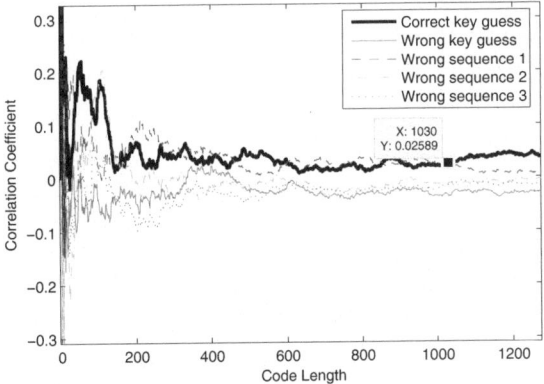

Fig. 3. Required number of power traces to extract key bits from TSC based on spread-spectrum technique

Detectability. The additional leakage of this spread-spectrum inspired TSC only depends on the key and a fixed code-generator. An evaluator can exploit this, by first averaging many traces using a fixed input and then, in a second step, averaging over another set of traces, where only a single key bit has been flipped. By calculating the difference of means of those traces he might get enough information to recover the code sequence that has been used to leak the toggled key bit. To harden the proposed TSC against this method we suggest to transfer only combination of bits, such that further interpretation by means of algebraic equations is required to understand and use the TSC. An alternative method to prevent unauthorized use of our TSC by an evaluator is to create an interdependence between plaintext input and initialization vector of the PRNG, i.e., either we generate the PRNG's initialization value based on a combinatorial circuit from the plaintext or we introduce a previous initialization phase and clock all plaintext bits through the PRNG using a separate shift register and an additional XOR gate (similar to the initialization of the A5/1 streamcipher). Although these approach requires additional logic and/or clock cycles, this involves a significantly higher effort for the evaluator to get access to the Trojan's communication channel.

3.2 TSCs Using Known Input Values

Our second proposal is a subtle TSC that leaks secret information obtained during the run of a block-cipher's key schedule. More precisely, we demonstrate an attack on the AES-128 block-cipher and its corresponding key schedule in this section. While the TSC following spread-spectrum theory used subsequent leakages to implement a CDMA channel, this design was inspired by conventional side-channel analysis. The idea is to artificially introduce leaking intermediate states in the key schedule that depend on known input bits and key bits, but that naturally would not occur during regular processing of the cipher. These values can then be exploited by differential power analysis attacks. The TSC proposed uses an artificial intermediate state consisting of only a single bit of generated leakage. We evaluated several functions for combining input bits with key bits to find methods that

1. theoretically allow a good discrimination of wrong and right key bit guesses in a differential power analysis using the correlation coefficient,
2. use only a few input bits (≤ 16),
3. require a few logic gates (< 100),
4. do not lead to any exploitable correlation in case intermediate values the crypto-core processes during his regular operation are attacked.

For our demonstration TSC, we selected a very straightforward AND-XOR combination that is based on up to 16 input bits, but promises easily detectable results. This function uses AND conjunctions to pairwise combine each key bit with another input bit. The output of the AND gates are then combined to the leaked intermediate value by XORing all of them.

The encryption property of this TSC results from the attackers choice to select which of the 128 input bits of the AES cipher are combined with which key bits. Note that the attacker's secret to access the TSC is based on a permuted choice; in other words it can be implemented solely by wiring. For the sake of simplicity, we propose to leak 1-2 bytes of the AES round key for each round of the key schedule. This could be, for example, the first and the third byte of each round key. Note that if only one byte is leaked, the key space is already reduced sufficiently ($128 - 10 \cdot 8 = 48$ bit) for an attacker to mount a brute-force attack on a known plaintext-ciphertext pair, e.g., by using special-purpose hardware as a COPACOBANA [4]. On the other hand, designing the TSC to extract two bytes simultaneously enables the attacker to reveal the whole key without the need to apply brute-force to any unknown bits.

Analyzing the available key space, we consider an intermediate value generated as described above using 8 key bits and 8 input bits. In AES-128, we have 16 possible key bytes that can be leaked by a TSC[2]. There are $128!/120!$ different choices for selecting a sequence of 8 different bits from 128 bits. Therefore, we estimate the effective key space to $128!/120! \cdot 16 \approx 9.22 \cdot 10^{17}$ possible keys, which corresponds to a key length of approximately 59 bits. For combination functions using 16 bits the keyspace is significantly larger.

Experimental Results. We evaluated this TSC by implementing an AES key schedule connected to the proposed TSC on a SASEBO standard side-channel evaluation board [18]. The target FPGA on this board is a Xilinx Virtex-2 PRO XC2VP7-5 FPGA. The implemented TSC is sketched in Figure 4 and is based only on 8 plaintext bits and 8 bits of the round key register within the AES-128 key schedule. Note that many real-world implementations store the plaintext

---

[2] It makes sense to use complete bytes instead of unrelated bits, due to the SBOX inside the g-function of the AES key schedule. This facilitates reconstruction of the AES key from the leaked round key snippets.

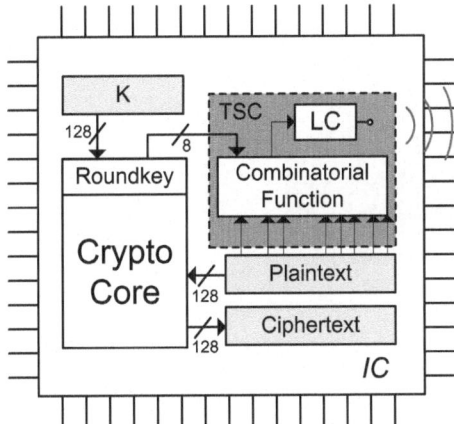

Fig. 4. Diagram showing an alternative TSC circuit attacking 8 bits of an AES-128 key schedule. It consists of a combinatorial circuit taking a set of plaintext and round key bits as inputs that are combined into a single output bit, finally leaked into the power signature using a leakage circuit (LC).

input in the state which is overwritten in subsequent rounds. Hence, the TSC might require an additional register to store the used plaintext bits for the entire runtime of the attack.

In this setup, we additionally employed a leakage circuit (LC) that can be implemented very efficiently with FPGAs (although our main focus are TSCs on ASICs, this could be a solution for the case that an FPGA is the target device). We configured a single LUT of the FPGA as 16-bit shift register (SRL16 feature) and loaded it with an initial alternating sequence of zeros and ones. The shift register is only clocked in case the input to the leakage circuit is one, which results in an additional dynamic power consumption. Including the logic from the combinatorial circuit, the overall size of the implemented TSC results to only 14 LUTs occupying a total of 7 slices (equivalent to 29 gates when implemented as ASIC).

Our experimental results demonstrate that a recovery of the TSC information can easily be achieved by means of a correlation power analysis. The plots of the correlation coefficient show very distinct peaks for correct key guesses. The example detection given in Figure 5 shows 10 peaks each indicating a covert transmission of another key byte. The figure to the right shows the correlation coefficients used to identify the content of the fourth transmitted byte as the value 65.

Detectability. This TSC approach can also be detected and exploited by diligent evaluation of the device. In this case an evaluator again uses variations of traces that differ only in single plaintext or key bits. This allows him to distinguish which key bits and plaintext bits were used in the combination function. While flipping combinations of bits, analysis of the corresponding behavior of the TSC will finally allow to reverse-engineer the used combination function.

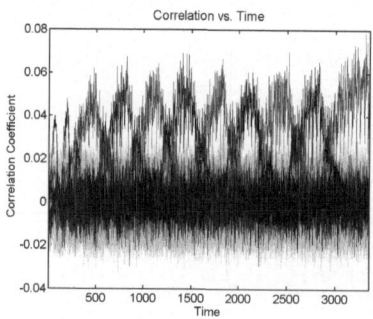 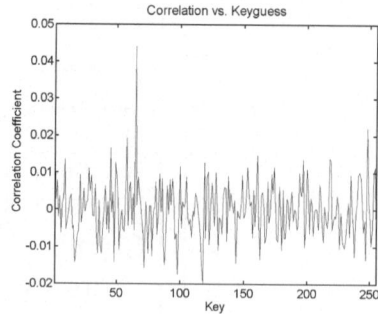

(a) Subsequent exposure of 10 key bytes (b) Discrimination of the key byte for the 4th transmission

Fig. 5. Recovery of the TSC Information by means of the correlation coefficient

To make our TSC less susceptible for these types of detection strategies, we suggest to extend the applied combination function by XORing additional plaintext bits, logic combinations of plaintext bits, key bits or even combinations of key bits. By introducing such complex linear or non-linear equations, the interpretation of the observed differences by the evaluator can be complicated to an unfeasible level. Such an improvement also assists the attacker to detect a key that only consists of zeros. Up to now, the detection of the zero key is based on the absence of any correlation for all other key guesses, which might be an undesired property.

4   Further Work

In this work, we introduce a general concept of Trojan side-channels and still let room for many improvements for both the implementation and the detection perspective. In this section we discuss aspects beyond the scope of this paper that are subject to (our) current research: the detectability and availability of TSCs by third parties and evaluators has not been sufficiently verified experimentally yet. Upcoming designs of TSCs in ASICs have to be tested with special attentions to these properties, so that their threats can be better understood and judged. Improvements to the TSCs include smaller and more subtle combination functions, better leakage circuits and more elegant methods for CDMA code generation. For example, the shift register of the input dependent TSC could be omitted by shifting the contents of the register containing the plaintext bits instead. The leakage circuit could even completely be omitted by using a combination function with well-designed leakage behavior. For a first idea on TSC performance in ASICs, we have already performed experiments using circuit-level simulations of the CDMA-based TSC implemented in a 45nm predictive transistor model. These experiments also indicate feasibility of our concept in real hardware implementations beyond the limited scope of FPGAs with its very coarse-grain logic elements. Therefore, the ASIC implementations of the TSCs require much less gates than the logic required for our preliminary FPGA experiments.

5   Conclusions

Trojan side-channels form a subtle class of hardware Trojans that are very promising to evade detection strategies applied by evaluators. The known methods for detection of "conventional" Trojan hardware circuits, such as optical inspection of metal layers or fingerprinting of circuits, will most likely fail on TSCs due to their extremely small size. TSCs also do not require a direct connection to I/O pins and do not effect the functional behavior of the pristine IC. Since TSCs are only dependent on inputs of the cryptographic implementation under attack, the design space of TSCs allows for a multitude of potential TSC properties. The two types of TSCs demonstrated in this work show how considerable these differences might be, resulting in completely different TSC applications and detection schemes. Moreover, the degrees of freedom include the selection of

- physical channels (power, EM radiation, timing, heat, etc.),
- combination functions,
- internal states determining also time and/or input dependence of the leakage
- leakage circuits depending on their input or on transitions of their input

TSCs implemented during the manufacturing process in untrusted semiconductor foundries pose a very serious threat to all kinds of security modules. We provided a short discussion on the detectability of our TSCs by side-channel evaluators. Given the flexibility of TSCs with better hiding methods, further research is required for evaluators to develop more practical detection schemes to recognize the next generation TSCs.

TSCs based on CDMA are an universal tool for leaking information independent of the cryptographic algorithm used. Moreover, information other than the mere key can be leaked through the CDMA side-channel. We would like to stress that CDMA-based TSCs can potentially also find applications in constructive uses.

- Since TSC can be viewed as a form of physical encryption, one can imagine other cryptographic protocols and applications using TSC as building blocks.
- TSC can be used for anti-counterfeiting in a straightforward manner: authentic ICs can identify themselves via the TSC by sending an ID, whereas illegal but functionally correct copies lack this capability.
- TSC can be used for conveying internal status information about a circuit to facilitate the testability of a circuit.

Acknowledgements

The work described in this paper has been supported in part by the European Commission through the ICT programme under contract ICT-2007-216676 ECRYPT II.

References

1. High Performance Microchip Supply, Annual Report by the Defense Science Board (2008), http://www.acq.osd.mil/dsb/
2. McCormack, R.: It's Like Putting A Band-Aid On A Bullet Hole, Manufacturing & Technology News (2008),
   http://www.manufacturingnews.com/news/08/0228/art1.html
3. King, S.T., Tucek, J., Cozzie, A., Grier, C., Jiang, W., Zhou, Y.: Designing and implementing malicious hardware. In: Proceedings of the 1st USENIX Workshop on Large-Scale Exploits and Emergent Threats (LEET), pp. 1–8 (2008)
4. Güneysu, T., Kasper, T., Novotný, M., Paar, C., Rupp, A.: Cryptanalysis with COPACOBANA. IEEE Transactions on Computers 57(11), 1498–1513 (2008)
5. Chen, Z., Guo, X., Nagesh, R., Reddy, A., Gora, M., Maiti, A.: Hardware Trojan Designs on BASYS FPGA Board. In: Embedded System Challenge Contest in Cyber Security Awareness Week, CSAW (2008)
6. Kiamilev, F., Hoover, R.: Demonstration of Hardware Trojans. In: DEFCON 16, Las Vegas (2008)
7. Wang, X., Tehranipoor, M., Plusquellic, J.: Detecting Malicious Inclusions in Secure Hardware: Challenges and Solutions. In: 1st IEEE International Workshop on Hardware-Oriented Security and Trust (HOST), pp. 15–19 (2008)
8. Soden, J.M., Anderson, R.E., Henderson, C.L.: IC Failure Analysis: Magic, Mystery, and Science. IEEE Design & Test of Computers 14, 59–69 (1997)
9. Banga, M., Hsiao, M.S.: A Region Based Approach for the Identification of Hardware Trojans. In: 1st IEEE International Workshop on Hardware-Oriented Security and Trust (HOST), pp. 40–47 (2008)
10. Chakraborty, R., Paul, S., Bhunia, S.: On-Demand Transparency for Improving Hardware Trojan Detectability. In: 1st IEEE International Workshop on Hardware-Oriented Security and Trust (HOST), pp. 48–50 (2008)
11. Agrawal, D., Baktir, S., Karakoyunlu, D., Rohatgi, P., Sunar, B.: Trojan Detection using IC Fingerprinting. In: IEEE Symposium on Security and Privacy, pp. 296–310 (2007)
12. Fukunaga, K.: Introduction to Statistical Pattern Recognition, 2nd edn. Computer Science and Scientific Computing Series. Academic Press, London (1990)
13. Rad, R.M., Wang, X., Tehranipoor, M., Plusquellic, J.: Power supply signal calibration techniques for improving detection resolution to hardware Trojans. In: International Conference on Computer-Aided Design (ICCAD), pp. 632–639 (2008)
14. Jin, Y., Makris, Y.: Hardware Trojan Detection Using Path Delay Fingerprint. In: 1st IEEE International Workshop on Hardware-Oriented Security and Trust (HOST), pp. 51–57 (2008)
15. Kocher, P., Jaffe, J., Jun, B.: Differential Power Analysis. In: Wiener, M. (ed.) CRYPTO 1999. LNCS, vol. 1666, pp. 388–397. Springer, Heidelberg (1999)
16. Kocher, P.: Timing Attacks on Implementations of Diffie-Hellman, RSA, DSS, and Other Systems. In: Koblitz, N. (ed.) CRYPTO 1996. LNCS, vol. 1109, pp. 104–113. Springer, Heidelberg (1996)
17. Shamir, A., Tromer, E.: Acoustic cryptanalysis, online proof of concept,
   http://people.csail.mit.edu/tromer/acoustic/
18. Resarch Center for Information Security (RCIS): Side-channel Attack Standard Evaluation Board (SASEBO),
   http://www.rcis.aist.go.jp/special/SASEBO/index-en.html

19. Proakis, J.: Digital communications, 4th edn. McGraw-Hill, New York (2000)
20. Gierlichs, B., Lemke-Rust, K., Paar, C.: Templates vs. Stochastic Methods. In: Goubin, L., Matsui, M. (eds.) CHES 2006. LNCS, vol. 4249, pp. 15–29. Springer, Heidelberg (2006)
21. Rajski, J., Tyszer, J.: Primitive polynomials over GF(2) of degree up to 660 with uniformly distributed coefficients. Journal of Electronic Testing: theory and applications, 645–657 (2003)
22. Standaert, F., Oldenzeel, L., Samyde, D., Quisquater, J.: Power analysis of FPGAs: how practical is the attack? In: Y. K. Cheung, P., Constantinides, G.A. (eds.) FPL 2003. LNCS, vol. 2778, pp. 701–711. Springer, Heidelberg (2003)
23. Ors, S., Oswald, E., Preneel, B.: Power-analysis attacks on an FPGA - first experimental results. In: Walter, C.D., Koç, Ç.K., Paar, C. (eds.) CHES 2003. LNCS, vol. 2779, pp. 35–50. Springer, Heidelberg (2003)

MERO: A Statistical Approach for Hardware Trojan Detection[*]

Rajat Subhra Chakraborty[**], Francis Wolff, Somnath Paul,
Christos Papachristou, and Swarup Bhunia

Department of Electrical Engineering and Computer Science,
Case Western Reserve University, Cleveland OH–44106, USA
rsc22@case.edu

Abstract. In order to ensure trusted in–field operation of integrated circuits, it is important to develop efficient low–cost techniques to detect malicious tampering (also referred to as *Hardware Trojan*) that causes undesired change in functional behavior. Conventional post–manufacturing testing, test generation algorithms and test coverage metrics cannot be readily extended to hardware Trojan detection. In this paper, we propose a test pattern generation technique based on multiple excitation of rare logic conditions at internal nodes. Such a statistical approach maximizes the probability of inserted Trojans getting triggered and detected by logic testing, while drastically reducing the number of vectors compared to a weighted random pattern based test generation. Moreover, the proposed test generation approach can be effective towards increasing the sensitivity of Trojan detection in existing *side–channel* approaches that monitor the impact of a Trojan circuit on power or current signature. Simulation results for a set of ISCAS benchmarks show that the proposed test generation approach can achieve comparable or better Trojan detection coverage with about 85% reduction in test length on average over random patterns.

1 Introduction

The issue of *Trust* is an emerging problem in semiconductor integrated circuit (IC) security [1,2,3,8] This issue has become prominent recently due to widespread outsourcing of the IC manufacturing processes to untrusted foundries in order to reduce cost. An adversary can potentially tamper a design in these fabrication facilities by the insertion of malicious circuitry. On the other hand, third-party CAD tools as well as hardware intellectual property (IP) modules used in a design house also pose security threat in terms of incorporating malicious circuit into a design [3]. Such a malicious circuit, referred to as a *Hardware Trojan*, can trigger and affect normal circuit operation, potentially with catastrophic consequences in critical applications in the domains of communications, space, military and nuclear facilities.

---

[*] The work is funded in part by a DoD seedling grant FA-8650-08-1-7859.
[**] Corresponding author.

C. Clavier and K. Gaj (Eds.): CHES 2009, LNCS 5747, pp. 396–410, 2009.

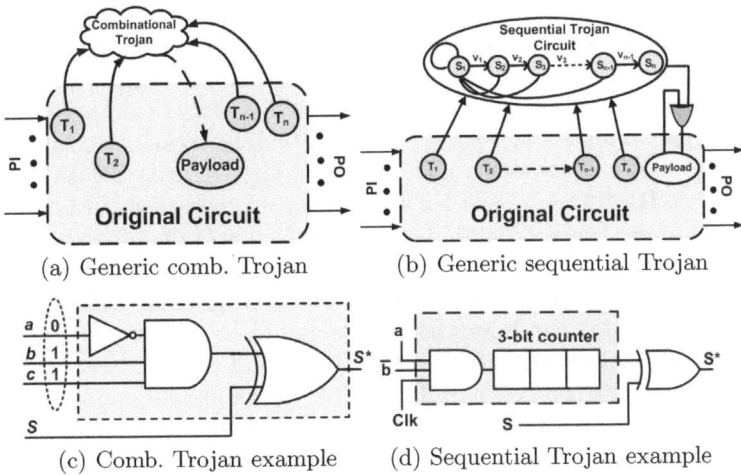

(a) Generic comb. Trojan   (b) Generic sequential Trojan

(c) Comb. Trojan example   (d) Sequential Trojan example

Fig. 1. Generic model for combinational and sequential Trojan circuits and corresponding examples

An intelligent adversary will try to hide such tampering of IC's functional behavior in a way that makes it extremely difficult to detect with conventional post–manufacturing test [3]. Intuitively, it means that the adversary would ensure that such a tampering is manifested or *triggered* under very rare conditions at the internal nodes, which are unlikely to arise during test but can occur during long hours of field operation [13]. Fig. 1 shows general models and examples of hardware Trojans. The *combinational Trojans* as shown in Fig. 1(a) do not contain any sequential elements and depend only on the simultaneous occurrence of a set of rare node conditions (e.g. on nodes T_1 through node T_n) to trigger a malfunction. An example of a combinational Trojan is shown in Fig. 1(c) where the node S has been modified to $S^\star$, and malfunction is triggered whenever the condition $a = 0, b = 1, c = 1$ is satisfied. The *sequential Trojans* shown in Fig. 1(b), on the other hand, undergo a sequence of state transitions (S_1 through S_n) before triggering a malfunction. An example is shown in Fig. 1(d), where the 3–bit counter causes a malfunction at the node S on reaching a particular count, and the count is increased only when the condition $a = 1, b = 0$ is satisfied at the positive clock–edge. We refer to the condition of Trojan activation as the *triggering condition* and the node affected by the Trojan as its *payload*.

In order to detect the existence of a Trojan using logic testing, it is not only important to trigger a rare event at a set of internal nodes, but also to propagate the effect of such an event at the payload to an output node and observe it. Hence, it is very challenging to solve the problem of Trojan detection using conventional test generation and application, which are designed to detect manufacturing defects. In addition, the number of possible Trojan instances has a combinatorial dependence on the number of circuit nodes. As an example, even with the assumption of maximum 4 trigger nodes and a single payload, a relatively small

ISCAS–85 circuit such as *c880* with 451 gates can have $\sim 10^9$ triggers and $\sim 10^{11}$ distinct Trojan instances, respectively. Thus, it is not practical to enumerate all possible Trojan instances to generate test patterns or compute test coverage. This indicates that instead of an exact approach, a statistical approach for test vector generation for Trojans can be computationally more tractable.

In this paper, we propose a methodology, referred to as *MERO* (**M**ultiple **E**xcitation of **R**are **O**ccurence) for statistical test generation and coverage determination of hardware Trojans. The main objective of the proposed methodology is to derive a set of test patterns that is *compact* (minimizing test time and cost), while maximizing the Trojan detection coverage. The basic concept is to detect low probability conditions at the internal nodes, select candidate Trojans triggerable by a subset of these rare conditions, and then derive an optimal set of vectors than can trigger each of the selected low probability nodes *individually to their rare logic values multiple times* (e.g. at least N times, where N is a given parameter). As analyzed in Section 3.1, this increases the probability of detection of a Trojan having a subset of these nodes as its trigger nodes. By increasing the toggling of nodes that are random–pattern resistant, it improves the probability of activating a Trojan compared to purely random patterns. The proposed methodology is conceptually similar to *N-detect test* [5,6] used in stuck-at ATPG (automatic test pattern generation), where test set is generated to detect each single stuck-at fault in a circuit by at least N different patterns, in the process improving test quality and defect coverage [6]. In this paper, we focus on digital Trojans [13], which can be inserted into a design either in a design house (e.g. by untrusted CAD tool or IP) or in a foundry. We do not consider the Trojans where the triggering mechanism or effect are analog (e.g. thermal).

Since the proposed detection is based on functional validation using logic values, it is robust with respect to parameter variations and can reliably detect very small Trojans, e.g. the ones with few logic gates. Thus, the technique can be used as *complementary to the side–channel Trojan detection approaches* [1,9,10,11] which are more effective in detecting large Trojans (e.g. ones with area $> 0.1\%$ of the total circuit area). Besides, the *MERO* approach can be used to increase the detection sensitivity of many side-channel Trojan detection techniques such as the ones that monitor the power/current signature, by increasing the activity in a Trojan circuit. Using an integrated Trojan coverage simulation and test generation flow, we validate the approach for a set of ISCAS combinational and sequential benchmarks. Simulation results show that the proposed test generation approach can be extremely effective for detecting arbitrary Trojan instances of small size, both combinational and sequential.

The rest of the paper is organized as follows. Section 2 describes previous work on Trojan detection. Section 3 describes the mathematical justification of the *MERO* methodology, the steps of the *MERO* test generation algorithm and the Trojan detection coverage estimation. Section 4 describes the simulation setup and presents results for a set of ISCAS benchmark circuits with detailed analysis. Section 5 concludes the paper.

2   Trojan Detection: Previous Work

Previously proposed Trojan detection approaches can be classified under two main classes: (1) destructive approaches and (2) non–destructive approaches. In the *destructive* approaches, the manufactured IC is de–metallized layer by layer, and chip microphotographs of the layers are integrated and analyzed by advanced software to detect any tampering [4]. However, the applicability of such approaches is limited by the fact that the hacker is most likely to modify only a small random sample of chips in the production line. This means that the success of detecting the Trojan depends totally on correctly selecting a manufactured IC instance that has actually been tampered. Also, destructive methods of validating an IC are extremely expensive with respect to time and cost and technology intensive, with validation of a single IC taking months [3]. Hence, it is important to investigate efficient non–destructive Trojan detection approaches.

Two non-destructive Trojan detection techniques can be categorized into two broad classes: (1) *Side-channel Analysis* based and (2) *Logic Testing* based techniques. The *side–channel* analysis based techniques utilize the effect of an inserted Trojan on a measurable physical quantity, e.g. the supply current [1,11] or path delays [10]. Such a measured circuit parameter can be referred as a *fingerprint* for the IC [1]. Side–channel approaches of detecting Trojans belong to a class of generic powerful techniques for IC authentication, and are conceptually applicable to Trojans of all operational modes and to designs of arbitrary size and complexity. Only local *activation* of the Trojans is sufficient to detect them, and methods have been proposed to maximize the possibility of locally activating Trojans [9]. However, there are two main issues with the side–channel based approaches that limit their practical applicability:

1. An intelligent adversary can craft a very small Trojan circuit with just a few logic gates which causes minimal impact on circuit power or delay. Thus it can easily evade side–channel detection techniques such as the ones described in [1,10].
2. The fingerprint is extremely vulnerable to process variations (i.e. *process noise*) and measurement noise. Even advanced de–noising techniques such as those applied in [1] fail to detect arbitrarily small Trojans under process variations.

Logic testing based approaches, on the other hand, are extremely reliable under process variations and measurement noise effects. An important challenge in these approaches is the inordinately large number of possible Trojans an adversary can exploit. Relatively few works have addressed the problem of Trojan detection using logic testing. In [12], a design methodology was proposed where special circuitry was embedded in an IC to improve the controllability and observability of internal nodes, thereby facilitating the detection of inserted Trojans by logic testing. However, this technique does not solve the problem of detecting Trojans in ICs which have not been designed following that particular design methodology.

3   Statistical Approach for Trojan Detection

As described in Section 1, the main concept of our test generation approach is based on generating test vectors that can excite candidate trigger nodes individually to their rare logic values multiple (at least N) times. In effect, the probability of activation of a Trojan by the simultaneous occurrence of the rare conditions at its trigger nodes increases. As an example, consider the Trojan shown in Fig. 1(c). Assume that the conditions $a = 0$, $b = 1$ and $c = 1$ are very rare. Hence, if we can generate a set of test vectors that induce these rare conditions at these nodes individually N times where N is sufficiently large, then a Trojan with triggering condition composed jointly of these nodes is highly likely to be activated by the application of this test set. The concept can be extended to sequential Trojans, as shown in Fig. 1(d), where the inserted 3–bit counter is clocked on the simultaneous occurrence of the condition $ab' = 1$. If the test vectors can sensitize these nodes such that the condition $ab' = 1$ is satisfied at least 8 times (the maximum number of states of a 3–bit counter), then the Trojan would be activated. Next, we present a mathematical analysis to justify the concept.

3.1   Mathematical Analysis

Without loss of generality, assume that a Trojan is triggered by the rare logic values at two nodes A and B, with corresponding probability of occurrence p_1 and p_2. Assume T to be the total number of vectors applied to the circuit under test, such that both A and B have been individually excited to their rare values *at least* N times. Then, the expected number of occurrences of the rare logic values at nodes A and B are given by $E_A = T{\cdot}p_1 {\geq} N$ and $E_B = T{\cdot}p_2 {\geq} N$, which lead to:

$$T \geq \frac{N}{p_1} \quad \text{and} \quad T \geq \frac{N}{p_2} \tag{1}$$

Now, let p_j be the probability of simultaneous occurrence of the rare logic values at nodes A and B, an event that acts as the trigger condition for the Trojan. Then, the expected number of occurrences of this event when T vectors are applied is:

$$E_{AB} = p_j{\cdot}T \tag{2}$$

In the context of this problem, we can assume $p_j > 0$, because an adversary is unlikely to insert a Trojan which would never be triggered. Then, to ensure that the Trojan is triggered at least once when T test vectors are applied, the following condition must be satisfied:

$$p_j{\cdot}T \geq 1 \tag{3}$$

From inequality (1), let us assume $T = c{\cdot}\frac{N}{p_1}$. where $c{\geq}1$ is a constant depending on the actual test set applied. Inequality (3) can then be generalized as:

$$S = c{\cdot}\frac{p_j}{p_1}{\cdot}N \tag{4}$$

where S denotes the number of times the trigger condition is satisfied during the test procedure. From this equation, the following observations can be made about the interdependence of S and N:

1. For given parameters c, p_1 and p_j, S is proportional to N, i.e. the expected number of times the Trojan trigger condition is satisfied increases with the number of times the trigger nodes have been individually excited to their rare values. This observation forms the main motivation behind the *MERO* test generation approach for Trojan detection.

2. If there are q trigger nodes and if they are assumed to be mutually independent, then $p_j = p_1 \cdot p_2 \cdot p_3 \cdots p_q$, which leads to:

$$S = c \cdot N \cdot \prod_{i=2}^{q} p_i \tag{5}$$

As $p_i < 1 \quad \forall i = 1, 2, \cdots q$, hence, with the increase in q, S decreases for a given c and N. In other words, with the increase in the number of trigger nodes, it becomes more difficult to satisfy the trigger condition of the inserted Trojan for a given N. Even if the nodes are not mutually independent, a similar dependence of S on q is expected.

3. The trigger nodes can be chosen such that $p_i \leq \theta \quad \forall i = 1, 2, \cdots q$, so that θ is defined as a *trigger threshold* probability. Then as θ increases, the corresponding selected rare node probabilities are also likely to increase. This will result in an increase in S for a given T and N i.e. the probability of Trojan activation would increase if the individual nodes are more likely to get triggered to their rare values.

All of the above predicted trends were observed in our simulations, as shown in Section 4.

3.2   Test Generation

Algorithm 1 shows the major steps in the proposed reduced test set generation process for Trojan detection. We start with the golden circuit netlist (without any Trojan), a random pattern set (V), list of rare nodes (L) and number of times to activate each node to its rare value (N). First, the circuit netlist is read and mapped to a *hypergraph*. For each node in L, we initialize the number of times a node encounters a rare value (A_R) to 0. Next, for each random pattern v_i in V, we count the number of nodes (C_R) in L whose rare value is satisfied. We sort the random patterns in decreasing order of C_R. In the next step, we consider each vector in the sorted list and modify it by perturbing one bit at a time. If a modified test pattern increases the number of nodes satisfying their rare values, we accept the pattern in the reduced pattern list. In this step we consider only those rare nodes with $A_R < N$. The process repeats until each node in L satisfies its rare value at least N times. The output of the test generation process is a minimal test set that improves the coverage for both combinational and sequential Trojans compared to random patterns.

Algorithm 1. Procedure *MERO*

Generate reduced test pattern set for Trojan detection

Inputs: Circuit netlist, list of rare nodes (L) with associated rare values, list of random patterns (V), number of times a rare condition should be satisfied (N)

Outputs: Reduced pattern set (R_V)

1: Read circuit and generate *hypergraph*
2: **for all** nodes in L **do**
3:     set number of times node satisfies rare value (A_R) to 0
4: **end for**
5: set $R_V = \Phi$
6: **for all** random pattern in V **do**
7:     Propagate values
8:     Count the # of nodes (C_R) in L with their rare value satisfied
9: **end for**
10: Sort vectors in V in decreasing order of C_R
11: **for all** vector v_i in decreasing order of C_R **do**
12:     **for all** bit in v_i **do**
13:         Perturb the bit and re-compute # of *satisfied rare values* (C_R')
14:         **if** ($C_R' > C_R$) **then**
15:             Accept the perturbation and form v_i' from v_i
16:         **end if**
17:     **end for**
18:     Update A_R for all nodes in L due to vector v_i
19:     **if** v_i' increases A_R for at least one rare node **then**
20:         Add the modified vector v_i' to R_V
21:     **end if**
22:     **if** ($A_R \geq N$) for all nodes in L **then**
23:         break
24:     **end if**
25: **end for**

3.3   Coverage Estimation

Once the reduced test vector set has been obtained, computation of Trigger and Trojan coverage can be performed for a given *trigger threshold* (θ) (as defined in Section 3.1) and a given number of trigger nodes (q) using a random sampling approach. From the Trojan population, we randomly select a number of q–trigger Trojans, where each trigger node has signal probability less than equal θ. We assume that Trojans comprising of trigger nodes with higher signal probability than θ will be detected by conventional test. From the set of sampled Trojans, Trojans with false trigger conditions which cannot be justified with any input pattern are eliminated. Then, the circuit is simulated for each vector in the given vector set and checked whether the trigger condition is satisfied. For an activated Trojan, if its effect can be observed at the primary output or scan flip-flop input, the Trojan is considered "covered", i.e. detected. The percentages of Trojans activated and detected constitute the *trigger coverage* and *Trojan coverage*, respectively.

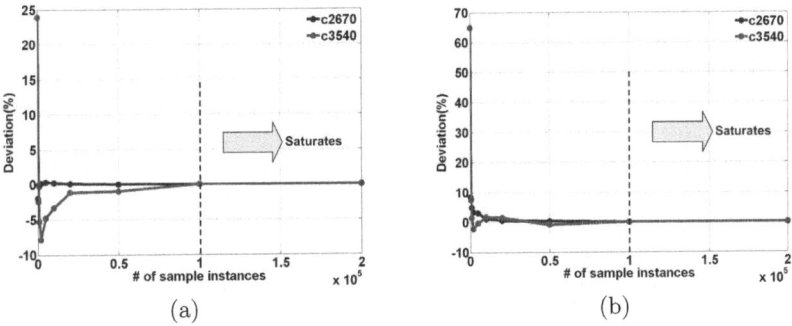

Fig. 2. Impact of sample size on trigger and Trojan coverage for benchmarks c2670 and c3540, $N = 1000$ and $q = 4$: (a) deviation of trigger coverage, and (b) deviation of Trojan coverage

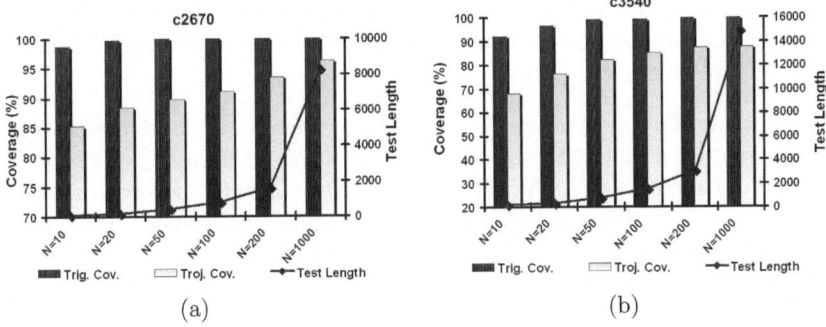

Fig. 3. Impact of N (number of times a rare point satisfies its rare value) on the trigger/Trojan coverage and test length for benchmarks (a) c2670 and (b) c3540

3.4 Choice of Trojan Sample Size

In any random sampling process an important decision is to select the sample size in a manner that represents the population reasonably well. In the context of Trojan detection, it means further increase in sampled Trojans, renders negligible change in the estimated converge. Fig. 2 shows a plot of percentage deviation of Trigger and Trojan coverage ($q = 4$) from the asymptotic value for two benchmark circuits with varying Trojan sample size. From the plots, we observe that the coverage saturates with nearly 100,000 samples, as the percentage deviation tends to zero. To compromise between accuracy of estimated coverage and simulation time, we have selected a sample size of 100,000 in our simulations.

3.5 Choice of N

Fig. 3 shows the trigger and Trojan coverage for two ISCAS–85 benchmark circuits with increasing values of N, along with the lengths of the corresponding testset. From these plots it is clear that similar to N–$detect$ tests for stuck-at fault where defect coverage typically improves with increasing N, the trigger and Trojan coverage obtained with the *MERO* approach also improves steadily with N, but then both saturate around $N = 200$ and remain nearly constant for larger values of N. As expected, the test size also increases with increasing N. We chose a value of $N = 1000$ for most of our experiments to reach a balance between coverage and test vector set size.

3.6 Improving Trojan Detection Coverage

As noted in previous sections, Trojan detection using logic testing involves simultaneous triggering of the Trojan and the propagation of its effect to output nodes. Although the proposed test generation algorithm increases the probability of Trojan activation, it does not explicitly target increasing the probability of a malicious effect at payload being observable. *MERO* test patterns, however, achieves significant improvement in Trojan coverage compared to random patterns, as shown in Section 4. This is because the Trojan coverage has strong correlation with trigger coverage. To increase the Trojan coverage further, one can use the following low-overhead approaches.

1. *Improvement of test quality*: We can consider number of nodes observed along with number of nodes triggered for each vector during test generation. This means, at step 13-14 of Algorithm 1, a perturbation is accepted if the sum of triggered and observed nodes improves over previous value. This comes at extra computational cost to determine the number of observable nodes for each vector. We note that for a small ISCAS benchmark *c432* (an interrupt controller), we can improve the Trojan coverage by 6.5% with negligible reduction in trigger coverage using this approach.
2. *Observable test point insertion*: We note that insertion of very few observable test points can achieve significant improvement in Trojan coverage at the cost of small design overhead. Existing algorithm for selecting observable test points for stuck-at fault test [14] can be used here. Our simulation with *c432* resulted in about 4% improvement in Trojan coverage with 5 judiciously inserted observable points.
3. *Increasing N and/or increasing the controllability of the internal nodes*: Internal node controllability can be increased by judiciously inserting few controllable test points or increasing N. It is well-known in the context of stuck-at ATPG, that scan insertion improves both controllability and observability of internal nodes. Hence, the proposed approach can take advantage of low-overhead design modifications to increase the effectiveness of Trojan detection.

4   Results

4.1 Simulation Setup

We have implemented the test generation and the Trojan coverage determination in three separate C programs. All the three programs can read a Verilog netlist

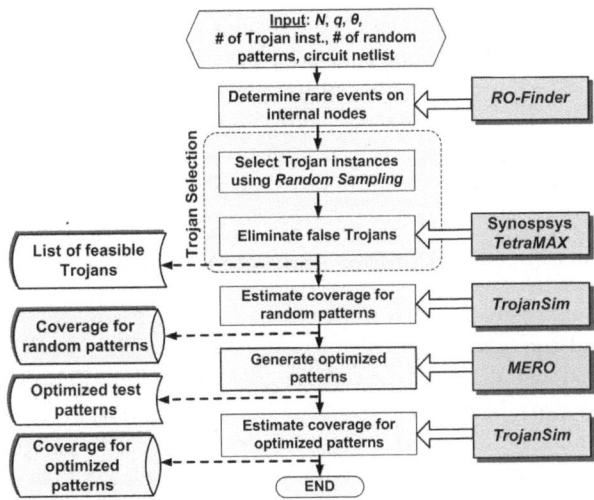

Fig. 4. Integrated framework for rare occurrence determination, test generation using *MERO* approach, and Trojan simulation

Table 1. Comparison of Trigger and Trojan coverage among ATPG patterns [7], Random (100K, input weights: 0.5), and *MERO* patterns for $q = 2$ and $q = 4$, $N = 1000$, $\theta = 0.2$

		ATPG patterns				Random (100K patterns)				*MERO* Patterns			
		q = 2		q = 4		q = 2		q = 4		q = 2		q = 4	
Ckt.	Nodes (Rare/ Tot.)	Trig. Cov. (%)	Troj. Cov. (%)	Trig. Cov. (%)	Troj. Cov. (%)	Trig. Cov. (%)	Troj. Cov. (%)	Trig. Cov. (%)	Troj. Cov. (%)	Trig. Cov. (%)	Troj. Cov. (%)	Trig. Cov. (%)	Troj. Cov. (%)
c2670	297/1010	93.97	58.38	30.7	10.48	98.66	53.81	92.56	30.32	100.00	96.33	99.90	90.17
c3540	580/1184	77.87	52.09	16.07	8.78	99.61	86.5	90.46	69.48	99.81	86.14	87.34	64.88
c5315	817/2485	92.06	63.42	19.82	8.75	99.97	93.58	98.08	79.24	99.99	93.83	99.06	78.83
c6288	199/2448	55.16	50.32	3.28	2.92	100.00	98.95	99.91	97.81	100.00	98.94	92.50	89.88
c7552	1101/3720	82.92	66.59	20.14	11.72	98.25	94.69	91.83	83.45	99.38	96.01	95.01	84.47
s13207‡	865/2504	82.41	73.84	27.78	27.78	100	95.37	88.89	83.33	100.00	94.68	94.44	88.89
s15850‡	959/3004	25.06	20.46	3.80	2.53	94.20	88.75	48.10	37.98	95.91	92.41	79.75	68.35
s35932‡	970/6500	87.06	79.99	35.9	33.97	100.00	93.56	100.00	96.80	100.00	93.56	100.00	96.80
Avg.	**724/2857**	**74.56**	**58.14**	**19.69**	**13.37**	**98.84**	**88.15**	**88.73**	**72.30**	**99.39**	**93.99**	**93.50**	**82.78**

‡These sequential benchmarks were run with 10,000 random Trojan instances to reduce run time of Tetramax.

and create a *hypergraph* from the netlist description. The first program, named as *RO-Finder* (**R**are **O**ccurence **Finder**), is capable of functionally simulating a netlist for a given set of input patterns, computing the signal probability at each node and identifying nodes with low signal probability as rare nodes. The second program *MERO* implements algorithm-1 described in Section 3.2 to generate the reduced pattern set for Trojan detection. The third program, *TrojanSim* (**Trojan Sim**ulator), is capable of determining both Trigger and Trojan coverage for a given test set using random sample of Trojan instances. A q-trigger random Trojan instance is created by randomly selecting the trigger nodes from the list of rare nodes. We consider one randomly selected payload node for each Trojan. Fig. 4 shows the flow-chart for the *MERO* methodology. Synopsys *TetraMAX* was used to justify the trigger condition for each Trojan and eliminate the false Trojans. All simulations and test generation were carried out on a Hewlett-Packard Linux workstation with a 2GHz dual-core Intel processor and 2GB RAM.

4.2   Comparison with Random and ATPG Patterns

Table 1 lists the trigger and Trojan coverage results for a set of combinational (ISCAS-85) and sequential (ISCAS-89) benchmarks using stuck-at ATPG patterns (generated using the algorithm in [7]), weighted random patterns and *MERO* test patterns. It also lists the number of total nodes in the circuit and the number of rare nodes identified by *RO-Finder* tool based on signal probability. The signal probabilities were estimated through simulations with a set of 100,000 random vectors. For the sequential circuits, we assume full-scan implementation. We consider 100,000 random instances of Trojans following the sampling policy described in Section 3.4, with one randomly selected payload node for each Trojan. Coverage results are provided in each case for two different trigger point count, $q = 2$ and $q = 4$, at $N = 1000$ and $\theta = 0.2$.

Table 2 compares reduction in the length of the testset generated by the *MERO* test generation method with 100,000 random patterns, along with the corresponding run-times for the test generation algorithm. This run-time includes the execution time for *Tetramax* to validate 100,000 random Trojan instances, as well as time to determine the coverage by logic simulation. We can make the following important observations from these two tables:

1. The stuck-at ATPG patterns provide poor trigger and Trojan coverage compared to *MERO* patterns. The increase in coverage between the ATPG and *MERO* patterns is more significant in case of higher number of trigger points.
2. From Table 2, it is evident that the reduced pattern with N=1000 and $\theta = 0.2$ provides comparable trigger coverage with significant reduction in test

Table 2. Reduction in test length with MERO approach compared to 100K random patterns along with runtime, $q = 2$, N=1000, θ=0.2

Ckt.	*MERO* test length	% Reduction	Run-time (s)
c2670	8254	**91.75**	30051.53
c3540	14947	**85.05**	9403.11
c5315	10276	**89.72**	80241.52
c6288	5014	**94.99**	15716.42
c7552	12603	**87.40**	160783.37
s13207[†]	26926	**73.07**	23432.04
s15850[†]	32775	**67.23**	39689.63
s35932[†]	5480	**94.52**	29810.49
Avg.	**14534**	**85.47**	**48641.01**

†These sequential benchmarks were run with 10,000 random Trojan instances to reduce run time of *Tetramax*.

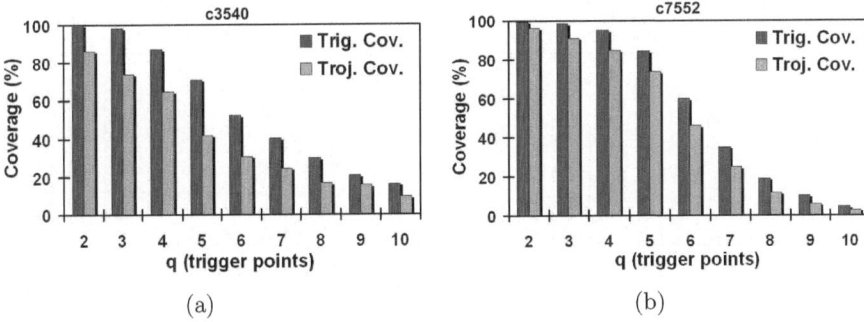

Fig. 5. Trigger and Trojan coverage with varying number of trigger points (q) for benchmarks (a) c3540 and (b) c7552, at $N = 1000$, $\theta = 0.2$

length. The average improvement in test length for the circuits considered is about 85%.

3. Trojan coverage is consistently smaller compared to trigger coverage. This is because in order to detect a Trojan by applying an input pattern, besides satisfying the trigger condition, one needs to propagate the logic error at the payload node to one or more primary outputs. In many cases although the trigger condition is satisfied, the malicious effect does not propagate to outputs. Hence, the Trojan remains triggered but undetected.

4.3 Effect of Number of Trigger Points (q)

The impact of q on coverage is evident from the Fig. 5, which shows the decreasing trigger and Trojan coverage with the increasing number of trigger points for two combinational benchmark circuits. This trend is expected from the analysis of Section 3.1. Our use of *TetraMAX* for justification and elimination of the false triggers helped to improve the Trojan coverage.

4.4 Effect of Trigger Threshold (θ)

Fig. 6 plots the trigger and Trojan coverage with increasing θ for two ISCAS-85 benchmarks, at $N = 1000$ and $q = 4$. As we can observe, the coverage values improve steadily with increasing θ while saturating at a value above 0.20 in both the cases. The improvement in coverage with θ is again consistent with the conclusions from the analysis of Section 3.1.

4.5 Sequential Trojan Detection

To investigate the effectiveness of the *MERO* test generation methodology in detecting sequential Trojans, we designed and inserted sequential Trojans modeled following Fig. 1(d), with 0, 2, 4, 8, 16 and 32 states, respectively (the case with zero states refers to a combinational Trojan following the model of Fig. 1(c)). A cycle-accurate simulation was performed by our simulator *TrojanSim*, and the

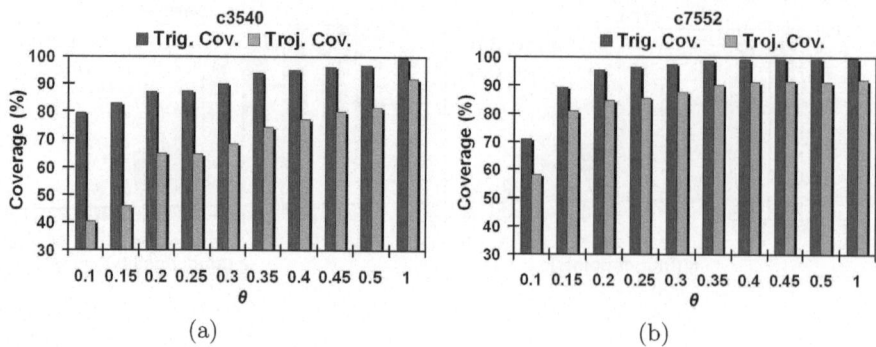

Fig. 6. Trigger and Trojan coverage with *trigger threshold* (θ) for benchmarks (a) c3540 and (b) c7552, for $N = 1000$, $q = 4$

Table 3. Comparison of sequential Trojan coverage between random (100K) and MERO patterns, $N = 1000$, $\theta = 0.2$, $q = 2$

Ckt.	Trigger Cov. for 100K Random Vectors (%)					Trigger Cov. for *MERO* Vectors (%)				
	Trojan State Count					Trojan State Count				
	0	2	4	8	16	32	0	2	4	8	16	32
s13207	100.00	100.00	99.77	99.31	99.07	98.38	100.00	100.00	99.54	99.54	98.84	97.92
s15850	94.20	91.99	86.79	76.64	61.13	48.59	95.91	95.31	94.03	91.90	87.72	79.80
s35932	100.00	100.00	100.00	100.00	100.00	100.00	100.00	100.00	100.00	100.00	100.00	100.00
Avg.	**98.07**	**97.33**	**95.52**	**91.98**	**86.73**	**82.32**	**98.64**	**98.44**	**97.86**	**97.15**	**95.52**	**92.57**

Ckt.	Trojan Cov. for 100K Random Vectors (%)					Trojan Cov. for *MERO* Vectors (%)				
	Trojan State Count					Trojan State Count				
	0	2	4	8	16	32	0	2	4	8	16	32
s13207	95.37	95.37	95.14	94.91	94.68	93.98	94.68	94.68	94.21	94.21	93.52	92.82
s15850	88.75	86.53	81.67	72.89	58.4	46.97	92.41	91.99	90.62	88.75	84.23	76.73
s35932	93.56	93.56	93.56	93.56	93.56	93.56	93.56	93.56	93.56	93.56	93.56	93.56
Avg.	**92.56**	**91.82**	**90.12**	**87.12**	**82.21**	**78.17**	**93.55**	**93.41**	**92.80**	**92.17**	**90.44**	**87.70**

Trojan was considered detectable only when the output of the golden circuit and the infected circuit did not match. Table 3 presents the trigger and Trojan coverage respectively obtained by 100,000 randomly generated test vectors and the *MERO* approach for three large ISCAS-89 benchmark circuits. The superiority of the *MERO* approach over the random test vector generation approach in detecting sequential Trojans is evident from this table.

Although these results have been presented for a specific type of sequential Trojans (counters which increase their count conditionally), they are representative of other sequential Trojans whose state transition graph (STG) has no "loop". The STG for such a FSM has been shown in Fig. 7. This is a 8-state FSM which changes its state only when a particular internal node condition C_i is satisfied at state S_i, and the Trojan is triggered when the FSM reaches state S_8. The example Trojan shown in Fig. 1(d) is a special case of this model, where the conditions C_1 through C_8 are identical. If each of the conditions C_i is as rare as the condition $a = 1, b = 0$ required by the Trojan shown in Fig. 1(d),

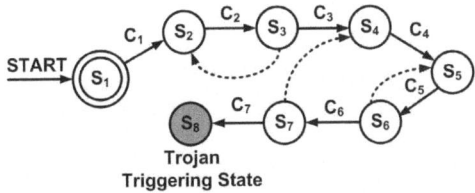

Fig. 7. FSM model with no loop in state transition graph

then there is no difference between these two Trojans as far as their rareness of getting triggered is concerned. Hence, we can expect similar coverage and test length results for other sequential Trojans of this type. However, the coverage may change if the FSM structure is changed (as shown with dotted line). In this case, the coverage can be controlled by changing N.

4.6 Application to Side-Channel Analysis

As observed from the results presented in this section, the *MERO* approach can achieve high trigger coverage for both combinational and sequential Trojans. This essentially means that the *MERO* patterns will induce activity in the Trojan triggering circuitry with high probability. A minimal set of patterns that is highly likely to cause activity in a Trojan is attractive in power or current signature based side-channel approach to detect hardware Trojan [1,9,11]. The detection sensitivity in these approaches depends on the induced activity in the Trojan circuit by applied test vector. It is particularly important to enhance sensitivity for the Trojans where the leakage contribution to power by the Trojan circuit can be easily masked by process or measurement noise. Hence, *MERO* approach can be extended to generate test vectors for side-channel analysis, which requires amplifying the Trojan impact on side-channel parameter such as power or current.

5 Conclusions

Conventional logic test generation techniques cannot be readily extended to detect hardware Trojans because of the inordinately large number of possible Trojan instances. We have presented a statistical Trojan detection approach using logic testing where the concept of multiple excitation of rare logic values at internal nodes is used to generate test patterns. Simulation results show that the proposed test generation approach achieves about 85% reduction in test length over random patterns for comparable or better Trojan detection coverage. The proposed detection approach can be extremely effective for small combinational and sequential Trojans with small number of trigger points, for which side-channel analysis approaches cannot work reliably. Hence, the proposed detection approach can be used as complementary to side-channel analysis based detection schemes. Future work will involve improving the test quality which will help in minimizing the test length and increasing Trojan coverage further.

References

1. Agrawal, D., Baktir, S., Karakoyunlu, D., Rohatgi, P., Sunar, B.: Trojan detection using IC fingerprinting. In: IEEE Symp. on Security and Privacy, pp. 296–310 (2007)
2. Ravi, S., Raghunathan, A., Chakradhar, S.: Tamper resistance mechanisms for secure embedded systems. In: Intl. Conf. on VLSI Design, pp. 605–611 (2006)
3. DARPA BAA06-40: TRUST for Integrated Circuits,
   http://www.darpa.mil/BAA/BAA06-40mod1/html
4. Kumagai, J.: Chip Detectives. IEEE Spectrum 37, 43–49 (2000)
5. Amyeen, M.E., Venkataraman, S., Ojha, A., Lee, S.: Evaluation of the Quality of N-Detect Scan ATPG Patterns on a Processor. In: Intl. Test Conf., pp. 669–678 (2004)
6. Pomeranz, I., Reddy, S.M.: A Measure of Quality for n-Detection Test Sets. IEEE. Trans. on Computers. 53, 1497–1503 (2004)
7. Mathew, B., Saab, D.G.: Combining multiple DFT schemes with test generation. IEEE Trans. on CAD. 18, 685–696 (1999)
8. Adee, S.: The Hunt for the Kill Switch. IEEE Spectrum 45, 34–39 (2008)
9. Banga, M., Hsiao, M.S.: A Region Based Approach for the Identification of Hardware Trojans. In: Intl. Workshop on Hardware-oriented Security and Trust, pp. 40–47 (2008)
10. Jin, Y., Makris, Y.: Hardware Trojan Detection Using Path Delay Fingerprint. In: Intl. Workshop on Hardware-oriented Security and Trust, pp. 51–57 (2008)
11. Rad, R.M., Wang, X., Tehranipoor, M., Plusqellic, J.: Power Supply Signal Calibration Techniques for Improving Detection Resolution to Hardware Trojans. In: Intl. Conf. on CAD, pp. 632–639 (2008)
12. Chakraborty, R.S., Paul, S., Bhunia, S.: On-Demand Transparency for Improving Hardware Trojan Detectability. In: Intl. Workshop on Hardware-oriented Security and Trust, pp. 48–50 (2008)
13. Wolff, F., Papachristou, C., Bhunia, S., Chakraborty, R.S.: Towards Trojan-Free Trusted ICs: Problem Analysis and Detection Scheme. In: Design, Automation and Test in Europe, pp. 1362–1365 (2008)
14. Geuzebroek, M.J., Van der Linden, J.T., Van de Goor, A.J.: Test Point Insertion that Facilitates ATPG in Reducing Test Time and Data Volume. In: Intl. Test Conf., pp. 138–147 (2002)

On Tamper-Resistance from a Theoretical Viewpoint
The Power of Seals*

Paulo Mateus[1] and Serge Vaudenay[2]

[1] SQIG /Instituto de Telecomunicações - IST/TULisbon
1049-001 Lisboa, Portugal
pmat@math.ist.utl.pt
http://sqig.math.ist.utl.pt
[2] EPFL
CH-1015 Lausanne, Switzerland
serge.vaudenay@epfl.ch
http://lasecwww.epfl.ch

Abstract. Tamper-proof devices are pretty powerful. They can be used to have better security in applications. In this work we observe that they can also be maliciously used in order to defeat some common privacy protection mechanisms. We propose the theoretical model of *trusted agent* to formalize the notion of programmable secure hardware. We show that protocols not using tamper-proof devices are not deniable if malicious verifiers can use trusted agents. In a strong key registration model, deniability can be restored, but only at the price of using key escrow. As an application, we show how to break invisibility in undeniable signatures, how to sell votes in voting schemes, how to break anonymity in group/ring signatures, and how to carry on the Mafia fraud in non-transferable protocols. We conclude by observing that the ability to put boundaries in computing devices prevents from providing full control on how private information spreads: the concept of sealing a device is in some sense incompatible with privacy.

1 Introduction

Tamper-proof hardware devices have been used quite massively in industrial and commercial applications. There exists a wide spectrum of tamper-proof devices, ranging in their price and security, from simple smartcards to the IBM 4758, which has several physical penetration sensors, including temperature, radiation, pressure, etc. Clearly, people are currently surrounded by devices (aimed at) instantiating trusted agents. People wear smart cards, secure tokens, their PCs have Trusted Computing Platforms, their media readers have secure hardware to deal with DRMs, their iPhones have a self-blocking secure hardware, passports have secure RFID tags, etc.

So far, secure hardware devices have been used to implement some strong security protocols with the hypothesis that they are tamper-resistant. The idea of using tamper-proof devices to realize cryptographic functionalities goes back (at least) to 1986 [10]. Due to existence of all side channel attacks, whether tamper resistance is possible in

---

* A preliminary version of this paper was presented as a poster at EUROCRYPT'09.

C. Clavier and K. Gaj (Eds.): CHES 2009, LNCS 5747, pp. 411–428, 2009.

practice is still an open question. Current allegedly tamper-resistant devices are (at least) trusted by banks, mobile telephone operators, companies selling access control devices, software companies, media content providers, hardware manufacturers, governments, and so on. It is unlikely that none of these organizations would ever try to take any malicious advantage out from their devices. So, assuming some adversaries would use tamper-proof devices for attacks is a legitimate assumption. In this paper we show how to make several privacy attacks using trusted tamper-proof devices.

In this work, we formalize the notion of programmable secure hardware by introducing the *trusted agent model*. Informally speaking, the trusted agent model consists in assuming that it is possible to acquire a trusted device (agent) that runs honestly a known program in a secure environment (tamper proof) without any way of running another program. At the first time it is switched on, we can load a code whose digest will be permanently displayed. Later, we can interact with the device through the interface defined by this program only. Quite importantly, every output will be appended to the digest of the original program so that someone looking at the display is ensured that the output is produced by a device having been set up with a program of displayed digest. We show that within this model, it is possible

- to transfer proofs of zero-knowledge protocols after completion (in particular: to transfer the verification of an invisible signature);
- to register rogue public keys and prove the ignorance of a secret key (which then can be used to break anonymity in ring signatures or non-transferability mechanisms);
- to sell ballots in e-voting systems.

In a nutshell, for any interactive proof protocol, we can load the verifier algorithm in a trusted agent and make a malicious verifier relay protocol messages between the prover and the trusted agent. Afterward completion, the agent ends up in a state which testifies that the proof protocol was correctly run and provide some kind of forensic evidence. Clearly, such a honest device could be used to defeat the invisible signature paradigm [6] when maliciously used. One could say that this trivial attack could be defeated by classical non-transferability techniques like having a Public Key Infrastructure (PKI) for verifiers [2,11,15,16]. However, this would work only if the prover is convinced that the verifier possesses himself a secret key. A trusted agent could still be maliciously used to register a key whose secret part would be ignored by the verifier. Later, the agent could prove that the verifier must ignore the secret key and continue to defeat non-transferability. Finally, the only key registration model which could fix this would imply some kind of key escrow: some information making the registrating authority able to impersonate the verifier would eventually have to leak in order to thwart the previous attack. Key escrow however leads us to other privacy concerns.

Another possible use of registering a public key whose secret component is sealed in a trusted tamper-proof hardware would be to break anonymity in group signatures or ring signatures [23]. Interestingly, it makes it possible to *prove ignorance*. It could also be used in voting systems and open the door to vote selling.

While it is debatable if the trusted agent model is realizable or not, assuming it cannot be used by adversaries is a much greater error than assuming that it can. For this reason, we believe that cryptographers should mind the proposed trusted agent model when designing future protocols.

Related work. Classical ZK proof systems fulfill a privacy property called *deniabil-ity* [21] stating that the verifier cannot prove knowledge to a third party after interact-ing with the prover. That is, the verifier cannot transfer the proof upon completion. The more general concept of *non-transferability* is also central in some cryptographic schemes, such as invisible signatures [6][1] that use interactive verification in order to prevent the signature to be authenticated to an unauthorized third party. A different way to enforce deniability of a signature is to use a group or ring signature [23] between the signing party and the verifier. In this case, the signer can deny the signature by claiming that it was computed by the other party.

Several flaws have been found to non-transferability protocols, and improvements have been proposed (see e.g. [2,11,15,16]). The attacks are focused in adversaries that are online with the verifier during the interaction period. For offline attacks, it is ac-cepted that the protocols are secure. However, herein we will present an offline attack that will render several non-transferability protocols useless under the assumption that the participants can trust tamper-proof devices.

The idea of using tamper-proof hardware to transfer proofs of ZK protocols was first introduced in the context of quantum memory [18,19].

In general, setup phases in cryptographic protocols is a critical issue. Participants are often assumed to securely register their public keys, although doing so is not trivial. Key setup is a problem for the Universal Composability (UC) framework by Canetti [3]. For instance, the key registration model by Barak, Canetti, Nielsen and Pass [1] assumes that the secret key of honest participants is safely stored by the key registration au-thority. In [22], Ristenpart and Yilek considered several variants of key registration protocols and have shown tricky interference with the security in several group sig-nature protocols. They noticed that security proofs often assume that all participants send their secret keys to a trusted authority in a KOSK model (as for *Knowledge Of Secret Key*) although some signature schemes could still be secure in a less demanding key registration process such as producing a self-signed certificate for the public key, what they call the POP (as for *Proof Of Possession*). Our results show that POP is ei-ther not enough in the trusted agent model, or compromises some other cryptographic property.

Katz [17] used another approach consisting in assuming the existence of tamper-proof hardware tokens. These tokens could be used to achieve commitment, thus any well-formed functionality. Contrarily to these hardware tokens, we assume that trusted agents are private (namely: their holders do not give them to another user) and display the initial code (or its digest) so that any other party can trust that it is in a state which is a consequence of having set it up with this code. The question whether a tamper-proof hardware can be trusted to run what it is supposed to is discussed e.g. in [14].

In [21], Pass introduced the notion of deniable zero-knowledge which is immune to offline proof transfer. ZK in the standard model is essentially deniable. However, zero-knowledge in the common reference string (CRS) model is not always deniable.

---

[1] As suggested by several authors, we use the term of *invisible signature* to designate what is more often called *undeniable signature* since the term *undeniable* is a little confusing, espe-cially when we introduce the notion of deniability.

Structure of the paper. The paper is organized as follows, in Section 2 we introduce the trusted agent model and the nested trusted agent model. In Section 3 we study deniability. We show that deniability is impossible if honest participants do not use trusted agents but the malicious verifier does. We show that deniability is possible when the prover uses trusted agents. In other cases, it is impossible in the nested trusted agent model, and possible in the trusted agent model. Section 4 studies a key registration model. It shows that key registration with key escrow makes non-transferability possible. We provide examples of malicious use of trusted agents in Section 5. Finally, we draw some conclusions in Section 6.

2   The Trusted Agent Model

Multiparty computation model. In a multiparty setting, several participants or function-alities[2] run different algorithms and can communicate using pairwise communication channels. Channels are assumed to be secure in the sense that leakage or corruption in transmitted messages can only be made by one of the two end participants on this channel. We consider a static adversarial model in which participants are either honest or corrupted. Honest participants run predefined algorithms whereas corrupted participants may run arbitrary algorithms and talk to an (imaginary) adversary to collude. We use calligraphic characters (e.g., $\mathcal{P}_V$ or $\mathcal{F}_{TA}$) to denote participants and functionalities and capital characters (e.g., V or M) to denote the algorithms they run. By convention we will denote with a star $*$ the corrupted participants in a static model. Sometimes, a participant $\mathcal{P}$ invoking a functionality O will be referred to $\mathcal{P}$ querying an *oracle* O and we will write $\mathcal{P}^O$ for this type of communication. Later, a trusted agent will be defined by a functionality and used as an oracle. At the beginning, an arbitrary environment $\mathcal{E}$ sends input to all participants (including the adversary and functionalities) and collect the output at the end.

We stress that we do not necessarily assume that malicious participants have the same privileges as honest participants, which means that they can have access to different sets of functionalities. For instance, a malicious participant may use a trusted agent as a tool for cheating while we would not want a honest one to need an extra device.

Recall that an interactive machine is a next-message deterministic function applied to a current *view*. The view of the algorithm is a list containing all inputs to the machine (including the random coins) and all messages which have been received by the machine (with a reference to the communication channel through which it was delivered so that they can see which ones come from a trusted agent). The view is time dependent and can always be expanded by adding more messages.

We denote by $P \leftrightarrow V$ two interactive algorithms P and V interacting with each other, following a given protocol. When there is a single message sent by e.g. P to V, we say that the protocol is non-interactive and we denote it by $P \rightarrow V$. If O_P (resp. O_V) is the list of functionalities that participant $\mathcal{P}_P$ (resp. $\mathcal{P}_V$) may invoke when running the algorithm P (resp. V) we denote by $P^{O_P} \leftrightarrow V^{O_V}$ the interaction. More precisely, we denote by $P^{O_P(r_{O_P})}(x_P; r_P) \leftrightarrow V^{O_V(r_{O_V})}(x_V; r_V)$ the experiment of running P with

---

[2] Following the traditional terminology of universal composability [3], a functionality is a virtual participant performing honestly a specific cryptographic task.

input x_P and random coins r_P with access to O_P initialized with random coins r_{OP} and interacting with V with input x_V and random coins r_V with access to O_V initialized with random coins r_{OV}. We denote by $\mathrm{View}_V(P^{O_P(r_{OP})}(x_P; r_P) \leftrightarrow V^{O_V(r_{OV})}(x_V; r_V))$ the *final view* of V in this experiment, i.e. x_V, r_V and the list of messages from either P or O_V.

The trusted agent model. We assume that it is possible to construct a trusted device (agent) that runs honestly a known program (a minimal boot loader) in a secure environment (tamper proof) without any way of running another program. Moreover we assume that the device's memory is private, and that the only way to interact with the device is by using the interface defined by the program. A device is attached to a participant called its *holder*. He entirely controls the communication with it. The holder may however show the display of the device to another participant which would give him some kind of evidence of the outcome produced by a trusted agent. Below, we model trusted agents in a similar way as Katz's secure tokens [17]. Differences will be discussed in the full version of the paper.

We consider (probabilistic) interactive Turing machines with four kinds of tapes: the input tape, the working tape, the output tape, and the random tape. We define their *state* by the state of the automaton and the content of the working tape. We consider a programming language to specify the *code* of the transition function of the Turing machine and its *initial state*. All trusted agents are modeled by a functionality $\mathcal{F}_{\mathsf{TA}}$. To access to a particular trusted agent, we use a sid value. For each used sid, $\mathcal{F}_{\mathsf{TA}}$ stores a tuple in memory of form $(\mathsf{sid}, \mathcal{P}, r, C, \mathsf{state}, \mathsf{out})$, where $\mathcal{P}$ identifies the holder of the trusted agent, r denotes its random tape, C the loaded code to be displayed, state its current state, and out its output tape. $\mathcal{F}_{\mathsf{TA}}$ treats the following queries.

Query SEND(sid, m) **from participant** $\mathcal{P}$: If there is a tuple $(\mathsf{sid}, \mathcal{P}', \dots)$ registered with a participant $\mathcal{P}' \neq \mathcal{P}$, ignore the query. Otherwise:
  - If there is a tuple with correct participant, parse it to $(\mathsf{sid}, \mathcal{P}, r, C, \mathsf{state}, \mathsf{out})$ and set in to the value of m.
  - If there is no tuple registered, interpret m as a code C. Extract from it the value state of its initial state. Then set in and out to the empty string. Pick a string r of polynomial length at random. Then, store a new tuple $(\mathsf{sid}, \mathcal{P}, r, C, \mathsf{state}, \mathsf{out})$.
  Then, define a Turing machine with code C and initial state state, random tape set to r, input tape set to in, and output tape set to out. Then, reset all head positions and run the machine until it stops, and at most a polynomial number of steps. Set state to the new state value, set out to the content of the output tape, and update the stored tuple with the new values of state and out.
  Note that the registered $(\mathsf{sid}, \mathcal{P}, r, C)$ are defined by the first query and never changed.
Query SHOWTO$(\mathsf{sid}, \mathcal{P}')$ **from participant** $\mathcal{P}$: If there is no tuple of form $(\mathsf{sid}, \mathcal{P}, r, C, \mathsf{state}, \mathsf{out})$ with the correct $(\mathsf{sid}, \mathcal{P})$, ignore. Otherwise, send to $\mathcal{P}'$ the pair formed by the code C and the content out.

Here, the holder $\mathcal{P}$ asks for the creation of a new trusted agent by invoking a fresh instance sid of the functionality which becomes an agent. The holder is the only participant who can send messages to the agent. The holder can define to whom to send response messages by the SHOWTO message. (Incidentally, the holder can ask to see the output message himself.) The response message (as *displayed* on the device) consists of the originally loaded code C and the current output out. Since the channel from

$\mathcal{F}_{TA}$ to $\mathcal{P}'$ is secure, $\mathcal{P}'$ is ensured that some instance of $\mathcal{F}_{TA}$ (i.e. some trusted agent) was run with code C and produced the result out. Showing a trusted agent may provide a simple way to authenticate some data. By convention, we denote by $[C : out]$ an incoming message from $\mathcal{F}_{TA}$ composed by a code C and a value out. The action of *checking* $[C : out]$ means that the receiver checks that it comes form $\mathcal{F}_{TA}$, that C matches the expected code, and that out matches the expected pattern from the context.

One important property of this functionality is that it is well-formed. We say that a list of oracles O is *well-formed* if for any pair $(\mathcal{P}_0, \mathcal{P}_1)$ of participants and any algorithm M with access to O, there exists an algorithm S with access to O such that the two following experiments are indistinguishable from the environment:

1. $\mathcal{P}_0$ runs M^O and $\mathcal{P}_1$ runs an algorithm doing nothing.
2. $\mathcal{P}_0$ runs and algorithm defined by
   - for any incoming message m from $\mathcal{P} \neq \mathcal{P}_1$, $\mathcal{P}_0$ sends $[\mathcal{P}, m]$ to $\mathcal{P}_1$;
   - for any incoming message $[\mathcal{P}, m]$ from $\mathcal{P}_1$, $\mathcal{P}_0$ sends m to $\mathcal{P}$;
   - upon message $[out, m]$ from $\mathcal{P}_1$, the algorithm ends with output out.
   The participant $\mathcal{P}_1$ runs S^O.

Typically, S emulates the algorithm M by treating all messages forwarded by $\mathcal{P}_0$ and by using $\mathcal{P}_0$ as a router. This means that the output of O is not modified if the algorithm M is run by $\mathcal{P}_0$ or by $\mathcal{P}_1$. Informally, being well-formed means that the distribution of roles among the participants does not affect the behavior of O. An example of a functionality for which this is *not* the case is a key registration functionality who registers the name of the sending participant and reports it to a directory. So, the adversary could check if the key was registered by $\mathcal{P}_0$ or by $\mathcal{P}_1$ and tell it to the environment. As for $\mathcal{F}_{TA}$, it checks that messages come from the same holder but his identity has no influence.

Relevance of the model in practice. Our model for trusted agent could easily be implemented (assuming that tamper-resistance can be achieved) provided that we could trust a manufacturer and that nobody could counterfeit devices. Obviously this is a quite strong assumption but this could make sense in applications where there is a liable entity which must be trusted. For instance, digital payment relies on trusted agents issued by a liable bank: credit cards have a tamper-proof embedded chips and e-banking is often based on trusted secure tokens such as secureID. Nation-wide e-governance could be based on protocols using trusted agents. It is already the case for passports, ID documents, or health cards with tamper-proof RFID chips. In this paper, we demonstrate that such devices can be used for malicious reasons and not only to protect the user against attacks. We show in Appendix A that our trusted agent model can perform bit commitment. Following Katz's reasoning [17], since we can realize commitments in the $\mathcal{F}_{TA}$-hybrid model we can also realize many protocols which suffer from being impossible to realize in the bare model. Namely, we can realize any well-formed functionality [5].

Nested trusted agents. Regular trusted agents run Turing machines which cannot interact with other functionalities during their computation time. We can consider more general trusted agents who can use external oracles. Typically, we will consider generalized trusted agents (which we call *nested* trusted agents) who can become the holder of another trusted agents. To take an example, a (human) holder may communicate with a trusted agent "of depth 2" modeled by the functionality $\mathcal{F}_{\mathsf{TA}}^2$. The participant may then receive $[C : \text{out}]$ messages from $\mathcal{F}_{\mathsf{TA}}^2$ (if the holder asked to) or from the functionality $\mathcal{F}_{\mathsf{TA}}^1$ of regular trusted agents (upon the request by the nested trusted agent).

Formally, if O is a list of functionalities, we consider the functionality $\mathcal{F}_{\mathsf{TA}}^O$ which looks like $\mathcal{F}_{\mathsf{TA}}$ with the difference that the running code C can now send messages to all functionalities in O. Note that if $O = \bot$ this means that no oracle is used and then, $\mathcal{F}_{\mathsf{TA}}^O$ is the regular $\mathcal{F}_{\mathsf{TA}}$ functionality. When O designates a trusted agent functionality, we assume that $\mathcal{F}_{\mathsf{TA}}^O$ keeps a record of which instance sid of $\mathcal{F}_{\mathsf{TA}}^O$ queries which instance sid$'$ of O so that only the holder device $\mathcal{F}_{\mathsf{TA}}^O(\text{sid})$ can communicate to a designated trusted agent $O(\text{sid}')$, just like for human holders. Equivalently, we can say that $\mathcal{F}_{\mathsf{TA}}^O$ works by cloning itself in clones $\mathcal{F}_{\mathsf{TA}}^O(\text{sid})$.

We define $\mathcal{F}_{\mathsf{TA}}^0 = \bot$ (this is a dummy functionality doing nothing) and $\mathcal{F}_{\mathsf{TA}}^n = \mathcal{F}_{\mathsf{TA}}^{\mathcal{F}_{\mathsf{TA}}^{n-1}}$ iteratively. We have $\mathcal{F}_{\mathsf{TA}}^1 = \mathcal{F}_{\mathsf{TA}}$. We further define $\mathcal{F}_{\mathsf{NTA}} = \mathcal{F}_{\mathsf{TA}}^{\mathcal{F}_{\mathsf{NTA}}}$. That is, instances of $\mathcal{F}_{\mathsf{NTA}}$ can invoke $\mathcal{F}_{\mathsf{NTA}}$. We obtain a hierarchy of functionalities starting with $\bot$ and $\mathcal{F}_{\mathsf{TA}}$ and ending with $\mathcal{F}_{\mathsf{NTA}}$. To simplify, we consider $\mathcal{F}_{\mathsf{TA}}^n$ as a restricted usage of $\mathcal{F}_{\mathsf{NTA}}$ for all n. That is, holders load nested trusted agents with codes which are clearly made for an agent of a given depth. A participant receiving a message $[C : \text{out}]$ from $\mathcal{F}_{\mathsf{NTA}}$ can see that it is from a trusted agent of depth bounded by n.

3   Forensic Attacks Based on a Trusted Witness (Deniability Loss)

We recall here the definitions of a hard predicate and a zero-knowledge argument of knowledge system [12,13]. We slightly adapt it so that the prover and the verifier can talk to a specific list of oracles (typically: trusted agents). If a list is specified as $\bot$ or unspecified, we consider that no oracle is used. Quite importantly, we do not necessarily assume that honest and malicious verifiers have access to the same oracles.

Definition 1 (Hard predicate). *Let $R(x,w)$ be a predicate relative to a statement x and a witness w. We say that R is a hard predicate if (1) there is a polynomial-time Turing machine A such that $A(x;w)$ yields 1 if and only if $R(x,w)$ and (2) there is no probabilistic polynomial-time Turing machine B such that for any x, $B(x;r)$ returns w such that $R(x,w)$ with non-negligible probability (over the random coins r).*

Definition 2 (Zero-knowledge argument). *Let $R(x,w)$ be a predicate relative to a statement x and a witness w, O_P, O_V, O_V^* be three lists of oracles initialized using a list of random coins r_I. An argument of knowledge for R relative to (O_P, O_V) is a pair (P^{O_P}, V^{O_V}) of polynomial-time interactive machines $P^{O_P}(x,w;r_P)$ and $V^{O_V}(x,z;r_V)$ such that: x is a common input; P has a secret input w; V has an auxiliary input z and produces a binary output (accept or reject); and, moreover, the system fulfills the following properties:*

- Completeness: *for any r_I, r_P, r_V, x, w, z such that $R(x,w)$ holds, the outcome of interaction $P^{O_P}(x,w;r_P) \leftrightarrow V^{O_V}(x,z;r_V)$ makes V accept.*

– Soundness: *there exists a polynomial-time algorithm E (called* extractor) *which is given full black-box access[3] to the prover such that for any x and z, any polynomial-time algorithm P^* with access to O_P, if the probability (over all random coins) that $P^{*O_P}(x; r_P) \leftrightarrow V^{O_V}(x, z; r_V)$ makes V accept is non-negligible, then $E^{P^*}(x; r)$ produces w such that $R(x, w)$ with non-negligible probability (over r).*

The argument system is called zero-knowledge (ZK) *relative to O_V^* (or O_V^*-ZK) if for any polynomial-time algorithm $V^{*O_V^*}$ with access to O_V^* there exists a polynomial-time algorithm $S^{O_V^*}$ (called* simulator), *which could be run by the verifier, such that for any x, w, and z such that $R(x, w)$, the experiments of either computing $\text{View}_V(P^{O_P}(x, w; r_P) \leftrightarrow V^{*O_V^*}(x, z; r_V))$ or running $S^{O_V^*}(x, z; r)$ produce two random (over all random coins) outputs with indistinguishable distributions.*

As shown by the following classical lemma, our definition of zero-knowledge is essentially *deniable* because the simulator can be run by the verifier [21]. This means that V^* cannot produce some y which could serve to feed a malicious prover P^*.[4]

Lemma 3. *Let (P^{O_P}, V^{O_V}) be an argument of knowledge system for R. The system is O_V^*-ZK if and only if for any polynomial-time algorithm $V^{*O_V^*}$ producing a final output y, there exists a polynomial-time algorithm $S^{O_V^*}$ which could be run by the verifier such that for any x, w and z such that $R(x, w)$, the experiments of either running $P^{O_P}(x, w; r_P) \leftrightarrow V^{*O_V^*}(x, z; r_V)$ and getting the final output y of V^* or running $S^{O_V^*}(x, z; r)$ produce two random outputs with indistinguishable distributions.*

In the next lemma, we show that if the honest verifier has access to O_V, a malicious verifier V^* holding a nested trusted agent N^{O_V} can defeat deniability.

Lemma 4 (Transference Lemma). *Let O_P, O_V be any oracle lists. We assume O_V is well-formed. Let N^{O_V} be a nested trusted agent with O_V embedded. Let (P^{O_P}, V^{O_V}) be an argument of knowledge for R. We assume that V only receives messages from either O_V or the prover $\mathcal{P}_P$. There exists a non-interactive argument of knowledge $(Q^{O_P, N^{O_V}}, W)$ for R and a malicious verifier $V^{*N^{O_V}}$ producing a final string y such that the random variables*

$$\text{View}_W \left(P^{O_P}(x, w; r_P) \leftrightarrow V^{*N^{O_V}}(x; r_V) \rightarrow W(x; r_W) \right) \quad and$$

$$\text{View}_W \left(Q^{O_P, N^{O_V}}(x, w; r_P) \rightarrow W(x; r_W) \right)$$

are indistinguishable.

Proof. We define a code C implementing the algorithm V^{O_V} to be run by N^{O_V}. The code terminates the protocol by yielding either "x accepted" or "abort".

---

[3] This means that E can call P^* as a subroutine, choose all inputs including the random tape (i.e. it has "rewindable access"), see all outputs including queries to the oracles invoked by P^* and simulate their responses.

[4] This notion of deniability is sometimes called *self-simulatability* [1] to avoid confusion with other notions of deniability which are used in encryption or signature.

To construct $Q^{O_P,N^{O_V}}$, we first load N^{O_V} with the same C. Then, we simulate the protocol between P^{O_P} and N^{O_V}. Finally, Q sends SHOWTO $\mathcal{P}_W$ where $\mathcal{P}_W$ is the participant running W. To define W, we just make it check that the message is $[C : x$ accepted$]$ with the correct C and x and that the message comes from a trusted agent. Clearly, $(Q^{O_P,N^{O_V}},W)$ is an argument of knowledge for R: it is complete, and for soundness we observe that if W accepts, then it must have received $[C : x$ accepted$]$ from a trusted agent who thus must have run the code C and complete the proof verification. This means that from the malicious $Q^{*O_P,N^{O_V}}$ interacting with N^{O_V} we can first extract an algorithm P^{*O_P} to complete the proof with V^{O_V} and then extract a witness.

To construct V^{*N}, we simply let V^* load N^{O_V} with C and relay messages between P^{O_P} and N^{O_V}. Finally, V^* sends SHOWTO $\mathcal{P}_W$. Clearly, $\text{View}_W(P^{O_P} \leftrightarrow V^{*N} \leftrightarrow W)$ and $\text{View}_W(Q^{O_P,N} \leftrightarrow W)$ are identically distributed. $\qquad \square$

For $O_V = \bot$, this result tells us that we can make any argument of knowledge non-interactive by using a trusted agent. In other words, a malicious verifier V^* equipped with a trusted agent O, after interacting with the prover P, can behave as a prover Q to transfer non-interactively the argument of knowledge to any verifier W offline. This is done by simply certifying a correct execution of V by O. Clearly, trusted agents make the whole notion of NIZK pretty simple to achieve. This further leads us to making deniable zero-knowledge collapse.

Theorem 5. *Let O_P, O_V be any oracle lists. We assume O_V is well-formed. Let N^{O_V} be a nested trusted agent with O_V embedded. Let R be any hard predicate. No argument of knowledge (P^{O_P}, V^{O_V}) for R such that V only receives messages from O_V or the prover $\mathcal{P}_P$ is N^{O_V}-ZK.*

In particular, if O is a trusted agent, no (P,V) argument for R is O-ZK. In clear, if a malicious verifier can use a trusted agent but the honest participants do not, the argument system is not zero-knowledge.

Proof. Let (P^{O_P}, V^{O_V}) be a N^{O_V}-ZK argument of knowledge for a relation R such that V only receives messages from O_V or the prover $\mathcal{P}_P$. We define V^*, Q, W by Lemma 4. Due to Lemma 3, there must exist a simulator $S^{N^{O_V}}$ making a string y without interacting with P^{O_P}. This string is indistinguishable from the one generated by V^{*N} after the interaction with P^{O_P}, so it must be accepted by W. Since $(Q^{O_P,N^{O_V}},W)$ is an argument of knowledge for R, we can use an extractor on $S^{N^{O_V}}$ to get a witness w for x. This contradicts that R is hard. $\qquad \square$

Our result shows the inadequacy of deniable zero-knowledge as soon as adversaries can use trusted agents. It does not mean that deniable zero-knowledge is impossible in this model since honest participants could also use trusted agents to protect against transference attacks.

Indeed, if the prover uses a trusted agent which can directly send messages to the verifier (which is excluded in the hypothesis of Theorem 5), then it is possible to realize deniable zero-knowledge as depicted in Figure 1.[5] The code C makes $\mathcal{F}_{TA}$ waits for

---

[5] More formally: we can realize, in the sense of the universal composability framework, a zero-knowledge functionality $\mathcal{F}_{ZK}$ in the $\mathcal{F}_{TA}$-hybrid model.

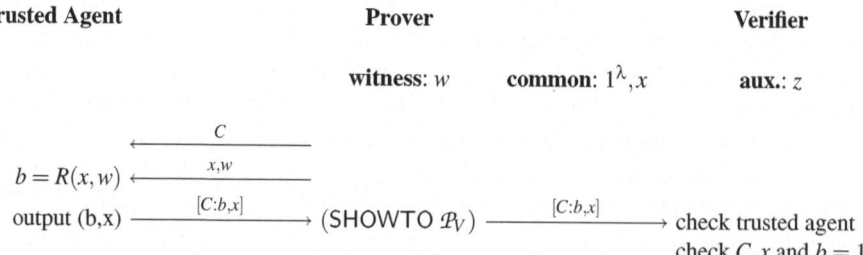

Fig. 1. UC-realization of the $\mathcal{F}_{ZK}$ in the $\mathcal{F}_{TA}$-hybrid model

(x,w), computes $b = R(x,w)$, and outputs (b,x). Clearly, for this protocol to work it is essential that in the last step, the message $[C:b,x]$ reaches V from the prover's TA via an authenticated channel. The protocol is sound, because if the TA yields $[C:1,x]$, it must be the case that it received w such that $R(x,w) = 1$ and the extractor can see messages from P to it. Moreover, it is deniable, because $\mathrm{View}_V(P^{\mathcal{F}_{TA}}(x,w;r_P) \leftrightarrow V^*(x,z;r_V)) = \langle x,z,r_V, \mathcal{F}_{TA} : [C:1,x] \rangle$ and this string could be forged from x,z,r_V by the verifier running $S(x,z;r_V)$.

If the prover uses no oracle, the situation is more complicated. Actually, in the asymmetric case where the verifier uses a nested trusted agent of depth n but the malicious verifier uses a nested trusted agent of higher depth, no zero-knowledge is feasible due to Th. 5. Note that the attack in Th. 5 requires an oracle N^{O_V} for V^* with higher depth than O_V for V. In fact, in symmetric cases where both verifiers use a nested TA $\mathcal{F}_{TA}^n$ with same depth n, zero-knowledge is possible.[6] In symmetric cases where verifiers use $\mathcal{F}_{NTA}$ (i.e. nested trusted agents of unbounded depth), no zero-knowledge is possible since $\mathcal{F}_{TA}^{\mathcal{F}_{NTA}} = \mathcal{F}_{NTA}$. Feasible ZK results are summarized in the following table.

oracle for P	oracle for V	oracle for V^*	feasibility	comment
none	none	none	yes	classical situation
none	$\mathcal{F}_{TA}^n$	$\mathcal{F}_{TA}^n$	yes	
none	$\mathcal{F}_{TA}^n$	$\mathcal{F}_{TA}^{n+1}$	no	Th. 5
none	$\mathcal{F}_{NTA}$	$\mathcal{F}_{NTA}$	no	Th. 5
$\mathcal{F}_{TA}$	any	any	yes	Fig. 1, V receives messages from the prover's TA

4   Attacks Based on Public Key Registration Shift

ZK protocols have the property that a verifier cannot simulate a prover *after* completion of the attack. Nevertheless, the verifier could still play the *Mafia fraud* attack [9]. Indeed, if a third party, say Eve, and the verifier V^* are online, V^* may just relay messages between the prover and Eve while Eve may play a role of a honest verifier. This type

---

[6] We can show this by having V make a commitment (using a TA) to a random challenge prior to a Σ-protocol so that we transform a honest-verifier zero-knowledge protocol into a full zero-knowledge one.

of attack is addressed by a stronger notion of non-transferability. More concretely, we define non-transferability based on [20].

Definition 6. *Let O_P, O_V, O_V^*, O_W^* be some lists of oracles. Let (P^{O_P}, V^{O_V}) be an interactive argument of knowledge for a relation R in the key registration model. We assume that the verifier $\mathcal{P}_V$ is designated by a reference given as an extra common input. We say that the argument is non-transferable relative to $O_V^* | O_W^*$ if, for any malicious (polynomial-time) verifier $V^{*O_V^*}$ run by $\mathcal{P}_V$, and any polynomial-time $W^{*O_W^*}$ interacting with V^* and not run by $\mathcal{P}_V$, there exists a simulator $S^{O_V^*}$ run by $\mathcal{P}_V$ such that for any x, w and z such that $R(x,w)$ the random variables (over all random coins)*

$$\text{View}_{W^*}(P^{O_P}(x, \mathcal{P}_V, w; r_P) \leftrightarrow V^{*O_V^*}(x, \mathcal{P}_V, z; r_V) \leftrightarrow W^{*O_W^*}(x; r_W))$$

and $\text{View}_{W^*}(S^{O_V^*}(x,z;r) \leftrightarrow W^{*O_W^*}(x;r_W))$ *are indistinguishable.*

Thanks to Lemma 3, by using a dummy W^* receiving a single message y and doing nothing else we can see that for any O_W^*, non-transferability relative to $O_V^* | O_W^*$ implies O_V^*-zero-knowledge. Hence, non-transferability is a stronger notion than zero-knowledge (thus deniability).

Interestingly, the protocol of Fig. 1 using no key registration is non-transferable since for any V^* we can simulate the fact that the algorithm receives the string $[C:1,x]$ without the help of the prover. But maybe this is not the ideal non-transferable protocol that we want to use because it requires a secure channel from the prover's TA to the verifier. So, in what follows we assume that the prover uses no trusted agent.

A classical technique to achieve such a strong non-transferability uses proofs to a designated verifier V. This verifier is designated by its public key. That is, the verifier holds a public/private key pair (k,s). One way to make interactive proofs non-transferable consists of replacing the proof of knowledge for secret w by a proof of knowledge of either w or s. This way, a malicious verifier trying to transfer the proof to someone else will not prove knowledge of w since the verifier is assumed to hold s. This works because the verifier is not able to deny knowing s.

Practically, non-transferability strongly relies on the key setup assumption. To formalize this, we use a key registration model. If $\mathcal{P}$ wants to register a public key k, he runs the (Reg, Dir) registration protocol with the registration authority. We model the key registration authority by a new functionality $\mathcal{F}_{CA}^{Dir}$ which registers $(\mathcal{P}, k)$ in a directory. We assume that this functionality is first set up with coins r_D. An instance of the functionality is referred to by sid.

Query REGISTER(sid) from $\mathcal{P}$: launch a Dir protocol session to interact with $\mathcal{P}$. If an output k is produced, store $(\mathcal{P}, k)$. Ignore any further REGISTER(sid) query.
Query CHECK(sid, $\mathcal{P}, k$) from $\mathcal{P}'$: if $(\mathcal{P}, k)$ is stored, sends (sid, yes) to $\mathcal{P}'$. Otherwise sends (sid, no) to $\mathcal{P}'$.

In [22], Ristenpart and Yilek define several key registration models. They consider an arbitrary key registration protocol (Reg, Dir) in which Reg$(k,s;r_R)$ is run by any registrant participant $\mathcal{P}$ with secret key s willing to register a public key k (e.g. generated by some algorithm G) and Dir(r_D) is run by a certificate authority which is assumed to be trusted. Several examples of key registration protocols are defined in [22]. The Plain protocol simply consists of sending a public key from Reg to Dir. The Kosk protocol (as for *Knowledge Of Secret Key*) consists of sending a public key joined with the secret key from Reg to Dir so that the directory authority can check that the registrant knows a secret key consistent with the registered public key. This quite strong model has a brother protocol consisting of making the authority generate the key pair and sending it to the registrant. This protocol is called KRK as for *Key Registration with Knowledge* in [1]. This is the model that could be used when discussing about identity-based protocols because of their intrinsic escrow property. The SPop protocol (as for *Signature-based Proof of Possession*) consists of sending a public key as a self-signed certificate. The Dir protocol first checks the signature before accepting the public key. This is indeed what is used in many practical cases to register a key in a Public Key Infrastructure (PKI). However, the SPop protocol is a pretty weak proof of possession while the Kosk protocol leads to important privacy concerns due to key escrow. We enrich this list with a ZKPop protocol which is an arbitrary zero-knowledge proof of knowledge for the secret key attached to the public key.

Our point is that if proving ignorance of s is doable for V^* then the transference attack could still apply with this construction. More formally, in a protocol where V receives messages from the prover and O_V only, with $O_W^* = O_V$, if V^* acts as a relay between the prover and W^* and is able to register a public key generated by W^* without knowing the secret key, then V^* literally transfers the proof to W^*. We have thus to check which registration model makes it possible for V^* to register a public key k while being able to prove ignorance of the secret key s. This is clearly the case of the Plain and ZKPop models: since V^* and W^* are colluding, V^* can just relay messages between the registering authority and W^* and learns nothing about s. In the SPop model (where V^* acting the same would have to learn more than the public key to register), we have to assume that a self-signed certificate does not provide any extra information to simulate a malicious prover to show that the proof is transfered. On the other side, this is clearly not the case of protocols based on key escrow such as the Kosk or KRK models. Indeed, key escrow surprisingly helps privacy by restoring non-transferability in the trusted agent model.

Theorem 7. *Let O_V^*, O_W^* be two lists of oracles. In the key registration model using* Kosk *there exists some argument of knowledge $(P, V^{\mathcal{F}_{CA}^{Dir}})$ which is non-transferable relative to $O_V^* | O_W^*$.*

Consequently, $(P, V^{\mathcal{F}_{CA}^{Dir}})$ is a O_V^*-ZK even when O_V^* includes trusted agents.

Proof. We use a Σ protocol defined by four algorithms $P_1(x, w; r_P) = (a, t)$, $P_2(t, e) = z$, $\mathsf{Extract}(x, a, e, z, e', z') = w$, and $\mathsf{Simulate}(x, e; r) = (a, z)$, a bitlength $\ell(\lambda)$ defining the domain for e, and a polynomially computable predicate $V_0(x, a, e, z)$. Following [7], the Σ protocol $P(x, w; r_P) \leftrightarrow V(x; r_V)$ works as follows: the prover runs $P_1(x, w; r_P) = (a, t)$ and sends a to V; The verifier picks a random bitstring e of $\ell(\lambda)$ bits and sends it to P; The prover runs $P_2(t, e) = z$ and sends z to V; The verifier accepts if and only if $V_0(x, a, e, z)$ holds. Following the definition of Σ-protocols, the verifier always accept if $R(x, w)$ holds and the protocol is correctly executed; $\mathsf{Extract}(x, a, e, z, e', z')$ must returns a witness w' such that $R(x, w')$ whenever the conditions $e \neq e'$, $V_0(x, a, e, z)$, and $V_0(x, a, e', z')$ are satisfied (this is the *special soundness* property); and for any x and e,

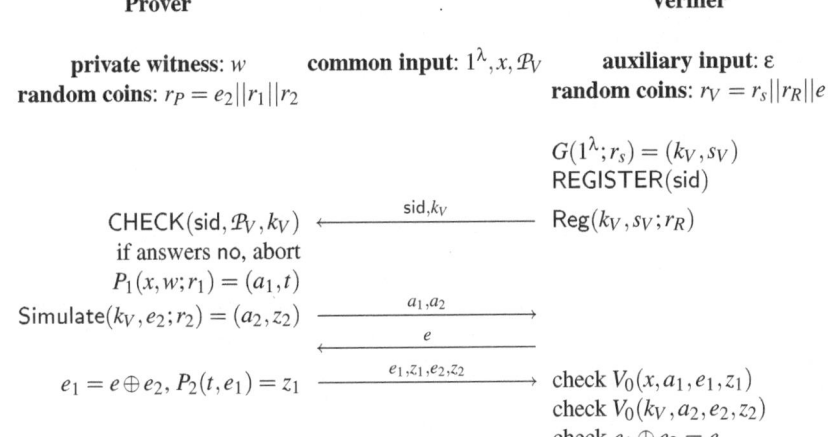

Fig. 2. A non-transferable ZK proof system

$\text{Simulate}(x, e; r) = (a, z)$ should define a random (a, e, z) triplet such that $V_0(x, a, e, z)$ holds with same distribution as the triplets generated by the honest run of the protocol (this is the *special zero-knowledge* property).

We modify the protocol of [8] as depicted on Fig. 2. Relative to the Kosk key registration model, obtain a non-transferable argument of knowledge. If V^* does not send any valid k_V binded to $\mathcal{P}_V$ to P, the simulation is trivial since it does not use w. Otherwise, V^* must have sent some k_V together with s_V to $\mathcal{F}_{CA}^{Dir}$. A simulator for V^* could then use s_V to simulate P in the OR proof. □

In this construction based on key escrow, the registering authority could abuse the protocol and make a cheating prover to the designated verifier. We further show that this (bad) property is necessary for any $(P, V^{\mathcal{F}_{CA}^{Dir}})$ protocol which is non-transferable.

Theorem 8. *Let O_P, O_V be two lists of oracles. We assume that O_V is well-formed and that $\mathcal{P}_V$ can only receive messages from either O_V, $\mathcal{F}_{CA}^{Dir}$, or $\mathcal{P}_P$. Let $\mathcal{P}_D$ be an authority who runs an emulator D for the $\mathcal{F}_{CA}^{Dir}$ functionality for a given protocol (Reg, Dir). Let $(P^{O_P}, V^{O_V, \mathcal{F}_{CA}^{Dir}})$ be an argument of knowledge for R which is non-transferable relative to $\perp | O_V$. We let $\tilde{V}^{O_V}(x, \mathcal{P}_V, z; r_V)$ denote the protocol who simulates $V^{O_V \mathcal{F}_{CA}^{Dir}}(x, \mathcal{P}_V, z; r_V)$ with all messages for $\mathcal{P}_D$ and $\mathcal{P}_P$ sent to the same counterpart. There exists an algorithm $D^{*O_P}(x, z; r)$ such that for any $r_I, x, z, D^{*O_P} \leftrightarrow \tilde{V}^{O_P}$ accepts with high probability.*

This means that if an interactive proof using well-formed oracles for the verifier is non-transferable and with the property that the registering authority cannot cheat with the verifier, then the verifier $\mathcal{P}_V$ must receive messages from an oracle O_P, e.g. using a TA.

Proof. We let $W^{*O_V} = \tilde{V}^{O_V}$ be run by a participant $\mathcal{P}_W$ with $\mathcal{P}_V$ as counterpart. Here, V^* is "router" who "reroutes" the requests by W^* to the registration authority to $\mathcal{P}_D$ and others to $\mathcal{P}_P$. Clearly, $P^{O_P} \leftrightarrow V^{*\mathcal{F}_{CA}^{Dir}} \leftrightarrow W^{*O_V}$ makes W^* always accept since O_V is well-formed. Thanks to the definition of non-transferability, there is a simulator S

such that the view from W^* to either $P^{O_P} \leftrightarrow V^{*\mathcal{F}_{CA}^{Dir}} \leftrightarrow W^{*O_V}$ or $S^{\mathcal{F}_{CA}^{Dir}} \leftrightarrow W^{*O_V}$ are indistinguishable. We can thus define $D^* = S^{\mathcal{F}_{CA}^{Dir}}$ so that $D^* \leftrightarrow \tilde{V}$ accepts with high probability.   □

5   Malicious Applications

5.1   Shedding Light on Invisible Signatures (Invisibility Loss)

Undeniable signatures (aka *invisible signatures*) were invented by Chaum and van Antwerpen in [6] and have two basic features: (i) *interactive verification*, that is, the verification process is interactive and so the signer can choose who can verify his signature; (ii) *disavowal protocol* which allows the signer to prove that a given signature is a forgery. The first feature enables the signer to restrict the verification of the signature to those he wishes to. If the document leaks, a third party would not be able to verify the signature alone.

More formally, an invisible signature scheme is defined by two algorithms and a relation R: algorithm $\mathsf{Setup}(1^\lambda; K_s) = K_p$ is making keys and algorithm $\mathsf{Sign}(m, K_s; r) = s$ is making signatures. The relation $R(x, w)$ with witness $w = K_s$ defines valid signatures $x = (m, s, K_p)$. The scheme also comes with two ZK proof of knowledge protocols

$$(P_{\mathsf{Confirm}}(x, K_s; r_P), V_{\mathsf{Confirm}}(x; r_V)) \quad \text{and} \quad (P_{\mathsf{Deny}}(x, K_s; r_P), V_{\mathsf{Deny}}(x; r_V))$$

for the relations R and $\neg R$, respectively. Besides the zero-knowledge proof of knowledge properties, the scheme requires signature to be *existentially unforgeable* and *invisible*. Several definitions for invisibility exist in the literature. The weakest one requires the existence of a simulator $S(m, K_p; r) = s$ that makes strings look like signatures, such that no algorithm based on K_p only can distinguish between Sign and S. This does not prevent from transferability issues. Clearly, a verifier V^* for (Confirm or Deny) equipped with a trusted agent O could transfer a proof universally from Lemma 4 to any offline W. Somehow, this malicious verifier would remove the "invisibility shield" on the signature which would then become visible.

There are some invisible signature schemes featuring non-transferability properties [20]. They however require verifiers to be given public and privates keys as well. We have seen how to defeat this protection in Section 4.

5.2   Selling Votes (Receipt-Freeness Loss)

Another application where transferring the protocol to a trusted agent would be dangerous is for e-voting: clearly, a trusted agent casting a ballot on behalf of a malicious elector could later testify the ballot content and receipt-freeness would be broken. E-democracy could collapse due to corruption with the help of trusted agents.

5.3   Denying Ring Signatures (Anonymity Loss)

Ring signatures were proposed by Rivest, Shamir and Tauman [23] to allow members of a certain group to sign a message without conveying any information on who inside the group signed the message. Informally, the ring signature works as follows. A signer creates a "ring" of members including himself. Each ring member $1 \leq i \leq n$ has a public k_i and secret key s_i. The public key specifies a trapdoor permutation and the secret key specifies the trapdoor information needed to compute its inverse. The signing process generates a ciphertext that could have been generated by anyone knowing at least one secret key. The verification process only requires the knowledge of the public keys. This way, the signer can hide in a ring that he created himself. Ring signature can be used e.g. for whistleblowing in order to protect the signer. Ring signatures can be used as a countermeasure to spamming. The idea is to have every email sent together with a ring signature of a ring consisting of the sender P and the receiver V. The reason to have email signed by the sender is to authenticate its origin and moreover, to make the email somehow binding. The reason to have the receiver in the ring is to prevent him from exhibiting the signed email to a third party W^*. In such a case, the email could have been forged by the receiver V^* so the sender can deny it.

If one member, say Victor, of the ring can prove the ignorance of his own secret key, then he can show that he was not able to sign any message with the ring signature, that is, he denies the signature. One way for Victor doing this in the Plain registration model is to take some pseudorandom generator π, some seed x and publish as public key $k = \pi(x)$. In this way he could present the pseudorandom generator and the seed to a third party W^* and convince him that he was not able to use the ring signature, since he did not know the secret key. To fix this, the key registration model should at least mandate the use of a proof of possession of a secret key, e.g. using self-signed certificates like the SPop or ZKPop protocol.

To defeat this, Victor owning a trusted agent could have his agent to simulate a key registration process so that only the agent would know the secret key. The attack could be more vicious here since the agent could still be used to sign messages in a ring. The only difference is that the agent would keep record of all signed messages and could, upon request, certify that a message was signed or not. The signature by the trusted agent of the certificate together with its code is an evidence to anyone trusting the agent.

In [22], Ristenpart and Yilek proved that ring signatures could guaranty anonymity for rings larger than 2 even when the adversary can select the keys under the key registration model using an SPop protocol. Clearly, this result is no longer valid in the trusted agent model.

Once again, the attack relies on the public key registration issue, and the only way to thwart it seems to use key escrow: the Kosk protocols. This however enables the registration authority to forge a ring signature with a ring consisting of honest P and V: having V's secret key makes it possible to impersonate P to V. Finally, it seems that we either have to choose between having signatures deniable or forgeable.

We could still fix this problem by making honest participants use trusted agents to help registering keys in a Kosk-like model still ensuring privacy: a trusted agent could simulate Dir running Kosk. The certificate from the trusted agent could then be sent to D to register. Again, this fix is void if malicious registrants use nested trusted agents.

6 Conclusions

We have defined the Trusted Agent Model. In the past, several cryptographic protocols requiring trusted agents have been proposed but researchers prefer to develop protocols without them. However it does not prevent to *maliciously use* such devices if they exist. We devised scenarii in which adversaries equipped with such devices could defeat several cryptographic properties, e.g. invisibility in invisible signatures, receipt-freeness in e-voting, or anonymity in ring signatures. Fundamentally, these failures come from the strange nature of deniability in protocols. We have shown that deniability is not possible for regular protocols (namely, protocols not using trusted agents) if adversaries can use trusted agents. Deniability becomes possible again if honest and corrupted verifiers can use trusted agents. It collapses again if the malicious verifier can use a nested trusted agent of depth higher than the one the honest verifier is using. It can be restored again in a key registration model. We can even achieve non-transferability which is a stronger form of deniability, but this comes at the price of key escrow: if a protocol is non-transferable, then the key registration authority has the privilege to create malicious provers. Namely, non-transferability requires giving the authority some piece of information which could be used to cheat with proofs, which is pretty bad for privacy. An ultimate solution consists of making the proving part trivial by having proofs (resp. signatures) assessed by a trusted agent instead of running a protocol with the prover.

Although our "attacks" are pretty trivial, we think the issue of malicious use of trusted devices in practice has been overlooked so far. Clearly, these devices could defeat some privacy protocols. To the authors it does not seem acceptable on one hand, to accept tamper-proof hardware, and on the other hand, assume that adversaries cannot use such hardware and its properties to perform attacks.

Probably, the most interesting question that this paper opens is whether privacy is a self-contradicting concept or not. As shown herein, as soon as we place boundaries around devices, we can no longer control how private data spreads, so boundaries are somehow harming privacy. On the other hand, privacy strongly relies on setting boundaries on data.

Acknowledgments

P. Mateus was partially supported by Instituto de Telecomunicações, FCT and EU FEDER through PTDC, namely via QSec PTDC/EIA/67661/2006 Project, IT project QuantTel, European Network of Excellence EURO-NF and IBM Portuguese scientific prize 2005.

References

1. Barak, B., Canetti, R., Nielsen, J.B., Pass, R.: Universally composable protocols with relaxed set-up assumptions. In: Annual ACM Symposium on Theory of Computing: FOCS 2004, pp. 186–195. IEEE Computer Society Press, Los Alamitos (2004)
2. Camenisch, J.L., Michels, M.: Confirmer signature schemes secure against adaptive adversaries (Extended abstract). In: Preneel, B. (ed.) EUROCRYPT 2000. LNCS, vol. 1807, pp. 243–258. Springer, Heidelberg (2000)

3. Canetti, R.: Obtaining universally composable security: Towards the bare bones of trust. In: Kurosawa, K. (ed.) ASIACRYPT 2007. LNCS, vol. 4833, pp. 88–112. Springer, Heidelberg (2007)

4. Canetti, R., Fischlin, M.: Universally composable commitments. In: Kilian, J. (ed.) CRYPTO 2001. LNCS, vol. 2139, pp. 19–40. Springer, Heidelberg (2001)

5. Canetti, R., Lindell, Y., Ostrovsky, R., Sahai, A.: Universally composable two-party and multi-party secure computation. In: Annual ACM Symposium on Theory of Computing: STOC 2002, pp. 494–503. ACM Press, New York (2002)

6. Chaum, D., Van Antwerpen, H.: Undeniable signatures. In: Brassard, G. (ed.) CRYPTO 1989. LNCS, vol. 435, pp. 212–216. Springer, Heidelberg (1990)

7. Cramer, R.: Modular design of secure, yet practical cryptographic protocols. Phd thesis, University of Amsterdam (1996)

8. Damgård, I.: On Σ-protocols. Lecture notes, University of Aahrus (2005), http://www.daimi.au.dk/%7Eivan/Sigma.pdf

9. Desmedt, Y.: Major security problems with the unforgeable (feige)-fiat-shamir proofs of identity and how to overcome them. In: The 6th Worldwide Congress on Computer and Communications Security and Protection: Securicom 1988, pp. 147–149. SEDEP (1988)

10. Desmedt, Y.G., Quisquater, J.-J.: Public-key systems based on the difficulty of tampering. In: Odlyzko, A.M. (ed.) CRYPTO 1986. LNCS, vol. 263, pp. 111–117. Springer, Heidelberg (1987)

11. Desmedt, Y.G., Yung, M.: Weaknesses of undeniable signature schemes. In: Davies, D.W. (ed.) EUROCRYPT 1991. LNCS, vol. 547, pp. 205–220. Springer, Heidelberg (1991)

12. Goldwasser, S., Micali, S., Rackoff, C.: The knowledge complexity of interactive proof-systems. In: Proceedings of the Seventeenth Annual ACM Symposium on Theory of Computing, STOC 1985, pp. 291–304. ACM Press, New York (1985)

13. Goldwasser, S., Micali, S., Rackoff, C.: The knowledge complexity of interactive proof systems. SIAM Journal of Computing 18(1), 186–208 (1989)

14. Gratzer, V., Naccache, D.: Alien vs. quine, the vanishing circuit and other tales from the industry's crypt. In: Vaudenay, S. (ed.) EUROCRYPT 2006. LNCS, vol. 4004, pp. 48–58. Springer, Heidelberg (2006)

15. Jakobsson, M.: Blackmailing using undeniable signatures. In: De Santis, A. (ed.) EURO-CRYPT 1994. LNCS, vol. 950, pp. 425–427. Springer, Heidelberg (1995)

16. Jakobsson, M., Sako, K., Impagliazzo, R.: Designated verifier proofs and their applications. In: Maurer, U.M. (ed.) EUROCRYPT 1996. LNCS, vol. 1070, pp. 143–154. Springer, Heidelberg (1996)

17. Katz, J.: Universally composable multi-party computation using tamper-proof hardware. In: Naor, M. (ed.) EUROCRYPT 2007. LNCS, vol. 4515, pp. 115–128. Springer, Heidelberg (2007)

18. Mateus, P.: Attacking zero-knowledge proof systems. Habilitation thesis, Department of Mathematics, Instituto Superior Técnico, 1049-001 Lisboa, Portugal, 2005. Awarded the Portuguese IBM Scientific Prize (2005) (in Portuguese)

19. Mateus, P., Moura, F., Rasga, J.: Transferring proofs of zero-knowledge systems with quantum correlations. In: Dini, P., et al. (eds.) Proceedings of the First Workshop on Quantum Security: QSec 2007, p. 9. IEEE Computer Society Press, Los Alamitos (2007)

20. Monnerat, J., Vaudenay, S.: Short 2-move undeniable signatures. In: Nguyen, P.Q. (ed.) VI-ETCRYPT 2006. LNCS, vol. 4341, pp. 19–36. Springer, Heidelberg (2006)

21. Pass, R.: On deniability in the common reference string and random oracle model. In: Boneh, D. (ed.) CRYPTO 2003. LNCS, vol. 2729, pp. 316–337. Springer, Heidelberg (2003)

22. Ristenpart, T., Yilek, S.: The power of proofs-of-possession: Securing multiparty signatures against rogue-key attacks. In: Naor, M. (ed.) EUROCRYPT 2007. LNCS, vol. 4515, pp. 228–245. Springer, Heidelberg (2007)
23. Rivest, R.L., Shamir, A., Tauman, Y.: How to leak a secret. In: Boyd, C. (ed.) ASIACRYPT 2001. LNCS, vol. 2248, pp. 552–565. Springer, Heidelberg (2001)

A   Secure Commitment Using Trusted Agents

We consider the $\mathcal{F}_{\mathsf{com}}$ functionality[7] defined by

Query COMMIT$(\mathsf{sid}, \mathcal{P}, \mathcal{P}', b)$ **from** $\mathcal{P}$:   record b, send the message $[\mathsf{receipt}, \mathsf{sid}, \mathcal{P}, \mathcal{P}']$ to $\mathcal{P}'$, and ignore any future COMMIT queries.

Query OPEN(sid) **from** $\mathcal{P}$: if no value b was recorded, ignore. Otherwise, send $[\mathsf{open}, \mathsf{sid}, b]$ to $\mathcal{P}'$.

Here is a protocol to realize this ideal functionality using trusted agents:

- To emulate COMMIT$(\mathsf{sid}, \mathcal{P}, \mathcal{P}', b)$ by $\mathcal{P}$, participant $\mathcal{P}$ sets a code C as detailed below, makes the queries SEND(sid, C) to $\mathcal{F}_{\mathsf{TA}}$, then SEND$(\mathsf{sid}, b)$, then SHOWTO $(\mathsf{sid}, \mathcal{P}')$. Participant $\mathcal{P}'$ gets a message $[C : \mathsf{receipt}, N]$, checks that it comes from $\mathcal{F}_{\mathsf{TA}}$, that C is correct (as defined below), and stores N.
- To emulate OPEN(sid), $\mathcal{P}$ queries SEND$(\mathsf{sid}, \emptyset)$ to $\mathcal{F}_{\mathsf{TA}}$ and finally SHOWTO $(\mathsf{sid}, \mathcal{P}')$. Participant $\mathcal{P}'$ gets a message $[C : \mathsf{open}, N', b']$, checks that it comes from $\mathcal{F}_{\mathsf{TA}}$, that C is still correct, and that $N = N'$. The value b' is revealed.

The code C takes a first message b as input, picks a random nonce N, stores N and b, and responds by $(\mathsf{receipt}, N)$. Then it waits for a dummy message and responds by (open, N, b).

Given a static adversary $\mathcal{A}$ interacting with $\mathcal{P}$, $\mathcal{P}'$ and $\mathcal{F}_{\mathsf{TA}}$ in the real world, we define a simulator $\mathcal{S}$ interacting with $\mathcal{P}$, $\mathcal{P}'$ and $\mathcal{F}_{\mathsf{TA}}$ in the ideal world so that for any environment $\mathcal{E}$ interacting with $\mathcal{P}$, $\mathcal{P}'$ and $\mathcal{A}$ resp. $\mathcal{S}$, the real and ideal views of $\mathcal{E}$ are indistinguishable. Therefore, the protocol UC-realizes $\mathcal{F}_{\mathsf{com}}$ in the $\mathcal{F}_{\mathsf{TA}}$-hybrid model.

Indeed, when $\mathcal{P}$ and $\mathcal{P}'$ are both honest or both corrupted, constructing the simulator is trivial. When $\mathcal{P}$ is honest but $\mathcal{P}'$ is corrupted, the $[C : \mathsf{receipt}, N]$ message to $\mathcal{P}'$ can be perfectly simulated from the $[\mathsf{receipt}, \mathsf{sid}, \mathcal{P}, \mathcal{P}']$ by picking a fresh nonce N and maintain a table of $\mathsf{sid} \leftrightarrow N$ pairs. The $[C : \mathsf{open}, N', b']$ message can be simulated upon message $[\mathsf{open}, \mathsf{sid}, b]$ from $\mathcal{F}_{\mathsf{com}}$ by looking up at the table for the correct N' and setting $b' = b$. Thanks to the trusted agent property and the C code there must be an appropriate pair. When $\mathcal{P}$ is corrupted and $\mathcal{P}'$ is honest, the messages to $\mathcal{F}_{\mathsf{TA}}$ with correct code C can be perfectly simulated from the message COMMIT$(\mathsf{sid}, \mathcal{P}, \mathcal{P}', b)$ resp. OPEN(sid) to $\mathcal{F}_{\mathsf{com}}$.

---

[7] The functionality $\mathcal{F}_{\mathsf{mcom}}$ [4] could be used as well.

Mutual Information Analysis:
How, When and Why?

Nicolas Veyrat-Charvillon* and François-Xavier Standaert**

UCL Crypto Group, Université catholique de Louvain, B-1348 Louvain-la-Neuve
{nicolas.veyrat,fstandae}@uclouvain.be

Abstract. The Mutual Information Analysis (MIA) is a generic side-channel distinguisher that has been introduced at CHES 2008. This paper brings three contributions with respect to its applicability to practice. First, we emphasize that the MIA principle can be seen as a toolbox in which different (more or less effective) statistical methods can be plugged in. Doing this, we introduce interesting alternatives to the original proposal. Second, we discuss the contexts in which the MIA can lead to successful key recoveries with lower data complexity than classical attacks such as, *e.g.* using Pearson's correlation coefficient. We show that such contexts exist in practically meaningful situations and analyze them statistically. Finally, we study the connections and differences between the MIA and a framework for the analysis of side-channel key recovery published at Eurocrypt 2009. We show that the MIA can be used to compare two leaking devices only if the discrete models used by an adversary to mount an attack perfectly correspond to the physical leakages.

1   Introduction

The most classical solutions used in non profiled side-channel attacks are Kocher's original DPA [14] and correlation attacks using Pearson's correlation coefficient, introduced by Brier *et al.* [5]. In 2008, another interesting side-channel distinguisher has been proposed, denoted as Mutual Information Analysis (MIA) [12]. MIA aims at genericity in the sense that it is expected to lead to successful key recoveries with as little assumptions as possible about the leaking devices it targets. In this paper, we confirm and extend the ideas of Gierlichs *et al.* and tackle three important questions with respect to this new distinguisher.

1. How to use MIA? In general, MIA can be viewed as the combination of two subproblems. In a first stage of the attack, an adversary has to *estimate* the leakage *probability density functions* for different key-dependent models. In a second stage of the attack, this adversary has to *test the dependence* of these models with actual measurements. In the original description of [12], the MIA is using histograms for the first stage and a Kullback-Leibler divergence for the

---

* Work supported in part by the Walloon Region research project SCEPTIC.
** Associate researcher of the Belgian Fund for Scientific Research (FNRS - F.R.S.).

C. Clavier and K. Gaj (Eds.): CHES 2009, LNCS 5747, pp. 429–443, 2009.

second stage. In this paper, we argue that in fact, the MIA can be seen as a toolbox in which different probability density estimation techniques and notions of divergence can be used. We show that these different solutions (some of them being introduced in [3,19]) yield different results for the attack effectiveness. We also introduce an alternative test that is at least as generic as the original MIA but does not require an explicit estimation of the leakage probability densities.

2. When to use MIA? In a second part of this paper, we analyze the contexts in which MIA can be necessary (*i.e.* when other side-channel attacks would not succeed). In [19], it is argued that MIA is particularly convenient in higher-order side-channel attacks because of its simple extension to multi-dimensional scenarios. In this paper, we show that MIA can also be useful in univariate side-channel attacks, if the models used by an adversary to mount an attack are not sufficiently precise. Hence, we complement the original experiment of [12] against a dual-rail pre-charged implementation. In order to further validate this intuition, we analyze an arbitrary degradation of the leakage models and show that after a certain threshold, MIA leads to a more effective key recovery than the corresponding correlation attack using Pearson's coefficient. We also discuss the effect of incorrect models theoretically and intuitively.

3. Why to use MIA? Eventually, in a third part of the paper, we investigate the relations between the MIA and the information theoretic *vs.* security model of [22]. We exhibit that although having similar foundations, MIA and this model have significantly different goals and are not equivalent in general. We also show that in certain idealized contexts (namely, when adversaries can exploit leakage predictions that perfectly correspond to the actual measurements), the MIA can be used as a metric to compare different cryptographic devices.

2   Background

2.1   Information Theoretic Definitions

Entropy. The entropy [7] of a random variable X on a discrete space $\mathcal{X}$ is a measure of its uncertainty during an experiment. It is defined as:

$$\mathrm{H}\left[X\right] = -\sum_{x \in \mathcal{X}} \Pr\left[X = x\right] \log_2(\Pr\left[X = x\right]).$$

The joint entropy of a pair of random variables X, Y expresses the uncertainty one has about the combination of these variables:

$$\mathrm{H}\left[X, Y\right] = -\sum_{x \in \mathcal{X}, y \in \mathcal{Y}} \Pr\left[X = x, Y = y\right] \log_2(\Pr\left[X = x, Y = y\right]).$$

The joint entropy is always greater than that of either subsystem, with equality only if Y is a deterministic function of X. The joint entropy is also sub-additive. Equality occurs only in the case where the two variables are independent.

$$\mathrm{H}\left[X\right] \leq \mathrm{H}\left[X, Y\right] \leq \mathrm{H}\left[X\right] + \mathrm{H}\left[Y\right].$$

Finally, the conditional entropy of a random variable X given another variable Y expresses the uncertainty on X which remains once Y is known.

$$H[X|Y] = - \sum_{x \in \mathcal{X}, y \in \mathcal{Y}} \Pr[X = x, Y = y] \log_2(\Pr[X = x|Y = y]).$$

The conditional entropy is always greater than zero, with equality only in the case where X is a deterministic function of Y. It is also less than the entropy of X. Equality only occurs if the two variables are independent.

$$0 \leq H[X|Y] \leq H[X].$$

All these measures can be straightforwardly extended to continuous spaces by differentiation. For example, the differential entropy is defined as:

$$H[X] = - \int_{x \in \mathcal{X}} \Pr[X = x] \log_2(\Pr[X = x]).$$

The differential entropy can be negative, contrary to the discrete entropy.

Mutual information. The mutual information is a general measure of the dependence between two random variables. On a discrete domain, the mutual information of two random variables X and Y is defined as:

$$I(X;Y) = \sum_{x \in \mathcal{X}, y \in \mathcal{Y}} \Pr[X = x, Y = y] \log_2 \left(\frac{\Pr[X = x, Y = y]}{\Pr[X = x] \cdot \Pr[Y = y]} \right).$$

It is directly related to Shannon's entropy, and can be expressed using entropies:

$$\begin{aligned} I(X;Y) &= H[X] - H[X|Y] \\ &= H[X] + H[Y] - H[X,Y] \\ &= H[X,Y] - H[X|Y] - H[Y|X] \end{aligned}$$

It can also be straightforwardly extended to the continuous case:

$$I(X;Y) = \int_{x \in \mathcal{X}} \int_{y \in \mathcal{Y}} \Pr[X = x, Y = y] \log_2 \left(\frac{\Pr[X = x, Y = y]}{\Pr[X = x] \cdot \Pr[Y = y]} \right).$$

2.2   Pearson's Correlation Coefficient

This coefficient is a simpler measure of dependence between two random variables X and Y. Computing it does not require the knowledge of the probability density functions of X and Y but it only measures the linear dependence between these variables (whereas mutual information is able to detect any linear or non-linear dependence). It is defined as follows (with $\overline{X}$ the mean value of X):

$$\rho(X,Y) = \frac{\sum_{x \in \mathcal{X}, y \in \mathcal{Y}} (x - \overline{X}) \cdot (y - \overline{Y})}{\sqrt{\sum_{x \in \mathcal{X}} (x - \overline{X})^2 \cdot \sum_{y \in \mathcal{Y}} (y - \overline{Y})^2}}.$$

2.3   Side-Channel Analysis

In a side-channel attack, an adversary tries to recover secret information from a leaking implementation, *e.g.* a software program or an IC computing a cryptographic algorithm. The core idea is to compare key-dependent models of the leakages with actual measurements. Typically, the adversary first defines the subkeys that he aims to recover. For example, in a block cipher implementation, those subkeys could be one byte of the master key. Then, for each subkey candidate, he builds models that correspond to the leakage generated by the encryption of different plaintexts. Eventually, he evaluates which model (*i.e.* which subkey) gives rise to the best prediction of the actual leakages, measured for the same set of plaintexts. As a matter of fact and assuming that the models can be represented by a random variable X and the leakages can be represented by a random variable Y, the side-channel analysis can simply be seen as the problem of detecting a dependence between those two variables. Pearson's coefficient and the mutual information can be used for this purpose.

In the following, we consider side-channel attacks restricted by two important assumptions. First, we investigate *univariate attacks*, *i.e.* attacks in which one compares the leakage models X with a single sample in the leakage traces. It means that the variable Y has only one dimension. Second, we consider *discrete leakage models*, *i.e.* we assume that the variable X is discrete (by contrast, the actual leakage variable Y can be continuous). We note that univariate attacks are typical scenarios in standard DPA attacks such as [14] and discrete leakage models are also a very natural assumption as long as the side-channel attacks cannot be enhanced with profiling and characterization [6]. Hence, these two assumptions can be seen as reasonable starting points for the analysis of MIA.

3   How to Use MIA: The Information Theoretic Toolbox

Following the previous informal description, let us denote the subkey candidates in a side-channel attack as k_j and the models corresponding to those subkeys as X_j. The distinguisher used in a mutual information analysis is defined as:

$$d_j = \hat{\mathrm{I}}(X_j; Y).$$

For simplicity, we will omit the j subscript in the following of the paper. The idea behind this procedure is that a meaningful partition of Y where each subset corresponds to a particular model value will relate to a side-channel sample distribution $\hat{\Pr}[Y|X = x]$ distinguishable from the global distribution of $\hat{\Pr}[Y]$. The estimated mutual information will then be larger than zero. By contrast, if the key guess is incorrect, the false predictions will form a partition corresponding to a random sampling of Y and therefore simply give scaled images of the total side-channel probability density function (pdf for short). Hence, the estimated mutual information will be equal (or close) to zero in this case.

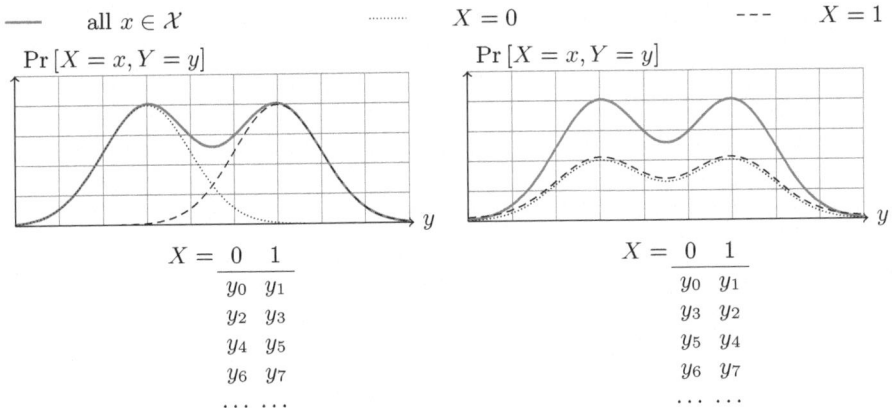

Fig. 1. Probability densities and associated leakage partitions for correct (left) and wrong (right) subkey hypotheses in the case of a single bit DPA attack

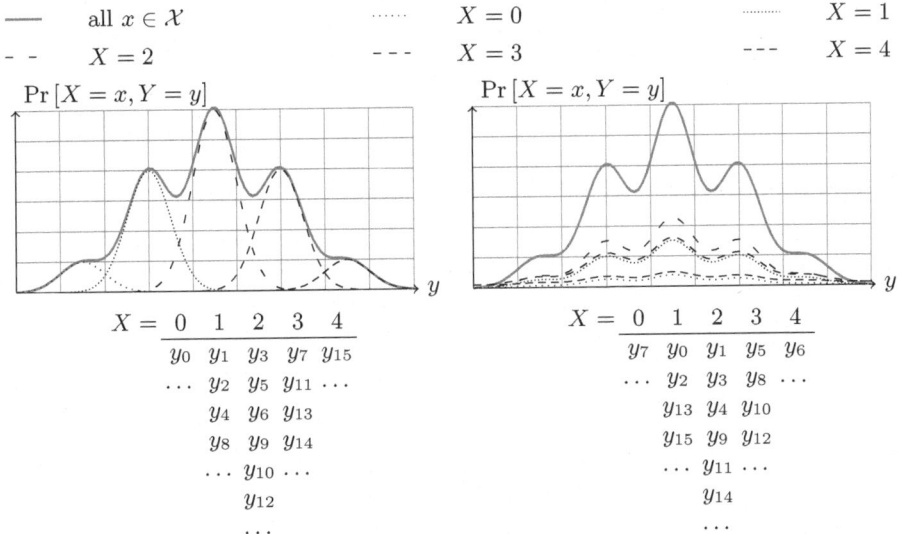

Fig. 2. Probability densities and associated leakage partitions for correct (left) and wrong (right) subkey hypotheses in the case of a 4-bit DPA attack

Example. Let us imagine a target implementation in which the adversary receives leakages of the form $y = H_W(S(p \oplus k)) + n$ where H_W is the Hamming weight function, S the 4-bit S-box of the block cipher Serpent, p a known plaintext, k the target subkey of the attack and n is a Gaussian noise. Let us also assume two different attacks: in the first one, the model X corresponds to a single bit of $S(p \oplus k)$; in the second one, the model X corresponds to $H_W(S(p \oplus k))$. Figures 1 and 2 illustrate what happens asymptotically to the correct and a wrong subkey hypotheses in the case these two attacks. They clearly show the higher dependence for the correct subkey (*i.e.* the left figures) that is expected by [12].

In theory, the MI distinguisher tests a *null hypothesis* stating that the predicted leakage values and the side-channel samples are independent if the subkey hypothesis is false. When this hypothesis is not verified, the adversary assumes that he found the correct subkey. However, in practice there may exist certain dependencies between a wrong subkey candidate and the actual leakages (*e.g.* ghost peaks as in [5]). Hence, the adversary generally selects the subkey that leads to the highest value of the distinguisher. This description underlines that a MIA is essentially composed of the two problems listed in introduction:

1. An estimation of some probability density functions, namely those of the global samples and of the samples corresponding to each modeled leakage.
2. The test of a null hypothesis stating that the predicted leakages and their actual side-channel values are independent.

As a matter of fact, different solutions can be considered for this purpose. Therefore, in the remainder of this section, we first review some possible techniques to estimate the probability density functions used in a side-channel attack. Then we present various probability-distance measures that can replace the usual relative entropy in mutual information analysis. Eventually, we discuss the possibility to compare two pdf without explicitly estimating them and briefly mention alternative attack techniques inspired from "all-or-nothing" multiple-bit DPA.

3.1   Probability Density Function Estimation

The problem of modeling a probability density function from random samples of this distribution is a well studied problem in statistics, referred to as density estimation. A number of solutions exist, ranging from simple histograms to kernel density estimation, data clustering and vector quantization. The authors of [12] used histograms for density estimation as a proof of concept for MIA. But in certain contexts, an attack can be greatly improved by using more advanced techniques. In the following, we summarize a few density estimation tools that have been initially suggested in [3] as relevant to side-channel attacks and then applied to MIA in [19]. They are detailed in Appendix A.

Non-parametric methods. One interesting feature of the MIA is that it does not rely on particular assumptions on the leakages. Hence, it is natural to consider non-parametric estimation techniques first since, *e.g.* assuming Gaussian leakages would again reduce the genericity of the distinguisher. In practice, two techniques can generally be used for this purpose:

- Histograms perform a partition of the samples by grouping them into bins. Each bin contains the samples of which the value falls into a certain range. The respective ranges of the bins have equal width and form a partition of the range between the extreme values of the samples. Using this method, one approximates a probability by dividing the number of samples that fall within a bin by the total number of samples (see Appendix A.1).

– Kernel density estimation is a generalization of histograms. Instead of bundling samples together in bins, it adds (for each observed sample) a small kernel centered on the value of the leakage to the estimated pdf. The resulting estimation is a sum of small "bumps" that is much smoother than the corresponding histogram. It usually provides faster convergence towards the true distribution. Note that although this solution requires to select a Kernel and a bandwidth (details are given in Appendix A.2), it does not assume anything more about the estimated pdf than histograms.

Parametric methods. Contrary to the previous techniques, parametric methods for density estimation require certain assumptions about the leakages. They consequently trade some of the genericity of the MIA for a hopefully better effectiveness, *i.e.* they are an intermediate solution between attacks using the correlation coefficient and the original MIA of [12]. In this context, a particularly interesting tool is the finite mixture estimation. A mixture density is a probability density function that consists in a convex combination of probability density functions. Given a set of densities $p_1(x), \ldots, p_n(x)$, and positive weights $w_1, \ldots, w_n$ verifying $\sum w_i = 1$, the finite mixture is defined as:

$$\hat{\Pr}[x] = \sum_{i=0}^{n-1} w_i \, p_i(x).$$

A typical choice is to assume a mixture of Gaussian densities (see, *e.g.* [15]), which leads to an efficient parametric estimation of the pdf (see Appendix A.3).

3.2 Probability-Distance Measures

Once the probability densities have been estimated, one has to test whether the predicted leakages are correlated with the actual measurements. This dependence is tested using a probability-distance measure which allows deciding which subkey is the most likely to be the correct one. As in the previous section, different solutions can be used, that we detail and connect to the original MIA.

Kullback-Leibler divergence. The Kullback-Leibler divergence, or relative entropy [7], is a measure of the difference between two probability density functions P and Q. It is not a distance, as it is non-commutative and does not satisfy the triangle inequality. The KL divergence of Q from P, where P and Q are two probability functions of a discrete random variable X, is defined as:

$$D_{\mathsf{KL}}\left(\mathsf{P}\|\mathsf{Q}\right) = \sum_{x \in \mathcal{X}} \Pr\left[X = x, X \sim \mathsf{P}\right] \log \frac{\Pr\left[X = x, X \sim \mathsf{P}\right]}{\Pr\left[X = x, X \sim \mathsf{Q}\right]},$$

where $\Pr\left[X = x, X \sim \mathsf{P}\right]$ denotes the probability that the random variable X equals x when it follows the density function P. The mutual information can be defined in terms of Kullback-Leibler divergence, as being the divergence between the joint distribution $\Pr\left[X = x, Y = y\right]$ and the product distribution $\Pr\left[X = x\right] \cdot$

$\Pr[Y = y]$, or as the expected divergence between the conditional distribution $\Pr[Y = y | X = x]$ and $\Pr[Y = y]$. In other words:

$$I(X; Y) = D_{\mathsf{KL}}(\Pr[X = x, Y = y] \parallel \Pr[X = x] \cdot \Pr[Y = y])$$
$$= E_{x \in \mathcal{X}}(D_{\mathsf{KL}}(\Pr[Y = y | X = x] \parallel \Pr[Y = y]))$$

Hence, it can be seen as the expected value of the divergence between the leakage distributions taken conditionally to the models and the marginal distribution.

F-divergences. The f-divergence [9] is a function of two probability distributions P and Q that is used to measure the difference between them. It was introduced independently by Csiszàr [8] and Ali and Silvey [1] and is defined as:

$$I_f(\mathsf{P}, \mathsf{Q}) = \sum_{x \in \mathcal{X}} \Pr[X = x, X \sim \mathsf{Q}] \times f\left(\frac{\Pr[X = x, X \sim \mathsf{P}]}{\Pr[X = x, X \sim \mathsf{Q}]}\right),$$

where f is a parameter function. Some classical examples include:

- Kullback-Leibler divergence: $f(t) = t \log t$
- Inverse Kullback-Leibler: $f(t) = -\log t$
- Pearson χ^2–divergence: $f(t) = (t - 1)^2$
- Hellinger distance: $f(t) = 1 - \sqrt{t}$
- Total variation: $f(t) = |t - 1|$

As detailed in [12], the qualitative motivation for using the mutual information as a metric of dependence is sound. But one can wonder about its effectiveness. That is, all the previous f functions ensure an asymptotically successful attack. But are there significant differences in the convergence of the corresponding distinguishers? We note that the previous list is not exhaustive. For example, one could consider the Jensen-Shannon divergence that is a popular method based on the Kullback-Leibler divergence, with the useful difference that it is always a finite value: $D_{\mathsf{JS}}(P \parallel Q) = \frac{1}{2}(D_{\mathsf{KL}}(P \parallel M) + D_{\mathsf{KL}}(Q \parallel M))$, where $M = \frac{1}{2}(P + Q)$. Similarly, the earth mover's or Mallow distances [4,17] could also be used.

3.3   Distinguishing without Explicit pdf Estimation

Interestingly, an explicit pdf estimation is not always necessary and there also exist statistical tools to compare two pdfs directly from their samples. The Kolmogorov-Smirnov test is typical of such non parametric tools. For different samples x_i and a threshold x_t, it first defines an empirical cumulative function:

$$F(x_t) = \frac{1}{n} \sum_{i=1}^{n} \chi_{x_i \le x_t}, \text{ where } \chi_{x_i \le x_t} = \begin{cases} 1 \text{ if } x_i \le x_t \\ 0 \text{ otherwise.} \end{cases}$$

Then, the Kolmogorov-Smirnov distance is defined by:

$$D_{\mathsf{KS}}(P \parallel Q) = \sup_{x_t} |F_P(x_t) - F_Q(x_t)|.$$

This distance can then be used to test a null hypothesis. Since it is based on a supremum rather than a sum as the previous distances, it is better integrated to the following (MIA-inspired) distinguisher:

$$E_{x \in \mathcal{X}} \left(D_{\mathsf{KS}} \left(\Pr\left[Y = y | X = x\right] \| \Pr\left[Y = y\right] \right) \right)$$

This is further improved by normalizing each KS distance with the number of samples used in its computation, taking into account the convergence:

$$E_{x \in \mathcal{X}} \left(\frac{1}{|Y|X = x|} D_{\mathsf{KS}} \left(\Pr\left[Y = y | X = x\right] \| \Pr\left[Y = y\right] \right) \right),$$

where $|Y|X = x|$ is the number of leakages samples with modeled value x. Finally, an even more efficient alternative to the KS test is the two sample Cramér-von-Mises test [2], which is also based on the empirical cumulative function.

$$D_{\mathsf{CvM}} \left(P \| Q \right) = \int_{-\infty}^{+\infty} \left(F_P(x_t) - F_Q(x_t) \right)^2 dx_t.$$

3.4 All-or-Nothing Comparisons

Eventually, we mention that the MIA is defined as the expected value of a divergence between the leakage distributions conditionally to the model values and the marginal leakage distribution, *i.e.* $E_{x \in \mathcal{X}} \left(D_{\mathsf{KL}} \left(\Pr\left[Y = y | X = x\right] \| \Pr\left[Y = y\right] \right) \right)$. But divergences between the conditional distributions could be considered as well, as in "all-or-nothing" DPA attacks (see, *e.g.* [3] for an example).

3.5 How Much Does It Matter? Experimental Results

The previous sections illustrate that MIA is in fact a generic tool in which different statistics can be plugged in. A natural question is to evaluate the extend to which different pdf estimations and definitions of divergence affect the effectiveness of the distinguisher. For this purpose, we carried out attacks based on the traces that are publicly available in the DPA Contest [10] and computed the success rate defined in [22] in function of the number of traces available to the adversary (*i.e.* encrypted messages), over 1000 independent experiments, using a Hamming weight leakage model. The results of these experiments are in Figure 3 from which we can extract different observations: First, classical attacks using the correlation coefficient are the most effective in this simple context. Second, the pdf estimation tools have a stronger impact than the notion of divergence on the MIA-like attacks. In particular and as far as non-parametric pdf estimations are concerned, the Kernel-based MIA performs significantly better than its counterpart using histograms. Eventually, it is worth noting the good behavior of the normalized KS and Cramér-von-Mises tests for which pdf estimation is not required. They are interesting alternatives to the other tests because of their simple implementation which makes them comparable to plain histograms in terms of processing workload. The Cramér-von-Mises criterion seems to behave

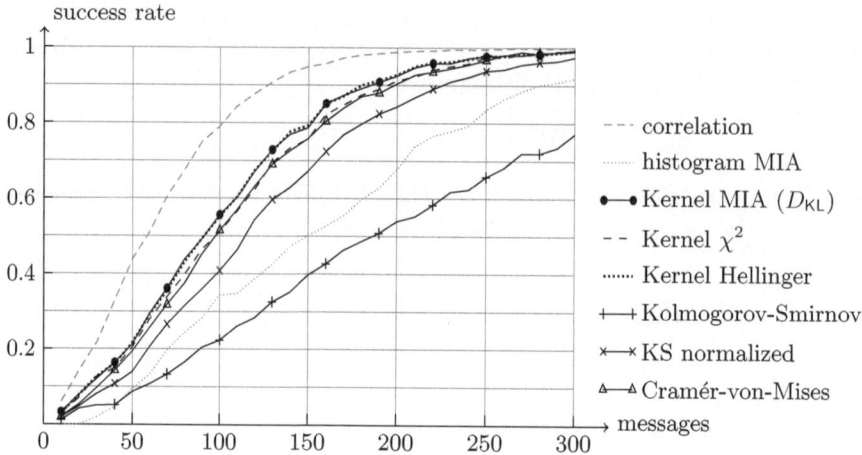

Fig. 3. Success rate of different attacks against the first DES S-box in the DPA Contest

as efficiently as the kernel-based methods, while avoiding the (hard) problem of choosing the kernel bandwidth. Hence, an intriguing open problem is to determine wether this test can be as efficient in more challenging contexts (e.g. implementations protected with masking or other countermeasures).

4   When To Use It: MIA versus Correlation

The experiments of Figure 3 suggest (as already emphasized by the authors in [12]) that when a reasonable leakage model is known by the adversary, standard DPA techniques such as using Pearson's correlation coefficient are more efficient than MIA. Hence, an obvious question is to determine the existence of contexts in which MIA would be necessary. With this respect, it is shown in [19] that higher-order attacks against masking schemes are good examples of such situations. This is essentially because MIA easily generalizes to multivariate statistics and hence does not need to worry about the combination of the leakages such as, *e.g.* [18]. In this section, we aim to show that MIA can even be useful in a univariate context, as soon as the adversary's leakage model is sufficiently imprecise.

Theoretically, this can be easily explained as follows. Let us assume that the leakages Y can be written as the sum of a deterministic part X_P (representing a perfect model) and a gaussian distributed random part R (representing some noise in the measurements): $Y = X_P + R$ and that a side-channel adversary exploits a leakage model $X_A = f(X_P)$. In ideal scenarios, we have $X_A = X_P$ but in practice, there generally exist deviations between the adversary's model and the perfect model, here represented by the function f. Correlation attacks are asymptotically successful as long as $\rho(X_A^g, Y) > \rho(X_A^w, Y)$, *i.e.* the correlation for the model corresponding to a correct subkey (with g superscript) is higher than the one for a wrong subkey candidate (with w superscript). If the adversary's model can again be written as $X_A = X_P + R'$ with R' another additive Gaussian

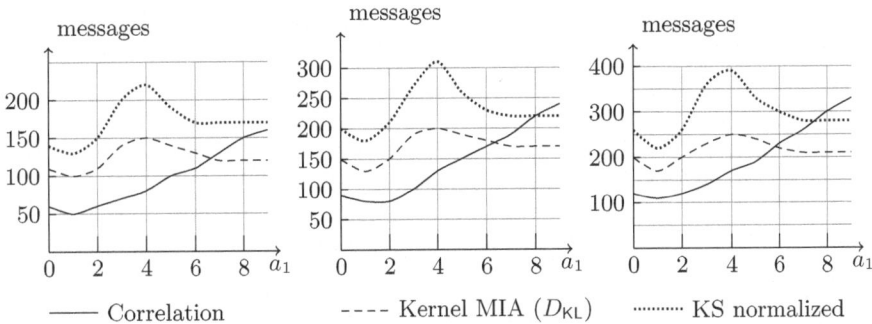

Fig. 4. Weight of the first leaking bit versus number of messages needed to reach a success rate of 50% (left), 75% (middle) and 90% (right), for different attacks

noise, then correlation attacks will obviously remain the best solution. But in general, imprecisions in the models can take any shape (not only additive). This may lead correlation attacks to fail where, *e.g.* MIA can still succeed.

As an illustration, an interesting case that is reasonably connected to practice is to assume a data bus in a micro-controller such that one bit (say the LSB) leaks significantly more than the others (*e.g.* because of a larger capacitance). Taking the example of Section 3, this time with the 8-bit AES S-box, we could imagine that the leakages equal: $y = \sum_{i=1}^{8} a_i \cdot [S(p \oplus k)]_i$. If the bit coefficients $a_i = 1$ for all i, we have Hamming weight leakages again. But by increasing a coefficient (*e.g.* a_1) and keeping the same Hamming weight model for the adversary, we can force this model to be arbitrarily wrong. Figure 4 illustrates the results of attacks that simulate this scenario. It shows that the number of messages required to reach a given success rate always increases with a_1 for the attacks using the correlation coefficient. By contrast, it stabilizes at some point for the MIA and KS test. Hence, for a sufficiently "wrong" leakage model, MIA-like attacks become useful. It is worth noting that the stabilization observed for the MIA and KS tests can be understood by looking at the pdf for a correct subkey candidate in Appendix B (again simplified to a 4-bit example): once a_1 is sufficiently heavy for the global pdf to be made of two disconnected pdf (one for $[S(p \oplus k)]_1 = 0$, one for $[S(p \oplus k)]_1 = 1$), the effectiveness of these distinguishers remains constant. Eventually, it is worth mentioning that while the MIA better resists to incorrect models than correlation attacks, it is not immune against them. One still requires that $I(X_A^g; Y) > I(X_A^w; Y)$. In other words, choosing random models will obviously not lead to successful attacks.

5   Why To Use It: MIA as an Evaluation Metric

Since the primary goal of the MIA is to distinguish subkeys, an adversary is not directly concerned with the value of $I(X_A^g; Y)$ but rather with the fact that it is higher than $I(X_A^w; Y)$. However, once a successful attack is performed, one can also wonder about the meaning of this value. In other words, can the mutual information $I(X_A^g; Y)$ additionally be used as an evaluation metric for side-channel attacks, as the information theoretic metric suggested in [22]?

In order to discuss this question, we can again take the simple example of the previous section in which the leakages are the sum of a perfect model and a Gaussian noise: $Y = X_P + R$. Say the target subkey in a side-channel attack is denoted by a variable K. The model in [22] suggests to evaluate a leaking implementation with $H[K|Y]$. Because of the additive noise, this can be written as: $H[K|Y] = H[K|X_P] + H[X_P|Y]$. Additionally assuming that $R = 0$, we find: $H[K|Y] = H[K|X_P]$. By contrast, the MIA does not directly apply to the subkey variable, but to subkey-dependent leakage models. That is, assuming that an adversary performs MIA with a perfect leakage model, it computes: $I(X_P; Y) = H[X_P] - H[X_P|Y]$ with $H[X_P|Y] = 0$ if $R = 0$. Using the relation:

$$I(K; X_P) = H[K] - H[K|X_P],$$

we have that if an adversary performs the MIA with a perfect leakage model and no noise (and a perfect pdf estimation tool), the following equation holds:

or similarly:
$$H[K|Y] = H[K|X_P] = H[K] - I(X_P; Y),$$
$$I(K; Y) = I(X_P; Y).$$

It implies that MIA and the metric of [22] can be used equivalently in this case. Adding additive noise R to the leakages will not change the situation since it will simply add a term $H[X_P|Y]$ to the previous equations. But as in Section 4, this equality does not hold anymore if the adversary's model is not perfect and the imperfections are not simply additive, i.e. if we have $Y = f(X_P) \neq X_P + R$. Then, the previous equality will turn into an inequality:

or similarly:
$$H[K|Y] \leq H[K] - I(X_P; Y),$$
$$I(K; Y) \geq I(X_P; Y).$$

That is, the mutual information computed by the MIA with an incorrect leakage model will tend to underestimate the amount of information leaked by the chip. In other words, MIA is a generic distinguisher while the conditional entropy $H[K|Y]$ is a generic evaluation metric for side-channel attacks. The reason of this genericity comes from the information theoretic nature of these tools. In practice, MIA can be used to approach a fair evaluation metric if a perfect leakage model is available to the adversary but it deviates from this metric as soon as this conditions is not respected anymore[1]. This deviation essentially comes from the need to use an intermediate variable (corresponding to an intermediate value in the target algorithm, e.g. an S-box output) in non profiled side-channel attacks rather than considering the subkey leakages directly. That is, MIA computes $I(X_P; Y)$ rather than $H[K|Y]$. Summarizing, the MIA and the model of [22] have different objectives, namely recovering keys for MIA and allowing fair evaluations of leaking devices for the model. They also generally exploit different adversarial contexts, namely non-profiled attacks for the MIA and profiled attacks for the model. But eventually, the reason for using these tools is similar since they both allow capturing any kind of dependencies in the physical leakages and consequently lead to generic attacks and evaluation of the attacks and leakages.

---

[1] When moving to multivariate statistics, perfect models should be considered for each sample which yields the open question of how to efficiently exploit multiple models.

References

1. Ali, S.M., Silvey, S.D.: A general class of coefficients of divergence of one distribution from another. Journal of the Royal Statistical Society, Series B (Methodological) 28(1), 131–142 (1966)
2. Anderson, T.W.: On the distribution of the two-sample cramér-von mises criterion. The Annals of Mathematical Statistics 33(3), 1148–1159 (1962)
3. Aumoniér, S.: Generalized correlation power analysis. In: Ecrypt Workshop on Tools For Cryptanalysis. Kra kòw, Poland (September 2007)
4. Bickel, P., Levina, E.: The earth's mover's distance is the mallows distance: some insights from statistics. In: Computer Vision 2001, vol. 2, pp. 251–256 (2001)
5. Brier, E., Clavier, C., Olivier, F.: Correlation power analysis with a leakage model. In: Joye, M., Quisquater, J.-J. (eds.) CHES 2004. LNCS, vol. 3156, pp. 16–29. Springer, Heidelberg (2004)
6. Chari, S., Rao, J., Rohatgi, P.: Template attacks. In: Kaliski Jr., B.S., Koç, Ç.K., Paar, C. (eds.) CHES 2002. LNCS, vol. 2523, pp. 13–28. Springer, Heidelberg (2003)
7. Cover, T.M., Thomas, J.A.: Elements of Information Theory. Wiley, Chichester (1991)
8. Csiszár, I.: Information-type measures of difference of probability distributions and indirect observation. Studia Sci. Math. Hungar. 2, 229–318 (1967)
9. Csiszár, I., Shields, P.C.: Information theory and statistics: a tutorial. Commun. Inf. Theory 1(4), 417–528 (2004)
10. DPA Contest 2008/2009, http://www.dpacontest.org/
11. Freedman, D., Diaconis, P.: On the histogram as a density estimator. Probability Theory and Related Fields 57(4), 453–476 (1981)
12. Gierlichs, B., Batina, L., Tuyls, P., Preneel, B.: Mutual information analysis. In: Oswald, E., Rohatgi, P. (eds.) CHES 2008. LNCS, vol. 5154, pp. 426–442. Springer, Heidelberg (2008)
13. Härdle, W.: Smoothing Techniques: With Implementation in S. Springer Series in Statistics (December 1990)
14. Kocher, P., Jaffe, J., Jun, B.: Differential power analysis. In: Wiener, M. (ed.) CRYPTO 1999. LNCS, vol. 1666, pp. 388–412. Springer, Heidelberg (1999)
15. Lemke, K., Paar, C.: Gaussian mixture models for higher-order side channel analysis. In: Nejdl, W., Tochtermann, K. (eds.) EC-TEL 2006. LNCS, vol. 4227, pp. 14–27. Springer, Heidelberg (2006)
16. Laird, N., Dempster, A., Rubin, D.: Maximum likelihood from incomplete data via the em algorithm. Journal of the Royal Statistical Society, Series B (Methodological) 39(1), 1–38 (1977)
17. Mallows, C.L.: A note on asymptotic joint normality. The Annals of Mathematical Statistics 43(2), 508–515 (1972)
18. Messerges, T.S.: Using second-order power analysis to attack DPA resistant software. In: Paar, C., Koç, Ç.K. (eds.) CHES 2000. LNCS, vol. 1965, pp. 238–251. Springer, Heidelberg (2000)
19. Prouff, E., Rivain, M.: Theoretical and practical aspects of mutual information based side channel analysis. In: Abdalla, M., Pointcheval, D., Fouque, P.-A., Vergnaud, D. (eds.) ACNS 2009. LNCS, vol. 5536, pp. 499–518. Springer, Heidelberg (2009)
20. Scott, D.W.: On optimal and data-based histograms. Biometrika 66(3), 605–610 (1979)
21. Silverman, B.W.: Density Estimation for Statistics and Data Analysis. Chapman & Hall/CRC, Boca Raton (1986)

22. Standaert, F.-X., Malkin, T.G., Yung, M.: A unified framework for the analysis of side-channel key recovery attacks (extended version). Cryptology ePrint Archive, Report 2006/139 (2006), http://eprint.iacr.org/
23. Turlach, B.A.: Bandwidth selection in kernel density estimation: a review. In: CORE and Institut de Statistique (1993)
24. Zhang, M.H., Cheng, Q.S.: Determine the number of components in a mixture model by the extended ks test. Pattern Recogn. Lett. 25(2), 211–216 (2004)

A Density Estimation Techniques

A.1 Histograms

For n bins noted b_i, the probability is estimated as:

$$\hat{\Pr}[\underline{b_i} \leq x \leq \overline{b_i}] = \frac{\#b_i}{q}, \text{ where } q = \sum_{0 \leq j \leq n} \#b_j$$

The optimal choice for the *bin width* h is an issue in Statistical Theory, as different bin sizes can greatly modify the resulting model. For relatively simple distributions, which is usually the case of side-channel leakages, reasonable choices are Scott's rule [20] ($h = 3.49 \times \hat{\sigma}(x) \times n^{-1/3}$) and Freedman-Diaconis' rule [11] ($d = 2 \times \mathrm{IQR}(x) \times n^{-1/3}$, $\mathrm{IQR} = $ interquartile range). While histograms are quite easy to implement, they generally provide a very slow convergence towards the target pdf, lack smoothness and heavily depend on bin width.

A.2 Kernel Density Estimation

The probability is estimated as:

$$\hat{\Pr}[X = x] = \frac{1}{nh} \sum_i K\left(\frac{x - x_i}{h}\right),$$

where the kernel function K is a real-valued integrable function satisfying $\int_{-\infty}^{\infty} K$ $(u)\, du = 1$ and $K(u) = -K(u)$ for all u. Some kernel functions are in Table 1. Similarly to histograms, the most important parameter is the *bandwidth* h. Its optimal value is the one minimizing the AMISE (Asymptotic Mean Integrated Squared Error), which itself usually depends on the true density. A number of approximation methods have been developed, see [23] for an extensive review. In our case , we used the modified estimator [21,13]:

$$h = 1.06 \times \min\left(\hat{\sigma}(x), \frac{\mathrm{IQR}(x)}{1.34}\right) n^{-\frac{1}{5}}$$

Table 1. Some kernel functions. i is defined as: $i(u) = 1$ if $|u| \leq 1$, 0 otherwise.

Kernel	$K(u)$	Kernel	$K(u)$
Uniform	$\frac{1}{2}i(u)$	Triweight	$\frac{35}{32}(1-u^2)^3 i(u)$
Triangle	$(1-|u|)i(u)$	Tricube	$\frac{70}{81}(1-|u|^3)^3 i(u)$
Epanechnikov	$\frac{3}{4}(1-u^2)i(u)$	Gaussian	$\frac{1}{\sqrt{2\pi}}exp\left(-\frac{1}{2}u^2\right)$
Quartic	$\frac{15}{16}(1-u^2)^2 i(u)$	Cosinus	$\frac{\pi}{4}\cos\left(\frac{\pi}{2}u\right)i(u)$

A.3 Gaussian Mixtures

This parametric method models the pdf as:

$$\hat{\Pr}(X = x) = \sum_{i=0}^{n-1} w_i \, \mathcal{N}(x, \mu_i, \sigma_i),$$

where the μ_i and σ_i are the respective means and deviations of each mixture component. This method can be thought of as a generalization of the kernel density estimation with gaussian kernels, where one is not restricted to $w_i = \frac{1}{nh}$ or $\sigma_i = \frac{1}{h}$. The main advantage of the finite mixture method is that it usually leads to a number of mixture elements significantly smaller than the number of samples used to form the model in a kernel density estimation. An efficient algorithm called the *Expectation Maximization* (EM) algorithm [16] allows one to give a good approximation of a pdf in the form of a finite mixture. Given the number of components in the mixture, it computes their weights and gaussian parameters. Some additional procedures have been proposed that help choosing the number of components to be used in a mixture, for example in [24].

B Effect of Incorrect Leakage Models

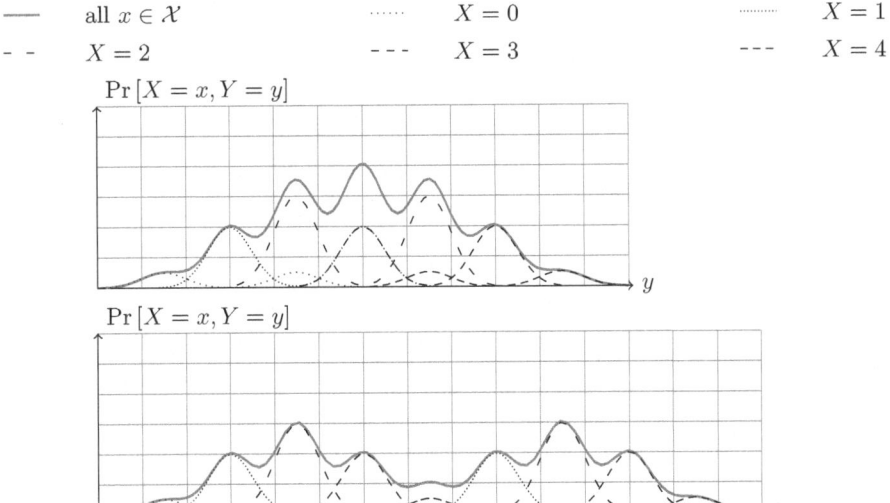

Fig. 5. Behavior of the probability densities for the correct subkey in a 4-bit DPA, assuming a Hamming weight leakage model and $a_1 = 3$ (up) and $a_1 = 5$ (down)

Fault Attacks on RSA Signatures with Partially Unknown Messages

Jean-Sébastien Coron[1], Antoine Joux[2], Ilya Kizhvatov[1], David Naccache[3], and Pascal Paillier[4]

[1] Université du Luxembourg
6, rue Richard Coudenhove-Kalergi, L-1359 Luxembourg, Luxembourg
{jean-sebastien.coron,ilya.kizhvatov}@uni.lu
[2] DGA and Université de Versailles
UVSQ PRISM 45 avenue des États-Unis, F-78035, Versailles CEDEX, France
antoine.joux@m4x.org
[3] École normale supérieure
Département d'informatique, Groupe de Cryptographie
45, rue d'Ulm, F-75230 Paris CEDEX 05, France
david.naccache@ens.fr
[4] Gemalto, Cryptography & Innovation
6, rue de la Verrerie, F-92447 Meudon sur Seine, France
pascal.paillier@gemalto.com

Abstract. Fault attacks exploit hardware malfunctions to recover secrets from embedded electronic devices. In the late 90's, Boneh, DeMillo and Lipton [6] introduced fault-based attacks on CRT-RSA. These attacks factor the signer's modulus when the message padding function is deterministic. However, the attack does not apply when the message is partially unknown, for example when it contains some randomness which is recovered only when verifying a *correct* signature.

In this paper we successfully extends RSA fault attacks to a large class of partially known message configurations. The new attacks rely on Coppersmith's algorithm for finding small roots of multivariate polynomial equations. We illustrate the approach by successfully attacking several randomized versions of the ISO/IEC 9796-2 encoding standard. Practical experiments show that a 2048-bit modulus can be factored in less than a minute given one faulty signature containing 160 random bits and an unknown 160-bit message digest.

Keywords: Fault attacks, digital signatures, RSA, Coppersmith's theorem, ISO/IEC 9796-2.

1 Introduction

1.1 Background

RSA [21] is undoubtedly the most common digital signature scheme used in embedded security tokens. To sign a message m with RSA, the signer applies an

C. Clavier and K. Gaj (Eds.): CHES 2009, LNCS 5747, pp. 444–456, 2009.

encoding (padding) function μ to m, and then computes the signature $\sigma = \mu(m)^d \bmod N$. To verify the signature, the receiver checks that $\sigma^e = \mu(m) \bmod N$. As shown by Boneh, DeMillo and Lipton [6] and others (e.g. [18]), RSA implementations can be vulnerable to fault attacks, especially when the Chinese Remainder Theorem (CRT) is used; in this case the device computes $\sigma_p = \mu(m)^d \bmod p$ and $\sigma_q = \mu(m)^d \bmod q$ and the signature σ is computed from σ_p and σ_q by Chinese Remaindering.

Assuming that the attacker is able to induce a fault when σ_q is computed while keeping the computation of σ_p correct, one gets $\sigma_p = \mu(m)^d \bmod p$ and $\sigma_q \neq \mu(m)^d \bmod q$ and the resulting (faulty) signature σ satisfies

$$\sigma^e = \mu(m) \quad \bmod p , \qquad \sigma^e \neq \mu(m) \quad \bmod q .$$

Therefore, given one faulty σ, the attacker can factor N by computing

$$\gcd(\sigma^e - \mu(m) \bmod N, N) = p . \qquad (1)$$

Boneh *et al.*'s fault attack is easily extended to any deterministic RSA encoding, *e.g.* the Full Domain Hash (FDH) [5] encoding where $\sigma = H(m)^d \bmod N$ and $H : \{0,1\}^* \mapsto \mathbb{Z}_N$ is a hash function. The attack is also applicable to probabilistic signature schemes where the randomizer used to generate the signature is sent along with the signature, *e.g.* as in the Probabilistic Full Domain Hash (PFDH) encoding [11] where the signature is $\sigma \| r$ with $\sigma = H(m \| r)^d \bmod N$. In that case, given the faulty value of σ and knowing r, the attacker can still factor N by computing $\gcd(\sigma^e - H(m \| r) \bmod N, N) = p$.

1.2   Partially-Known Messages: The Fault-Attacker's Deadlock

However, if the message is not entirely given to the attacker the attack is thwarted, *e.g.* this may occur when the signature has the form $\sigma = (m\|r)^d \bmod N$ where r is a random nonce. Here the verifier can recover r only after completing the verification process; however r can only be recovered when verifying a *correct* signature. Given a faulty signature, the attacker cannot retrieve r nor infer $(m\|r)$ which would be necessary to compute $\gcd(\sigma^e - (m\|r) \bmod N, N) = p$.

In other words, the attacker faces an apparent deadlock: recovering the r used in the faulty signature σ seems to require that σ is a correctly verifiable signature. Yet, obviously, a correct signature does not factor N. These conflicting constraints cannot be conciliated unless r is short enough to be guessed by exhaustive search. Inducing faults in many signatures does not help either since different r values are used in successive signatures (even if m remains invariant). As a result, randomized RSA encoding schemes are usually considered to be inherently immune against fault attacks.

1.3   The New Result

We overcome the deadlock by showing how to extract *in some cases* the unknown message part (UMP) involved in the generation of faulty RSA signatures. We develop several techniques that extend Boneh *et al.*'s attack to a large class of partially known message configurations. We nonetheless assume that certain conditions on the unknown parts of the encoded message are met; these conditions may depend on the encoding function itself and on the hash functions used. To illustrate our attacks, we have chosen to consider the ISO/IEC 9796-2 standard [16]. ISO/IEC 9796-2 is originally a deterministic encoding scheme often used in combination with message randomization (e.g. in EMV [13]). The encoded message has the form:

$$\mu(m) = \mathtt{6A_{16}} \parallel m[1] \parallel H(m) \parallel \mathtt{BC_{16}}$$

where $m = m[1] \parallel m[2]$ is split into two parts. We show that if the unknown part of $m[1]$ is not too large (e.g. less than 160 bits for a 2048-bit RSA modulus), then a single faulty signature allows to factor N as in [6][1]. The new method is based on a result by Herrmann and May [12] for finding small roots of linear equations modulo an unknown factor p of N; [12] is itself based on Coppersmith's technique [7] for finding small roots of polynomial equations using the LLL algorithm [19]. We also show how to extend our attack to multiple UMPs and to *scenarii* where more faulty signatures can be obtained from the device.

1.4   The ISO/IEC 9796-2 Standard

ISO/IEC 9796-2 is an encoding standard allowing partial or total message recovery [16,17]. The encoding can be used with hash functions $H(m)$ of diverse digest sizes k_h. For the sake of simplicity we assume that k_h, the size of m and the size of N (denoted k) are all multiples of 8. The ISO/IEC 9796-2 encoding of a message $m = m[1] \parallel m[2]$ is

$$\mu(m) = \mathtt{6A_{16}} \parallel m[1] \parallel H(m) \parallel \mathtt{BC_{16}}$$

where $m[1]$ consists of the $k - k_h - 16$ leftmost bits of m and $m[2]$ represents the remaining bits of m. Therefore the size of $\mu(m)$ is always $k - 1$ bits. Note that the original version of the standard recommended $128 \le k_h \le 160$ for partial message recovery (see [16], §5, note 4). In [9], Coron, Naccache and Stern introduced an attack against ISO/IEC 9796-2; the authors estimated that attacking $k_h = 128$ and $k_h = 160$ would require respectively 2^{54} and 2^{61} operations. After Coron et al.'s publication, ISO/IEC 9796-2 was amended and the current official requirement (see [17]) is now $k_h \ge 160$. In a recent work Coron, Naccache, Tibouchi and Weinmann successfully attack the currently valid version of ISO/IEC 9796-2 [10].

To illustrate our purpose, we consider a message $m = m[1] \parallel m[2]$ of the form

$$m[1] = \alpha \parallel r \parallel \alpha', \qquad m[2] = \mathrm{DATA}$$

where r is a message part unknown to the adversary, α and α' are strings known to the adversary and DATA is some known or unknown string[2]. The size of r is

---

[1] In our attack, it does not matter how large the unknown part of $m[2]$ is.

[2] The attack will work equally well in both cases.

denoted k_r and the size of $m[1]$ is $k - k_h - 16$ as required in ISO/IEC 9796-2. The encoded message is then

$$\mu(m) = 6A_{16} \| \alpha \| r \| \alpha' \| H(\alpha \| r \| \alpha' \| \text{DATA}) \| BC_{16} \tag{2}$$

Therefore the total number of unknown bits in $\mu(m)$ is $k_r + k_h$.

2   Fault Attack on Partially-Known Message ISO/IEC 9796-2

This section extends [6] to signatures of partially known messages encoded as described previously. We assume that after injecting a fault the opponent is in possession of a faulty signature σ such that:

$$\sigma^e = \mu(m) \mod p, \qquad \sigma^e \neq \mu(m) \mod q. \tag{3}$$

From (2) we can write

$$\mu(m) = t + r \cdot 2^{n_r} + H(m) \cdot 2^8 \tag{4}$$

where t is a known value. Note that both r and $H(m)$ are unknown to the adversary. From (3) we obtain:

$$\sigma^e = t + r \cdot 2^{n_r} + H(m) \cdot 2^8 \mod p.$$

This shows that $(r, H(m))$ must be a solution of the equation

$$a + b \cdot x + c \cdot y = 0 \mod p \tag{5}$$

where $a := t - \sigma^e \mod N$, $b := 2^{n_r}$ and $c := 2^8$ are known. Therefore we are left with solving equation (5) which is linear in the two variables x, y and admits a small root $(x_0, y_0) = (r, H(m))$. However the equation holds modulo an unknown divisor p of N and not modulo N. Such equations were already exploited by Herrmann and May [12] to factor an RSA modulus $N = pq$ when some blocks of p are known. Their method is based on Coppersmith's technique for finding small roots of polynomial equations [7]. Coppersmith's technique uses LLL to obtain two polynomials $h_1(x, y)$ and $h_2(x, y)$ such that

$$h_1(x_0, y_0) = h_2(x_0, y_0) = 0$$

holds over the integers. Then one computes the resultant between h_1 and h_2 to recover the common root (x_0, y_0). To that end, we must assume that h_1 and h_2 are algebraically independent. This ad hoc assumption makes the method heuristic; nonetheless it turns out to work quite well in practice. Then, given the root (x_0, y_0) one recovers the randomized encoded message $\mu(m)$ and factors N by GCD.

Theorem 1 (Herrmann-May [12]). *Let N be a sufficiently large composite integer with a divisor $p \geq N^{\beta}$. Let $f(x,y) = a + b \cdot x + c \cdot y \in \mathbb{Z}[x,y]$ be a bivariate linear polynomial. Assume that $f(x_0, y_0) = 0 \bmod p$ for some (x_0, y_0) such that $|x_0| \leq N^{\gamma}$ and $|y_0| \leq N^{\delta}$. Then for any $\varepsilon > 0$, under the condition*

$$\gamma + \delta \leq 3\beta - 2 + 2(1 - \beta)^{3/2} - \varepsilon \tag{6}$$

one can find $h_1(x,y), h_2(x,y) \in \mathbb{Z}[x,y]$ such that $h_1(x_0, y_0) = h_2(x_0, y_0) = 0$ over $\mathbb{Z}$, in time polynomial in $\log N$ and ε^{-1}.

We only sketch the proof and refer the reader to [12] and [8] for more details. Assume that $b = 1$ in the polynomial f (otherwise multiply f by $b^{-1} \bmod N$) and consider the polynomial

$$f(x,y) = a + x + c \cdot y$$

We look for (x_0, y_0) such that $f(x_0, y_0) = 0 \bmod p$. The basic idea consists in generating a family $\mathcal{G}$ of polynomials admitting (x_0, y_0) as a root modulo p^t for some large enough integer t. Any linear combination of these polynomials will also be a polynomial admitting (x_0, y_0) as a root modulo p^t. We will use LLL to find such polynomials with small coefficients. To do so, we view any polynomial $h(x,y) = \sum h_{i,j} x^i y^j$ as the vector of coefficients $\left(h_{i,j} X^i Y^j\right)_{i,j}$ and denote by $\|h(xX, yY)\|$ this vector's Euclidean norm. Performing linear combinations on polynomials is equivalent to performing linear operations on their vectorial representation, so that applying LLL to the lattice spanned by the vectors in $\mathcal{G}$ will provide short vectors representing polynomials with root $(x_0, y_0) \bmod p^t$.

We now define the family $\mathcal{G}$ of polynomials as

$$g_{k,i}(x,y) := y^i \cdot f^k(x,y) \cdot N^{\max(t-k,0)}$$

for $0 \leq k \leq m$, $0 \leq i \leq m - k$ and integer parameters t and m. For all values of indices k, i, it holds that $g_{k,i}(x_0, y_0) = 0 \bmod p^t$. We first sort the polynomials $g_{k,i}$ by increasing k values and then by increasing i values. Denoting $X = N^{\gamma}$ and $Y = N^{\delta}$, we write the coefficients of the polynomial $g_{k,i}(xX, yY)$ in the basis $x^{k'} \cdot y^{i'}$ for $0 \leq k' \leq m$ and $0 \leq i' \leq m - k'$. Let L be the corresponding lattice; L's dimension is

$$\omega = \dim(L) = \frac{m^2 + 3m + 2}{2} = \frac{(m+1)(m+2)}{2}$$

and we have

$$\det L = X^{s_x} Y^{s_y} N^{s_N}$$

where

$$s_x = s_y = \sum_{k=0}^{m} \sum_{i=0}^{m-k} i = \frac{m(m+1)(m+2)}{6}$$

and

$$s_N = \sum_{i=0}^{t} (m+1-i) \cdot (t-i) \,.$$

We now apply LLL to the lattice L to find two polynomials $h_1(x, y)$ and $h_2(x, y)$ with small coefficients.

Theorem 2 (LLL [19]). *Let L be a lattice spanned by $(u_1, \dots, u_\omega)$. Given the vectors $(u_1, \dots, u_\omega)$, the LLL algorithm finds in polynomial time two linearly independent vectors b_1, b_2 such that*

$$\|b_1\|, \|b_2\| \leq 2^{\omega/4} (\det L)^{1/(\omega-1)} .$$

Therefore using LLL we can get two polynomials $h_1(x, y)$ and $h_2(x, y)$ such that

$$\|h_1(xX, yY)\|, \|h_2(xX, yY)\| \leq 2^{\omega/4} \cdot (\det L)^{1/(\omega-1)} . \tag{7}$$

Using Howgrave-Graham's lemma (below), we can determine the required bound on the norms of h_1 and h_2 to ensure that (x_0, y_0) is a root of both h_1 and h_2 over the integers:

Lemma 1 (Howgrave-Graham [14]). *Assume that $h(x, y) \in \mathbb{Z}[x, y]$ is a sum of at most ω monomials and assume further that $h(x_0, y_0) = 0 \mod B$ where $|x_0| \leq X$ and $|y_0| \leq Y$ and $\|h(xX, yY)\| < B/\sqrt{\omega}$. Then $h(x_0, y_0) = 0$ holds over the integers.*

Proof. We have

$$|h(x_0, y_0)| = \left| \sum h_{ij} x_0^i y_0^i \right| = \left| \sum h_{ij} X^i Y^j \left(\frac{x_0}{X} \right)^i \left(\frac{y_0}{Y} \right)^j \right|$$

$$\leq \sum \left| h_{ij} X^i Y^j \left(\frac{x_0}{X} \right)^i \left(\frac{y_0}{Y} \right)^j \right| \leq \sum |h_{ij} X^i Y^j|$$

$$\leq \sqrt{\omega} \|h(xX, yY)\| < B$$

Since $h(x_0, y_0) = 0 \mod B$, this implies that $h(x_0, y_0) = 0$ over the integers. □

We apply Lemma 1 with $B := p^t$. Using (7) this gives the condition:

$$2^{\omega/4} \cdot (\det L)^{1/(\omega-1)} \leq \frac{N^{\beta t}}{\sqrt{\omega}} . \tag{8}$$

[12] shows that by letting $t = \tau \cdot m$ with $\tau = 1 - \sqrt{1 - \beta}$, we get the condition:

$$\gamma + \delta \leq 3\beta - 2 + 2(1 - \beta)^{3/2} - \frac{3\beta(1 + \sqrt{1 - \beta})}{m}$$

Therefore we obtain as in [12] the following condition for m:

$$m \geq \frac{3\beta(1 + \sqrt{1 - \beta})}{\varepsilon} .$$

Since LLL runs in time polynomial in the lattice's dimension and coefficients, the running time is polynomial in $\log N$ and $1/\varepsilon$.

2.1   Discussion

For a balanced RSA modulus ($\beta = 1/2$) we get the condition:

$$\gamma + \delta \leq \frac{\sqrt{2} - 1}{2} \cong 0.207 \tag{9}$$

This means that for a 1024-bit RSA modulus N, the total size of the unknowns x_0 and y_0 can be at most 212 bits. Applied to our context, this implies that for ISO/IEC 9796-2 with $k_h = 160$, the size of the UMP r can be as large as 52 bits. Section 3 reports practical experiments confirming this prediction. In [8] we provide a Python code for computing the bound on the size of the unknown values ($k_r + k_h$) as a function of the modulus size.

2.2   Extension to Several Unknown Bits Blocks

Assume that the UMP used in ISO/IEC 9796-2 is split into n different blocks, namely

$$\mu(m) = 6\mathsf{A}_{16} \,\|\, \alpha_1 \,\|\, r_1 \,\|\, \alpha_2 \,\|\, r_2 \,\|\, \cdots \,\|\, \alpha_n \,\|\, r_n \,\|\, \alpha_{n+1} \,\|\, H(m) \,\|\, \mathsf{BC}_{16} \tag{10}$$

where the UMPs $r_1, \ldots, r_n$ are all part of the message m. The α_i blocks are known. In [8], we show how to recover p from one faulty signature, using the extended result of Herrmann and May [12]. It appears that if the total number of unknown bits plus the message digest is less than 15.3% of the size of N, then the UMPs can be fully recovered from the faulty signature and Boneh *et al.*'s attack will apply again. However the number of blocks cannot be too large because the attack's runtime increases exponentially with n.

2.3   Extension to Two Faults Modulo Different Factors

Assume that we can get two faulty signatures, one incorrect modulo p and the other incorrect modulo q. This gives the two equations

$$a_0 + b_0 \cdot x_0 + c_0 \cdot y_0 = 0 \quad \mathrm{mod}\ p$$
$$a_1 + b_1 \cdot x_1 + c_1 \cdot y_1 = 0 \quad \mathrm{mod}\ q$$

with small unknowns x_0, y_0, x_1, y_1. We show in [8] that by multiplying the two equations, we get a quadri-variate equation modulo N which can be solved by linearization under the following bound:

$$\gamma + \delta \leq \frac{1}{6} \cong 0.167 \,.$$

This remains weaker than condition (9). However the attack is significantly faster because it works over a lattice of constant dimension 9. Moreover, the 16.7% bound is likely to lend itself to further improvements using Coppersmith's technique instead of plain linearization.

2.4   Extension to Several Faults Modulo the Same Factor

To exploit single faults, we have shown how to use lattice-based techniques to recover p given N and a bivariate linear equation $f(x, y)$ admitting a small root (x_0, y_0) modulo p. In this context, we have used Theorem 1 which is based on approximate GCD techniques from [15]. In the present section we would like to generalize this to use ℓ different polynomials of the same form, each having a small root modulo p. More precisely, let ℓ be a fixed parameter and assume that as the result of ℓ successive faults, we are given ℓ different polynomials

$$f_u(x_u, y_u) = a_u + x_u + c_u y_u \tag{11}$$

where each polynomial f_u has a small root (ξ_u, ν_u) modulo p with $|\xi_u| \leq X$ and $|\nu_u| \leq Y$. Note that, as in the basic case, we re-normalized each polynomial f_u to ensure that the coefficient of x_u in f_u is equal to one. To avoid double subscripts, we hereafter use the Greek letters ξ and ν to represent the root values. We would like to use a lattice approach to construct new multivariate polynomials in the variables $(x_1, \cdots, x_\ell, y_1, \cdots, y_\ell)$ with the root $R = (\xi_1, \cdots, \xi_\ell, \nu_1, \cdots, \nu_\ell)$. To that end we fix two parameters m and t and build a lattice on a family of polynomials $\mathcal{G}$ of degree at most m with root R modulo $B = p^t$. This family is composed of all polynomials of the form

$$y_1^{i_1} y_2^{i_2} \cdots y_\ell^{i_\ell} f_1(x_1, y_1)^{j_1} f_2(x_2, y_2)^{j_2} \cdots f_\ell(x_\ell, y_\ell)^{j_\ell} N^{\max(t-j,0)} ,$$

where each i_u, j_u is non-negative, $i = \sum_{u=1}^\ell i_u$, $j = \sum_{u=1}^\ell j_u$ and $0 \leq i + j \leq m$. Once again, let L be the corresponding lattice. Its dimension ω is equal to the number of monomials of degree at most m in 2ℓ unknowns, i.e.

$$\omega = \binom{m + 2\ell}{2\ell} .$$

Since we have a common upper bound X for all values $|\xi_u|$ and a common bound for all $|\nu_u|$ we can compute the lattice's determinant as

$$\det(L) = X^{s_x} Y^{s_y} N^{s_N} ,$$

where s_x is the sum of the exponents of all unknowns x_u in all occurring monomials, s_y is the sum of the exponents of the y_u and s_N is the sum of the exponents of N in all occurring polynomials. For obvious symmetry reasons, we have $s_x = s_y$ and noting that the number of polynomials of degree exactly d in ℓ unknowns is $\binom{d+\ell-1}{\ell-1}$ we find

$$s_x = s_y = \sum_{d=0}^m d \binom{d+\ell-1}{\ell-1} \binom{m-d+\ell}{\ell} .$$

Likewise, summing on polynomials with a non-zero exponent v for N, where the sum of the j_u is $t - v$ we obtain

$$s_N = \sum_{v=1}^t v \binom{t-v+\ell-1}{\ell-1} \binom{m-t+v+\ell}{\ell} .$$

As usual, assuming that $p = N^\beta$ we can find a polynomial with the correct root over the integers under the condition of formula (8).

Concrete Bounds: Using the notation of Theorem 1, we compute effective bounds on $\gamma + \delta = \log(XY)/\log(N)$ from the logarithm of condition (8), dropping the terms $\sqrt{\omega}$ and $2^{\omega/4}$ which become negligible as N grows. For concrete values of N, bounds are slightly smaller. Dividing by $\log(N)$, we find

$$s_x \cdot (\gamma + \delta) + s_N \leq \beta t \omega .$$

Thus, given k, t and m, we can achieve at best

$$\gamma + \delta \leq \frac{\beta t \omega - s_N}{s_x}.$$

In [8], we provide the achievable values of $\gamma + \delta$ for $\beta = 1/2$, for various parameters and for lattice dimensions $10 \leq \omega \leq 1001$.

Recovering the Root: With 2ℓ unknowns instead of two, applying usual heuristics and hoping that lattice reduction directly outputs 2ℓ algebraically independent polynomials with the prescribed root over the integers becomes a wishful hope. Luckily, a milder heuristic assumption suffices to make the attack work. The idea is to start with K equations instead of ℓ and iterate the lattice reduction attack for several subsets of ℓ equations chosen amongst the K available equations. Potentially, we can perform $\binom{K}{\ell}$ such lattice reductions. Clearly, since each equation involves a different subset of unknowns, they are all different. Note that this does not suffice to guarantee algebraic independence; in particular, if we generate more than K equations they cannot be algebraically independent. However, we only need to ascertain that the root R can be extracted from the available set of equations. This can be done, using Gröbner basis techniques, under the heuristic assumption that the set of equations spans a multivariate ideal of dimension zero *i.e.* that the number of solutions is finite.

Note that we need to choose reasonably small values of ℓ and K to be able to use this approach in practice. Indeed, the lattice that we consider should not become too large and, in addition, it should be possible to solve the resulting system of equations using either resultants or Buchberger's algorithm which means that neither the degree nor the number of unknowns should increase too much.

Asymptotic Bounds: Despite the fact that we cannot hope to run the multi-polynomial variant of our attack when parameters become too large, it is interesting to determine the theoretical limit of the achievable value of $\gamma + \delta$ as the number of faults ℓ increases. To that end, we assume as previously that $\beta = 1/2$, let $t = \tau m$ and replace ω, s_x and s_N by the following approximations:

$$\omega \cong \frac{m^{2\ell}}{(2\ell)!}, \quad s_x \cong \sum_{d=0}^{m} \frac{d^\ell (m-d)^\ell}{(\ell-1)! \, \ell!}, \quad s_N \cong \sum_{v=1}^{t} v \frac{(t-v)^{\ell-1}(m-t+v)^\ell}{(\ell-1)! \, \ell!} .$$

Table 1. Bound for the relative size $\gamma + \delta$ of the unknowns as a function of the number of faults ℓ

ℓ	1	2	3	4	5	6	7	8	9	10
$\gamma + \delta$	0.207	0.293	0.332	0.356	0.371	0.383	0.391	0.399	0.405	0.410

For small ℓ values we provide in Table 1 the corresponding bounds on $\gamma + \delta$. Although we do not provide further details here due to lack of space, one can show that the bound $\gamma + \delta$ tends to $1/2$ as the number of faults ℓ tends to infinity and that all $\gamma + \delta$ values are algebraic numbers.

3   Simulation Results

Assuming that fault injection can be performed on unprotected devices (see Section 4), we simulated the attack. In the experiment we generated faulty signatures (using the factors p and q) and applied to them the attack's mathematical analysis developed in the previous sections to factor N. For our experimental results of physical fault injection see Section 4.

3.1   Single-Fault Attack Simulations

We first consider a single-UMP, single-fault attack when $H = $ SHA-1 i.e. $k_h = 160$. Using the SAGE library LLL implementation, computations were executed on a 2GHz Intel notebook.

Experimental results are summarized in Table 2. We see that for 1024-bit RSA, the randomizer size k_r must be quite small and the attack is less efficient than exhaustive search[3]. However for larger moduli, the attack becomes more efficient. Typically, using a single fault and a 158-bit UMP, a 2048-bit RSA modulus was factored in less than a minute.

Table 2. Single fault, single UMP 160-bit digests ($k_h = 160$). LLL runtime for different parameter combinations.

modulus size k	UMP size k_r	m	t	lattice dim. ω	runtime
1024	6	10	3	66	4 minutes
1024	13	13	4	105	51 minutes
1536	70	8	2	45	39 seconds
1536	90	10	3	66	9 minutes
2048	158	8	2	45	55 seconds

---

[3] Exhausting a 13-bit randomizer took 0.13 seconds.

3.2   Multiple-Fault Simulations

To test the practicality of the approach presented in Section 2.4, we have set $(\ell, t, m) = (3, 1, 3)$ i.e. three faulty signatures. This leads to a lattice of dimension 84 and a bound $\gamma + \delta \leq 0.204$. Experiments were carried out with 1024, 1536 and 2048 bit RSA moduli. This implementation also relied on the SAGE library [20] running on a single PC. Quite surprisingly, we observed a very large number of polynomials with the expected root over the integers. The test was run for three random instances corresponding to the parameters in Table 3.

Table 3. Three faults, single UMP, 160-bit digests ($k_h = 160$). LLL runtime for different parameter combinations.

modulus size k	UMP size k_r	runtime
1024	40	49 seconds
1536	150	74 seconds
2048	250	111 seconds

Three faults turn-out to be more efficient than single-fault attacks (Table 3 vs. Table 2). In particular for a 1024-bit RSA modulus, the three-fault attack recovered a 40-bit UMP r in 49 seconds[4], whereas the single-fault attack only recovered a 13-bit UMP in 51 minutes.

4   Physical Fault Injection Experiments

We performed fault injection on an unprotected device to demonstrate the entire attack flow. We obtain a faulty signature from a general-purpose 8-bit microcontroller running an RSA implementation and factor N using the mathematical attack of Section 2.

Our target device is an Atmel ATmega128 [3], a very pupular RISC microcontroller (μC) with an 8-bit AVR core. The μC was running an RSA-CRT implementation developed in C using the BigDigits multiple-precision arithmetic library [4]. The μC was clocked at 7.3728 MHz using a quartz crystal and powered from a 5V source.

We induced faults using voltage spikes (*cf.* to [1] and [2] for such attacks on similar μCs). Namely, we caused brief power cut-offs (spikes) by grounding the chip's V_{cc} input for short time periods. Spikes were produced by an FPGA-based board counting the μC's clock transitions and generating the spike at a precise moment. The cut-off duration was variable with 10ns granularity and the spike temporal position could be fine-tuned with the same granularity. The fault was heuristically positioned to obtain the stable fault injection in one of the RSA-CRT branches (computing σ_p or σ_q). A 40ns spike is presented in Figure 1. Larger spike durations caused a μC's reset.

---

[4] We estimate that exhaustive search on a 40-bit UMP would take roughly a year on the same single PC.

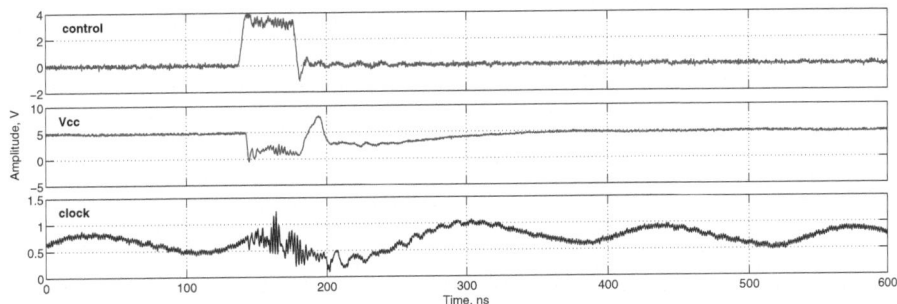

Fig. 1. Spike captured with a DSO: control signal from FPGA, power supply cut-off, and induced glitch in the clock signal

[8] provides more details on a 1536-bit RSA signature experiment conducted using our setup.

5   Conclusion

The paper introduced a new breed of partially-known message fault attacks against RSA signatures. These attacks allow to factor the modulus N given a single faulty signature. Although the attack is heuristic, it works well in practice and paradoxically becomes more efficient as the modulus size increases. As several faulty signatures are given longer UMPs and longer digests become vulnerable.

References

1. Schmidt, J.-M., Herbst, C.: A practical fault attack on square and multiply. In: Proceedings of FDTC 2008, pp. 53–58. IEEE Computer Society Press, Los Alamitos (2008)
2. Kim, C.H., Quisquater, J.-J.: Fault attacks for CRT based RSA: New attacks, new results, and new countermeasures. In: Sauveron, D., Markantonakis, K., Bilas, A., Quisquater, J.-J. (eds.) WISTP 2007. LNCS, vol. 4462, pp. 215–228. Springer, Heidelberg (2007)
3. ATmega128 datasheet, http://www.atmel.com/dyn/resources/prod_documents/doc2467.pdf
4. BigDigits multiple-precision arithmetic source code, Version 2.2., http://www.di-mgt.com.au/bigdigits.html
5. Bellare, M., Rogaway, P.: The exact security of digital signatures - how to sign with RSA and rabin. In: Maurer, U.M. (ed.) EUROCRYPT 1996. LNCS, vol. 1070, pp. 399–416. Springer, Heidelberg (1996)
6. Boneh, D., DeMillo, R.A., Lipton, R.J.: On the importance of checking cryptographic protocols for faults. Journal of Cryptology 14(2), 101–119 (2001)
7. Coppersmith, D.: Small solutions to polynomial equations, and low exponent vulnerabilities. Journal of Cryptology 10(4), 233–260 (1997)
8. Coron, J.S., Joux, A., Kizhvatov, I., Naccache, D., Paillier, P.: Fault Attacks on Randomized RSA Signatures. Full version of this paper, http://eprint.iacr.org

9. Coron, J.-S., Naccache, D., Stern, J.P.: On the security of RSA padding. In: Wiener, M. (ed.) CRYPTO 1999. LNCS, vol. 1666, pp. 1–18. Springer, Heidelberg (1999)

10. Coron, J.-S., Naccache, D., Tibouchi, M., Weinmann, R.P.: Practical cryptanalysis of ISO/IEC 9796-2 and EMV signatures. In: Halevi, S. (ed.) CRYPTO 2009. LNCS, vol. 5677, pp. 428–444. Springer, Heidelberg (2009), eprint.iacr.org/2009/203.pdf

11. Coron, J.-S.: Optimal security proofs for PSS and other signature schemes. In: Knudsen, L.R. (ed.) EUROCRYPT 2002. LNCS, vol. 2332, pp. 272–287. Springer, Heidelberg (2002)

12. Herrmann, M., May, A.: Solving linear equations modulo divisors: On factoring given any bits. In: Pieprzyk, J. (ed.) ASIACRYPT 2008. LNCS, vol. 5350, pp. 406–424. Springer, Heidelberg (2008)

13. EMV, Integrated circuit card specifications for payment systems, Book 2. Security and Key Management. Version 4.2 (June 2008), http://www.emvco.com

14. Howgrave-Graham, N.A.: Finding small roots of univariate modular equations revisited. In: Darnell, M.J. (ed.) Cryptography and Coding 1997. LNCS, vol. 1355, pp. 131–142. Springer, Heidelberg (1997)

15. Howgrave-Graham, N.A.: Approximate integer common divisors. In: CALC, pp. 51–66 (2001)

16. ISO/IEC 9796-2, Information technology - Security techniques - Digital signature scheme giving message recovery, Part 2: Mechanisms using a hash-function (1997)

17. ISO/IEC 9796-2:2002 Information technology – Security techniques – Digital signature schemes giving message recovery – Part 2: Integer factorization based mechanisms (2002)

18. Joye, M., Lenstra, A., Quisquater, J.-J.: Chinese remaindering cryptosystems in the presence of faults. Journal of Cryptology 21(1), 27–51 (1999)

19. Lenstra, A., Lenstra Jr., H., Lovász, L.: Factoring polynomials with rational coefficients. Mathematische Annalen 261, 513–534 (1982)

20. SAGE, Mathematical Library, http://www.sagemath.org

21. Rivest, R., Shamir, A., Adleman, L.: A method for obtaining digital signatures and public key cryptosystems. Communications of the ACM 21, 120–126 (1978)

Differential Fault Analysis on DES Middle Rounds

Matthieu Rivain

Oberthur Technologies & University of Luxembourg
m.rivain@oberthur.com

Abstract. Differential Fault Analysis (DFA) is a powerful cryptanalytic technique that disturbs cryptographic computations and exploits
erroneous results to infer secret keys. Over the last decade, many works
have described and improved DFA techniques against block ciphers thus
showing an inherent need to protect their implementations. A simple and
widely used solution is to perform the computation twice and to check
that the same result is obtained. Since DFA against block ciphers usually
targets the last few rounds, one does not need to protect the whole ciphering thus saving computation time. However the number of rounds to
protect must be chosen very carefully in order to prevent security flaws.
To determine this number, one must study DFA targeting middle rounds
of the cipher. In this paper, we address this issue for the Data Encryption
Standard (DES) algorithm. We describe an attack that breaks DES by
introducing some faults at the end of round 9, 10, 11 or 12, more or less
efficiently depending on the fault model and the round number.

1 Introduction

Fault analysis is a class of implementation attacks that consists in disturbing
cryptographic computations to recover secret keys. Among these attacks, one
merely identifies two families which differ in the information exploited to recover the key. *Differential Fault Analysis* (DFA) [3] exploits the difference between correct and faulty results while other attacks focus on the behavior of the
corrupted computation, namely on whether the induced fault effectively pro-
vokes an erroneous result or not. Among them, one lists *safe-error attacks* [26]
on exponentiation algorithms as well as *Ineffective Fault Analysis* (IFA) [9, 23]
and *Collision Fault Analysis* (CFA) [16] against block ciphers implementations.

IFA and CFA consider an adversary that is able to set an intermediate variable
of its choice to a known value (usually to 0). If the result is erroneous or if a fault
is detected, the attacker knows that the intermediate variable was different from
the induced value. Obtaining this information for several encryptions enables
key-recovery. A simple way to thwart this kind of attack is to use data masking [15, 2] which is often applied to protect embedded implementation against
power analysis [18]. Indeed, masking ensures that no single intermediate variable
provides information on the secret key. However, masking does not ensure the
result integrity and is hence ineffective against DFA [5].

C. Clavier and K. Gaj (Eds.): CHES 2009, LNCS 5747, pp. 457–469, 2009.
© International Association for Cryptologic Research 2009

DFA on block ciphers was first introduced by Biham and Shamir against DES [3]. Since then, several DFA were proposed on AES [11,4,13,22,7,25,17] as well as on other block ciphers such as IDEA [10] and CLEFIA [8,24]. These different works demonstrate the vulnerability of block ciphers towards DFA and the subsequent need of including countermeasures to embedded implementations. A straightforward way to protect any algorithm against DFA is to compute it twice and check whether the obtained results are equal or not. Another similar solution is to verify the integrity of an encryption by a decryption and *vice versa*. It is also possible to include redundancy and coherence checking at the operation level; the complexity-security ratio of such schemes is usually of the same order than the one of computation doubling [19]. An advantage of computation doubling is the scalability on the number of rounds to protect. In fact, most of DFA techniques target the last few rounds of the block cipher. To thwart these attacks, one only need to double the computation of these last few rounds thus saving computation time. However, a question remains: how many rounds should be protected to obtain a good security level towards DFA? To answer this question, we need to investigate DFA on middle rounds of the cipher.

This issue has been addressed in [21] by Phan and Yen for the AES block cipher. They apply block cipher cryptanalysis techniques to improve DFA on AES and exhibit some attacks against rounds 7, 6 and 5. Concerning DES, the original work by Biham and Shamir [3] described an attack that exploits a fault corrupting either round 16, 15 or 14 (and equivalently the end of round 15, 14 or 13). In his PhD thesis [1], Akkar investigates the application of differential cryptanalysis techniques to attack earlier rounds of DES. In a first place, the considered attacker is assumed to be able to induce a differential of its choice in the DES internal value at the end of some round. The last round key is recovered by guessing every 6-bit parts independently and by selecting, for each subkey, the candidate that produces the expected differential at the S-box output the more frequently. The obtained attacks are quite efficient but, as mentioned by the author, the fault model is not realistic. Akkar then applies this attack under two more realistic fault models: a single bit switch at a fixed position (in the left part of the DES internal state) and a single bit switch at a random position (in the right part of the DES internal state). For the fixed position bit error model, the attack needs a few hundred fault injections at the end of round 11 and it fails on round 9 (the attack on round 10 is not considered). For the random position bit error model, the attack needs a few dozen fault injections at the end of round 12 and it fails on round 11.

In this paper, we generalize and improve the attack described by Akkar in [1]. We consider various realistic fault models for an error induced in the left part of the DES internal state, including the bit error model and the byte error model with chosen error position or random error position. As we will argue, disturbing the left part leads to better attacks than disturbing the right part. Moreover, we use more accurate distinguishers than the one proposed in [1]. In the usual (chosen position) byte error model, our attack recovers the whole last round key with a 99% success rate using 9 faults on round 12, 210 faults on round 11 and

13400 faults on round 10. In the (chosen position) bit error model, these numbers are reduced to 7, 11 and 290, respectively.

2   Data Encryption Standard

The Data Encryption Standard (DES) [12] is a block cipher that was selected by the US National Bureau of Standards in 1976 as an official standard for data encryption. DES uses a 56-bit key (usually represented on 64 bits including 8 parity check bits) and it operates on 64-bit blocks. It has an iterative structure applying 16 times the same round transformation F which is preceded by a bit-permutation IP and followed by the inverse bit-permutation IP^{-1}. Every round transformation is parameterized by a 48-bit round key k_r that is derived from the secret key through a key schedule process. To summarize, a ciphertext C is computed from a plaintext P as follows:

$$C = IP^{-1} \circ \left(\bigcirc_{r=1}^{16} F_{k_r} \right) \circ IP(P) \ .$$

The round transformation follows a *Feistel scheme*, namely, the block is split into two 32-bit parts L (the left part) and R (the right part), and F is defined as:

$$F_{k_r}(L, R) = (R, L \oplus f_{k_r}(R)) \ ,$$

where f is a function parameterized with a 48-bit key and operating on a 32-bit block. This structure is illustrated on Fig. 1. In the sequel, the output block of the r-th round shall be denoted as (L_r, R_r). Defining $(L_0, R_0) = IP(P)$, we have $(L_r, R_r) = F_{k_r}(L_{r-1}, R_{r-1})$ for every $r \le 16$ and $C = IP^{-1}(L_{16}, R_{16})$.

The function f of the DES first applies an *expansion layer* E that expands the 32 input bits into 48 output bits by duplicating 16 of them. The round key is then introduced by bitwise addition afterward the block is split into eight 6-bit blocks, each entering into a different *substitution box* (S-box) S_i producing a 4-bit output. Finally, the 32 bits from the eight S-box outputs are permuted through a bit-permutation P which yields the 32-bit output block.

In the sequel, E_i and P_i^{-1} denote the i-th 6-bit coordinate of the expansion layer E and the i-th 4-bit coordinate of the bit-permutation P^{-1}, respectively.

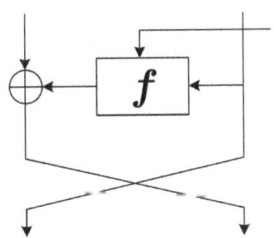

Fig. 1. Round transformation in the Feistel scheme

Similarly, $\mathbf{k}_{r,i}$ shall denote the i-th 6-bit part of a round key $\mathbf{k}_r$. We hence have the equality:

$$P_i^{-1}(f_{\mathbf{k}_r}(\cdot)) = S_i(E_i(\cdot) \oplus \mathbf{k}_{r,i}) . \tag{1}$$

3   Fault Models

Our attack consists in corrupting some bits of the left part of the DES internal state at the end of the r-th round with $r \in \{9, 10, 11, 12\}$. We shall consider different fault models depending on the statistical distribution of the induced error. We first consider the *bit error model* : one and one single bit of the left part is switched. We also consider the *byte error model* : one byte of the left part is switched to a random and uniformly distributed value. Furthermore, the fault position may be either *chosen* by the attacker or *random* among the 32 bit-positions or the 4 byte-positions of the left part.

In the sequel, $\widetilde{L}_i$ and $\widetilde{R}_i$ will respectively denote the corrupted value of the left part L_i and the right part R_i at the end of round i and $\widetilde{C} = \mathsf{IP}^{-1}(\widetilde{L}_{16}, \widetilde{R}_{16})$ will denote the faulty ciphertext. We shall further denote by ε the induced error that is defined as $\varepsilon = L_r \oplus \widetilde{L}_r$.

4   Attack Description

4.1   General Principle

Let us denote by Δ the bitwise difference between the correct value and the corrupted value of the left part at the end of the fifteenth round: $\Delta = L_{15} \oplus \widetilde{L}_{15}$. Due to the Feistel scheme, we have the following relation:

$$R_{16} \oplus \widetilde{R}_{16} = f_{\mathbf{k}_{16}}(L_{16}) \oplus f_{\mathbf{k}_{16}}(\widetilde{L}_{16}) \oplus \Delta . \tag{2}$$

Based on (2), an adversary that knows Δ can mount a key recovery attack. The principle is to make a guess on the value of the round key $\mathbf{k}_{16}$. Then, given a pair of ciphertexts $(C, \widetilde{C})$, the attacker checks whether (2) is consistent for this guess. If not, the guess is discarded. In this way, $\mathbf{k}_{16}$ is uniquely determined using few pairs of ciphertexts. Due to the structure of f (see (1)), the attacker does not need to guess the whole round key $\mathbf{k}_{16}$ but he can guess and check each subkey $\mathbf{k}_{16,i}$ independently. When an error is induced in the final rounds, the differential Δ (or at least a part of it) can be predicted according to the pair $(C, \widetilde{C})$ which enables the attack [3]. This is no more the case for an error induced in a middle round; in that case the attack must be extended.

As noted in [1], if an error ε is induced in the left part at the end of the thirteenth round then Δ equals ε. Therefore, an attacker that is able to induce a chosen (or at least known) error in L_{13} can apply the previous attack. For a fault induced in the left part during an earlier round, the equality $\Delta = \varepsilon$ does not hold anymore. However the statistical distribution of Δ may be significantly biased (depending on the fault model and the round number). Indeed, as illustrated in Fig. 2, a fault injection in the left part skips one round before propagating through the function f. Besides, the error propagation path from L_r to L_{15} sticks through the function f only once for $r = 12$, twice for $r = 11$, etc. This is quite low considering the slow diffusion of the function f. As a result, a fault induced in L_r may produce a differential Δ with a distribution that is significantly biased. As described hereafter, this bias enables a key recovery attack based on a statistical distinguisher.

Remark 1. From Fig. 2, it can be noticed that the injection of an error ε in L_r is equivalent to the injection of ε in R_{r+1}. This demonstrates the relevance of attacking the left part rather than the right one. Besides, this explains why the attack on the right part described in [1] is inefficient compared to the one on the left part on the same round.

Let us define, for every $i \in \{1, \cdots, 8\}$, the function g_i as the prediction of the i-th 4-bit coordinate of $\mathsf{P}^{-1}(\Delta)$ according to a pair $(C, \widetilde{C})$ and to a guess k on the value of $\mathbf{k}_{16,i}$:

$$g_i(C, \widetilde{C}, k) = \mathsf{S}_i\big(\mathsf{E}_i(L_{16}) \oplus k\big) \oplus \mathsf{S}_i\big(\mathsf{E}_i(\widetilde{L}_{16}) \oplus k\big) \oplus \mathsf{P}_i^{-1}\big(R_{16} \oplus \widetilde{R}_{16}\big) \ .$$

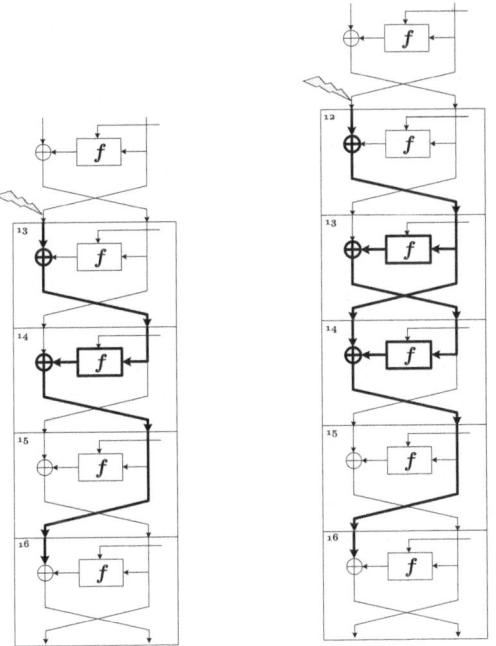

(a) From L_{12} to L_{15}.   (b) From L_{11} to L_{15}.

Fig. 2. Error propagation paths

From (1) and (2), it can be checked that, for the correct key guess, $g_i(C, \widetilde{C}, k)$ equals $\mathsf{P}_i^{-1}(\Delta)$. On the other hand, for a wrong key guess, $g_i(C, \widetilde{C}, k)$ can be assumed to have a uniform distribution. This is a classical assumption in block cipher cryptanalysis known as the *wrong-key assumption*.

Let us define, for every $i \in \{1, \cdots, 8\}$ and for every $\delta \in \{0, \cdots, 15\}$, the probability $p_i(\delta)$ as:

$$p_i(\delta) = \Pr\left[\mathsf{P}_i^{-1}(\Delta) = \delta\right] .$$

To summarize, according to the wrong-key assumption, we have:

$$\Pr\left[g_i(C, \widetilde{C}, k) = \delta\right] = \begin{cases} p_i(\delta) & \text{if } k = \mathbf{k}_{16,i} \\ \frac{1}{16} & \text{otherwise} \end{cases} \tag{3}$$

Provided that the distribution $p_i(\cdot)$ is significantly biased, (3) clearly exhibits a wrong-key distinguisher for $\mathbf{k}_{16,i}$.

4.2   Wrong-Key Distinguishers

We define hereafter two possible distinguishers $d(k)$ for a key candidate k which are expected to be maximal for the correct key candidate $k = \mathbf{k}_{16,i}$. These distinguishers take as input a set of N pairs $(C_n, \widetilde{C}_n)$, $1 \leq n \leq N$. The choice of the distinguisher to use depends on the attacker's knowledge of the fault model.

Likelihood distinguisher. The attacker is assumed to have an exact knowledge of the fault model, namely he knows the distribution of ε. In that case, he can compute (or at least estimate) the distribution $p_i(\cdot)$ in order to use a maximum likelihood approach. The likelihood of a key candidate k is defined as the product of the probabilities $p_i\big(g_i(C_n, \widetilde{C}_n, k)\big)$ for $n = 1, \cdots, N$. For practical reasons, we make the classical choice to use the logarithm of the likelihood, namely $d(k)$ is defined as:

$$d(k) = \sum_{n=1}^{N} \log\big(p_i\big(g_i(C_n, \widetilde{C}_n, k)\big)\big) .$$

Squared Euclidean Imbalance (SEI) distinguisher. The attacker does not have a precise knowledge of the fault model and is hence not able to estimate the distribution $p_i(\cdot)$. In that case, an alternative strategy is to look for the strongest bias in the distribution of $g_i(C_n, \widetilde{C}_n, k)$. This is done by computing the squared Euclidean distance to the uniform distribution (known as *squared Euclidean imbalance*), namely $d(k)$ is defined as:

$$d(k) = \sum_{\delta=0}^{15} \left(\frac{\#\{n; g_i(C_n, \widetilde{C}_n, k) = \delta\}}{N} - \frac{1}{16}\right)^2 .$$

4.3   Chosen Error Position Strategies

In a chosen error position fault model scenario, we further have to define a strategy to choose the positions where to induce the errors.

Table 1. Destination S-boxes for the input bits of f

bits	1,32	2,3	4,5	6,7	8,9	10,11	12,13	14,15
S-boxes	1,8	1	1,2	2	2,3	3	3,4	4

bits	16,17	18,19	20,21	22,23	24,25	26,27	28,29	30,31
S-boxes	4,5	5	5,6	6	6,7	7	7,8	8

Bit error model. In the bit error model, ε has a single bit to 1 which implies that the function f in round $r+2$ has one or two *active S-boxes*. That is, the correct output and the corrupted output of f only differ for one or two S-boxes. Indeed, as shown in Table 1, the expansion layer sends every input bit of f in one or two S-boxes. In order to maximize the bias in the distribution of Δ, the bit-positions should be chosen among the ones entering in a single S-box hence slowing the error propagation. Our strategy is simply to first choose a bit-position entering in S-box 1 only, then in S-box 2 only, and so on until S-box 8 and start over with S-box 1, etc.

Remark 2. Relation (3) implicitly assumes that the i-th S-box in the sixteenth round is active, otherwise $g_i(C, \widetilde{C}, k)$ equals 0 for every k. For a chosen position bit error attack on round 12, each selected bit-position implies that two S-boxes are *inactive* in round 16. However, the pair of inactive S-boxes differs for each bit-position which ensures the soundness of the attack.

Byte error model. Concerning the byte error model, every input byte of f is spread over four S-boxes. This can be checked from Table 2 that gives the destination S-boxes of every input byte of f. As a result, a byte error in L_r always implies four active S-boxes in the output differential of f in round $r+2$. For the attacks in the chosen position byte error model, the four byte-positions are hence equivalently chosen since they all induce the corruption of exactly four S-boxes in round $r+2$.

Remark 3. In a chosen error position attack, several fault models are involved hence, for a given i, different distributions $p_i(\cdot)$ are induced. Consequently, the SEI distinguisher shall not be directly applied but the SEI of $p_i(\cdot)$ shall be estimated for every error position independently. The SEI distinguisher is then defined as the sum of the SEIs for the different error positions.

Remark 4. In our attack simulations, we tried more specific strategies taking into account the bias in the $(p_i(\cdot))_i$ distributions resulting from the different bit-error positions. These strategies did not yield substantial improvements.

Table 2. Destination S-boxes for the input bytes of f

bytes	1	2	3	4
S-boxes	8,1,2,3	2,3,4,5	4,5,6,7	6,7,8,1

5 Attack Simulations

This section presents some experimental results. We performed attack simulations for each of the fault models introduced in Sect. 3 with a fault induced at the end of round 12, 11, 10 or 9. For every round number and every fault model, we applied the likelihood distinguisher and the SEI distinguisher (see Sect. 4.2). For the likelihood distinguisher, we empirically computed the distributions $(p_i(\cdot))_i$ based on several[1] ciphertexts pairs, each obtained from the correct and faulty encryptions of a random plaintext[2].

In what follows, we consider an attack successful when the whole last round key is determined with a 99% success rate. This strong requirement is motivated by the fact that, for a triple DES, too many key bits remain to perform an exhaustive search once the last round key has been recovered. Therefore, one shall fully determine the sixteenth round key before reiterating the attack on the fifteenth and so on. Every subsequent attack on a previous round key can be performed by using the same set of ciphertexts pairs and is expected to be substantially more efficient since the error propagates on fewer rounds. This way, if the last round key is recovered with a 99% success rate then the cipher can be considered fully broken.

Fig. 3 shows the success rate (over 1000 simulations) of the different attacks (chosen/random position bit/byte error, likelihood/SEI distinguishers) on round 12, 11 and 10. Fig. 4 shows the success rate (over 10 to 100 simulations) for the attacks on round 9 in the bit error model. Attacks on round 9 in the byte error model all required more than 10^8 faults. The numbers of faults required for a 99% success rate are summarized in Table 3.

Attack efficiency vs. round number. The attacks on rounds 11 and 12 are very efficient: less than 25 faults are sufficient on round 12 while, on round 11, less than 100 faults are sufficient in a bit error model and less than 1000 faults are sufficient in a byte error model. On round 10, the attacks are still fairly efficient: the best attack (chosen position bit error model, likelihood distinguisher) requires 290 faults whereas the least efficient attack (chosen position byte error model, SEI distinguisher) requires 26400 faults. It is on round 9 that the attacks become quite costly since the most efficient attack in the bit error model (chosen position, likelihood distinguisher) requires around $3.4 \cdot 10^5$ faults and all the attack in the byte error model require more than 10^8 faults[3].

Attack efficiency vs. fault model. As expected, we observe that, for a given setting (random/chosen position, likelihood/SEI distinguisher), a bit error model always leads to more efficient attacks than a byte error model. Similarly, a chosen position usually leads to more efficient attacks than a random position. Some exceptions are observed for the SEI distinguisher for which a random position sometimes leads to more efficient attacks than a chosen position. The reason of

---

[1] 10^7 for bit errors models and 10^8 for byte errors models.

[2] Note that the value of the key does not change the $(p_i(\cdot))_i$ distributions.

[3] The most efficient one (chosen position byte error model, likelihood distinguisher) yielded a 0% success rate (over 10 attack simulations) for 10^8 faults.

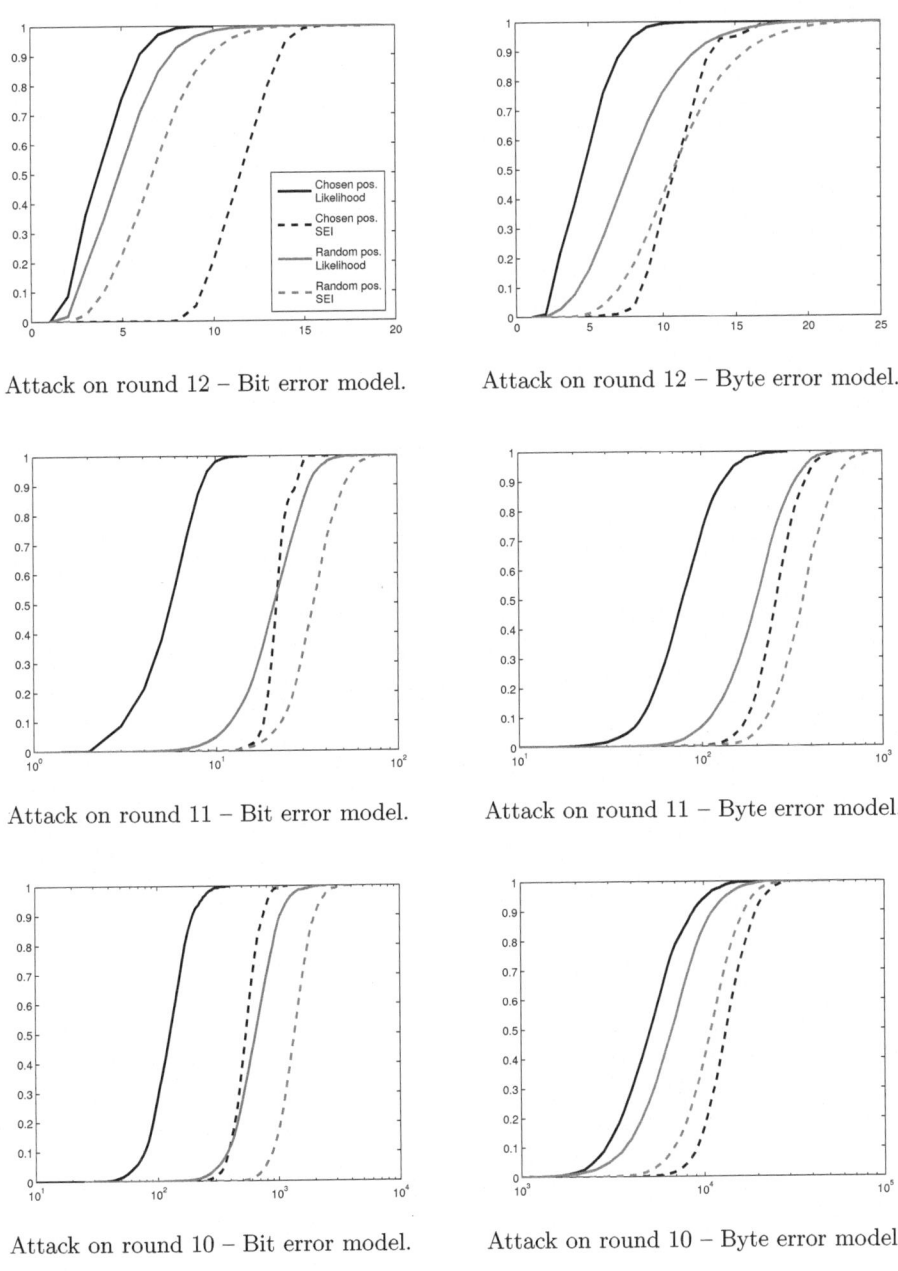

Attack on round 12 – Bit error model.   Attack on round 12 – Byte error model.

Attack on round 11 – Bit error model.   Attack on round 11 – Byte error model.

Attack on round 10 – Bit error model.   Attack on round 10 – Byte error model.

Fig. 3. Attacks on rounds 10, 11 and 12: success rate w.r.t. number of faults

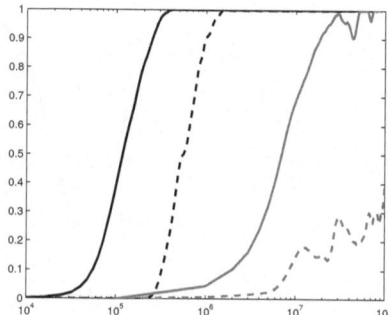

Fig. 4. Attacks on rounds 9, bit error model: success rate w.r.t. number of faults

Table 3. Number of faults to recover the 16-th round key with a 99% success rate

round	distinguisher	bit error		byte error	
		chosen pos.	random pos.	chosen pos.	random pos.
12	Likelihood	7	11	9	17
	SEI	14	12	17	21
11	Likelihood	11	44	210	460
	SEI	30	71	500	820
10	Likelihood	290	1500	13400	18500
	SEI	940	2700	26400	23400
9	Likelihood	$3.4 \cdot 10^5$	$2.2 \cdot 10^7$	$> 10^8$	$> 10^8$
	SEI	$1.4 \cdot 10^6$	$> 10^8$	$> 10^8$	$> 10^8$

this phenomenon may be that in a chosen position bit (resp. byte) error model, 8 (resp. 4) different SEIs are estimated based on 8 (resp. 4) times less faults than in the random position model where a single SEI is estimated (see Remark 3). As a result, these estimations are less precise which may render the attack less efficient than in the random position model. In these cases, the attacker can compute a single SEI based on all the faults, which amounts to perform the attack in the random position model.

To summarize, we naturally have that a bit error is better than a byte error and a chosen position is better than a random position. What was not *a priori* straightforward is the superiority of the random position bit error model compared to the chosen position byte error model. Except on round 12 where both cases are almost equivalent, our results show that the attacks in the random position bit error model are significantly more efficient than the ones in the chosen position byte error model.

Another interesting observation is that, in the bit error model, the ability to choose the error position is more advantageous than in the byte error model.

This phenomenon results from the strategy for the choice of the bit-positions (see Sect. 4.3) which selects 8 positions over 32 leading to more important bias in the distributions $p_i(\cdot)$ than the average case, whereas, in the chosen position byte error model, the 4 byte-positions are equivalently used.

Likelihood vs. SEI. As expected, the likelihood distinguisher always leads to more efficient attacks than the SEI distinguisher. It is interesting to note that this difference of efficiency is always greater in a chosen position model than in a random position model. Once again, this phenomenon results from the fact that, for a chosen position model, several different SEIs are estimated based on 4 or 8 times less faults compared to the random position model where a single SEI is estimated.

6 How Many Rounds To Protect ?

The question of the number of rounds to protect does not have a unique answer. Indeed, the answer to this question depends on the ability of an attacker to induce faults and on the number of correct and faulty ciphertexts pairs that he can collect. Besides, more efficient attacks that those described in this paper may exist.

What provide our paper are some lower bounds on the number of rounds to protect. We have shown that in a realistic fault model, efficient DFA attacks can be performed by inducing some faults until round 10. It seems therefore reasonable to protect at least the last seven rounds of the cipher. However, this may not suffice while considering a strong adversary model. We have shown that in a chosen position bit error model, $3.4 \cdot 10^5$ faults induced at the end of round 9 are sufficient to recover the last round key with a 99% confidence. Consequently, in order to thwart an adversary able to induce a single bit fault at a chosen position and to gather about 10^5 ciphertexts pairs, one shall at least protect the last eight rounds.

Attacks on initial rounds. As noted in [14], if an attacker has access to a decryption oracle then any DFA attack can be transposed on the initial rounds of the cipher. In fact, the attacker may obtain a faulty ciphertext $\widetilde{C}$ from a plaintext P by inducing a fault at the end of the first round. The plaintext P can then be viewed as the faulty result of a decryption of $\widetilde{C}$ for which a fault has been induced at the beginning of the last round. The attacker then asks for the decryption of $\widetilde{C}$ which provides him with a plaintext $\widetilde{P}$. The pair $(\widetilde{P}, P)$ thus constitutes a pair of correct and faulty results of the decryption algorithm with respect to an error induced at the beginning of the last round. According to this principle, any fault attack on an initial round of an encryption can be transposed to a fault attack on a final round of a decryption, provided that the attacker has access to a decryption oracle. In that case, the same number of rounds should be protected at the beginning and at the end of the cipher in order to obtain an homogenous security level. For a simple DES, based on our study, we recommend to protect the whole cipher. For a triple DES, one can only protect some rounds at the beginning of the first DES computation and some rounds at the end of the last DES computation; the number of protected rounds being at least seven according to our study.

7   Conclusion

In this paper, we have investigated differential fault analysis on DES middle rounds. We have described a generic attack and we have demonstrated its efficiency under various realistic fault models. We have shown that DES can be broken by inducing some faults at the end of rounds 12, 11, 10 and 9, more or less efficiently depending on the round number and the fault model. Although we focused on DES, our attack could be applied on any Feistel scheme.

References

1. Akkar, M.-L.: Attaques et méthodes de protections de systèmes cryptographiques embarqués. PhD thesis, Université de Versailles Saint-Quentin (October 1, 2004)
2. Akkar, M.-L., Giraud, C.: An Implementation of DES and AES, Secure against Some Attacks. In: Koç, Ç.K., Naccache, D., Paar, C. (eds.) CHES 2001. LNCS, vol. 2162, pp. 309–318. Springer, Heidelberg (2001)
3. Biham, E., Shamir, A.: Differential Fault Analysis of Secret Key Cryptosystem. In: Kaliski Jr., B.S. (ed.) CRYPTO 1997. LNCS, vol. 1294, pp. 513–525. Springer, Heidelberg (1997)
4. Blömer, J., Seifert, J.-P.: Fault Based Cryptanalysis of the Advanced Encryption Standard. In: Wright, R.N. (ed.) FC 2003. LNCS, vol. 2742, pp. 162–181. Springer, Heidelberg (2003)
5. Boscher, A., Handschuh, H.: Masking Does Not Protect Against Differential Fault Attacks. In: Breveglieri et al. [6], pp. 35–40
6. Breveglieri, L., Gueron, S., Koren, I., Naccache, D., Seifert, J.-P. (eds.): Fault Diagnosis and Tolerance in Cryptography – FDTC 2008. IEEE Computer Society Press, Los Alamitos (2008)
7. Chen, C.-N., Yen, S.-M.: Differential Fault Analysis on AES Key Schedule and Some Countermeasures. In: Safavi-Naini, R., Seberry, J. (eds.) ACISP 2003. LNCS, vol. 2727, pp. 118–129. Springer, Heidelberg (2003)
8. Chen, H., Wu, W., Feng, D.: Differential Fault Analysis on CLEFIA. In: Qing, S., Imai, H., Wang, G. (eds.) ICICS 2007. LNCS, vol. 4861, pp. 284–295. Springer, Heidelberg (2007)
9. Clavier, C.: Secret External Encodings Do Not Prevent Transient Fault Analysis. In: Paillier, Verbauwhede [20], pp. 181–194
10. Clavier, C., Gierlichs, B., Verbauwhede, I.: Fault Analysis Study of IDEA. In: Malkin, T.G. (ed.) CT-RSA 2008. LNCS, vol. 4964, pp. 274–287. Springer, Heidelberg (2008)
11. Dusart, P., Letourneux, G., Vivolo, O.: Differential Fault Analysis on A.E.S. In: Zhou, J., Yung, M., Han, Y. (eds.) ACNS 2003. LNCS, vol. 2846, pp. 293–306. Springer, Heidelberg (2003)
12. FIPS PUB 46-3. Data Encryption Standard (DES). National Institute of Standards and Technology, October 25 (1999)
13. Giraud, C.: DFA on AES. In: Dobbertin, H., Rijmen, V., Sowa, A. (eds.) AES 2005. LNCS, vol. 3373, pp. 27–41. Springer, Heidelberg (2005)

14. Giraud, C.: Attaques de cryptosystémes embarqués et contre-mesures associées. Thése de doctorat, Université de Versailles, Oberthur Card Systems (October 2007)

15. Goubin, L., Patarin, J.: DES and Differential Power Analysis – The Duplication Method. In: Koç, Ç.K., Paar, C. (eds.) CHES 1999. LNCS, vol. 1717, pp. 158–172. Springer, Heidelberg (1999)

16. Hemme, L.: A Differential Fault Attack against Early Rounds of (Triple-)DES. In: Joye, M., Quisquater, J.-J. (eds.) CHES 2004. LNCS, vol. 3156, pp. 254–267. Springer, Heidelberg (2004)

17. Kim, C.H., Quisquater, J.-J.: New Differential Fault Analysis on AES Key Schedule: Two Faults Are Enough. In: Grimaud, G., Standaert, F.-X. (eds.) CARDIS 2008. LNCS, vol. 5189, pp. 48–60. Springer, Heidelberg (2008)

18. Kocher, P., Jaffe, J., Jun, B.: Differential Power Analysis. In: Wiener, M. (ed.) CRYPTO 1999. LNCS, vol. 1666, pp. 388–397. Springer, Heidelberg (1999)

19. Malkin, T., Standaert, F.-X., Yung, M.: A Comparative Cost/Security Analysis of Fault Attack Countermeasures. In: Breveglieri, L., Koren, I., Naccache, D., Seifert, J.-P. (eds.) FDTC 2006. LNCS, vol. 4236, pp. 159–172. Springer, Heidelberg (2006)

20. Paillier, P., Verbauwhede, I. (eds.): CHES 2007. LNCS, vol. 4727. Springer, Heidelberg (2007)

21. Phan, R., Yen, S.-M.: Amplifying Side-Channel Attacks with Techniques from Block Cipher Cryptanalysis. In: Domingo-Ferrer, J., Posegga, J., Schreckling, D. (eds.) CARDIS 2006. LNCS, vol. 3928, pp. 135–150. Springer, Heidelberg (2006)

22. Piret, G., Quisquater, J.-J.: A Differential Fault Attack Technique against SPN Structures, with Application to the AES and KHAZAD. In: Walter, C.D., Koç, Ç.K., Paar, C. (eds.) CHES 2003. LNCS, vol. 2779, pp. 77–88. Springer, Heidelberg (2003)

23. Robisson, B., Manet, P.: Differential Behavioral Analysis. In: Paillier, Verbauwhede [20], pp. 413–426

24. Takahashi, J., Fukunaga, T.: Improved Differential Fault Analysis on CLEFIA. In: Breveglieri et al. [6], pp. 25–34

25. Takahashi, J., Fukunaga, T., Yamakoshi, K.: DFA Mechanism on the AES Key Schedule. In: Breveglieri, L., Gueron, S., Koren, I., Naccache, D., Seifert, J.-P. (eds.) Fault Diagnosis and Tolerance in Cryptography – FDTC 2007, pp. 62–74. IEEE Computer Society Press, Los Alamitos (2007)

26. Yen, S.-M., Joye, M.: Checking Before Output Not Be Enough Against Fault-Based Cryptanalysis. IEEE Transactions on Computers 49(9), 967–970 (2000)

Author Index